BLAZAR CONTINUUM VARIABILITY

A SERIES OF BOOKS ON RECENT DEVELOPMENTS IN ASTRONOMY AND ASTROPHYSICS

Printed by BookCrafters, Inc.

First published 1996

Library of Congress Catalog Card Number: 96-78616
ISBN 1-886733-30-9

D. Harold McNamara, Managing Editor of Conference Series
206 KMB
Brigham Young University
Provo, UT 84602-4463
801-378-2298

pasp@astro.byu.edu
Fax 801-378-2265

A SERIES OF BOOKS ON RECENT DEVELOPMENTS IN ASTRONOMY AND ASTROPHYSICS

Vol. 1-Progress and Opportunities in Southern Hemisphere Optical Astronomy: The CTIO 25th Anniversary Symposium
ed. V. M. Blanco and M. M. Phillips ISBN 0-937707-18-X

Vol. 2-Proceedings of a Workshop on Optical Surveys for Quasars
ed. P. S. Osmer, A. C. Porter, R. F. Green, and C. B. Foltz ISBN 0-937707-19-8

Vol. 3-Fiber Optics in Astronomy
ed. S. C. Barden ISBN 0-937707-20-1

Vol. 4-The Extragalactic Distance Scale: Proceedings of the ASP 100th Anniversary Symposium
ed. S. van den Bergh and C. J. Pritchet ISBN 0-937707-21-X

Vol. 5-The Minnesota Lectures on Clusters of Galaxies and Large-Scale Structure
ed. J. M. Dickey ISBN 0-937707-22-8

Vol. 6-Synthesis Imaging in Radio Astronomy: A Collection of Lectures from the Third NRAO Synthesis Imaging Summer School
ed. R. A. Perley, F. R. Schwab, and A. H. Bridle ISBN 0-937707-23-6

Vol. 7-Properties of Hot Luminous Stars: Boulder-Munich Workshop
ed. C. D. Garmany ISBN 0-937707-24-4

Vol. 8-CCDs in Astronomy
ed. G. H. Jacoby ISBN 0-937707-25-2

Vol. 9-Cool Stars, Stellar Systems, and the Sun. Sixth Cambridge Workshop
ed. G. Wallerstein ISBN 0-937707-27-9

Vol. 10-The Evolution of the Universe of Galaxies. The Edwin Hubble Centennial Symposium
ed. R. G. Kron ISBN 0-937707-28-7

Vol. 11-Confrontation Between Stellar Pulsation and Evolution
ed. C. Cacciari and G. Clementini ISBN 0-937707-30-9

Vol. 12-The Evolution of the Interstellar Medium
ed. L. Blitz ISBN 0-937707-31-7

Vol. 13-The Formation and Evolution of Star Clusters
ed. K. Janes ISBN 0-937707-32-5

Vol. 14-Astrophysics with Infrared Arrays
ed. R. Elston ISBN 0-937707-33-3

Vol. 15-Large-Scale Structures and Peculiar Motions in the Universe
ed. D. W. Latham and L. A. N. da Costa ISBN 0-937707-34-1

Vol. 16-Atoms, Ions and Molecules: New Results in Spectral Line Astrophysics
ed. A. D. Haschick and P. T. P. Ho ISBN 0-937707-35-X

Vol. 17-Light Pollution, Radio Interference, and Space Debris
ed. D. L. Crawford ISBN 0-937707-36-8

Vol. 18-The Interpretation of Modern Synthesis Observations of Spiral Galaxies
ed. N. Duric and P. C. Crane ISBN 0-937707-37-6

Vol. 19-Radio Interferometry: Theory, Techniques, and Application, IAU Colloquium 131
ed. T. J. Cornwell and R. A. Perley ISBN 0-937707-38-4

Vol. 20-Frontiers of Stellar Evolution, celebrating the 50th Anniversary of McDonald Observatory
ed. D. L. Lambert ISBN 0-937707-39-2

Vol. 21-The Space Distribution of Quasars
ed. D. Crampton ISBN 0-937707-40-6

Vol. 22-Nonisotropic and Variable Outflows from Stars
ed. L. Drissen, C. Leitherer, and A. Nota ISBN 0-937707-41-4

Vol. 23-Astronomical CCD Observing and Reduction Techniques
ed. S. B. Howell ISBN 0-937707-42-4

Vol. 24-Cosmology and Large-Scale Structure in the Universe
ed. R. R. de Carvalho ISBN 0-937707-43-0

Vol. 25-Astronomical Data Analysis Software and Systems I
ed. D. M. Worrall, C. Biemesderfer, and J. Barnes ISBN 0-937707-44-9

Vol. 26-Cool Stars, Stellar Systems, and the Sun, Seventh Cambridge Workshop
ed. M. S. Giampapa and J. A. Bookbinder ISBN 0-937707-45-7

Vol. 27-The Solar Cycle
ed. K. L. Harvey ISBN 0-937707-46-5

Vol. 28-Automated Telescopes for Photometry and Imaging
ed. S. J. Adelman, R. J. Dukes, Jr., and C. J. Adelman ISBN 0-937707-47-3

Vol. 29-Workshop on Cataclysmic Variable Stars
ed. N. Vogt ISBN 0-937707-48-1

Vol. 30-Variable Stars and Galaxies, in honor of M. S. Feast on his retirement
ed. B. Warner ISBN 0-937707-49-X

Vol. 31-Relationships Between Active Galactic Nuclei and Starburst Galaxies
ed. A. V. Filippenko ISBN 0-937707-50-3

Vol. 32-Complementary Approaches to Double and Multiple Star Research, IAU Collouquium 135
ed. H. A. McAlister and W. I. Hartkopf ISBN 0-937707-51-1

Vol. 33-Research Amateur Astronomy
ed. S. J. Edberg ISBN 0-937707-52-X

Vol. 34-Robotic Telescopes in the 1990s
ed. A. V. Filippenko ISBN 0-937707-53-8

Vol. 35-Massive Stars: Their Lives in the Interstellar Medium
ed. J. P. Cassinelli and E. B. Churchwell ISBN 0-937707-54-6

Vol. 36-Planets and Pulsars
ed. J. A. Phillips, S. E. Thorsett, and S. R. Kulkarni ISBN 0-937707-55-4

Vol. 37-Fiber Optics in Astronomy II
ed. P. M. Gray ISBN 0-937707-56-2

Vol. 38-New Frontiers in Binary Star Research
ed. K. C. Leung and I. S. Nha ISBN 0-937707-57-0

Vol. 39-The Minnesota Lectures on the Structure and Dynamics of the Milky Way
ed. Roberta M. Humphreys ISBN 0-937707-58-9

Vol. 40-Inside the Stars, IAU Colloquium 137
ed. Werner W. Weiss and Annie Baglin ISBN 0-937707-59-7

Vol. 41-Astronomical Infrared Spectroscopy: Future Observational Directions
ed. Sun Kwok ISBN 0-937707-60-0

Vol. 42-GONG 1992: Seismic Investigation of the Sun and Stars
ed. Timothy M. Brown ISBN 0-937707-61-9

Vol. 43-Sky Surveys: Protostars to Protogalaxies
ed. B. T. Soifer ISBN 0-937707-62-7

Vol. 44-Peculiar Versus Normal Phenomena in A-Type and Related Stars
ed. M. M. Dworetsky, F. Castelli, and R. Faraggiana ISBN 0-937707-63-5

Vol. 45-Luminous High-Latitude Stars
ed. D. D. Sasselov ISBN 0-937707-64-3

Vol. 46-The Magnetic and Velocity Fields of Solar Active Regions, IAU Colloquium 141
ed. H. Zirin, G. Ai, and H. Wang ISBN 0-937707-65-1

Vol. 47-Third Decinnial US-USSR Conference on SETI
ed. G. Seth Shostak ISBN 0-937707-66-X

Vol. 48-The Globular Cluster-Galaxy Connection
ed. Graeme H. Smith and Jean P. Brodie ISBN 0-937707-67-8

Vol. 49-Galaxy Evolution: The Milky Way Perspective
ed. Steven R. Majewski ISBN 0-937707-68-6

Vol. 50-Structure and Dynamics of Globular Clusters
ed. S. G. Djorgovski and G. Meylan ISBN 0-937707-69-4

Vol. 51-Observational Cosmology
ed. G. Chincarini, A. Iovino, T. Maccacaro, and D. Maccagni ISBN 0-937707-70-8

Vol. 52-Astronomical Data Analysis Software and Systems II
ed. R. J. Hanisch, J. V. Brissenden, and Jeannette Barnes ISBN 0-937707-71-6

Vol. 53-Blue Stragglers
ed. Rex A. Saffer ISBN 0-937707-72-4

Vol. 54-The First Stromlo Symposium: The Physics of Active Galaxies
ed. Geoffrey V. Bicknell, Michael A. Dopita, and Peter J. Quinn ISBN 0-937707-73-2

Vol. 55-Optical Astronomy from the Earth and Moon
ed. Diane M. Pyper and Ronald J. Angione ISBN 0-937707-74-0

Vol. 56-Interacting Binary Stars
ed. Allen W. Shafter ISBN 0-937707-75-9

Vol. 57-Stellar and Circumstellar Astrophysics
ed. George Wallerstein and Alberto Noriega-Crespo ISBN 0-937707-76-7

Vol. 58-The First Symposium on the Infrared Cirrus and Diffuse Interstellar Clouds
ed. Roc M. Cutri and William B. Latter ISBN 0-937707-77-5

Vol. 59-Astronomy with Millimeter and Submillimeter Wave Interferometry
ed. M. Ishiguro and Wm. J. Welch ISBN 0-937707-78-3

Vol. 60-The MK Process at 50 Years: A Powerful Tool for Astrophysical Insight
ed. C. J. Corbally, R. O. Gray, and R. F. Garrison ISBN 0-937707-79-1

Vol. 61-Astronomical Data Analysis Software and Systems III
ed. Dennis R. Crabtree, R. J. Hanisch, and Jeannette Barnes ISBN 0-937707-80-5

Vol. 62-The Nature and Evolutionary Status of Herbig Ae / Be Stars
ed. P. S. Thé, M. R. Pérez, and E. P. J. van den Heuvel ISBN 0-937707-81-3

Vol. 63-Seventy-Five Years of Hirayama Asteroid Families: The role of Collisions in the Solar System History
ed. R. Binzel, Y. Kozai, and T. Hirayama ISBN 0-937707-82-1

Vol. 64-Cool Stars, Stellar Systems, and the Sun, Eighth Cambridge Workshop
ed. Jean-Pierre Caillault ISBN 0-937707-83-X

Vol. 65-Clouds, Cores, and Low Mass Stars
ed. Dan P. Clemens and Richard Barvainis ISBN 0-937707-84-8

Vol. 66- Physics of the Gaseous and Stellar Disks of the Galaxy
ed. Ivan R. King ISBN 0-937707-85-6

Vol. 67-Unveiling Large-Scale Structures Behind the Milky Way
ed. C. Balkowski and R. C. Kraan-Korteweg ISBN 0-937707-86-4

Vol. 68-Solar Active Region Evolution: Comparing Models with Observations
ed. K. S. Balasubramaniam and George W. Simon ISBN 0-937707-87-2

Vol. 69-Reverberation Mapping of the Broad-Line Region in Active Galactic Nuclei
ed. P. M. Gondhalekar, K. Horne, and B. M. Peterson ISBN 0-937707-88-0

Vol. 70-Groups of Galaxies
ed. Otto G. Richter and Kirk Borne ISBN 0-937707-89-9

Vol. 71-Tridimensional Optical Spectroscopic Methods in Astrophysics
ed. G. Comte and M. Marcelin ISBN 0-937707-90-2

Vol. 72-Millisecond Pulsars—A Decade of Surprise, ed. A. A. Fruchter
M. Tavani, and D. C. Backer ISBN 0-937707-91-0

Vol. 73-Airborne Astronomy Symposium on the Galactic Ecosystem: From Gas to Stars to Dust
ed. M. R. Haas, J. A. Davidson, and E. F. Erickson ISBN 0-937707-92-9

Vol. 74-Progress in the Search for Extraterrestrial Life,
ed. G. Seth Shostak ISBN 0-937707-93-7

Vol. 75-Multi-Feed Systems for Radio Telescopes
ed. D. T. Emerson and J. M. Payne ISBN 0-937707-94-5

Vol. 76-GONG '94: Helio- and Astero-Seismology from the Earth and Space
ed. Roger K. Ulrich, Edward J. Rhodes, Jr., and Werner Däppen ISBN 0-937707-95-3

Vol. 77-Astronomical Data Analysis Software and Systems IV
ed. R. A. Shaw, H. E. Payne, and J. J. E. Hayes ISBN 0-937707-96-1

Vol. 78-Astrophysical Applications of Powerful New Databases
ed. S. J. Adelman and W. L. Wiese ISBN 0-937707-97-X

Vol. 79-Robotic Telescopes: Current Capabilities, Present Developments, and Future Prospects for Automated Astronomy
ed. Gregory W. Henry and Joel A. Eaton ISBN 0-937707-98-8

Vol. 80-The Physics of the Interstellar Medium and Intergalactic Medium
ed. A. Ferrara, C. F. McKee, C. Heiles, and P. R. Shapiro ISBN 0-937707-99-6

Vol. 81-Laboratory and Astronomical High Resolution Spectra
ed. A. J. Sauval, R. Blomme, and N. Grevesse ISBN 1-886733-01-5

Vol. 82-Very Long Baseline Interferometry and the VLBA,
ed. J. A. Zensus, P. K. Diamond, and P. J. Napier ISBN 1-886733-02-3

Vol. 83-Astrophysical Applications of Stellar Pulsation, IAU Colloquium 155
ed. R. S. Stobie and P. A. Whitelock ISBN 1-886733-03-1

Vol. 84-The Future Utilisation of Schmidt Telescopes, IAU Colloquium 148
ed. Jessica Chapman, Russell Cannon, Sandra Harrison, and Bambang Hidayat ISBN 1-886733-05-8

Vol. 85-Cape Workshop on Magnetic Cataclysmic Variables
ed. D. A. H. Buckley and B. Warner ISBN 1-886733-06-6

Vol. 86-Fresh Views of Elliptical Galaxies
ed. Alberto Buzzoni, Alvio Renzini, and Alfonso Serrano ISBN 1-886733-07-4

Vol. 87-New Observing Modes for the Next Century
ed. Todd Boroson, John Davies, and Ian Robson ISBN 1-886733-08-2

Vol. 88-Clusters, Lensing, and the Future of the Universe
ed. Virginia Trimble and Andreas Reisenegger ISBN 1-886733-09-0

Vol. 89-Astronomy Education: Current Developments, Future Coordination
ed. John R. Percy ISBN 1-886733-10-4

Vol. 90-The Origins, Evolution, and Destinies of Binary Stars in Clusters
ed. E. F. Milone and J. -C. Mermilliod ISBN 1-886733-11-2

Vol. 91-Barred Galaxies, IAU Colloquium 157
ed. R. Buta, D. A. Crocker, and B. G. Elmegreen ISBN 1-886733-12-0

Vol. 92-Formation of the Galactic Halo--Inside and Out
ed. H. L. Morrison and A. Sarajedini ISBN 1-886733-13-9

Vol. 93-Radio Emission from the Stars and the Sun
ed. A. R. Taylor and J. M. Paredes ISBN 1-886733-14-7

Vol. 94-Mapping, Measuring, and Modelling the Universe
ed. Peter Coles, Vicent Martinez, and Maria-Jesus Pons-Borderia ISBN 1-886733-15-5

Vol. 95-Solar Drivers of Interplanetary and Terrestrial Disturbances
ed. K.S. Balasubramaniam, S. L. Keil, and R. N. Smartt ISBN 1-886733-16-3

Vol. 96-Hydrogen-Deficient Stars
ed. C. S. Jeffery and U. Heber ISBN 1-886733-17-1

Vol. 97-Polarimetry of the Interstellar Medium
ed. W. G. Roberge and D. C. B. Whittet ISBN 1-886733-18-X

Vol. 98-From Stars to Galaxies: The Impact of Stellar Physics on Galaxy Evolution
ed. Claus Leitherer, Uta Fritze-von Alvensleben, and John Huchra ISBN 1-886733-19-8

Vol. 99-Cosmic Abundances
ed. Stephen S. Holt and George Sonneborn ISBN 1-886733-20-1

Vol. 100-Energy Transport in Radio Galaxies and Quasars
ed. P. E. Hardee, A. H. Bridle, and J. A. Zensus ISBN 1-886733-21-X

Vol. 101-Astronomical Data Analysis Software and Systems V
ed. George H. Jacoby and Jeannette Barnes ISSN 1080-7926

Vol. 102-The Galactic Center, 4th ESO/CTIO Workshop
ed. Roland Gredel ISBN 1-886733-22-8

Vol. 103-The Physics of Liners in View of Recent Observations
ed. M. Eracleous, A. Koratkar, C. Leitherer, and L. Ho ISBN 1-886733-23-6

Vol. 104-Physics, Chemistry, and Dynamics of Interplanetary Dust, IAU Colloquium 150
ed. Bo Å. S. Gustafson and Martha S. Hanner ISBN 1-886733-24-4

Vol. 105-Pulsars: Problems and Progress, IAU Colloquium 160
ed. M. Bailes, S. Johnston, and M.A. Walker ISBN 1-886733-25-2

Vol. 106-Minnesota Lectures on Extragalactic Neutral Hydrogen
ed. Evan D. Skillman ISBN 1-886733-26-0

Vol. 107-Completing the Inventory of the Solar System
ed. Terrence W. Rettig and Joseph Hahn ISBN 1-886733-27-9

Vol. 108-Model Atmospheres and Stellar Spectra
ed. S. J. Adelman, F. Kupka, and W. W. Weiss ISBN 1-886733-28-7

Vol. 109-Cool Stars, Stellar Systems, and the Sun, Ninth Cambridge Workshop
ed. Roberto Pallavicini and Andrea K. Dupree ISBN 1-886733-29-5

Vol. 110-Blazar Continuum Variability
ed. H. R. Miller, J. R. Webb, and J. C. Noble ISBN 1-886733-30-9

Inquiries concerning these volumes should be directed to the:
Astronomical Society of the Pacific
CONFERENCE SERIES
390 Ashton Avenue
San Francisco, CA 94112-1722 USA
415-337-1100
asp@stars.sfsu.edu

ASTRONOMICAL SOCIETY OF THE PACIFIC
CONFERENCE SERIES

Volume 110

BLAZAR CONTINUUM VARIABILITY

**An International Workshop held at
Florida International University
4-7 February 1996**

Edited by

H. R. Miller, J. R. Webb, and J. C. Noble

CONTENTS

II Radio Variability

III High-Energy Variability

IV Multi-frequency Variability

V Conference Summary

Preface

The idea to hold this conference was the result of several factors, the primary one being the need for a major conference on Blazar continuum variability where the latest results could be discussed. New satellite observations, recent Blazar outbursts, the rash of cooperative programs, and the completion of the OJ-94 project made it timely to hold a forum for the discussion of these results.

Another important reason for holding the conference was our desire to recognize the contributions of one of the most active and productive astronomers in this area of investigation, Professor Alex. G. Smith. A look at Alex's formidable resume leaves one with a sense of awe. His resume contains hundreds of papers, including seminal work on the decametric emission from Jupiter, radar, and, of course, his work on the variability of quasars from Rosemary Hill Observatory. In the midst of all this original research work was the important task of teaching and training others who would go on to work in various fields of study. His contageous curiosity, his persistence, his ability to teach and communicate, and his willingness to share his observations for the sake of knowledge has inspired us and has set the standard in this field. Alex has a large number of students who have received PhDs under his direction; most of which are either faculty members or involved in some type of research. He has also contributed to the astronomical community as Director of Cerro Tololo Interamerican Observatory and as editor of the AAS Photo-Bulletin for many years. It is to honor this wealth of contributions in astronomy that we decided to hold this conference and dedicate it to Dr. Alex G. Smith.

Miami, as the venue for this meeting during the winter time, seemed an ideal choice with its beautiful winter weather and abundant beaches. Also, perfectly timed was the construction of a new conference center on the campus of Florida International University. Unfortunately, the weather did not cooperate. The four days of the conference were the coldest, windiest days in recent memory. The silver lining inside this dark cloud, however, was that no one was tempted to skip sessions to go to the beaches! The conference center, although not quite finished, provided a comfortable and beautiful place for holding the meeting.

Dr. Hugh Aller, himself a pioneer in the field of quasar variability, was the guest speaker during the conference banquet. He recounted some interesting and enjoyable stories which revealed the lighter side of Alex Smith. I believe there was mention of speeding Jaguars and the police, as well as scientific innovation. We all wish to thank Dr. Aller for sharing his insights with us. After the dinner, Alex was presented with a picture of Rosemary Hill Observatory signed by all of his PhD students who attended the conference.

We would also like to thank Dr. Roger Blandford for summarizing the conference in the energetic and insightful way that only someone with his physical intuition and theoretical prowess could do.

Finally, we would like to thank the entire staff at the new FIU Conference Center for expertly handling all the logistical details that allowed us to concentrate on the science presented at the conference; and, of course, thanks to all the participants who shared their valuable work with us at the meeting and in this volume.

H. R. Miller and J. R. Webb

Participant List

Hugh Aller	hugh@astro.lsa.umich.edu
Margo Aller	margo@astro.lsa.umich.edu
Thomas Balonek	tbalonek@center.colgate.edu
Gang Bao	gbao@chara.gsu.edu
Roger Blandford	rdb@tapir.caltech.edu
Paul Boltwood	ae772@freenet.carleton.ca
Rolando Branly	rolando@alpha.nova.edu
Sandra Clements	cy$1de7@nervm.bitnet
Elizabeth Corbett	eac@star.herts.ac.uk
Luis Del Rio	delriol@servax.fiu.edu
Oliver Dreissigacker	odreissi@lsw.uni-heidelberg.de
José Fernandez	Fernandj@servax.fiu.edu
Markos Georganopoulos	markos@qjet.bu.edu
José Gómez	gomez@qjet.bu.edu
Karen Hackney	ksgc@wkuvx1.wku.edu
Richard Hackney	ksgc@wkuvx1.wku.edu
Robert Hartman	rch@puccini.gsfc.nasa.gov
Jochen Heidt	jheidt@hp3.lsw.uni-heidelberg.de
David Helfand	djh@carmen.phys.columbia.edu
James Hough	jhh@star.herts.ac.uk
Rene Hudec	rhudec@asu.cas.cz
Pamela Jenkins	jenkins@phyast.nhn.uoknor.edu
Minas Kafatos	mkafatos@compton.gmu.edu
Hidetosi Kubo	Kubo@astro.isas.ac.jb
Anthony Lawson	ajl@phastr.soton.ac.uk
Harry Lehto	hlehto@utu.fi
Ying-chi Lin	lin@egret0.stanford.edu
Laura Maraschi	maraschi@astmiu.mi.astro.it
Svetlana Marchenko	jotstad@bell.aispbu.spb.su
Alan Marscher	marscher@buast0.bu.edu
John Mattox	mattox@astro.umd.edu
H. Richard Miller	miller@chara.gsu.edu
Reshmi Mukerjee	muk@egret.gsfc.nasa.gov
Ian McHardy	imh@sousun.phys.soton.ac.uk
Damo Nair	nair@astro.ufl.edu
John Noble	noble@chara.gsu.edu
Franco Pacini	pacini@arcetri.astro.it
Stefani Paltani	paltani@Sigma-0.cesr.cnes.fr
Eric Perlman	perlman@baley.gsfc.nasa.gov
Joseph Pesce	pesce@stsci.edu
Elena Pian	pian@stsci.edu
Glenn Piner	glenn@egret.gsfc.nasa.gov

Martin Pohl	mkp@mpe-garching.mpg.de
Joseph Pollock	pollockjt@conrad.appstate.edu
Andrew Robinson	ar@star.herts.ac.uk
Ian Robson	eir@jach.hawaii.edu
Gustavo Romero	romero@irma.edu.ar
Alberto Sadun	asadun@ScottLAN.Edu
Rita Sambruna	sambruna@ros1.gsfc.nasa.gov
Roger Scott	ksgc@wkuvx1.wku.edu
Alexander Shcherbanovsky	webbj@servax.fiu.edu
Chris Shrader	shrader@grossc.gsfc.nasa.gov
Aimo Sillanpää	aimosill@sara.cc.utu.fi
Caroline Simpson	simpsonc@servax.fiu.edu
Mark Sincell	sincell@ux1.cso.uiuc.edu
Alex Smith	asmith@astro.ufl.edu
Niall Smith	nsmith@compserv.rtc-cork.ie
Paul Smith	psmith@as.arizona.edu
Tadayuki Takahasi	takahasi@tkynext.astro.isas.ac.jp
Leo Takalo	takalo@sara.cc.utu.fi
Harri Teräsranta	hte@vipunen.hut.fi
Steven Unwin	scu@astro.caltech.edu
C. Megan Urry	cmu@stsci.edu
Mauri Valtonen	mavalto@sara.cc.utu.fi
Walter Van Hamme	vanhamme@servax.fiu.edu
Federick Vrba	fjv@nofs.navy.mil
Stefan Wagner	S.wagner@lsw.uni-heidelberg.de
James Webb	webbj@servax.fiu.edu
Ann Wehrle	aew@ipac.caltech.edu
Paul Wiita	wiita@chara.gsu.edu
Wenge Wu	wx@deimos.caltech.edu
Ying Xiong	ying@chara.gsu.edu

Part I

UV-Optical-IR Variability

Blazar Continuum Variability
ASP Conference Series, Vol. 110, 1996
H. R. Miller, J. R. Webb, and J. C. Noble, eds.

The Multiple Timescales of AGN Variability

Alex. G. Smith

Department of Astronomy, University of Florida

Abstract. Variability in blazars and other active galactic nuclei (AGN) has now been studied for more than a third of a century. As in other fields of astrophysics, variability furnishes an important clue to the physical processes responsible for the emissions. Observers have recorded an astonishing gamut of timescales of variability, ranging in the optical from minutes to decades. In recent years, analyses of very long-term optical light curves have shown that many AGN undergo marked, quasi-cyclical changes in the base level from which shorter-term events erupt. The characteristic timescales of the cycles, which seldom suggest real periodicity, range from two or three years to decades. There are distinct differences in the mean timescales of the several classes of AGN: classical quasars, radio-quiet quasars, BL Lacertids, compact galaxies. Ongoing studies show that, similarly, there are characteristic differences among AGN types in the amplitudes and durations of the rapid optical flares that are superimposed on these long-term base-level meanderings.

1. History

Studies of variability go back to the very earliest days of quasar science. In 1960 Tom Mathews used the Caltech radio interferometer to obtain a precise position for the Cambridge radio source 3C 48. In an unscheduled paper at the December, 1960, AAS meeting, Allan Sandage announced that a 90-minute exposure with the Palomar 200-inch reflector showed a star-like object precisely at the radio position. We now know that this was the first optical identification of a quasar, but, quoting from a note in Sky and Telescope (1961), speculation at the time was "...that it was a relatively nearby star with most peculiar properties. It could be a supernova remnant." Spectra by Sandage, Munch, and Greenstein had already shown an outlandish mix of unidentified emission and absorption lines.

Almost immediately, Harlan Smith and Dorrit Hoffleit (1961), aided by five graduate students, scanned 2500 Harvard patrol plates covering the 3C 48 region. Their stated "working hypothesis" was that 3C 48 is a supernova remnant—or, less likely, a nova. Images of 3C 48 were identified on 75 of these plates, some going back as far as 1899. The rather negative conclusion from this first search for quasar variability was that "...no evidence has been found for continuous or sporadic change of brightness greater than 0.3 magnitude," which was about the estimated error. It is interesting that a dozen years of monitoring 3C 48 at Rosemary Hill showed a total range of 0.69 mag or $\pm\frac{1}{3}$ mag, just about the size of the error in the Smith-Hoffleit study. 3C 48 has not been a spectacular performer for us, nor was it listed as a blazar by Burbidge and Hewitt (1992).

A major break-through occurred in 1963 when the Australians Hazard, Mackey, and Shimmins (1963) used lunar occultations to pinpoint the position of the Cambridge radio source 3C 273. This quickly led to the discovery of a much brighter optical counterpart, and to the identification of its spectrum and redshift by Maarten Schmidt. Harlan Smith and Dorrit Hoffleit (1963) then repeated their archival search, this time with success, largely because 3C 273 is three magnitudes brighter than 3C 48. They found that over a 70-year period the optical flux from 3C 273 had varied by a factor of more than two. There appeared to be both long-term changes and more rapid fluctuations on timescales of weeks or months. Similar results were reported in 1963 by Sharov and Effremov (1963) from studies of the Pulkova plate collection.

Shortly afterwards, Sandage (1964) reported that photoelectric observations of 3C 48, 3C 196, and 3C 273 all showed significant changes in brightness, a result he called "exceedingly surprising." Noting that the change in 3C 48 amounted to 4×10^{44} ergs/second (based on the by-then known redshift), Sandage commented that this was "...a value so large that the reality of the variation might well be questioned." Sandage went on to describe "special observations" made during three nights in 1962 and 1963 "...to monitor very short-term fluctuations of the order of minutes or hours." While the result was negative, we can regard this as the first effort to detect what we now call "microvariability." Sandage concluded that "Because the first three objects studied have shown variability, it now becomes important to monitor all objects of the class..." and he noted that nine such objects were known at the time.

By the time our 76-cm Tinsley reflector saw first light at Rosemary Hill in 1968, the number of recognized quasars had grown dramatically, but I was still quite conscious of Sandage's prophetic words. We immediately began a program of regular photographic monitoring of a steadily growing list of QSO's on a schedule of a dozen nights per month of dark time—a schedule that has been pretty much maintained to the present day. Since Penston and Cannon (1970) had already noticed a correlation between optical activity and small radio spectral indices, most of the sources I added to the list were selected from the Parkes Catalog on the basis of their spectral index. Our list was not, and never has been, a random sample of QSO's. Instead, objects were added because they appeared to be "interesting."

By late 1970 we had enough data to publish a brief statistical summary in Nature (Folsom et al. 1971). Of 26 quasars and 5 compact galaxies that had been monitored intensively, 74% showed clear-cut optical variations. Of these, 5 objects fell in the optically violent (OVV) category that we now call "blazars." Figure 1 is from this early paper in Nature. It demonstrates the correlation between large-amplitude optical fluctuations and a small radio spectral index (averaged from 408 to 2650 MHz). We defined "large amplitude" as half-a-magnitude or more. Of the 53 quasars in the figure, 26 are from Rosemary Hill and 27 from observations I made with the 70-cm Soviet Maksutov at Cerro El Roble in Chile. Our result here clearly confirms Penston and Cannon's prediction based on a much smaller sample.

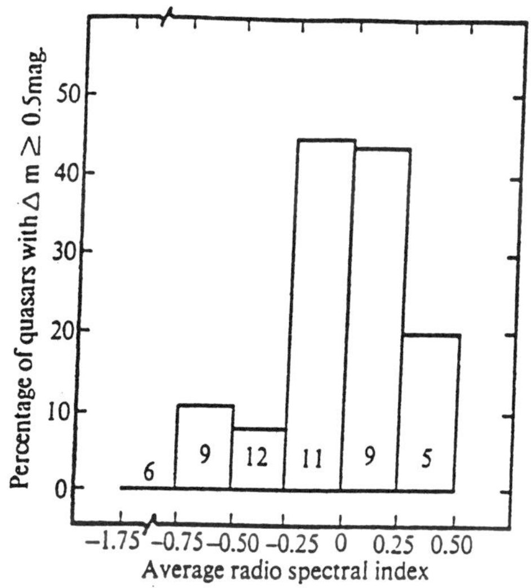

Figure 1. Optical variation as a function of radio spectral index.

2. Long-term studies

In 1987, Dick Miller, who is one of our present hosts, asked me to participate in a similar conference in Atlanta. As I debated over what to present, it seemed to me that the time had come to take a broad, long-term look at our uniquely large data base. As I began this, I became aware of a phenomenon I had really not seen discussed in the literature. Many of the long-term light curves seemed to display slow, quasi-cyclic changes in what we had come to call the "baseline" or "base level" from which the familiar and often spectacular "flares" erupt. Although there was rarely any suggestion of real periodicity, there did seem to be a characteristic timescale of brightening and fading that varied from source to source. A rather clean-cut illustration of this behavior is shown in Figure 2. A preliminary analysis of the phenomenon was presented at the Atlanta workshop (Smith et al. 1991).

By 1992, we had accumulated some 26,000 observations on 9000 photographic plates. With the help of Damo Nair, I made a detailed analysis of 60 classical radio-loud quasars for which we had the most complete light curves (Smith et al. 1993). We used three different techniques to estimate timescales from the light curves, some of which extended over 20 years. The most obvious technique was of course simple visual inspection, which was applied with great care. Probably the most powerful technique was the first-order structure function, which is defined as

$$\mathrm{SF}(t) = < [\mathrm{mag}(t) - \mathrm{mag}(t + dt)]^2 >,$$

where $\mathrm{mag}(t)$ and $\mathrm{mag}(t + dt)$ are observed magnitudes separated by the time interval dt, and the angular brackets denote an ensemble average. As is illustrated in Figure 3, if a time series contains cycles of period P, the structure

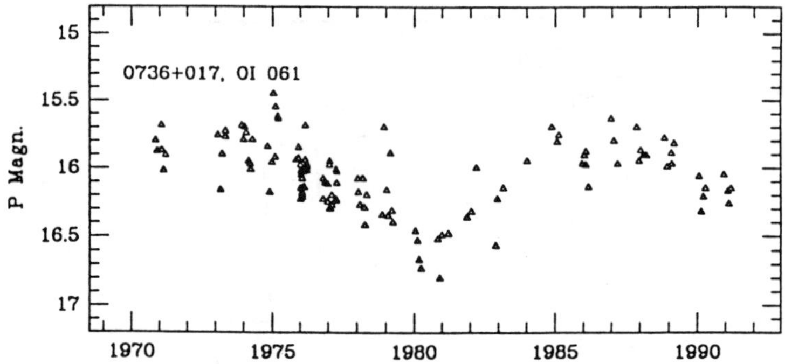

Figure 2. An illustration of rapid flares superimposed on long-term, quasi-cyclic changes in the base level of the quasar 0736 + 017.

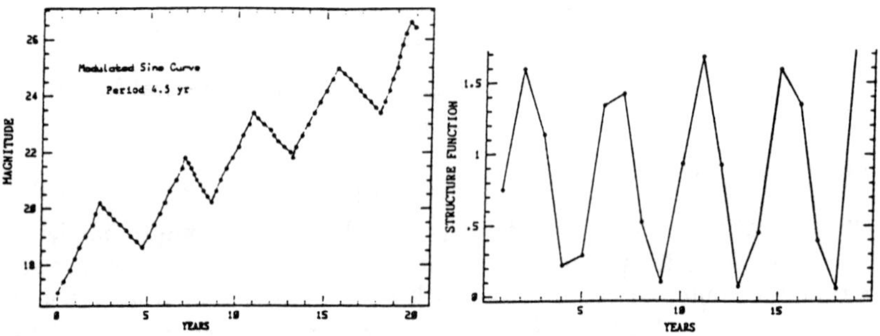

Figure 3. Response of the first-order structure function to cyclic variations in an artificial time series.

function rises to a maximum at $dt = P/2$, and then drops to a minimum at $dt = P$. If dt is extended to multiples of P, the structure function itself goes through a series of cycles.

A third technique for estimating timescales was the unequal-interval Fourier transform method of Deeming (1975), as modified by Horne and Baliunas (1986). This method is most useful when it produces a single strong peak, as illustrated in Figure 4. When cycles of unequal length are present, as is often the case, the power spectrum tends to be complex and difficult to interpret. In the 18 cases (out of 60) where the technique did produce a strong, clean peak, we used it largely to reinforce timescales estimated from the other methods. I should explain that by "timescale" we refer to a complete cycle of brightening and fading, say from a maximum through a minimum and back to maximum. Where there were cycles of unequal length, these were averaged to yield an estimate of the "characteristic timescale."

Of the 60 classical quasars, 42 (or 70%) showed convincing evidence of excursions in their base levels. Figure 5 (top) shows a histogram of the distribution of the observed timescales as estimated by the three methods just discussed. The

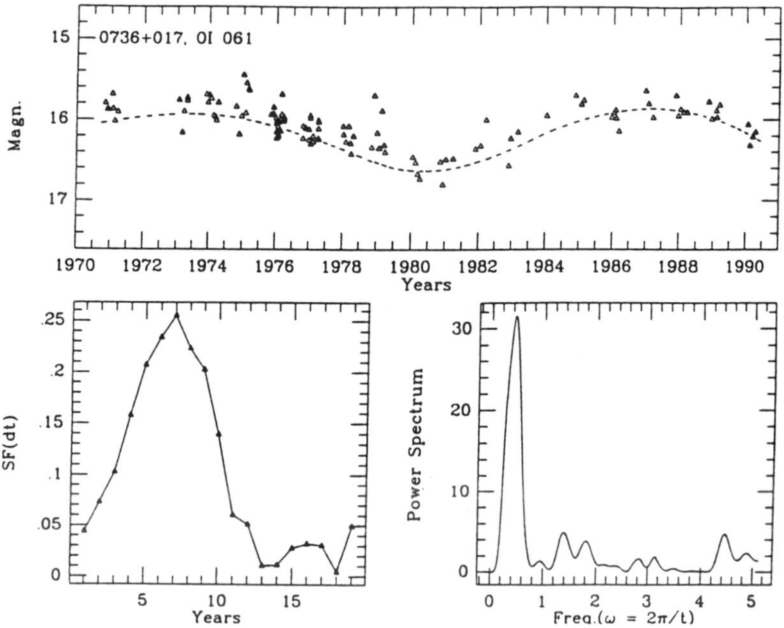

Figure 4. Structure function (lower left) and power spectrum (lower right) of the light curve (top panel) of the quasar 0736 + 017.

distribution peaks in the 6–9 year interval, but it is broad and rather symmetrical. The mean of the 42 timescales is 8.9 years. I should note that of these objects, 16 are listed as blazars or OVV's by Burbidge and Hewitt (1992). Since redshifts are known for all 42 quasars, the observed timescales are easily converted to timescales in the rest frames of the sources. The result is seen in the lower half of Figure 5. Now the peak is in the 4–6 year interval, and the mean falls to 5.0 years.

Obviously it's interesting to see if there might be a relationship between the rest-frame timescales and the redshifts. This is investigated in Figure 6, which is a scatter diagram for the 42 quasars. The line is a linear least-squares regression, for which the correlation coefficient is a modest 0.4. If the apparent negative correlation is real, it may imply that as quasars age, their base-level fluctuations become slower. An alternative interpretation is that the trend might be a luminosity effect. If the high-redshift quasars are generally more luminous, as seems reasonable, the observed trend could represent a tendency for high-luminosity objects to vary more slowly in agreement with the "universal" AGN model proposed by Malkan (1991).

Some later work I did in collaboration with Damo Nair extended the same type of study to three other groups of AGN: the BL Lacertids, the compact galaxies, and the radio-quiet quasars (Smith & Nair 1995). The same methods were used to estimate timescales for the base-level excursions of these objects. Figure 7 illustrates the application of the techniques to the Seyfert galaxy 3C 120.

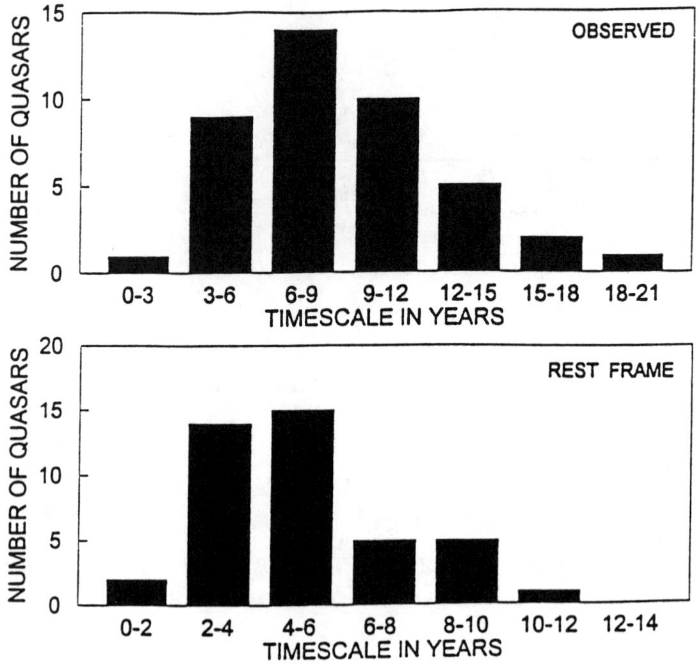

Figure 5. Baseline timescales of 42 classical quasars.

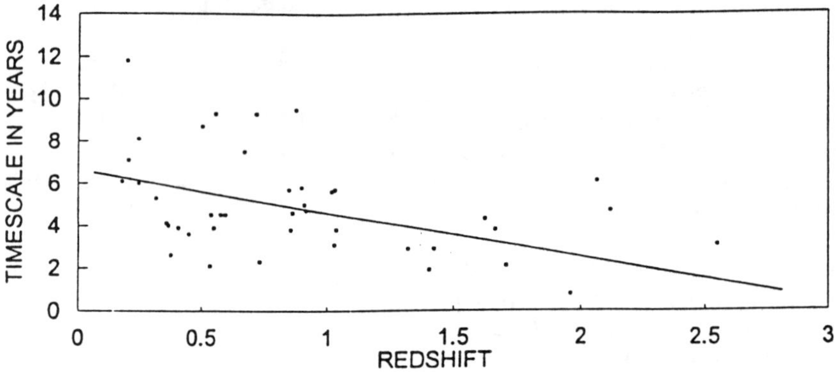

Figure 6. Relationship between timescales and redshift for the 42 classical quasars of Figure 5.

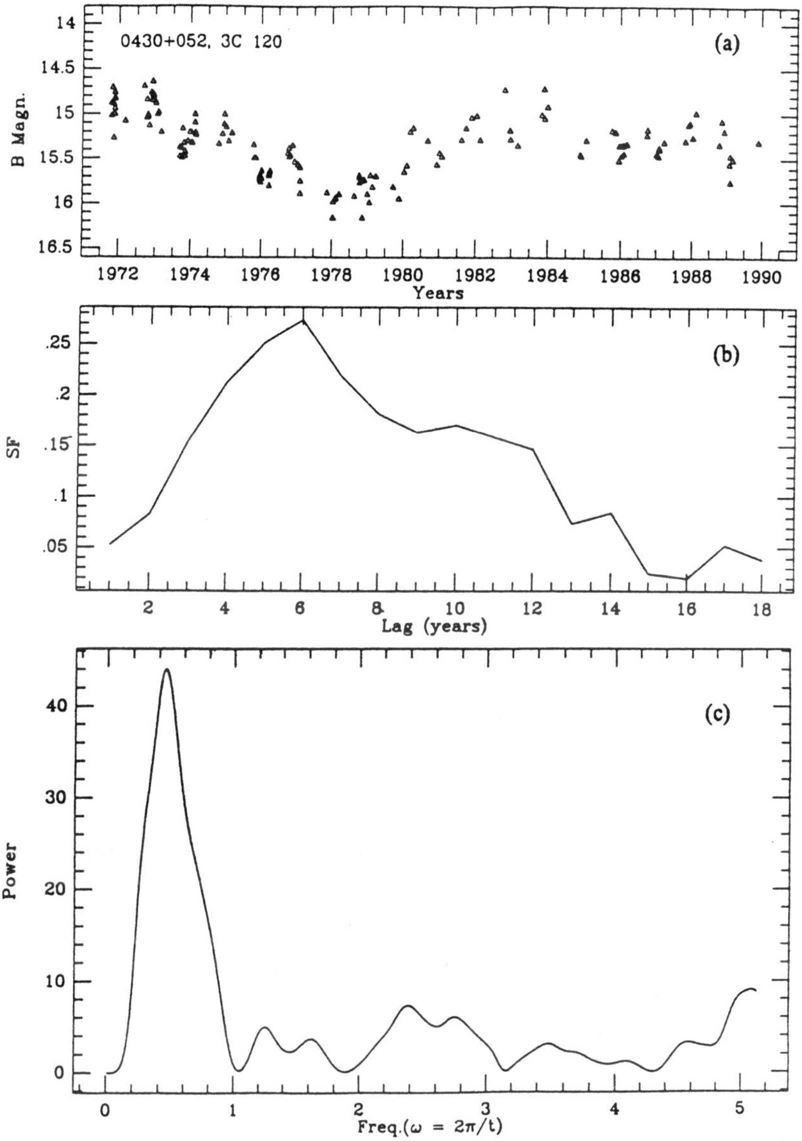

Figure 7. Structure function (middle panel) and power spectrum (bottom) of the optical light curve (upper panel) of 3C 120.

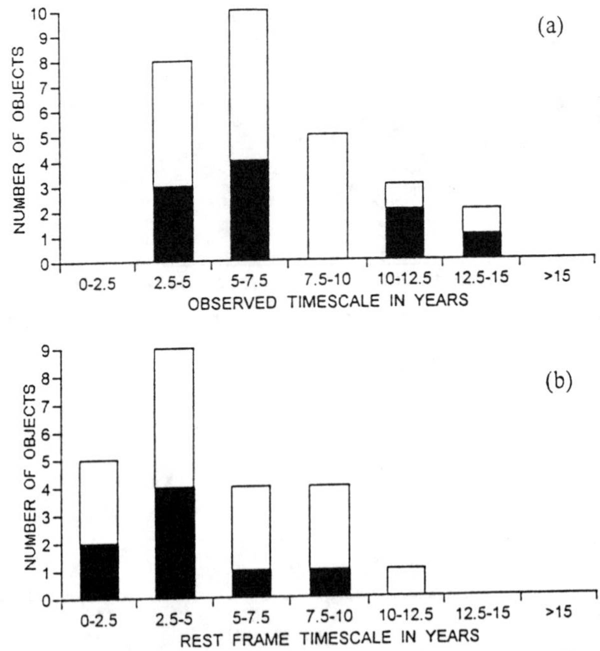

Figure 8. Distribution of base-level timescales of 28 BL Lacertids. (a) The observed timescales. (b) Rest-frame timescales.

Figure 8 shows the results obtained for 28 BL Lacertids (which is to say, blazars) for which we had reasonable runs of data ranging from 6 to 22 years. The upper histogram is for the observed timescales, while the lower plot shows rest-frame timescales. Not surprisingly, we were unable to find redshifts for 5 BL Lacs, so only 23 are plotted in the rest-frame histogram. The shaded parts of the bars represent timescales in which we have especially high confidence.

Valtaoja et al. (1992) suggested that there may be two classes of BL Lacs. One is comprised of moderate-strength FR I radio galaxies at small redshifts, with their radiation beamed toward us because of a relatively favorable orientation. The other class is powerful FR II galaxies at large redshifts, but very accurately aligned with our line-of-sight, so we see them as violent, high-polarization variables. A similar dichotomy was suggested by Burbidge and Hewitt (1989) with the break occurring near $z = 0.3$. Our results imply that BL Lacs at large redshifts have shorter rest-frame timescales. If we divide our sample at $z = 0.4$, those with $z > 0.4$ have a mean rest-frame timescale of 2.4 years, while the mean timescale for those with $z < 0.4$ is 6.6 years, nearly three times larger. There is indeed some implication that we might be dealing with two groups of objects.

Similar analyses were applied to our much smaller samples of compact galaxies and radio-quiet quasars. Only a dozen of our well-monitored sources are classified as compact galaxies. Of these, we were able to assign characteristic baseline timescales to 9, with the result illustrated in Figure 9. With so few ob-

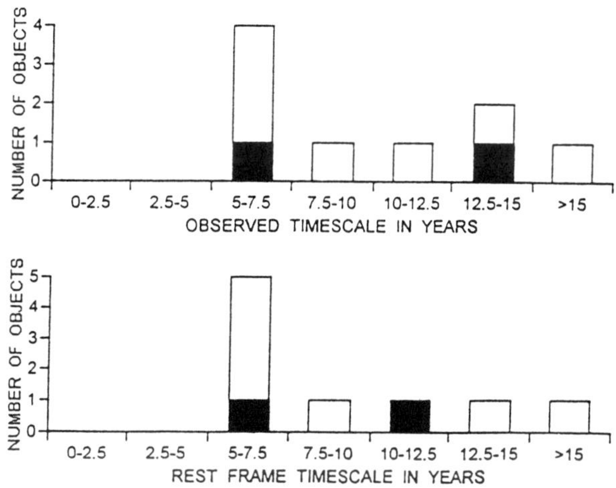

Figure 9. Distribution of base-level timescales of 9 compact galaxies. Upper panel: observed timescales. Lower panel: rest-frame timescales.

jects, one can scarcely expect a "smooth" distribution. The mean of the observed timescales is 10.4 years, while that of the rest-frame timescales is 9.6 years; because the redshifts of these sources are generally small, the difference in the two timescales is also small.

The light curve of the Seyfert galaxy 3C 120, seen in Figure 10, displays a phenomenon we have noted in a number of sources, including BL Lacertae itself—namely, an abrupt change in character. For 15 years 3C 120 followed a near-sinusoidal pattern with superimposed rapid small-amplitude flares. Then, starting around 1985, the base level became nearly constant while the amplitudes of the flares increased. Prior to 1985 the flare amplitudes were rather uniform at around 0.4 magnitude. By the end of 1995 the amplitudes had grown to over one magnitude. Such a transition seems quite unlikely if the phenomena were purely stochastic. The transition in BL Lac, ca. 1980, was in the opposite sense, from rather chaotic behavior to a smoothly varying baseline. It is suggestive that the sudden optical change in each of these blazars followed immediately after radio outbursts described by Aller et al. (1992) and attributed to shocked relativistic jets.

Of the 18 radio-quiet quasars in our program, we were able to identify base-level excursions in only 7. This may be due in part to less intensive monitoring of the radio-quiets since Patricia Edwards (1981) completed her dissertation on these objects. Five of the radio-quiets displayed secular brightening or fading of 0.2 to 0.5 magnitude during our 15 years of monitoring, hinting that there may be cycles much too long to be defined by the existing data. The rather meager results for this group are shown in Figure 11.

Finally, I have collected in Table 1 the average base-level timescales of all four groups of AGN. We see that both the classical quasars and the BL Lacertids have rest-frame timescales of about 5 years. This agreement might

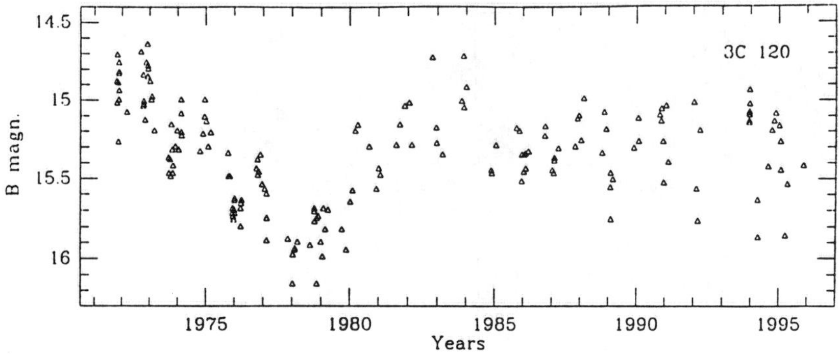

Figure 10. Extended long-term Rosemary Hill light curve of the much-studied Seyfert galaxy 3C 120. Note the marked change in behavior that began around 1985 and has contiued until the present.

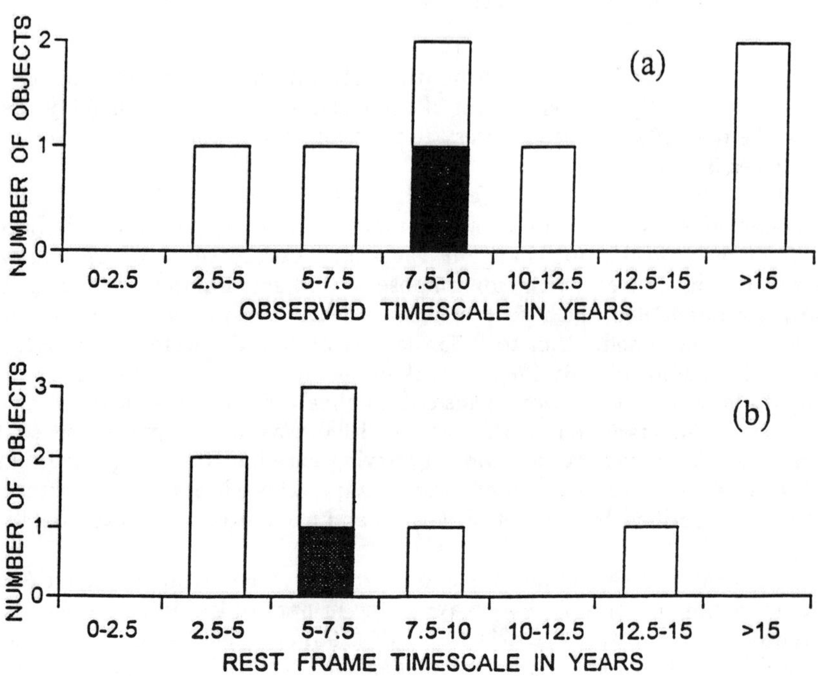

Figure 11. Distribution of base-level timescales of 7 radio-quiet quasars. Upper panel: observed timescales. Lower panel: rest-frame timescales.

be regarded as encouraging the so-called "unified" model of these sources, in which apparent differences in behavior are attributed to differences in the angle at which we happen to view them. The baselines of the radio-quiets seem to fluctuate somewhat more slowly; we are inclined to believe this difference is real even though the statistics are not very good. If the slow, secular changes we see in some radio-quiets are indeed indicative of very long cycles, the mean timescale for the group could become extremely large. And finally, the average base-level timescale for the compact galaxies is almost double that of the classical quasars and BL Lacs. This is consistent with the assumption that these are in fact a different type of object.

Table 1. Parameters of flares for four classes of AGN.

	MEAN REST-FRAME DURATION		MEAN AMPLITUDE	
	Base Level	Flares	Base Level	Flares
QUASARS	5.0 years	71 days	1.49 mag.	0.86 mag.
BL LACERTIDS	5.1	93	.88	1.31
RADIO QUIETS	6.5	108	.63	.48
GALAXIES	9.6	100	.55	.82

3. Data on flares

Thus far I have confined my analysis to the long cycles of the base-level excursions. I would like to conclude by describing some as-yet unpublished work in progress on the short-term flares of AGN. To do this, we identified numerous sections of light curves with substantial flare activity and replotted these with greatly expanded time axes so that I could make accurate measurements of the amplitudes and durations of the flares. So far, I have done this for about 300 flares of 86 objects of all four AGN classes.

The distribution of flare amplitudes for all of the groups is shown in Figure 12. To avoid giving undue weight to a single atypical event, the plotted amplitudes here and in subsequent figures are the means of the three largest flares of each object. The overall mean for each group is shown in the figure. Not surprisingly, the BL Lacertids have the largest amplitudes and the radio-quiets the least. The quasars and galaxies fall between these extremes. The upper tails of the distributions are of course dominated by the blazars.

Is there any relationship between flare amplitudes and the amplitudes of the baseline excursions? This is illustrated for the quasar group in Figure 13, where there appears to be a modest positive correlation. The line is a linear least-squares regression, which has a correlation coefficient R of 0.55.

We can also ask if there might be a relationship between the amplitudes of the flares and their durations. Somewhat surprisingly, the correlation shown in Figure 14 for the quasars appears to be negative. The tendency is for the larger flares (blazar activity) to have shorter durations, although the linear regression line represents a correlation coefficient of only 0.3. A similar result obtains for

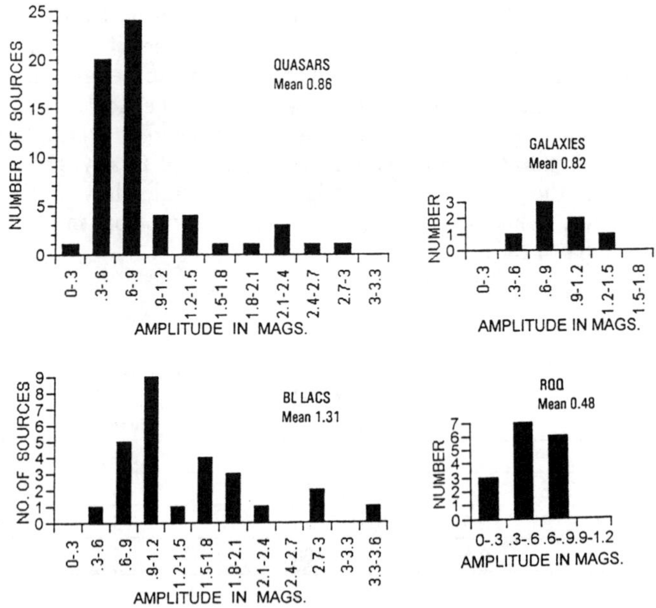

Figure 12. Distribution of flare amplitudes for four classes of AGN.

the BL Lacs (i.e., blazars) seen in Figure 15. Here the correlation coefficient is a little higher, 0.4. One might argue that we are seeing a kind of conservation law; large-amplitude flares tend to have shorter durations, while smaller flares last longer, tending to equalize the energy expended in each case.

Table 1 compares the mean durations of the baseline excursions of each type of object with the mean durations of their flares. Obviously the flares are far more rapid events. Also, the flares typically appear very irregular in outline—"jagged" is the description that comes to mind—whereas the baseline excursions are relatively smooth, sometimes even quasi-sinusoidal. The flares and the baseline excursions appear to be distinctly different phenomena. The last two columns of the table list the mean amplitudes of the flares and baseline excursions. For the quasars, the base-level meanderings substantially exceed the average flare amplitudes; this situation is reversed for the BL Lacertids, which of course display the largest flares.

In the 36 years that have elapsed since the first detection of a quasar, a scattering of patient optical observers have recorded timescales of variation ranging over six orders of magnitude. At one extreme are microvariations occurring in as little as 10 minutes; we ourselves have seen the flux of 3C 345 drop 34% in 19 minutes. At the other extreme are our base-level excursions with typical timescales of 5 to 10 years. It seems quite unlikely that any single mechanism could be responsible for such a gamut of phenomena. Hopefully, as the observations continue to accumulate, ingeneous modellers can invent an array of mechanisms worthy of such complexity.

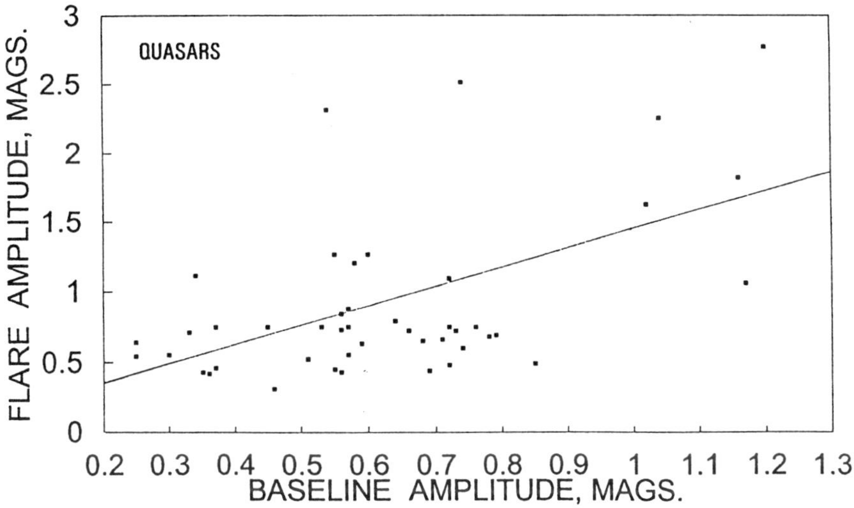

Figure 13. Flare amplitudes vs. the rest-frame durations of the base-level cycles for classical quasars.

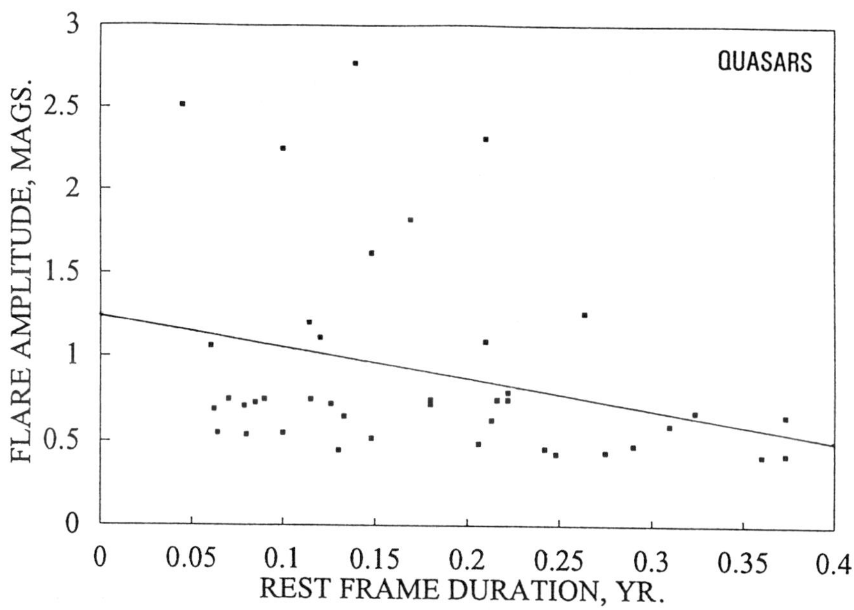

Figure 14. Flare amplitudes vs. the rest-frame durations of the flares for classical quasars.

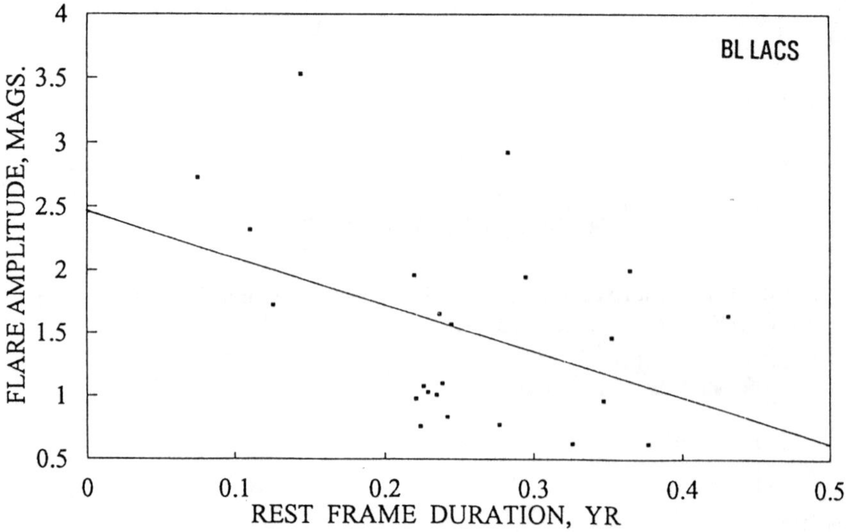

Figure 15. Flare amplitudes vs. the rest-frame durations of the flares for BL Lacertids.

Blazar Continuum Variability
ASP Conference Series, Vol. 110, 1996
H. R. Miller, J. R. Webb, and J. C. Noble, eds.

The Microvariability of Blazars and Related AGN

H. Richard Miller

Dept. of Physics and Astronomy, Georgia State University, Atlanta, GA 30303

John C. Noble

Dept. of Physics and Astronomy, Georgia State University, Atlanta, GA 30303

Center for Automated Space Sciences, Dept. of Physics and Astronomy, Western Kentucky University, Bowling Green, KY 42101

Abstract. The microvariability of blazars has been found to exhibit a wide range of morphological structure from simple linear trends to very complex variations. The morphology of the variability structure provides constraints on the geometry of the emitting region and on the physical processes responsible for the emissions. Radio-selected BL Lac objects have been observed to exhibit larger amplitude, more rapid microvariability than X-ray-selected BL Lac objects. This is consistent with the radio-selected BL Lac objects being more highly beamed than the X-ray-selected BL Lac objects. Microvariability was found to be present for a sample of radio-quiet AGN much less frequently than for a similar sample of radio-loud AGN. This is consistent with models that suggest that the microvariability is produced by shocks in a relativistic jet.

1. Introduction

Variability has been observed over a wide range of time scales and wavelengths for most classes of AGN. It can provide an important tool to probe of the size and structure of the central engine as well as the physical processes which are found at the heart of these intriguing objects. Variability has been detected on timescales ranging from decades to less than a minute. The major goals of AGN monitoring programs are to time-resolve the structure of the variability observed for these obects and to study the evolution of these variations in both time and frequency.

Blazars, as a class, are among the most violently variable AGN investigated and, as such, they provide a true challenge for monitoring programs. Variability timescales have been derived for many blazars from monitoring programs which attain a time resolution of days to years (e.g., Smith, 1996) and from several hours to days (e.g., Urry et al. 1993; Mattox & Wagner 1996). The latter variations have been denoted as intraday variability (e.g., Wagner & Witzel 1995). In many instances, blazars vary on timescales substantially shorter than the sampling interval in these previously noted monitoring programs. Recognizing the rapidity of the intrinsic variations, we embarked upon a monitoring program in 1977 in an effort to resolve the temporal structure of the optical variations limited only by the minimum possible sampling time, which may be as short as a few seconds. We identify those variations which exhibit variability structure on timescales ranging from a few seconds to several hours as *microvariability* in order to distinguish them from the intraday variability noted above.

For many years, despite early observational evidence to the contrary, e.g., Racine (1970), Miller et al. (1974), and Miller (1981), astronomers were reluctant to believe that AGN varied on time scales much shorter than a day. The observational difficulties associated with single-channel photoelectric photometry and variations in atmospheric transparency were cited as explanations for apparent variations.

It was the discovery of microvariability for a number of blazars in the optical (Miller 1988; Miller et al. 1989; Carini 1990; Carini & Miller 1992; Miller et al. 1992; Carini et al. 1990, 1991, 1992; Noble 1995; Noble & Miller 1996; Kidger & de Diego 1992) utilizing modern CCD detectors, and the UV (Miller et al. 1992; Urry et al. 1993) that finally convinced astronomers of the reality of this effect. At about the same time, Wagner and co-workers (Wagner et al. 1990; Wagner et al. 1993) detected correlated radio-optical intraday variability for a number of blazars which confirmed the existence of optical variability on time scales significantly shorter that a day. At the present time, the overwhelming majority of blazars which have been monitored for significant lengths of time have been found to exhibit microvariability.

2. Morphology of the variations

The luminosity of individual blazars may vary by as much as two or three orders of magnitude on the time scale of months or longer. These variations may appear to be erratic, but may also appear to be quasi-periodic or eposodic (e.g., Smith 1996). Day-to-day variations of a factor of two or more have also been observed for a number of blazars on many occasions (e.g., Webb et al. 1988). However, the variations detected on significantly shorter time scales may exhibit a character in one instance that is significantly different from the character observed at another time.

In the simpliest cases, the microvariations may appear to be a linear extension between the observations obtained on adjacent nights. We have observed this type of variation frequently for a number of the blazars which we have monitored. An excellent example of this type of behavior is shown in Figure 1 where we detect an approximate linear trend of 0.05 mag. for 3C 66A with a duration of more than 9 hours. These types of events are the most common type of variation which we detect, and they appear to be simply an extension of the large amplitude trends which are present in the long term light curves for these objects.

In other instances, very chaotic, erratic variations are observed in which the trends seen in the variations during one night can not be smoothly extended to the variations detected on subsequent nights. An excellent example of this type of variability is displayed in Figure 2 for the blazar AO 0235+164. In this example, we observe a linear brightening of almost 0.25 mag. on the first

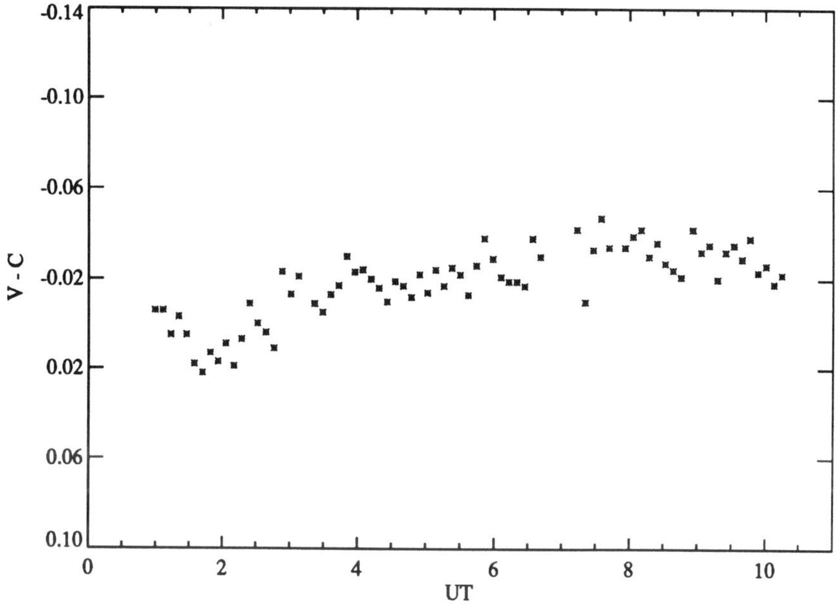

Figure 1. The light curve of 3C 66A for 89 November 07.

night, only to find a well-defined decline and rebound detected on the following night. However, on the third night we detect the onset of a rise in brightness which reaches a peak after several hours and then declines for the remainder of the night. This rise begins at a level that is considerably fainter than that last observed on the previous night. The observations on the fourth night find the source approximately 0.5 mag. brighter than the previous night.

In another case, the variations exhibit a complex character. An outstanding example of this is displayed on the fourth night shown in Figure 2. Erratic variations are observed with an amplitude of 0.05 mag for about 6 hours at a brightness level that is 0.5 mag. brighter than last observed on the previous night. This is followed by an abrupt plummeting of the brightness by approximately 0.25 mag. in less than 2 hours.

In rare instances, one occasionally detects well-defined events which have a duration of only a few hours. One of the best examples of this is displayed in Figure 3 where an outburst with substructure was observed for 3C 279. A well-observed rise and fall of approximately 0.15 mag. with a total duration of 3 hours is followed in detail by the observations. In a second example (Figure 4), a linear decline is observed for B2 1308+326 which has a well-defined excursion superposed on this decline. The duration of this excursion is approximately 45 minutes with an amplitude of approximately 0.03 mag.

These variations provide an important clue into the size and structure, physics, and dynamics of the radiating source region. This small sample of the types of variability structure detected implies that there may be a large

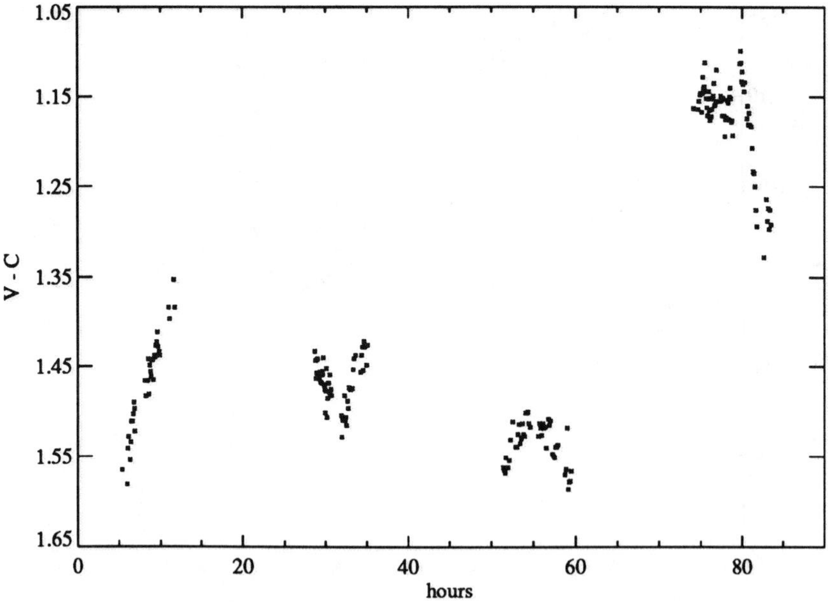

Figure 2. The light curve of AO 0235+164 for 91 November 05–08.

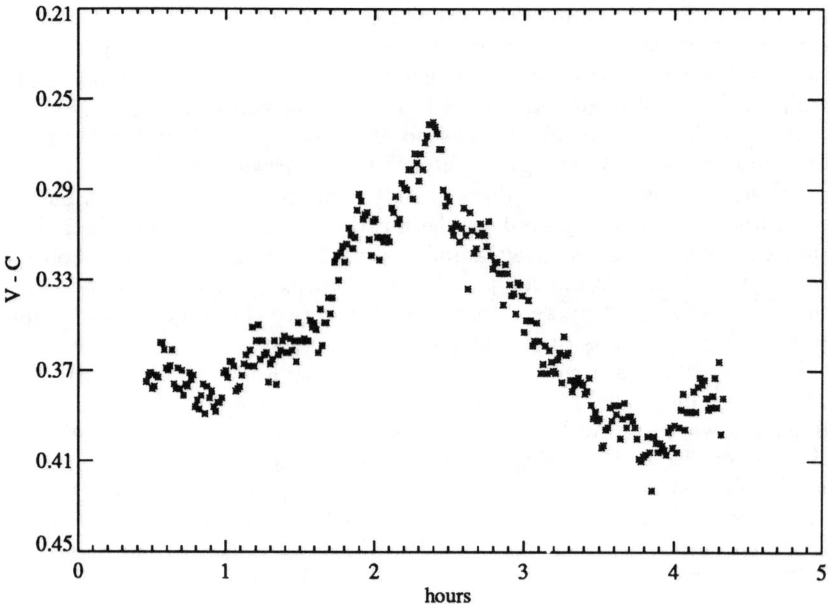

Figure 3. The light curve of 3C 279 for 89 May 10.

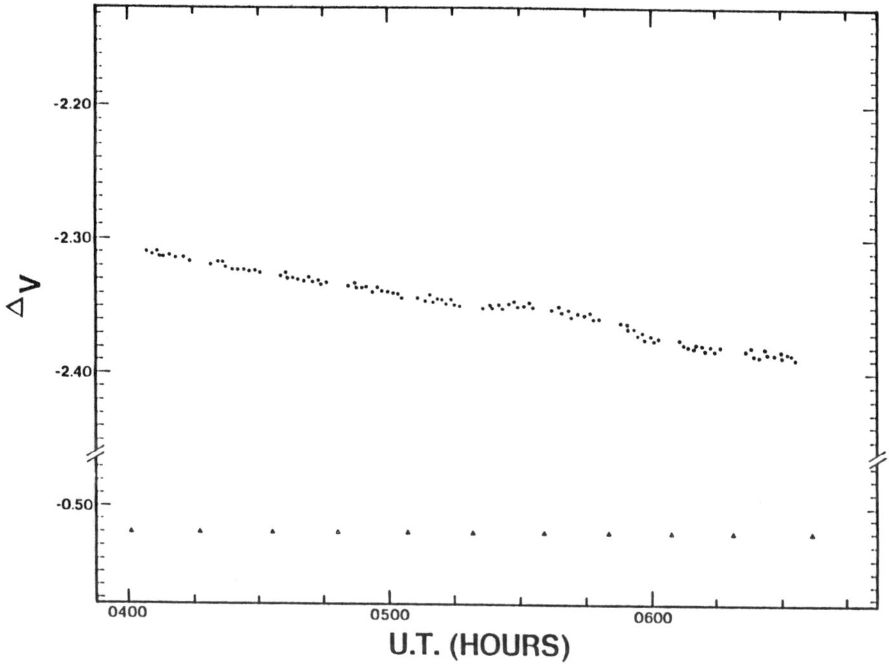

Figure 4. The light curve of B2 1308+326 and a comparison star for 78 May 13.

range in scale for the size of the emitting region found in the same or different objects. These observations also suggest that either the conditions under which these processes arise can have a large range of parameters discribing them, or there may be a variety of competing physical processes which are responsible for the variations that we detect. In either of these instances, it is obvious that additional, long term, high time resolution monitoring for many of these objects is necessary in order to further our understanding of the contraints provided by the variability on the physical mechanisms and on the physical conditions in the source region.

3. X-ray-selected versus radio-selected BL Lac objects

Padovani and Urry (1990) and Urry et al. (1991) have suggested that BL Lac objects can be separated into two subcatagories, X-ray-selected BL Lacs (XBLs) and radio-selected BL Lacs (RBLs). They point out that the differences in these two classes of BL Lacs can be accounted for by viewing the objects in these two classes from slightly different perspectives. In this senario, the RBLs are viewed with the axis of the relativistic jet oriented much closer to the observer's line-of-sight and are more narrowly beamed than are the XBLs. In addition, the RBLs are found to have significantly higher Lorentz factors than those found for the XBLs.

An alternative picture has been proposed by Padovanni and Giommi (1995) in which they suggest dividing the RBL-like and XBL-like objects into low energy-cutoff (LBL) and high energy cutoff (HBL) BL Lacs respectively, where the identification depends on the frequency of the synchrotron peak (Sambruna, Maraschi, & Urry 1996) provide a very recent discussion of SEDs of blazars.). In this case, the LBLs have the peak of their spectral energy distribution in the infrared and the HBLs have the peak of their spectral energy distribution in the UV.

In the first picture, the RBLs would, on the average, have significantly larger Lorentz factors than would be the case for the XBLs. However, in the second case, there is no reason to believe that there exists any significant difference in the typical Lorentz factors for the HBLs and LBLs. Therefore the variability may be used to discriminate between these two pictures. Since the RBLs, on the average, have significantly higher Lorentz factors than do the XBLs, one would expect that RBLs would exhibit larger amplitude, more rapid microvariability than do XBLs.

In order to examine these two hypotheses, we have investigated the amplitude of the microvariations observed for a sample of 32 BL Lac objects, 12 of which are XBLs and 20 of which are RBLs. We have measured the maximum amplitude of the variations which were detected during each continuous observing period for each object. We have then plotted a histogram of the number of observations during which variations of a given amplitude were observed. The histograms for the samples of RBLs and the XBLs are shown in Figures 5 and 6 respectively. In both samples, the distribution for small amplitude variations are quite similar. However, there appears to be an anomalous number of large amplitude events which are present in the sample of RBLs that is not found in the histogram for the XBLs. These large-amplitude events which are present in the RBL distribution are not due to only a few sources contributing all the large variations. From the detailed observations, it can be seen that 11 different sources are responsible for contributing to the large amplitude events of 0.10 mag or larger to the RBL histogram. There is also a conspicuous absence of small amplitude variations in both histograms, but this is due to the criterion that only 2.6σ changes or larger were considered statistically significant, and therefore, should not be construed as the non-existence of low amplitude events. The large-amplitude component detected in the histogram of the RBL variations is also suggested by the observations of Heidt and Wagner (1995) in their analysis of 28 RBLs.

A Kolmogorov-Smirnov (KS) test was performed on the RBL and XBL distributions shown in Figures 5 and 6 in order to determine the likelihood that these results might be obtained from sampling the same parent population of objects. The results of this test indicate that there is only a 2.2% chance that the two distributions are drawn from the same population of objects, i.e., if these two samples of objects *were* from the same parent population, there is only a 1 in 45 chance that similar observations would produce the two histograms shown. This implies that the large-amplitude variations observed for the RBLs, but not the XBLs, are probably unique to the RBLs. This result is consistent with the viewing angle hypothesis in which the RBLs have higher Lorentz factors than

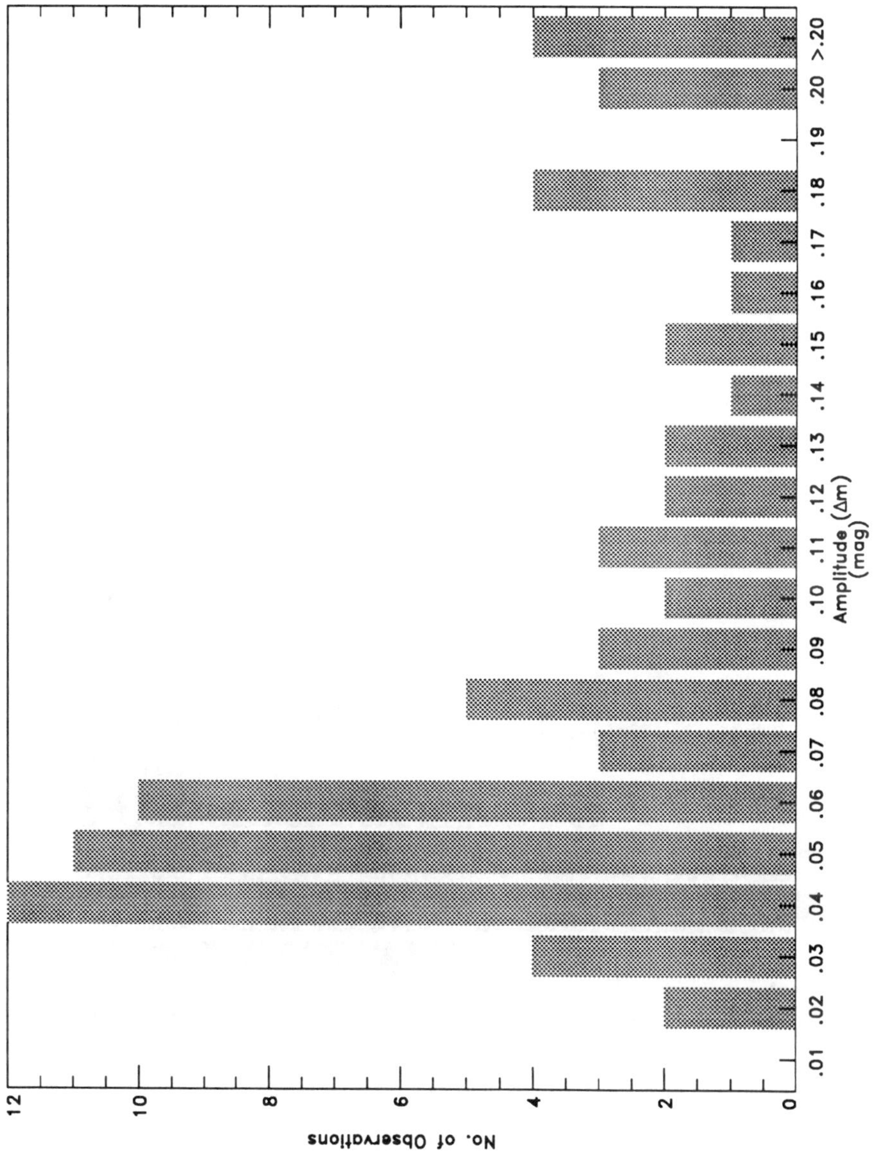

Figure 5. Histogram of varibility amplitudes for RBLs.

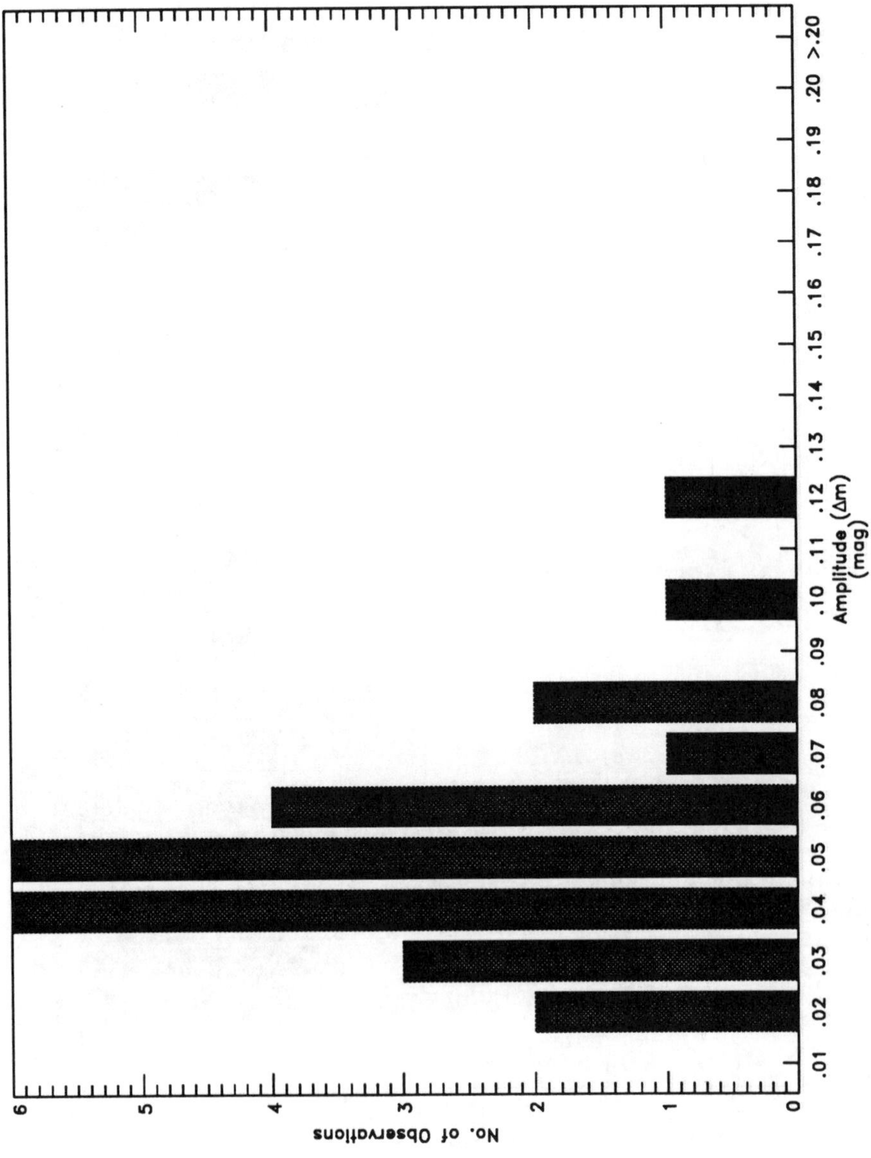

Figure 6. Histogram of varibility amplitudes for XBLs.

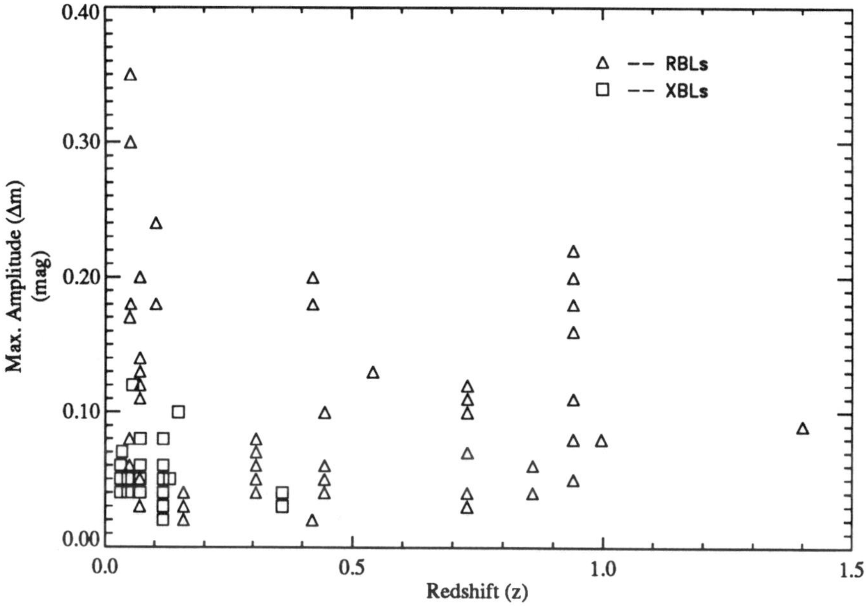

Figure 7. Plot of redshift vs. maximum amplitude of variation.

the XBLs, i.e., the larger Lorentz factors imply that the luminosity variations will be amplified more for the RBLs than for the XBLs.

We have also examined this sample of BL Lac objects to investigate if there exists any relationship between redshift and the amplitudes of the microvariations seen for the XBLs and RBLs. If such a relationship exists, it could suggest that the intrinsic luminosity of the central source might scale with the amplitude of the microvariations or the amount of galactic contamination due to an underlying galaxy might affect the observed brightness changes. The plot of redshift vs. maximum Δm is shown in Figure 7. There is no conspicuous trend present in the data displayed in this diagram. The columnar aspect of the data is indicative of there often being several different variations associated with one a single object. The RBLs and XBLs are both plotted in Figure 7, but with separate symbols in order to distinguish between them. Taken either separately or as a group, neither the RBLs nor XBLs exhibit any correlation between between the amplitude of the variations and redshift.

4. Radio-loud versus radio-quiet AGN

The difference between radio-quiet and radio-loud AGN has long been discussed, particularly in the context of the unified theories (e.g., Antonucci 1993). Radio-loud sources are expected to have relativistic jets which dominate their emission (Bregman 1992). In contrast, it is believed that most radio-quiet AGN do not have relativistic jets (Antonucci et al. 1990) or show only a very weak jet (Miller et al. 1993; Kellermann et al. 1994).

Even though the origin of the difference is an open question, the difference in radio emission makes the test of viable theoretical explanations of microvariability possible by comparing variability characteristics of radio-quiet and radio-loud AGN. Wiita et al. (1991, 1992), Chakrabarti and Wiita (1993), and Mangalam and Wiita (1993) have proposed a model that is based on the formation of disturbances in accretion disks. They suggest that numerous flares or hot spots on the accretion disk surrounding the central source can produce the microvariations observed in AGN. If these explanations are correct, one would expect that microvariability should be present for both radio-loud and radio-quiet AGN. On the other hand, the shocked jet based model which was proposed by Marscher and Gear (1985), Hughes et al. (1985), Valtaoja et al. (1988) and Qian et al. (1991), predicts the presence of microvariability only in radio-loud AGN. Therefore, if microvariability is observed for both radio-loud and radio-quiet AGN, this indicates that the variations are independent of the radio properties, and thus independent of the jet. In this case, the microvariations are consistent with an accretion disk model. However, if the microvariations are found only for the radio-loud AGN and not for the radio-quiet AGN, this suggests that they would most likely originate in the jet.

In order to test these hypotheses, we have investigated a sample of 30 bright quasars (Jang & Miller 1995; Jang & Miller 1996) selected from the PG-survey. A total of 11 of the objects studied were radio-loud, and the remaining 19 were radio-quiet. For the 11 radio-loud objects, 9 exhibited clear evidence of microvariations, which is similar to the results of Heidt and Wagner (1995), who found 82% of their sample of 34 blazars exhibited intraday variability, and Noble (1995), who found 84% of his sample of 32 blazars exhibited microvariability.

In contrast with these results, only 2 out of 19 objects in the radio-quiet sample exhibited microvariability. This very low detection rate for RQQSOs is consistent with the work of Gopal-Krishna et al. (1993, 1995) who reported two highly probable detections of microvariation in their sample of 12 RQQSOs. The best example for the detection of microvariability for a radio-quiet AGN is TON 951 and is shown in Figure 8. Observations of this object produced a clear detection of intranight variability for a radio-quiet AGN. Rapid variations are observed on January 25 and 30. When these nights' observations are compared in order to search for night to night variations, a change of 0.12 mag in the mean brightness level between the two nights was also found.

These results strongly indicate that microvariability is very likely associated with radio-loud AGN. This association is consistent with the models based on jets being primarily responsible for observed variations. However, it is believed that most AGN have accretion disks around their central regions. Thus if the AGN are viewed close to face-on, as unified schemes indicate for both the case of radio-quiet QSOs and radio-loud QSOs, the intrinsic variations in the disk are expected to be visible to the observer. Our results show a clear contrast between these two classes. It could imply that the strengths of disturbances, if they are present on the accretion disks, are so low as not to be detected in most cases. Otherwise, the weak disk disturbances would be swamped by jet-based disturbances for radio-loud AGN. However, clear detection of microvariations in only a few RQQSOs is not fully understood in the context of the present models

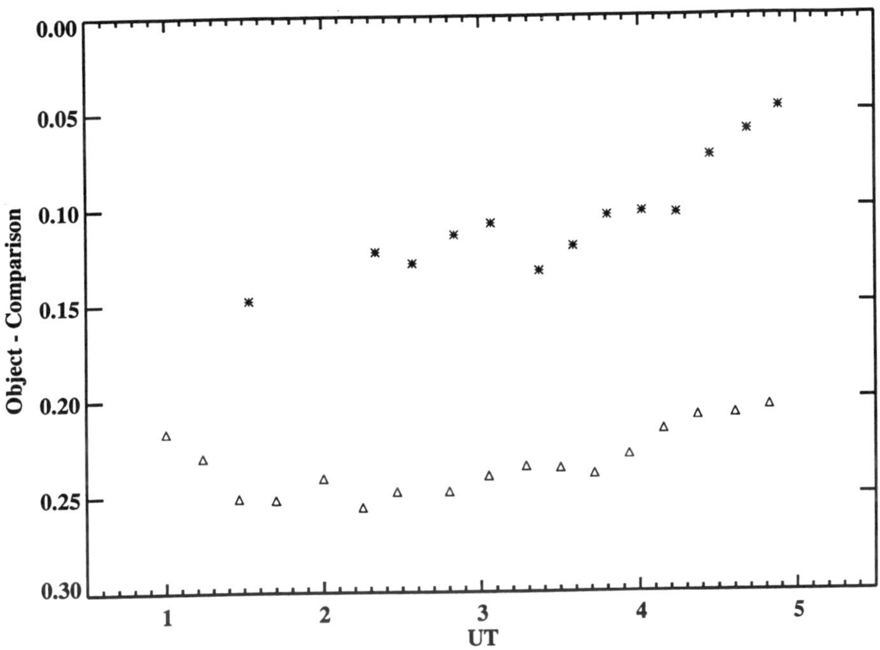

Figure 8. The light curves of TON 951 for 95 January 25 ($*$) and 30 ($\triangle$).

unless one assumes the selective presence of strong hot spots on accretion disks in some RQQSOs. The present observations do not allow one to ascertain if this is truly the case.

5. Summary

All classes of AGN which have been studied have exhibited microvariability, indicating that it is a characteristic of the AGN phenomena. However the nature, morphology, and duty cycle for the microvariability is quite different for different classes of AGN.

Radio-quiet AGN have duty cycles of $\sim 10-20\%$ and exhibit microvariations that, for the most part, appear to be linear trends that are merely smooth extensions of the small night-to-night variations that are well-established for these objects. Radio-loud AGN have duty cycles of $\sim 80\%$ and appear indistinguishable, in this respect, from blazars. However, the extreme complex variations observed for many blazars appears to be absent or quite infrequent for the radio-loud AGN.

Radio-selected BL Lac objects have been found to systematically exhibit larger-amplitude, more complex microvariations than X-ray-selected BL Lac objects. This characteristic is consistent with the RBLs having systematically higher Lorentz factors and being more narrowly beamed than the XBLs.

While it is clear that considerable progress has been made in understanding AGN microvariability, there are still a number of questions that remain to be answered. Do the various morphological classifications for the microvariations evolve in time in an orderly sequence or are the different morphological classes associated with different classes of AGN? How well is the morphology of the microvariations preserved and how does it evolve across several decades in frequency in the elctromagnetic spectrum? What is the duty cycle for the different classes of AGN? and how well has the duty cycle been determined to date? While some of these questions have answers in which we are beginning to have some degree of confidence, there are others to which we have only hints or clues.

In summary, it seems to me that the most important series of observations that must be carried out in the future is a continuous, high-time resolution monitoring program simultaneously at several frequencies and which has a duration of several months. It is only through programs such as this that we will truly understand the origin of the variability/microvariability of AGN.

Acknowledgments. The authors wish to thank the University of Florida and Lowell Observatory for providing allocations of observing time to this project. This work has been supported in part by an award from GSU's Chancellor's Initiative Fund to PEGA, and by grants from the Research Corporation and NASA (NAGW-4397). JCN was supported, in part, by a grant from NASA (NCCW-0085).

References

Antonucci, R. 1993, ARA&A, 31, 473

Antonucci, R., Barvainis, R., Alloin, D. 1990, ApJ, 353, 416

Bregman, J. N. 1992, in Variability of Active Galactic Nuclei, ed. H. R. Miller & P. J. Wiita (Cambridge, Cambridge University Press), p. 1

Carini, M. T. 1990, Ph.D. thesis, Georgia State University

Carini, M. T., Miller, H. R., & Goodrich, B. D. 1990, AJ, 100, 347

Carini, M. T., et al. 1991, AJ, 101, 1196

Carini, M. T., & Miller, H. R. 1992, ApJ, 385, 146

Carini, M. T., Miller, H. R., Noble, J. C., & Goodrich, B. D. 1992, AJ, 104, 15

Chakrabati, S. K., & Wiita, P. J. 1993, ApJ, 411, 602

Gopal-Krishna, Sagar, R., & Wiita, P. J. 1993, MNRAS, 262, 963

Gopal-Krishna, Sagar, R., & Wiita, P. J. 1995, MNRAS, 274,701

Heidt, J., & Wagner, S. J. 1995, A&A, 305, 42

Hughes, P. A., Aller, H. D., & Aller, M. F. 1985, ApJ, 298, 301

Jang, M., & Miller, H. R. 1995, ApJ, 452, 582

Jang, M., & Miller, H. R. 1996, AJ, in press

Kellermann, K. I., et al. 1994, AJ, 108, 1163

Kidger, M. R., & de Diego, J. A., 1992, A&AS, 93, 1

Mangalam, A. V., & Wiita, P. J. 1993, ApJ, 406, 420

Marscher, A. P., & Gear, W. K. 1985, ApJ, 298, 114

Mattox, J., & Wagner, S. J. 1996, these proceedings

Miller, H. R., Clonts, S. L., & Folsom, G. H. 1974, AJ, 79, 1352

Miller, H. R. 1981, ApJ, 244, 426

Miller, H. R. 1988, in Active Galactic Nuclei, ed. H. R. Miller & P. J. Wiita (Berlin, Springer-Verlag), p. 146

Miller, H. R., Carini, M. T., & Goodrich, B. D. 1989, Nature, 337, 627

Miller, H. R.., Carini, M. T., Noble, J. C., Webb, J. R., and Wiita, P. J. 1992, in Variability of Blazars, ed. E. Valtaoja & M. Valtonen (Cambridge, Cambridge University Press), p. 320

Miller, P., Rawlings, S., & Saunders, R. 1993, MNRAS, 263, 425

Noble, J. C. 1995, Ph.D. thesis, Georgia State University

Noble, J. C., & Miller, H. R. 1996, these proceedings

Padovani, P., & Giommi, P. 1995, ApJ, 444, 567

Padovani, P., & Urry, C. M. 1990, ApJ, 356, 75

Qian, S.-J., Quirrenbach, A., Witzel, A., Kirchbaum, T. P., Hummel, C. A., & Zensus, J. A. 1991, A&A, 241, 15

Racine, R. 1970, ApJ, 159, L99

Sambruna, R. M., Maraschi, L., & Urry, C. M. 1996, ApJ, 463, 444

Smith, A. G. 1996, these proceedings

Urry, C. M., Padovani, P., & Stickel, 1991, ApJ, 382, 501

Urry, C. M., et al. 1993, ApJ, 395, 187

Urry, C. M., & Padovani, P. 1995, PASP, 107, 803

Valtaoja, E., et al. 1988, A&A, 203, 1

Wagner, S. J., Sanchez-Pons, K., Anton, K., Quirrenbach, A., & Witzel, A. 1990, A&A, 235, L1

Wagner, S. J., Sanchez-Pons, K., Anton, K., Quirrenbach, A., & Witzel, A. 1991, in Variability of Active Galactic Nuclei, ed. H. R. Miller & P. J. Wiita (Cambridge, Cambridge University Press), p. 120

Wagner, S. J., et al. 1993, A&A, 271, 344

Wagner, S. J., & Witzel, A. 1995, ARA&A, 33, 163

Webb, J. R., et al. 1988, AJ, 95, 374

Wiita, P. J., Miller, H. R., Carini, M. T., & Rosen, A. 1991, in Structure and Emission Properties of Accretion Disks, IAU Colloquium 129, ed. J. P. Lasota et al., p. 557

Wiita, P. J., Miller, H. R., Gupta, N., & Chakrabarti, S. K. 1992, in Variability of Blazars, ed. E. Valtaoja & M. Valtonen (Cambridge, Cambridge University Press), p. 311

Blazar Continuum Variability
ASP Conference Series, Vol. 110, 1996
H. R. Miller, J. R. Webb, and J. C. Noble, eds.

Extreme Optical Microvariability in Three Selected Blazars

John C. Noble

Dept. of Physics and Astronomy, Georgia State University, Atlanta, GA 30303

Center for Automated Space Sciences, Dept. of Physics and Astronomy, Western Kentucky University, Bowling Green, KY 42101

H. Richard Miller

Dept. of Physics and Astronomy, Georgia State University, Atlanta, GA 30303

Abstract. Optical microvariability (flux variations on timescales of minutes to hours) is a well established phenomenon for the blazar class of active galactic nuclei (AGN). It has direct implications to the physical mechanisms that can produce such variations. By making a few hundred individual observations per night, a more detailed look at the specific character of these microvariations can be obtained. These specific characterizations can be then be analyzed in order to more accurately understand the microvariability phenomenon in blazars. This paper presents a study of some of the most extreme optical microvariability ever observed for this class of objects as depicted in observations of the three blazars 0109+22, AO 0235+164, and 4C 29.45. Observing runs of several days each have allowed for the acquisition of several hundred individual observations per object in an attempt to characterize the microvariability observed. The light-curves of the objects reveal both enormously rapid flux variations as well as interesting structure in the variations. Both of these aspects of the variations have direct implications on the physical mechanisms purported to produce them, namely shocks in jets or flares on an accretion disk. Only through detailed modeling of these types of variations will the controversy over which mechanism is responsible for them be resolved.

1. Introduction

The evidence for microvariability for the blazar class of AGN has been well-established (Miller 1988; Miller et al. 1989; Carini 1990; Carini et al. 1990, 1992; Carini & Miller 1992; Miller et al. 1992; Noble 1995). The connection between this variability and the underlying mechanism producing it remains a mystery. Two viable mechanisms for explaining the microvariability are: hot spots or flares from the accretion disk which surrounds the central super-massive black hole, or disturbances in the jets emanating from the core source. Three blazars, 4C 29.45 (1156+295), AO 0235+164, and 0109+22, monitored for microvariability, may shed some light on this mystery. These three blazars were chosen for optical monitoring of microvariability based on their historical brightness variations, which are some of the largest ranges in brightness for any blazar. Wills et al. (1983), show V-magnitude values for 4C 29.45 that range from around 13.3–18.2. Investigations by Miller (private communication) into the historical brightness of 0109+22 show this object to have a range in V-magnitude of between 12 and 16.5–17.0. AO 0235+164 has exhibited a range of around 5 magnitudes in brightness: the minimum brightness near V = 19.5 was reported by Spinrad and Smith (1975), and an outburst of V = 14.3 was observed by Rieke et al. (1976). Because of these large, long-term changes in brightness, it was believed that they would make good candidates for exhibiting large, short-term variations as well, as opposed to some of the more moderately varying blazars. Indeed this turned out to be the case during night-long optical monitoring of these objects.

2. Observations

Using the 0.9-m telescope at Kitt Peak National Observatory and the 1.07-m Hall telescope at Lowell Observatory, CCD images were taken of the respective blazars using Johnson V- and R-band filters. Several hundred successive images were taken each night, over several nights, in order to be able to capture even the most rapid events that might occur within a night. Processing of the images was done using standard IRAF[1] software packages which implemented bias-frame and flat-field averaging, and bias-subtraction and flat-field division of the object frames in order to redue background noisė sources. Aperture photometry was performed on the blazar and comparison stars of known magnitude in each frame using IRAF as well. Differences were taken between the instrumental magnitudes, computed from the aperture photometry, of both the object and any comparison stars in each frame in order to further eliminate magnitude fluctuations due to atmospheric conditions. Plots of the night-long light-curves for each blazar were then created using these instrumental magnitude-differences.

3. Results

Over each observing run, all of the objects displayed large brightness fluctuations similar to those seen over the long-term, year-to-year observations. Ignoring the details of the fluctuations that occurred during each night for the moment, each object shows changes in brightness of several tenths of a magnitude. Figure 1 shows 0109+22 undergoing an outburst event during which it brightened by $\sim$ 0.16 mag. and then faded by $\sim$ 0.34 mag. For the blazar 4C 29.45, Figure 2 reveals an outburst event over the last three days of observations of $\sim$ 0.30 mag., and a decrease in overall brightness of $\sim$ 0.73 mag. Similar week-long fluctuations can be seen for AO 0235+164 in Figures 3 and 4. This large-scale,

---

[1]IRAF is distributed by the National Optical Astronomy Observatories, which is operated by the Association of Universities for Research in Astronomy, Inc. under contract to the National Science Foundation.

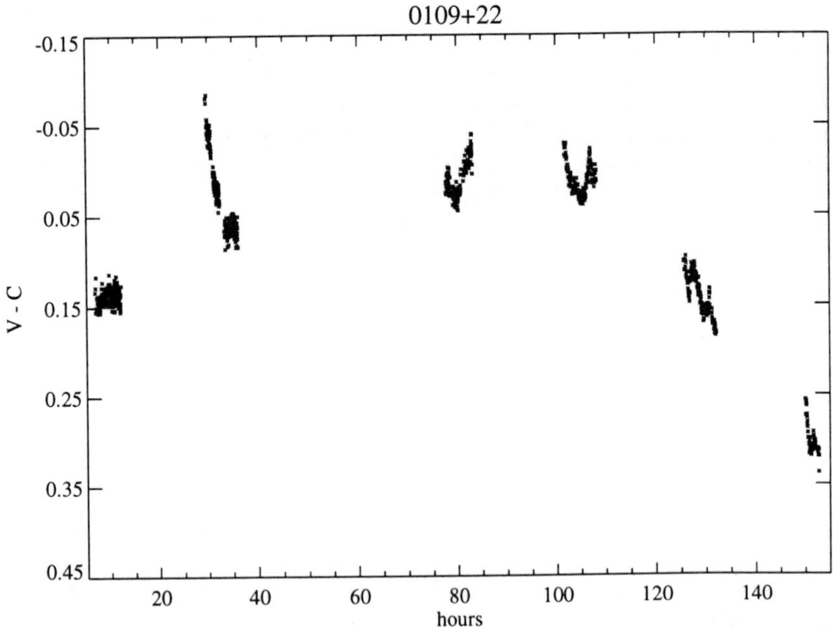

Figure 1. The light curve of 0109+22 from 94 September 23, 24, 26–29.

week-long variability is exactly what characterizes blazars, but it is also a signal that there must be shorter-term variability present as well.

These three blazars are dramatic examples of this class of object which display this microvariability, and; the microvariability exhibited is itself of a quite dramatic nature.

Some of the most rapid microvariability events ever observed are depicted in the observations reported here. The blazar 0109+22 shows a rate of change in brightness of $\sim$ 0.05 mag/hr over the first three-and-a-half hours of observations on the night of 94 September 24 (Fig. 1, 2nd group of data), and; on 94 September 26 (Fig. 1, 3rd group of data), the object shows an $\sim$ 0.07-mag/hr change in brightness. Such rapid variability is also seen in 4C 29.45 on 94 January 21, where a decline of $\sim$ 0.04 mag/hr occurred during the last 1.1 hours of observations (Fig. 2, 3rd group of data), and again on 94 January 22 (Fig. 2, 4th group of data), where a brightening of $\sim$ 0.05 mag/hr occurred during the middle of the observations. The blazar AO 0235+164 shows some of the *fastest* changes in brightness on the intra-day timescale: on 90 October 30 it underwent an $\sim$ 0.10-mag/hr decline during the first hour and then an $\sim$ 0.07-mag/hr decline during the last two hours of observations (Fig. 3, 4th group of data); similarly, on 91 November 08 (Fig. 4, 4th group of data), during a two-hour span, it declined in brightness by $\sim$ 0.19 mag!

In addition to the rapidity of the variability, the nightly observations of these three blazars also revealed very well-defined microvariability *events*. For example, on 94 September 27 (Fig. 1, 4th group of data), 0109+22 shows an out-

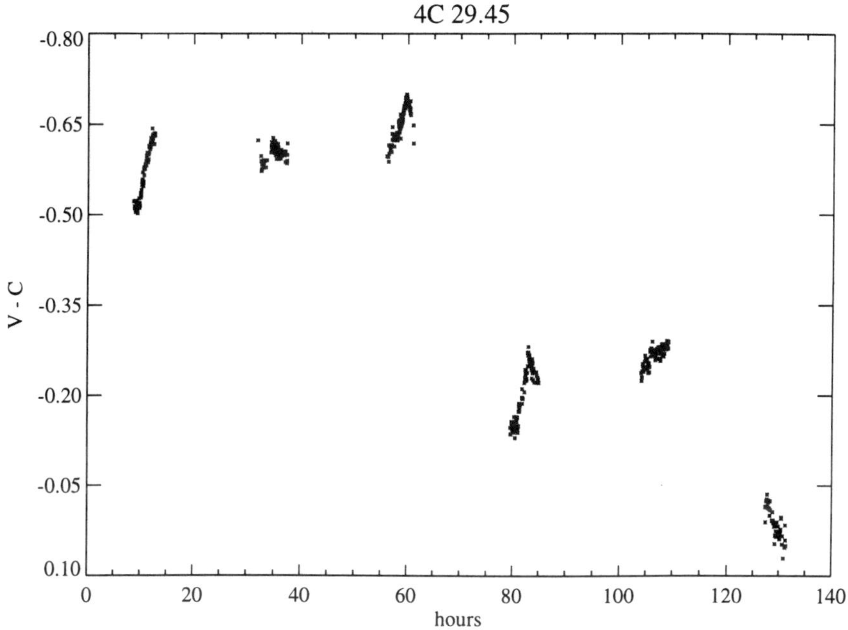

Figure 2. The light curve of 4C 29.45 from 94 January 19–24.

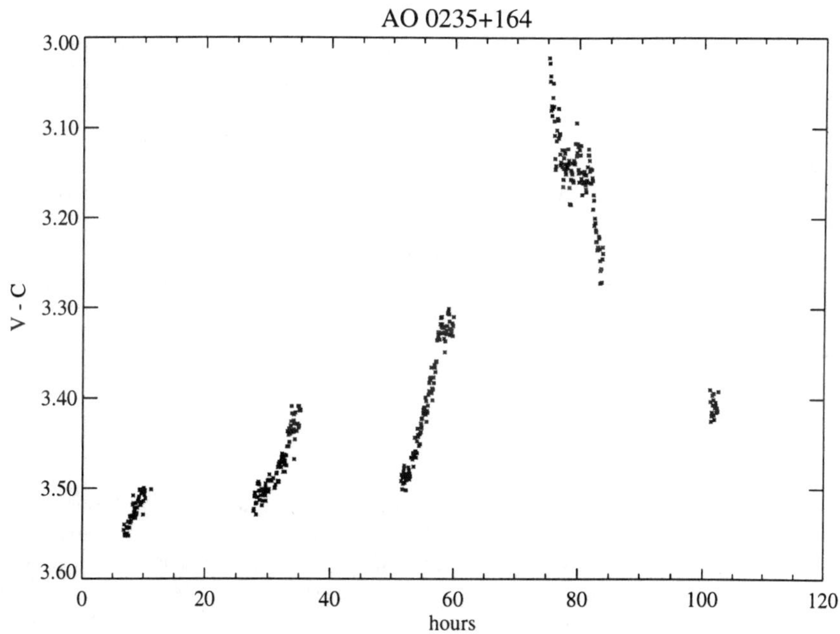

Figure 3. The light curve of AO 0235+164 from 90 October 27–31.

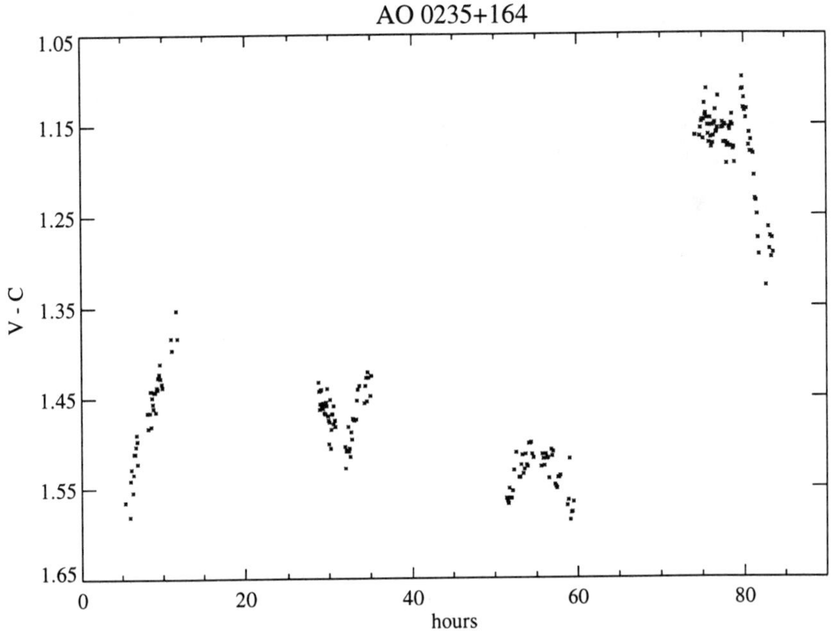

Figure 4. The light curve of AO 0235+164 from 91 November 05–08.

burst event near the end of the observations, and on 94 September 28 (Fig. 1, 5th group of data), it shows a well-defined event during the *middle* of the observations, and possibly a second near the end of the night. The blazar 4C 29.45 shows an outburst event at the end of observations on 94 January 21 (Fig. 2, 3rd group of data), and what looks like most of another event on 94 January 22 (Fig. 2, 4th group of data). AO 0235+164 exhibits a clearly defined outburst event on 91 November 07 (Fig. 4, 3rd group of data). This behavior suggests that some of the other variations in brightness captured for these blazars, like that for 4C 29.45 on 94 January 19 and AO 0235+164 on 90 October 29 and 30 and again on 91 November 05, are part of larger microvariability events. If so, they represent some of the largest changes in brightness ever observed on these timescales for the blazar class of objects.

4. Conclusions

It is interesting to note that all the objects were in a relatively faint phase with regard to the known ranges of their historical variability. This aspect could prove helpful in defining the type of mechanism that is responsible for the microvariability phenomenon. In one scenario, since these objects are all radio-loud, it might be reasonable to assume that during a quiescent period in the overall brightness of the object, the dominant radio jet is at a low level of radiation production. The implication here is that there must then be something other than the jet which is responsible for the microvariability events. The only other likely candidate for radiation production around blazars is an accretion disk of material surrounding the putative supermassive black hole central source. Therefore, if the jet radiation has been reduced considerably with respect to the radiation coming from the disk, then the microvariability phenomenon reported here may be indicative of the kind of activity going on in the disk. Of course, these results deal only with three objects, so a more detailed, statistical analysis of a larger set of blazars into the kind of microvariability activity that occurs during faint stages would be more appropriate to the discussion of this idea.

Acknowledgments. The authors wish to thank Lowell Observatory and Kitt Peak National Observatory for providing allocations of observing time to this project. This work was supported in part by an award from GSU's Chancellor's Initiative Fund to PEGA, and by grants from NASA (NAGW-4397) and the Research Corporation. JCN was also supported in part by NASA grant NCCW-0085.

References

Carini, M. T. 1990, Ph.D. thesis, Georgia State University

Carini, M. T., Miller, H. R., & Goodrich, B. D. 1990, AJ, 100, 347

Carini, M. T., & Miller, H. R. 1992, ApJ, 385, 146

Carini, M. T., Miller, H. R., Noble, J. C., & Goodrich, B. D. 1992, AJ, 104, 15

Miller, H. R. 1988, in Active Galactic Nuclei, ed. H. R. Miller & P. J. Wiita (Berlin, Springer-Verlag), p. 146

Miller, H. R., Carini, M. T., & Goodrich, B. D. 1989, Nature, 337, 627

Miller, H. R.., Carini, M. T., Noble, J. C., Webb, J. R., & Wiita, P. J. 1992, in Variability of Blazars, ed. E. Valtaoja, & M. Valtonen (Cambridge, Cambridge University Press), p. 320

Noble, J. C. 1995, Ph.D. thesis, Georgia State University

Rieke, G. H., et al. 1976, Nature, 260, 754

Spinrad, H., & Smith, H. E. 1975, ApJ, 201, 275

Wills, B. J., et al. 1983, ApJ, 274, 62

Blazar Continuum Variability
ASP Conference Series, Vol. 110, 1996
H. R. Miller, J. R. Webb, and J. C. Noble, eds.

Very Short-term Variability of PKS 2155–304 in the Optical Domain

S. Paltani[1]

Geneva Observatory, ch. des Maillettes 51, CH-1290 Sauverny, Switzerland

Institut d'Astronomie de l'Université de Lausanne, CH-1290 Chavannes-des-Bois, Switzerland

T.J.-L. Courvoisier

Geneva Observatory, ch. des Maillettes 51, CH-1290 Sauverny, Switzerland

INTEGRAL Science Data Centre, ch. d'Ecogia 16, CH-1290 Versoix, Switzerland

P. Bratschi, and A. Blecha

Geneva Observatory, ch. des Maillettes 51, CH-1290 Sauverny, Switzerland

Abstract. We have performed a two-week campaign with an optical telescope in La Silla, Chile, in order to quantify the properties of the very short-term variability of the well-known BL Lac PKS 2155–304. Variations on time scales as short as one hour have been observed. We find also some evidence for hardening of the spectrum when the flux increases. We estimate the variability properties using structure functions. They are compatible with one stochastic process with a powerlaw-type power-spectrum, the index being larger than 2.

1. Introduction

Many BL Lac objects are known to vary on very short time scales, as short as 1 hour (Wagner & Witzel 1995), especially those belonging to the "radio-selected" sub-class. One of them, PKS 2155–304, has been observed for four consecutive nights by Miller and Carini (1991). Although clear intra-day (or intra-night) variability has been detected, they found no evidence for variability on time scales significantly smaller than one day. Indeed the whole light curve really looks like one period of a 90-hour period sine function. This behavior is very different from the one described, for instance, by Wagner et al. (1991) for 0716+71, whose light curve shows much more complicated structures.

---

[1]Present address: CESR, 9, av. du Colonel-Roche, F-31029 Toulouse, France

Table 1. Statistics of the observations

	Filters				
	UG	BG	VG	RG	IC
Number of observations	28	114	268	226	133
Uncertainty (mJy)	–	0.07	0.06	0.05	0.07
Mean flux (mJy)	13.3	15.5	18.9	21.8	24.6
Standard deviation (mJy)	2.1	2.3	2.7	2.7	3.1
Standard deviation (%)	15.4	14.7	14.0	12.2	12.6

We describe here a much longer optical campaign on PKS 2155–304, which aimed at quantifying the amount of very short-term variability in this object. The tool that we chose for this purpose is the structure function, which is a very convenient way to obtain rough properties of a time series.

The campaign was also designed to allow us to get (limited) spectral information, which has been used to relate the change in flux with eventual hardening or softening of the spectrum.

2. Observations

We observed PKS 2155–304 for 15 consecutive nights between July 26 and August 9, 1995, with the 70-cm Swiss Telescope at La Silla, European Southern Observatory, Chile. Photometric observations have been obtained with a CCD camera in 5 filters: UG, BG, VG, RG and IC. UG, BG and VG are Geneva passbands, comparable to Johnson's U, B and V passbands respectively, while RG and IC are close to the Gunn R and Cousins I passbands respectively (Blecha & Bratschi 1996). A comparison star in the CCD field has been used for differential photometry.

The exposure time has been calculated to achieve an accuracy of the order of 1 % or better. This led us to set the exposure time to 2 min. in the VG band. The number of data points (and hence the time sampling) differs strongly from filter to filter; it exceeds 250 in the VG band (Table 1).

The shortest time scale sampled is 2 min. in VG and RG (27 data pairs with $\Delta\tau < 3$ min. in VG). The uncertainties are derived from the plateau of the structure functions at very small $\Delta\tau$ (see Table 1 and Section 5.). They are of the order of 0.5 %.

3. Light curves

Figure 1 shows the light curves in the 5 filters. Details of the light curves are given in Figure 2. Intra-day variations are clearly visible during almost all nights (principally in the VG and RG bands, due to the better sampling). Moreover the behavior of the light curves corresponds much better to what is usually referred to as intra-day variability.

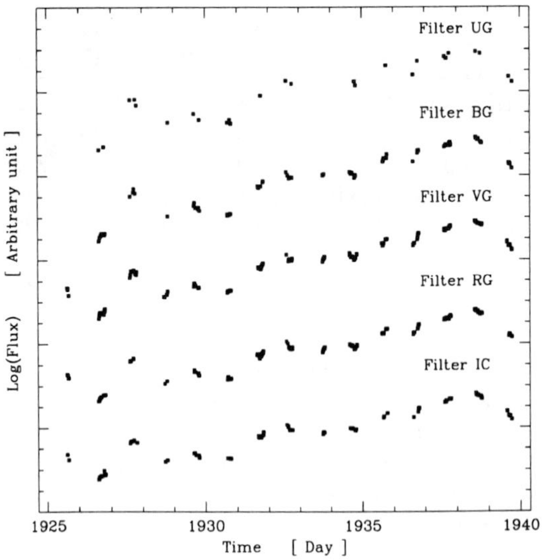

Figure 1. Light curves in the 5 optical bands

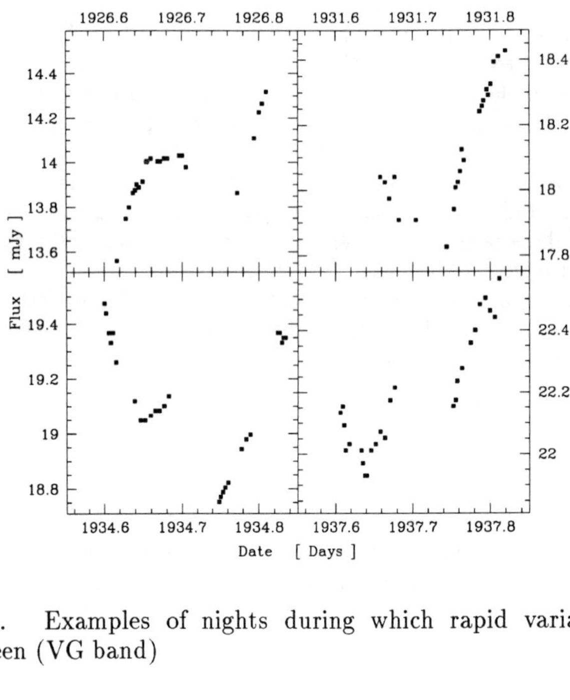

Figure 2. Examples of nights during which rapid variations are clearly seen (VG band)

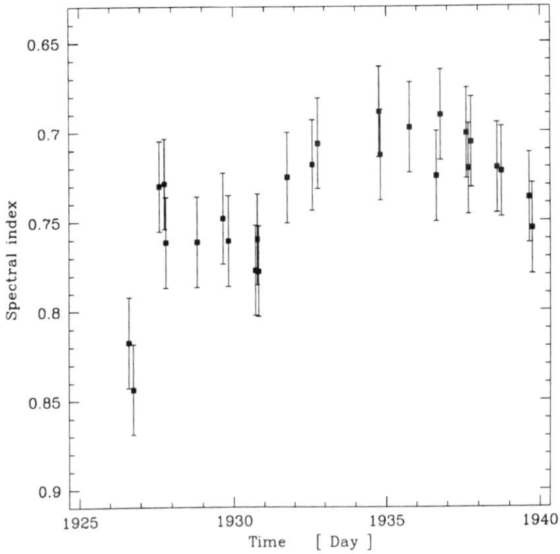

Figure 3. Spectral index as a function of time

Basic statistics of the data are given in Table 1. The variability is rather small. The 1991 campaign (Courvoisier et al. 1995), which lasted 1 month, has seen a variability of 18.3 % in the V band. The "long-term" amplitude of variability ($\Delta\tau > 10$ yr) obtained from the study of the IUE database (Paltani & Courvoisier 1994) gives 23.7 ± 0.2 % at 2 000 Å, which puts an upper limit to the "long-term" amplitude of variability in the V band, since the percentage of variability decreases whith the wavelength in all kinds of active galactic nuclei (Paltani & Courvoisier 1994). This decrease of variability is also marginally visible in this campaign.

4. Spectral index variation

Figure 3 shows the variation of the spectral index as a function of time. Estimations of the spectral index have been made every time that an observation in the UG filter has been obtained.

Comparing Figure 1 and Figure 3, we find evidence that the spectrum becomes harder when the object becomes brighter. During the November 1991 campaign (Courvoisier et al. 1995) the spectral index was found to be compatible with a constant, with a mean value of 0.66, which can be compared to our mean value of 0.74. However, the mean flux during the November 1991 campaign was about 50 % higher than during this one. This observation may thus confirm the existence of a correlation between the flux and the spectral index in the manner described above.

5. Structure function analysis

Figure 4 shows the structure functions (SFs) of the VG and RG light curves with a 2-min. time interval up to $\Delta\tau = 5$ hours and Figure 5 shows the structure functions of the BG, VG, RG and IC light curves with a 5-min. time interval up to $\Delta\tau = 10$ days.

The main feature of all SFs is a power-law whose low-$\Delta\tau$ part is contaminated by a white noise due to the measurement uncertainties (hence their estimations). We cannot exclude the presence of other features (for instance, pseudo-period); however we see no clear evidence for any of them.

The power-law extends down to 15 min. (eye estimation) without any evidence of steepening before merging into the white noise. As the SF has always a slope of 2 for $\Delta\tau$ significantly smaller than the minimum time scale of variation, this means that this minimum time scale is shorter than 15 min.

No clear plateau is found at "large" $\Delta\tau$, which means that the maximum time scale has not been reached. The slopes of the SFs are all compatible with 1.25 ± 0.1, and do not seem to be compatible with a slope of 1 (Figure 5). A power-law shaped SF indicates that the Fourier power spectrum of the light curve is also power-law shaped, the index of this new power-law being related to the one of the SF. In this case, the index of the power spectrum is about 2.3, very different from the index of a flicker noise (index = 1), and probably significantly larger the one of a random walk (index = 2). It means that the power in the light curve decreases very quickly toward the high frequencies. Therefore very short-term variability cannot be detected in PKS 2155–304, unless highly accurate observations can be obtained. This may explain why this behavior has not been detected in this object in the previous study of Miller and Carini (1991).

We can infer from the SF the variability in the V band observed after a delay $\Delta\tau$ (in days):

$$\sigma(\Delta\tau) = 1.06 \cdot \Delta\tau^{1.25/2} \text{ mJy}$$

To detect a 1-min. variability, an accuracy better than 0.01 mJy is required. On the other hand, it means that the variability observed in the "long"-term IUE campaign is reached when $\Delta\tau$ is of the order of 1 month. Thus the low-frequency ($1/f > 1$ month) part of the power spectrum has a negligible power (unless this power is distributed at frequencies below $f \simeq (10 \text{ years})^{-1}$).

References

Blecha A., & Bratschi P. 1996, A&A in press

Courvoisier T. J.-L., Blecha A., Bouchet P., et al. 1995, ApJ 438, 108

Miller H. R., & Carini M. T. 1991 in Variability of AGN, ed. H. R. Miller & P. J. Wiita (Cambridge, Cambridge University Press)

Paltani S., & Courvoisier T. J.-L. 1994, A&A 291, 74

Wagner S. J., Sanchez-Pons F., Anton K., Quirrenbach A., & Witzel A. 1991 in Variability of AGN, ed. H. R. Miller & P. J. Wiita (Cambridge, Cambridge University Press)

Wagner S. J., & Witzel A. 1991, ARA&A 33, 163

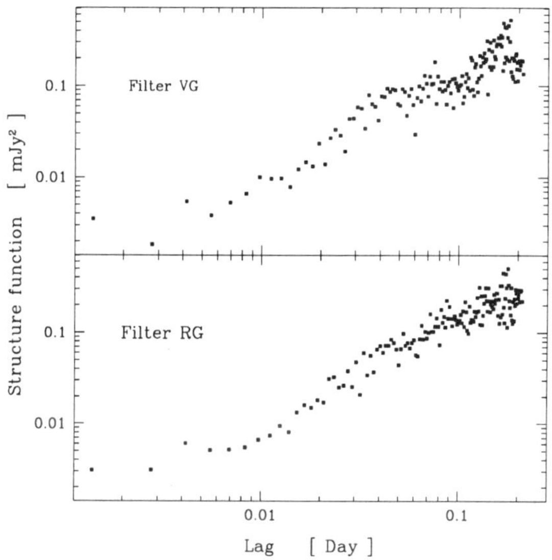

Figure 4. Structure functions of the VG and RG light curves with a 2-min. time interval

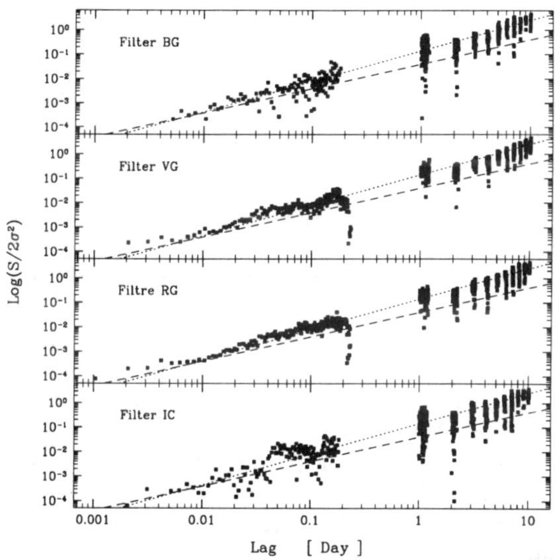

Figure 5. Structure functions of the BG, VG, RG and IC light curves with a 5-min. time interval (The dashed line is the expected structure function of a random walk; the dotted line has a slope of 1.25.)

Blazar Continuum Variability
ASP Conference Series, Vol. 110, 1996
H. R. Miller, J. R. Webb, and J. C. Noble, eds.

Accretion Disk Models for Rapid Variability

Paul J. Wiita

Department of Physics and Astronomy, Georgia State University, University Plaza, Atlanta GA 30303-3083

Abstract. Instabilities in accretion disks must produce some fluctuations in emission from blazars and other active galactic nuclei. Phenomenological models based upon generic hot-spots or flares on accretion disks have been shown to be capable of providing good matches to light-curves in the X-ray and optical bands. These models will be reviewed, along with some of the physical processes that might lead to such irregularities. Among the proposed processes are: the formation and break-up of spiral shocks within the disk; strong turbulence or vortices; plasma and MHD instabilities, some perhaps analogous to stellar flares. While some of these are only capable of producing variability through essentially thermal mechanisms, and would thus be concentrated in particular bands, others could be effective over wide ranges of frequencies.

1. Introduction

The tremendous number of instabilities to which accretion disks are subject guarantees that there will be fluctuations in emission from these disks which will yield some degree of variability in blazars and other AGN. Key questions to be addressed to such disk based models for variability include: Do such instabilities involve enough energy to be detectable in various bands? Do they operate on timescales short enough to contribute to microvariability? Are the expected differences in strength at various wavelengths, and possible time-lags between bands, compatible with observations? As of now, it is fair to say that there are no certain answers to any of these questions, but in this review we will begin to address some of them.

Shock-in-jet models (e.g., Marscher & Gear 1985; Hughes, Aller, & Aller 1985) certainly work well for relatively long time scale variations in blazars, and can probably also contribute to intraday variations in radio loud sources (e.g., Marscher 1996). Because the emission from blazars is dominated by a non-thermal (almost certainly synchrotron plus self-Compton) beamed continuum, it may be difficult or impossible to detect the direct effects of disk variability in blazars. But the recent discovery of optical microvariability in radio quiet AGN (Gopal-Krishna, Sagar, & Wiita 1993, 1995; Jang & Miller 1995, 1996; Miller 1996) as well as the X-ray variations in Seyferts (e.g., Lawrence & Papadakis 1993) seem more compatible with models based upon fluctuations produced in, or just above, accretion disks.

We begin with a listing of some of the types of instabilities that can affect accretion disks in AGN (§2), and then summarize phenomenological models that have illustrated how the observed rapid optical/UV and X-ray fluctuations can be accommodated within this paradigm (§3). Models based upon spiral shocks (§4) and various plasma effects (§5) are considered. Some key evidence of rapid optical variability in radio quiet QSOs is presented (§6) before our conclusions and suggestions for future efforts are given (§7).

2. Summary of disk instability mechanisms

Standard models for AGN of all types invoke accretion disks surrounding supermassive black holes (e.g., Rees 1984). These disks are subject to many instabilities and *will* certainly produce variations in emitted flux. A key preliminary question must be: which instabilities could be relevant to variability on different timescales?

Early attempts in this direction tried to make connections between some observed variability timescales and fundamental timescales arising from basic accretion disk models. The viscous accretion timescale is given by $t_{\rm acc} \simeq (v_\phi/c_{\rm s})(r/\alpha c_{\rm s})$, where r is the disk radius, $c_{\rm s}$ is the sound speed, and α is the standard viscosity parameter. The addition of a bolus of matter to the disk would naturally yield luminosity variations over this length of time. This timescale is typically $> 10^3$ yr for parameters expected for AGN accretion disks, even in systems which probably have relatively low central supermassive black hole (SMBH) masses such as Seyfert galaxies (Alloin et al. 1985).

A limit-cycle instability involving convective disks (as is almost certainly seen in dwarf novae) (Minishige & Shields 1990; Cannizzo 1992) may well explain long term on-off cycles ($t_{\rm recur} \sim 10^4 - 10^8$ yr). This mechanism is based on the fact that the $T_{\rm eff}$ vs. Σ (column density) curves have an **S** shape for convective disks. During a quiescent period, matter can pile up in the disk as the local mass infall rate is less than the steady-state supply at higher radii, and the entire disk resides on the lower portion of that **S**-curve. When $\Sigma_{\rm max}$ (where the curve reverses to the left) is exceeded, a heating front travels out from that annulus, thereby transforming nearby regions to the hot state (upper portion of the **S**-curve), thereby yielding much greater local viscosity (Cannizzo 1992). After some material has accreted, a cooling front, which begins at large r where $\Sigma < \Sigma_{\rm max}$, propagates back inward and shuts off the inflow, typically allowing only ~20% of the stored mass to fall into the central object (Mineshige & Shields 1990). The time during which the higher luminosity is seen is determined by the rate at which the cooling transition front goes through a significant fraction of the disk, and is typically $\approx 10^{-2} t_{\rm recur}$. By stretching all parameters to their limits it is conceivable that these limit-cycle instabilities may be extendible down to the decadal time-scales seen in many types of AGN (e.g., Smith 1996), but this point requires careful investigation. Limit-cycle instabilities are not associated with microvariability.

It was noted by Alloin et al. (1985) that the fast rise times seen in dwarf novae are much shorter than viscous timescales and could be produced by local thermal instabilities. These would travel with the sonic speed through the disk and yield timescales of order $t_{\rm s} \approx 40 R_{\rm out} M_6 T_4^{-1/2}$ hr, where $R_{\rm out}$ is the outer edge of the disk radius in units of the Schwarzschild radius, $M_6 \equiv M_{\rm SMBH}/10^6 M_\odot$, and T_4 is the temperature of the disk in units of 10^4 K. For typical disk sizes and temperatures, these timescales vary from months to years and therefore are not applicable to microvariability, though they may well be related to the multi-year variations that have been noted (Smith 1996).

Overall pulsational modes of the disk could, in principle, be excited and can probably produce periods ranging from days to many years (e.g., Vila 1979). However, such modes would be expected to produce different dominant periods in different wavelength bands, and nothing of this type has been observed. A recent analysis of vertical modes in accretion disks driven by the κ-mechanism in specific temperature ranges has argued that they could be fast enough ($\sim 10^3$ s) to account for small fluctuations in low mass, low accretion rate disks (Kanetake, Takeuti, & Fukue 1995).

We now turn our discussion towards more localized instabilities that could indeed occur over sub-day timescales and thus yield microvariability.

Hydrodynamical vortices (Abramowicz et al. 1992) may exist in accretion disks, driven by shear instabilities. While the physics is very uncertain, the presence of Jupiter's Great Red Spot inspires this idea. The existence of such vortices might allow us to directly view hotter portions of a disk closer to its equatorial plane. It could provide a suitable explanation for very fast X-ray fluctuations (Abramowicz et al. 1992). This concept also provides a way to allow for substantial quantities of "excess" flux to emerge from a fairly wide range of wavelengths on very similar timescales.

Stars (or lower-mass BHs) passing through or near accretion disks will induce gravitational perturbations which will engender spiral shocks in the disks (e.g., Chakrabarti & Wiita 1993). Such non-axisymmetric perturbations will lead to fluctuations in the emission from a disk, and will be discussed below (§4).

The possibility that magnetic fields could be dramatically enhanced by the shear in an accretion disk and could then rapidly release large amounts of stored energy was considered long ago by Shields and Wheeler (1976). While no definitive conclusions concerning the importance of magnetic fields in energy storage and release have yet been reached, this possibility remains intriguing. In analogy with solar flares producing photons from the radio through X-ray bands, we note that magnetic flares are possibly the only way in which accretion disk based fluctuations might be able to provide quasi-simultaneous emission over a wide range in wavelengths. Most authors who have examined magnetized accretion disks have assumed them to be not dissimilar to the sun, with a low $\beta_{\rm p}(\equiv p_{\rm gas}/p_{\rm mag})$ interior, and a coronal region above and below the disk, where $\beta_{\rm p} > 1$ is possible. For these disks magnetic flux probably escapes through magnetic buoyancy (e.g., Stella & Rosner 1984) into the corona. Magnetoacoustic oscillations of coronal loops can yield fluctuations in emission on timescales ranging from days to years (Krishan & Wiita 1994), and this and some other implications of these coronal structures will be considered below (§5).

Other possibilities exist, and a strong case has been made that both low $\beta_{\rm p}$ and high $\beta_{\rm p}$ accretion disks are physically realizable (Shibata, Tajima, & Matsumoto 1990). These low $\beta_{\rm p}$ disks would become magnetically cataclysmic, in that the flux builds up to explosive levels because the escape time is longer than the timescale for amplification through Keplerian shear motions (c.f., Balbus & Hawley 1992). The azimuthal fields could grow until the Alfvén velocity, V_A, reaches the Keplerian velocity (or $B^2/8\pi \simeq \rho GM/r$) over some 10–100 orbital periods (P). But, the growth of a sausage type instability that could expel the flux could occur on a timescale $\sim 0.1P$. Once outside the disk, collapse of the flux tube could occur through a pinch instability, also over short timescales ($< P$), thereby compressing the poloidal component of the large-scale magnetic field penetrating both the tube and the disk and accelerating particles to high energies (Shibata et al. 1990). If the accretion disks ar thick in their inner regions (e.g., Paczyński & Wiita 1980), then much of the magnetic flux generated in the inner portions of the disks is probably advected inward, and can only escape into the funnel region near the SMBH rotation axis (Chakrabarti & D'Silva 1994). These flux tubes can be anchored inside the accretion disk, thereby creating an active corona inside the chimney (D'Silva & Chakrabarti 1994). Again, timescales comparable to P are possible, and such processes could well contribute to rapid variability in blazars.

3. Phenomenological models for microvariability

Assuming that some of the mechanisms outlined above are capable of engendering variations in emission on or above the surface of accretion disks, we must then ask if such disk-based models can reproduce some of the basic features of actual observations of rapid variability in AGN. In this section we summarize results from simulations based upon phenomenological models for flares randomly emerging at different locations and different times on accretion disks. These are expected to be relevant for variations in the optical and UV bands (Wiita et al. 1991, 1992; Krolik et al. 1991), but better accretion disk atmosphere models should be used if comparisons with X-ray observations are to be made in detail. Nonetheless, these phenomenological models were actually first applied to X-ray emission from Seyfert galaxies (Abramowicz et al. 1989; Zhang & Bao 1991, hereafter ZB), and are quite successful at both reproducing the basic look of time series that are observed and in yielding power-density spectra (PDS) with appropriate slopes. Of course, much more work must be done to obtain careful physical arguments leading to natural magnitudes and timescales for individual flares, and steps in this direction are indicated in §§4 and 5 below.

The most detailed phenomenological models applied to optical/UV emission were presented in Mangalam and Wiita (1993, hereafter MW). The components contributing to variability included in our phenomenological model are:

1. Eclipsing of the hot spot by parts of the disk between that individual spot and the observer. There often exists a range of azimuthal positions at a given radius within which the spot is eclipsed, and this is directly dependent on the disk geometry and the observer's viewing angle.

2. Gravitation and rotation induced (Doppler) frequency shifts which vary with spot location and viewing angle.

3. Time delays due to differences in path length to the detector; this also varies with spot coordinates.

4. Intrinsic variability which is phenomenologically modeled via spot lifetimes and spot luminosities. If, as expected, most blazars are viewed close to face-on, then this intrinsic variability will be dominant for the fluctuations emerging from accretion disks in blazars, though this also means that such variations may be swamped by relativistically boosted emission from the approaching jet expected to be present in radio-loud AGN.

Similar effects were considered by ZB, Bao (1992), and Karas and Bao (1992) in the context of variations in X-ray emission. Additional general relativistic corrections, including multiple images, which can be important for X-ray emission from the inner part of the disk, were included in the work of Bao, Hadrava, and Østgaard (1994).

The MW simulations were done for both the more familiar α-disks, where the viscous shear, $t_{r\phi} = \alpha P_{\rm total}$ (e.g., Shakura & Sunyaev 1973), as well as the β-disk geometry, where $t_{r\phi} = \beta P_{\rm gas}$ (Eardley & Lightman 1975; Wandel & Petrosian 1988). For simplicity, we treated only Schwarzschild black holes, though Sun and Malkan (1989) have shown that similar spectra can be produced from Kerr black holes with suitably scaled M and accretion rates, $\dot{M}$, at least for α-disks. The standard α-disks are more prone to thermal instabilities, and we use the well known formulae in Novikov and Thorne (1973) and Page and Thorne (1974) for the disk structure. Because the viscous stresses couple to the entire pressure, they are more efficient in the case of α-disks than in the case of β-disks with otherwise identical parameters; the additional heating in the former case makes those disks thicker, but the consequently expanded disk implies a lower surface temperature.

The disk shape influences all aspects of the variability, most obviously component (1) above. The minimum inclination angle that produces a grazing eclipse of a spot in the interior was geometrically deduced (MW) in the case of β-disks to be dependent only on M and $\dot{M}$ (MW): $i_{\rm min} = \arctan(6.84 M_8/\dot{M}) = 81.7°$ when $M_8 = 1 = \dot{M}$ (in terms of solar masses/yr). This value is algebraically difficult to extract for α-disks but it can be graphically verified to be $< 1°$ for the same values of M and $\dot{M}$ and $\alpha = 1$. Clearly one expects eclipsing to influence variability much more for the thicker α-disks, and particularly for the X-ray producing inner region. Eclipsing is also more important at larger inclinations; in both of these situations the applicability of our model is weaker as we have not included the light-bending effects of general relativity. These are considered in the work of Bao (1992) and Bao et al. (1994), and can be significant for X-ray emission.

Two heuristic models for the variable components were considered by MW. The first is analogous to a ***hot-spot***, where one imagines a viscosity glitch or extra pulse of mass moving through the disk, brightening that region but just marginally raising its temperature; for this model we multiply each $dI_{\nu,{\rm em}}$ arising from the hot-spot zone by a constant factor. The second model, more similar to a stellar *flare*, assumes that the temperature of each randomly selected flare zone is raised by a constant factor. In all cases the flares or hot-spots were assumed to turn on abruptly at randomly chosen times and are located at uniformly weighted randomly chosen positions within the disks (which were divided into 200 radial and 50 azimuthal zones, stretching from $r_{\rm in} = 6$ to $r_{\rm out} = 2000$), where r is measured in units of $GM_{\rm BH}/c^2$. The perturbations are taken to live

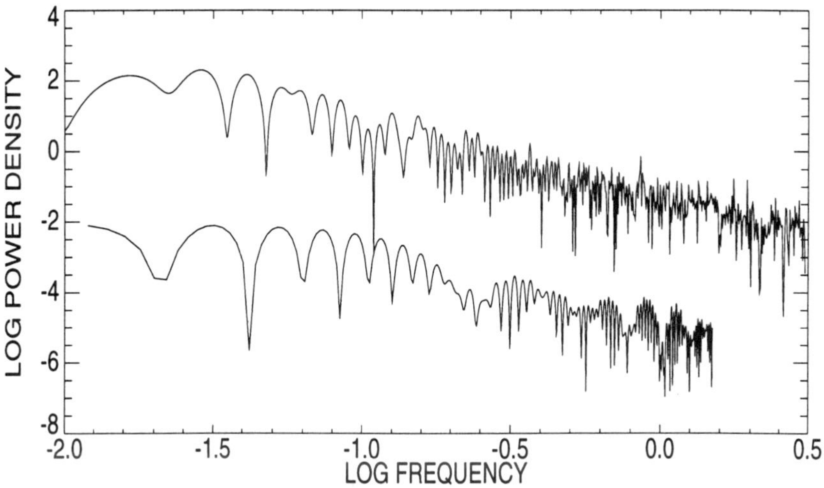

Figure 1. PDS of: (lower curve) observations of OJ 287 from 1–4 April 1988; (upper curve) β-disk model with M_8=30, $\dot{M}$=1, i=60° for 100 flares (with arbitrary offset). For both, $s \approx -1.9$.

either for fixed times or for times proportional to the local orbital periods. The numbers of extant hot-spots varied from 12 to 200.

Quiescent disk spectra are produced by integrating over the surface of the disk, assuming each annulus emits as a blackbody; this is reasonable as these disks are optically thick. Then the variable contribution to the spectrum is found by computing the radiation emanating from the spots that can be observed at a particular orientation for a set of frequencies corresponding to usual observational bands. The variable flux is normalized and the average is subtracted out, before the PDS is computed. A comparison of observational (data from Carini 1990) and phenomenological PDS's is given in Fig. 1.

The results of an analysis of some 30 of these simulations with various parameters can be summarized as follows. These models are capable of reproducing the observed PDS, which have slopes, s, defined by PDS amplitude $\propto f^s$, varying from ~ -1.2 to ~ -2.4 (M. Carini, private communication). We note that the X-ray PDS cluster around $s = -1.6$ (Lawrence & Papadakis 1993) and that these are also nicely fit by hot-spot models (ZB). The lightcurves and PDS's do not differ much for different filter bands centered in the range 1500Å–8290Å for the same run; typically the effect is for the slope to be *slightly* steeper for shorter wavelengths. The number of flares changes the slope, but also only marginally. The effect of having the flux, instead of the temperature, enhanced at the sites of the flares steepens the PDS slope a little bit. Naturally, color variations differ for "hot-spot" or "flare" models, with the latter being bluer. Increasing the inclination effectively linearly scales the contributions from all components contributing to the variability with the exception of eclipses. Taking the flare lifetimes proportional to the local Keplerian period as opposed to being independent of radius introduces additional power at lower frequencies, which steepens the slope. Differences in accretion rate and black hole mass scale the disk temperatures which affect the PDS in a complicated fashion because of the convolution of the variability with the sum of blackbody functions. This is the main reason for the differences between the α and β-disk PDS (β-disks almost always have steeper slopes than α-disks with identical parameters) which otherwise follow the same trends described above.

For high inclination angles of the line of sight to the disk normal, general relativistic effects can yield very strongly modulated fluxes for even single spots (Bao 1992). Assuming all hot-spots have the same extra intensity can still yield X-ray light curves and PDS's that look very similar to those of Seyfert galaxies (ZB). Breaks in the PDS characteristic of the inner edge of the disk can provide an important diagnostic of the last stable orbit of a disk, and thence the SMBH mass. In the absence of light bending, such breaks would be very sharp (ZB), but fully relativistic calculations yield more gentle shoulders (Bao 1992). Bao and Østgaard (1994) argued that such a shoulder was visible in an EXOSAT light curve (Lawrence et al. 1987) of the Seyfert galaxy NGC 4051, and used this feature to diagnose a characteristic timescale of ~ 2000 s, corresponding to last stable orbits around a SMBH of $\sim 4 \times 10^6 M_{\odot}$. While more recent data do not seem to show this shoulder (I. McHardy, private communication), this approach is a very promising one for both essentially proving the presence of a SMBH and of determining its mass.

Another diagnostic of the presence and mass of a central BH could come from future measurements of polarization variability for X-rays (Bao, Wiita, & Hadrava 1996). Preliminary calculations for accretion disk based models indicate that substantially larger swings in both fractional polarization and polarization angle are seen for harder X-rays than for softer ones if electron scattering from thermal emission dominates (Bao et al. 1996).

4. Spiral shocks in accretion disks

Non-axisymmetric tidal perturbations are induced by masses orbiting within, or passing through, an accretion disk. They will induce spiral waves that can steepen into shocks and can be quite efficient in transporting angular momentum; such waves were first investigated in the context of binary stars (e.g., Sawada, Matsuda, & Hachisu 1986; Matsuda et al. 1990; Różyczka & Spruit 1989). These shocks, producing regions of enhanced density and temperature, will naturally cause the perturbed accretion disk to emit more radiation than a standard steady-state disk, and we must ask if these spiral shock induced fluctuations are relevant for the variability observed within AGN.

Chakrabarti and Wiita (1993) used a 2-D hydrocode (Chakrabarti & Matsuda 1992) to simulate this situation for a range of disk types. We have been able to investigate a wide range of perturbations, involving mass ratios down to $q \equiv M_{\rm pert}/M_{\rm BH} = 10^{-3}$, which should be plausible for some AGNs. While such mass ratios are likely to be rare, and perturbers with $q \leq 10^{-5}$ would be more common, numerical dissipation does not allow us to investigate these cases with reasonable accuracy. We have also been able to examine disks whose inner edges are close enough to the central black hole so that perturbations in those

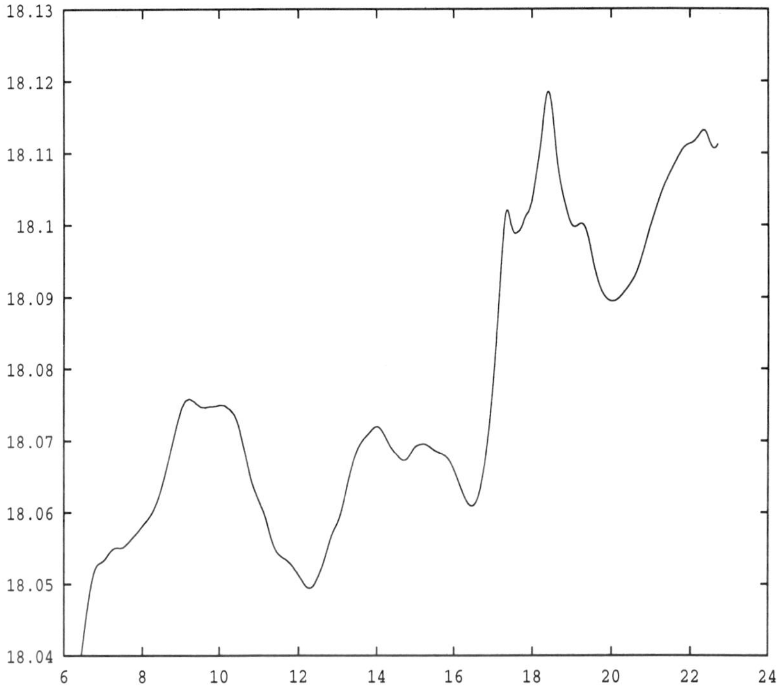

Figure 2. Light-curve (in nominal magnitudes) for a simulation with a perturbing mass with $q = 0.1$ within a constant angular momentum disk with $\gamma = 1.4$; a time difference of 2π corresponds to the period of the perturber. Adapted from Chakrabarti and Wiita (1993); ©American Astronomical Society

regions correspond to variations occurring on timescales of less than a week. We are extending simulations to finer grids so as to investigate shorter timescales characteristic of microvariability, but I now summarize the results of the extant simulations.

Our computations assume the excess emission is radiated thermally, with the higher densities and temperatures in compressed portions of the disk providing a possible physical underpinning for the hot-spots discussed above. Regardless of q, disk adiabatic index, γ, or initial angular momentum distributions, externally applied tidal forces induce spiral shocks in disks around the central massive object. Typically, the perturbations grow rapidly for roughly half of the orbital period of the perturbing mass, yielding increases in luminosity which one can identify as major flares. Subsequently, a quasi-steady state, with fluctuations on the order of 5%–10%, is usually established (Fig. 2). Spiral shocks are found to form, fragment, and reform on relatively short timescales, which can agree with those of microvariability if $M_{\rm SMBH} < 10^8\, M_\odot$ and the perturber penetrates to within a few hundred $R_{\rm S}$.

Simulations with initially Keplerian angular momentum distributions show at first one, and later usually two, well defined spiral structures. These spiral features are similar to those seen in binary star disk models, where the perturber is completely outside the disk; however, here the spiral structure can continue well outside the orbit of the perturber. Simulations with constant angular momentum tend to produce more violent disks, and at first well defined spirals do not form. Later, with the dissipation of angular momentum, spiral shock structure emerges. In most cases, quasi-stable, denser and hotter blobs (supported by pressure gradient and Coriolis forces) appear. These hot-spots form in the inner portions of the disk and slowly propagate outwards before dissipating. It is interesting to note that they live substantially longer than one local orbital period, $P(r)$, in the inner portions of the disk. But in the outer regions, where the Keplerian periods are longer, the blobs usually live $< P(r)$. These differences may be reflected in the different structures of the light curves for optical/UV and X-ray variabilities, which, in disk-based models, are dominated by outer and inner portions, respectively. When the perturber follows a parabolic orbit through the disk spiral patterns are never well defined until very late in the simulation, by which time the perturber has moved far away.

The initial conditions also have strong influences on the behavior of the intensity of the emitted radiation. The rise time in the flaring phase decreases monotonically with the mass of the perturber. This trend is roughly linear in Keplerian disks, but the dependence of rise time on q is much weaker in constant angular momentum disks. In the later fluctuating phase, larger amplitude (and more rapid) variability is induced by more massive perturbers, at least for the Keplerian cases, where the relationship between the rapid fluctuation timescale and q again appears to be roughly linear. In the constant angular momentum disks, this is not observed; in fact, the trend is opposite. For very low q, the dissipation of angular momentum through non-axisymmetric instabilities is insignificant. This causes higher frequency, but lower amplitude fluctuations. For high q, the equivalent viscosity is stronger, leading to larger amplitude, but slower, intensity variations. For the parabolic cases, the intensity of the emitted radiation rises rapidly, and the peak appears at a time somewhat after that when the perturber crosses the turning point. Subsequently, the intensity decays monotonically with very little fluctuation, on a much longer timescale.

5. Plasma effects and coronal models

Under certain circumstances, coherent plasma processes, such as Stimulated Raman Scattering (SRS) or Stimulated Brillouin scattering, could play a role in rapid variability (e.g., Krishan & Wiita 1990; Lesch & Pohl 1992; Gangadhara & Krishan 1995; Levinson & Blandford 1995). Coherent radiation processes may be absolutely necessary to explain extraordinarily high brightness temperatures ($T_B = I_\nu c^2/2k\nu^2 \simeq 10^{21}\,$K) that were recently apparently observed in PKS 0537$-$441 (Romero, Combi, & Colomb 1994), although more modest values of T_B can be covered by extreme forms of jet models if interstellar scintillation or microlensing are considered (Wiita 1993).

For high T_B sources such as pulsars and many blazars, the amplitude of the electric field is large enough to create non-linear effects in the plasma. The SRS process is particularly interesting; it involves the scattering of transverse electrostatic or Langmuir waves to produce coherent radiation over a wide range of frequencies. Transitions between this coherent radiation and individual scattering via the Inverse Compton process can be extremely fast (Gangadhara & Krishan 1993; Krishan & Wiita 1994; hereafter KW94). The fundamental case of monochromatic or narrow-band pump radiation was considered first (Krishan 1988; Krishan & Wiita 1990) and recently the more realistic situation of broad-band exciting radiation has been examined (Thompson et al. 1994; Levinson & Blandford 1995). Although the conditions necessary for SRS to occur are certainly possible in accretion flows onto SMBHs in AGN, since scattering is so severe for reasonable densities, it is difficult for this radiation to escape to be observed (Levinson & Blandford 1995), and there is no convincing evidence that it has been detected.

One class of plasma processes that could play a role in rapid variability involves oscillations in the plasma itself. If accretion disks around supermassive black holes have coronae not too different from the sun's then magnetoacoustic oscillations of coronal loops can yield fluctuations on timescales from hours to years (KW94). These loops are subject to both sausage and kink instabilities; the fast-modes of these magnetoacoustic waves can produce non-thermal emission that could range from the mm to UV bands and would show very weak polarization. For example, if the loop magnetic field is substantially stronger than the field outside the loop, then the timescales for the fast kink and fast sausage modes are approximately:

$$\tau_f \approx \frac{L}{\pi^{1/2} j V_{\rm A}} = 8.2 \times 10^5 \frac{L_{15} n_9^{1/2}}{j B_2} \ {\rm s}, \tag{1}$$

$$\tau_f' \approx \frac{\pi^{1/2} a}{V_{\rm A}} = 2.6 \times 10^4 \frac{a_{13} n_9^{1/2}}{B_2} \ {\rm s}, \tag{2}$$

where L is the length of the loop, a its radius, n the number density of electrons, B the magnetic field strength, $V_{\rm A}$ the Alfvén velocity, j is an integer, and subscripted variables are in scaled cgs units. For quantities to be expected in AGN disk coronae, variability on timescales of hours to months are expected, though the amplitude may be only marginally detectable (KW94).

Such coronal loops could also produce flares with substantial emission variations. Following the treatment of anomalous resistivity in Spicer's (1981) current interruption model for solar flares, KW94 find that this microinstability driven mechanism is likely to yield powerful flares, but with timescales ranging from weeks to years for parameters expected in AGN coronae; such emission would be concentrated in the X-ray band. Double-tearing modes (equivalent to the reconnection of the field between two neutral sheets) relying upon macroinstabilities and normal resistivities are likely to yield timescales longer than those arising directly from Alfvénic waves, and are thus unlikely to be important for *rapid* variability (KW94).

A second class of processes depends upon electromagnetic waves acting to modulate themselves. These waves modify the plasma properties and, in turn, the plasma acts back upon the "pump" wave to produce a non-linear active medium and additional EM waves which induce temporal variability in the emerging radiation. Various classes of instabilities, such as the Langmuir modulational instability, the oscillating two-stream instability, and the self-modulation of EM waves in a plasma are all types of modulational instabilities that can be described in terms of four-wave interactions (Bingham 1990). Although these instabilities have difficulties analogous to those mentioned above for SRS, they can operate on sub-day timescales for parameters relevant to AGN (KW94).

Another plasma process that can be relevant to observed fluctuations in radiation intensity and polarization is resonance absorption, though the chance of actually observing this phenomenon is probably small. This process occurs when an electromagnetic wave enters an ionized plasma, and describes the partial conversion of the photonic energy into electron plasma waves. When obliquely incident radiation whose electric vector lies in the plane of incidence (p-polarized light) penetrates such a plasma, part of the wave is absorbed, but the reflected wave will undergo a change in polarization. For AGN this polarization change can be significant and rapid for radio wavelengths penetrating broad line emission clouds or reprocessed by accretion disks in the vicinity of black holes (Krishan & Wiita 1996).

6. Search for microvariability in Radio Quiet QSOs

Very important evidence for the existence of disk-based microvariability would arise from the convincing discovery of intra-day variability in the optical/UV emission from Radio Quiet QSOs (RQQSOs). Optical microvariability has only been established for radio loud AGN, in particular the blazars, and it is very likely that some or all of it originates in the same relativistic jet that is naturally assumed to yield the synchrotron emission detected in the radio band. X-rays do vary rapidly in Seyferts, but most typically in the minority with radio cores (e.g., Lawrence & Papadakis 1993). The probable detection of optical microvariability in Seyferts (Dultzin-Hacyan et al. 1992; Noble et al. 1994) are also of objects with measurable radio power, and is not fully convincing because of the relatively large contamination in flux from the galaxy's stellar light caused by variable seeing and small shifts of the centers of the apertures through which the brightness is measured.

While it has been argued that even radio-quiet objects house weak jets (Miller, Rawlings, & Saunders 1993) perhaps of a different physical nature (Falcke & Biermann 1995), the evidence that they are unimportant overall in RQQSOs is strong (Gopal-Krishna et al. 1995). Therefore optical microvariability seen in radio quiet (preferably silent) QSOs, which are both nucleus dominated (as opposed to Seyferts) and probably lack jets (since they have no significant radio emission) would provide the very best evidence for disk based fluctuations (Gopal-Krishna et al. 1993).

Jang and Miller (1995, 1996) have searched for fast optical fluctuations in AGN with relatively weak radio emission (some Seyferts and some QSOs) and have found it to be rare, while optical microvariability from radio loud QSOs is about as common as it is in blazars (Miller 1996).

We have been looking at luminous ($M_V < -25$), and moderately bright ($m_V \approx 16$) RQQSOs over the past five years, mostly using the Vainu Bappu (2.34-m) telescope at Kavalur, India (Gopal-Krishna, Wiita, & Altieri 1992; Gopal-Krishna et al. 1993, 1995; Sagar, Gopal-Krishna, & Wiita 1996). We have used CCDs as N-star photometers, following the technique pioneered by

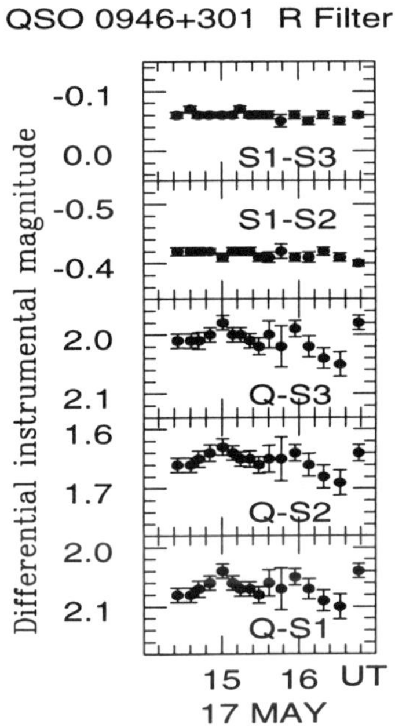

Figure 3. R-band observations of QSO 0946+301 on 17 May 1993. Starting from the bottom panel, differential light curves are presented for the QSO with respect to comparison stars 1, 2, and 3; similar curves are also given for the differences between stars 1 & 2 and 1 & 3. This figure was adapted from Gopal-Krishna, Sagar, and Wiita (1995); ©Royal Astronomical Society

Miller and collaborators (e.g., Miller, Carini, & Goodrich 1989; Miller & Carini 1991). This program has now observed 13 RQQSOs on multiple nights as well as one that we later discovered to have a weak radio core.

A clear detection of microvariability was made for the radio moderate QSO (0838+359) and a $>$ 98% probable (though less than 99.9% probable, which we chose as our conservative criterion) detection was found for the RQQSO 0946+301, part of which is shown in Fig. 3. Two other RQQSOs are probably variable (1049−006, 1444+408) and five others show intriguing hints of rapid changes (0117+213, 1206+459, 1254+047, 1352+011, 1630+377). More observations are needed to examine all of the possible candidates to confirm detections, characterize this microvariability, and contrast it with that of RLQSOs.

Nonetheless, we conclude that it is very likely that disk fluctuations *contribute* to optical/UV microvariability, even if those rapid changes are more common in sources where the brightness is dominated by non-thermal emission from jets.

7. Conclusions and future prospects

Our conclusions are easily summarized:

- Accretion disks are present in AGN and they must produce some level of variability in the observed intensities, as they are certainly subject to instabilities.
- Phenomenological models including general relativistic effects provide excellent matches for X-ray fluctuations and good matches for optical (and UV) microvariability.
- Physical mechanisms (shocks, vortices, magnetic flares, ...) that are plausible underpinnings for such phenomenological models exist. Most of these are not applicable to radio and γ-ray variations, and can thus only *contribute* to the observations.
- Observations of intraday variability in RQQSOs support the hypothesis that disk-based fluctuations are detectable.

The key theoretical task in this area is to improve the physical basis for various disk-based mechanisms. By doing so we could constrain the magnitudes of variations and time-scales at different wave-bands. These efforts have begun, but are at a preliminary state. It is almost certainly true that more complete calculations will allow us to eliminate many of the mechanisms mentioned above as plausible contributors to the observed microvariability. The theoretical models must be developed to the point that they provide solid estimates for the time-lags between emission in different bands, for these are important observational constraints. Predictions of frequency dependent polarization changes (e.g., Bao et al. 1996) could provide another key discriminant between models.

If additional observations show clear, tight correlations between widely separated bands (e.g., 0716+174; Wagner 1992) to be a reasonably common phenomenon, then this would be a severe difficulty for most disk-based models, though those based on coronal flares might still be of relevance.

The thrust of my argument is that disk-based emission fluctuations cannot be ignored and may have been seen. Nonetheless, I wish to stress that, at least in BL Lacertae objects, the strong dominance of the power-law continuum indicates that the rapid changes seen over the entire electromagnetic spectrum are not predominantly due to accretion disks, and are somehow generated in the relativistic jets.

Acknowledgments. I am most grateful for the insights and efforts of my colleagues, Bao Gang, Gopal-Krishna, Minhwan Jang, Vinod Krishan, Arun Mangalam, Dick Miller, John Noble, Ram Sagar, and Xiong Ying. This work was supported in part by NASA Grant NAG 5-3098, NSF Grant AST 91-02106, and Chancellor's Initiative Funds at Georgia State University.

References

Abramowicz, M. A., Bao, G., Lanza, A. & Zhang, X.-H. 1989, in Proceedings of the 23rd ESLAB Symposium, ed. J. Hunt & B. Battrick (Noordvijk, ESA), 871

Abramowicz, M. A., Lanza, A., Spiegel, E. A., & Szuszkiewicz, E. 1992, Nature, 356, 41

Alloin, D., Pelat, D., Phillips, M., & Whittle, M. 1985, ApJ, 288, 205

Balbus, S. A. & Hawley, J. F. 1996, ApJ, 400, 610

Bao, G. 1992, A&A, 257, 594

Bao, G., Hadrava, P., & Østgaard, E. 1994, ApJ, 425, 63

Bao, G., & Østgaard, E. 1994, ApJ, 422, L51

Bao, G., Wiita, P. J., & Hadrava, P. 1996, these proceedings

Bingham, R. 1990, Physica Scripta, T30, 24

Cannizzo, J. H. 1992, ApJ, 385, 94

Carini, M. T. 1990, Ph.D. dissertation, Georgia State University

Chakrabarti, S. K., & D'Silva, S. 1994, ApJ, 424, 138

Chakrabarti, S. K., & Matsuda, T. 1992, ApJ, 390, 639

Chakrabarti, S. K., & Wiita, P. J. 1993, ApJ, 411, 602

D'Silva, S., & Chakrabarti, S. K. 1994 ApJ, 424, 149

Dultzin-Hacyan, D., Schuster, W. J., Parrao, L., Peña, J. H., Peniche, R., Benitez, E., & Costero, R. 1992, AJ, 103, 1769

Eardley, D. M., & Lightman, A. P. 1975, ApJ 200, 189

Falcke, H., & Biermann, P. L. 1995, A&A, 293, 665

Gangadhara, R. T., & Krishan, V. 1993, ApJ, 415, 505

Gangadhara, R. T.,& Krishan, V. 1995, ApJ, 440, 116

Gopal-Krishna, Sagar, R., & Wiita, P. J. 1993, MNRAS, 262, 963

Gopal-Krishna, Sagar, R., & Wiita, P. J. 1995, MNRAS, 274, 701

Gopal-Krishna, Wiita, P. J., & Altieri, B. 1993, A&A, 271, 89

Hughes, P. A., Aller, H. D., & Aller, M. F. 1985, ApJ, 298, 301

Jang, M., & Miller, H. R. 1995, ApJ, 452, 582

Jang, M., & Miller, H. R. 1996, ApJ, submitted

Kanetake, R., Takeuti, M., & Fukue, J. 1995, MNRAS, 276, 971

Karas, V., & Bao, G. 1992, A&A, 257, 531

Krishan, V. 1988, MNRAS, 230, 183

Krishan, V., & Wiita, P. J. 1990, MNRAS, 246, 597

Krishan, V., & Wiita, P. J. 1994, ApJ, 423, 116 (KW94)

Krishan, V., & Wiita, P. J. 1996, MNRAS, submitted

Krolik, H. H., Horne, K., Kallman, T. R., Malkan, M. A., Edelson, R. A., & Kriss, G. A. 1991, ApJ, 371, 541

Lawrence, A., & Papadakis, I. 1993, ApJ, 414, L85

Lawrence, A., Watson, W. G., Pounds, K. A., & Elvis, M. 1987, Nature, 325, 694

Lesch, H., & Pohl, M. 1992, A&A, 254, 29

Levinson, A., & Blandford, R. 1995, MNRAS, 274, 717

Mangalam, A. V., & Wiita, P. J. 1993, ApJ, 406, 420 (MW)

Marscher, A. P. 1996, these proceedings

Marscher, A. P., & Gear, W. K. 1985, ApJ 298, 114

Matsuda, T., Sekino, N., Shima, E., Sawada, K., & Spruit, H. 1990, A&A, 235, 211

Miller, H. R. 1996, these proceedings

Miller, H. R., & Carini, M. T. 1991, in Variability of Active Galactic Nuclei, ed. H. R. Miller & P. J. Wiita (Cambridge, Cambridge Univ. Press), 111

Miller, H. R., Carini, M. T., & Goodrich, B. D. 1989, Nature 337, 627

Miller, P., Rawlings, S., & Saunders, R. 1993, MNRAS, 263, 425

Mineshige, S., & Shields, G. A. 1990, ApJ, 351, 47

Noble, J. C., Miller, H. R., Carini, M. T., Weinstein, D. 1994, in Frontiers of Space and Ground-Based Astronomy. ed. W. Wamsteker, M. S. Longair, & Y. Kondo (Dordrecht, Kluwer), 691

Novikov, I., & Thorne, K. S. 1973, in Black Holes, ed. C. DeWitt & B. DeWitt (New York, Gordon & Breach), 343

Paczyński, B., & Wiita, P. J. 1980, A&A, 88, 23

Page, D. N., & Thorne, K. S. 1974, ApJ 191, 499

Rees, M. J. 1984, ARA&A, 22, 471

Romero, G. E., Combi, J. A., & Colomb, F. R. 1994, A&A, 288, 731

Różyczka, M., & Spruit, H. C. 1989, in Theory of Accretion Disks, ed. F. Meyer et al. (Dordrecht, Kluwer), 341

Sagar, R., Gopal-Krishna, & Wiita, P. J. 1996, MNRAS, in press

Sawada, K., Matsuda, T., & Hachisu, I. 1986, MNRAS, 219, 75

Shakura, N. I., & Sunyaev, R. A. 1973, A&A 24, 337

Shibata, K., Tajima, T., & Matsumoto, R. 1990, ApJ, 350, 295

Shields, G. A., & Wheeler, J. C. 1976, ApLetters 17, 69

Smith, A. G. 1996, these proceedings

Spicer, D. S. 1981, Solar Phys, 70, 149

Stella, L., & Rosner, R. 1984, ApJ, 277, 312

Sun, W.-H., & Malkan, M. A. 1989, ApJ, 346, 68

Thompson, C., Blandford, R., Evans, C., & Phinney, E. S. 1994, ApJ, 422, 304

Vila, S. C. 1979, ApJ 234, 636

Wagner, S. J. 1992, in Variability of Blazars, ed. E. Valtaoja & M. Valtonen (Cambridge, Cambridge Univ. Press), 346

Wandel, A., & Petrosian, V. 1988, ApJ 329, L11

Wiita, P. J. 1993, in Accretion and Jets in Astrophysics, eds. Q. Li et al. (Wuhan, Huazhong Normal Univ. Press), 1

Wiita, P. J., Miller, H. R., Carini, M. T., & Rosen, A. 1991, in Structure and Emission Properties of Accretion Disks, IAU Colloq. 129, ed. C. Bertout et al. (Gif-sur-Yvette, Editions Frontières), 557

Wiita, P. J., Miller, H. R., Gupta, N., & Chakrabarti, S. K. 1992, in Variability of Blazars, ed. E. Valtaoja & M. Valtonen (Cambridge, Cambridge Univ. Press), 311

Zhang, X.-H., & Bao, G. 1991, A&A, 246, 21

Blazar Continuum Variability
ASP Conference Series, Vol. 110, 1996
H. R. Miller, J. R. Webb, and J. C. Noble, eds.

AGN Time Series and Stochastic Resonance

A. D. Nair
Dept. of Astronomy, University of Florida, Gainesville, FL 32611
nair@astro.ufl.edu

Abstract. The phenonmenon of stochastic resonance is explored as a possible explanation for the origin of variability in Active Galactic Nuclei (AGN). It is shown that stochastic resonance can produce sudden flares that rise and decay rapidly. The occurrence of these flares may be low-frequency quasi-periodic in nature. Examples of simulated light curves that approximately mimic AGN light curves are given.

1. Introduction

Most AGN are thought to be nonlinear and non-stationary systems (Vio et al., 1991, Nair, 1996). AGN light curves display a wide range of characteristics and morphologies, meaning that some AGN show frequent, but not strictly periodic flares (e.g., 1156+295) as shown in Fig. 1; in others the duration of flares usually do not last for the same amount of time (e.g., 3C 345). Some have shown a large initial outburst followed by a very long period of quiescence (e.g., 2345–167, Fig. 1). A common feature to all AGN light curves is that the flares, which seem to be superimposed on a noisy background, are of varying amplitudes and morphologies and are unpredictable.

Some AGN display multiple peaks during a flare (e.g., 3C 345, 3C 279). Here a generic model is proposed; these features may be explained as a result of stochastic resonance (SR) where noise-induced switching is the crucial ingredient. The most important property of SR is that, an optimum amount of noise applied at the input results in coherence at the output. This requirement is met only by a nonlinear system that is also a non-stationary system (Moss, 1992).

2. A physical model for variability

The assumption here is that all or at least a major portion of the variability arises due to a jet issuing from the putative central black hole. Birth of new knots in radio jets have been associated with flares in the optical and throughout the spectrum. The model adopted here is the simple model of Villermaux and Hopfinger (1994). They experimentally simulated a jet issuing forth from a nozzle at velocity u. It was found that for intermediate Reynolds number, ranging from 10^3 to about 4000, the jet displays oscillations. The self-sustained oscillations are due to the confinement of the jet. This confined jet is subject to shear (Kelvin-Helmholtz) instabilities, the jet diameter increases as the flow expands in the cavity (or the ISM) and the decrease of velocity sets up pressure gradi-

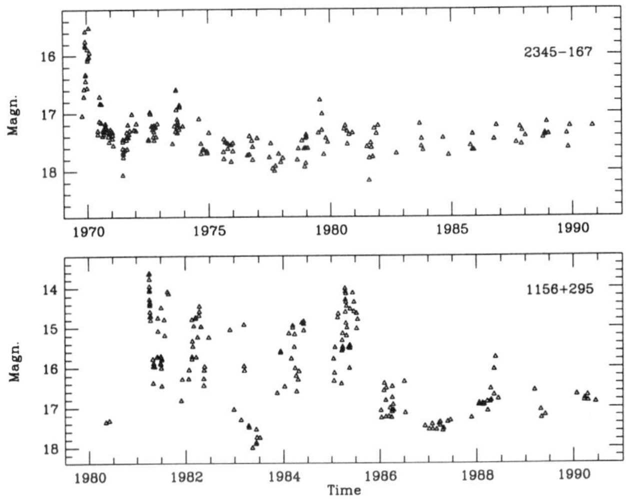

Figure 1. Rosemary Hill Observatory time-series of OVV quasars 2345−167 and 1156+295.

ents. These pressure gradients give rise to fluid particles at the upstream flow. The interaction of the fluid particles with the jet's shear instability is delayed by time τ. If y is the amplitude of the jet's shear instability, Villermaux and Hopfinger (1994) arrive at an expression for a time evolution of the amplitude as

$$\frac{d}{dt}y(t) = ry(t) - \mu(\int_0^\infty \mid y(t-t') \mid^2 \; f(t')dt')y(t) \tag{1}$$

where r and μ are fixed parameters and $f(t')$ is a general function which modulates the retarded action of the term $y(t-t')$.

Consider a particularly simple form of $f(t')$—the Dirac delta function—located at $t' = \tau$, then Eqn. (1) can be simplified to

$$\frac{d}{dt}y(t) = ry(t) - \mu(\mid y(t-\tau) \mid^2 y(t) \tag{2}$$

3. Stochastic resonance

The phenomenon of stochastic resonance (SR) arises as a result of the combined effect of periodic and stochastic forcing in a nonlinear system. SR was first discovered by Benzi et al. (1981) and has since found many interesting applications in many fields. For example, SR has been suggested as a cause for the periodicities observed in the occurrence of the earth's ice ages, in laser experiments McNamara et al. (1988) and many others. It is intended to show that noise induced switching may occur in the presence of jets and produce similar looking light curves in terms of the general features.

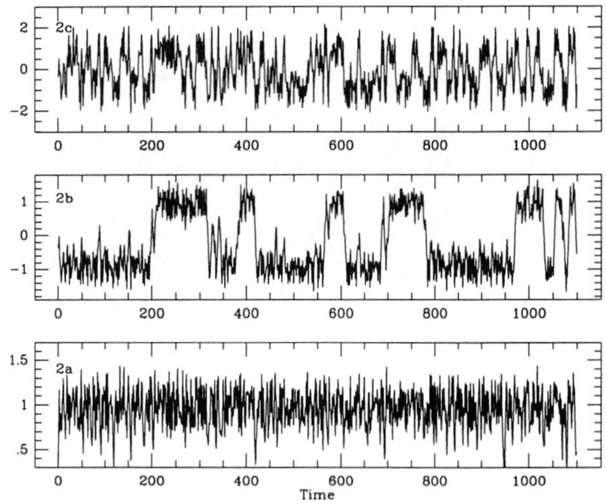

Figure 2. Time-series obtained by integrating Eqn. 3 shows coherence for intermediate values of noise variance (middle figure).

Finally, the last ingredient for SR is a low amplitude periodic modulation. Magnetic fields are thought to be present around AGN jets. The periodic modulation can be thought to represent a weak amplitude, low frequency magnetic field entwined perpendicular to the axis of the jet flow. There appears to be observational evidence supporting this scenario in the case of a variable radio-loud quasar 0805+046 (Feigelson & Isobe, 1984). Magnetic fields parallel to the jet flow have a stabilizing effect on the system and inhibit the growth of instabilities and therefore cannot be considered. The above nonlinear differential equation now reads

$$\frac{d}{dt}y(t) = ry(t) - \mu(\mid y(t-\tau) \mid^2 y(t) \ + \dot{c}e(t) + A\cos(\omega t + \phi) \tag{3}$$

where $e(t)$ is gaussian noise, $\dot{c}$ is a multiplying factor, ω is the frequency and A is amplitude of the periodic modulation. This is a Langevin equation and may be integrated using special techniques employed for stochastic differential equations. For this study a simple fixed step 4th-order Runge-Kutta scheme (Press et al. 1992) was used to integrate Eqn. 3.

The properties of SR may be defined as follows. The physical quantity that characterizes SR is the signal-to-noise ratio, this is defined as the ratio of the amplitude of the signal peak to the amplitude of the noise background in the power spectrum. For computational convenience only the power spectrum is considered. SR can be observed as the power spectrum rises to a maximum value from low values and goes to low values. The behavior of the time series can be thought of as being in two states, the low state and the high state. Switching between states occurs in the presence of noise due to an interaction between the unmodulated system, noise and the modulating frequency. When the noise is small the switching rate is almost zero, as the noise increases the

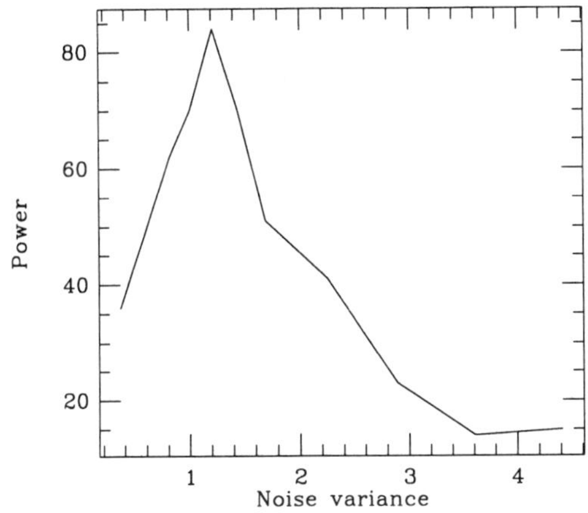

Figure 3. Maximum of the power spectra shows the optimum value of noise variance at which SR occurs.

interaction increases and there is coherence for an optimum value of the noise. When the noise level is increased further the coherence is lost and the time series is dominated by noise.

The maximum value in the power spectrum occurs at an optimum value for the noise intensity. At this optimum noise intensity the system is driven by noise induced switching that was not available at low noise amplitudes. The introduction of noise induces random switching to different states, the weak periodic signal makes switching coherent with the periodic function (Moss, 1992).

4. Simulations

The parameters used were $r = \mu = 0.5, \tau = 1, A = 0.1, \omega = 7.0$ and $\phi = 0.0$. It is customary to examine the effect of increasing the noise variance, $\dot{c}$ in Eqn. 3, from very small values to large values. As examples three time series are displayed in Fig. 2; they were derived from integrating Eqn. 3. The time series in Fig. 2a has a low noise level with variance 0.36, Fig. 2b has an intermediate noise level of variance 1.04. In Fig. 2c the noise variance is high 4.0.

Clearly, the induced coherence can be seen in Fig. 2b for an intermediate noise variance. This is depicted in another way in Fig. 3. The maximum power of the power spectra as a function of the noise variance shows an optimum value for the noise variance.

A variety of light curves can be generated, primarily, by varying the noise variance and the parameters in Eqn. 3. The system described by Eqn. 3 is very sensitive to the initial conditions and values of the parameters and as a result it is rare to find two simulated light curves that are alike. As examples consider Fig. 4, where the value of $\tau = 2$ and all parameters are unchanged from those to generate the light curves in Fig. 2. The light curve shown in Fig. 4a has much lower noise than the one in Fig. 4b. Fig. 4a has some similarity to Fig. 1a in the

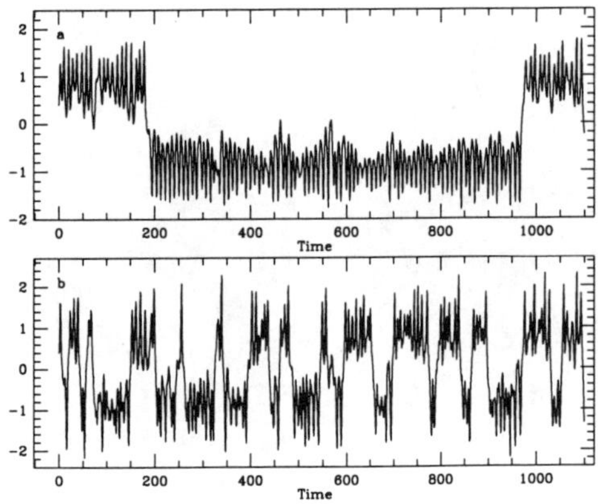

Figure 4. Simulated light curves with different noise variance; compare with Fig. 1.

sense that the light curve for the quasar 2345–167 shows an initial peak followed by a long period of quiescence. A similar phenomenon can be seen in Fig. 4a where there is peak (time 0 to about 200) followed by a period of quiescence. The light curve in Fig. 4b displays sporadic bursts not unlike that of the quasar light curve, 1156+295, of Fig. 1b. However, due to the rapid occurrence of flares in 1156+295 the light curve may be undersampled and the precise details of of morphology of the flares may be lost.

In Fig. 5 (top) the 3C 345 light curve extending over roughly 30 years is displayed, in the bottom part of Fig. 5 a light curve was obtained by integrating Eqn. 3 with a low level of noise. There is good deal of similarity in the first half of the two light curves. The main feature that the simulated light curve lacks and is present in 3C 345 is a secular trend of brightening and fading. This secular trend is not present in most AGN and it is not entirely clear if this trend arises from the sampling of the time series or whether it is intrinsic to the AGN.

In addition microvariability can be explained by this model as originating from random fluctuations of the background flux. These fluctuations are clearly evident in all of the simulated time series. This means that no clear periodicities will be present in all microvariability time series data. Hitherto, no clear case for periodicity have been reported in the literature.

5. Discussion

Several of the characteristics displayed by AGN light curves can be explained by SR. Some of these are, a general noisy background upon which the flares are superimposed, the noise content in the time series, the irregularity of the occurrence of flares, the occurrence of intermediate sized spikes at irregular intervals and the rapid rise and decline displayed by most AGN flares. SR has the ability to produce arbitrary shapes with changes of noise variance and initial conditions. The difference in individual AGN light curves can be thought of as arising due to differing amounts of noise. Some AGN, e.g., OJ 287, 3C 345,

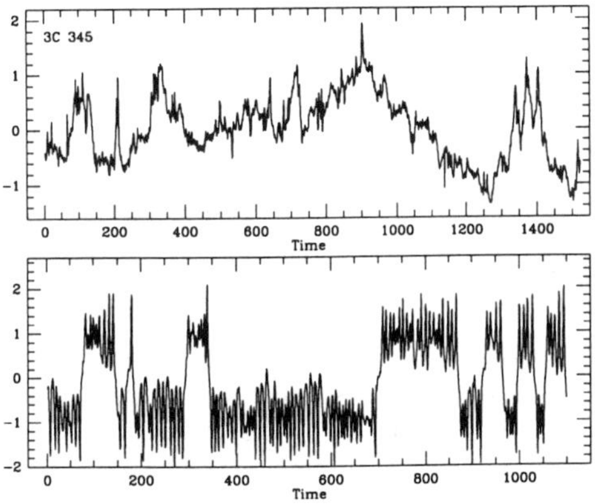

Figure 5. The 3C 345 light curve (top), and a simulated light curve obtained by varying the noise variance (bottom).

3C 279, display multiple peaks during flares. The above SR-simulated time series do depict such multiple peaks during an outburst.

Additive noise is the primary distinguishing property for the various AGN time series. According to SR, an AGN time series showing a larger number of flares (e.g., 1156+295) has more additive noise than one with 1 or 2 flares (e.g., 2345–167).

In summary, several predictions can be drawn as a consequence of this model. SR predicts that the system will be nonlinear and nonstationary; this is a property of almost all AGN time series. No two AGN light curves will be alike since all variations are noise driven. The morphology and duration of individual flares will not be identical from flare to flare; also that the occurrence of large flares will not be strictly periodic. Microvaribility arises from random background fluctuations and is not periodic. Finally, most AGN time series can be explained as arising from differing levels of noise.

References

Benzi, R., Sutera, A., & Vulpiani J. 1981, J. Phys. A, A, 14, L453

Feigelson, E., & Isobe, T. 1984, AJ, 89, 1464

McNamara, B., Wiesenfeld, K., & Roy, R. 1988, Phys.Rev.Lett, 60, 2626

Moss, F. 1992, Frontiers in Applied Mathematics, ed. G. H. Weiss (Philadelphia, SIAM)

Nair, A. D. 1996, in preparation

Press, W. H., Teukolsky, S. A., Vetterling, W. T., & Flannery, B. P. 1992, Numerical Recipes 2nd edition (Cambridge, Cambridge University Press)

Blazar Continuum Variability
ASP Conference Series, Vol. 110, 1996
H. R. Miller, J. R. Webb, and J. C. Noble, eds.

Intraday Variability Statistics of BL Lac Objects in the Optical Domain

J. Heidt

Landessternwarte, Königstuhl, 69117 Heidelberg, Germany

Abstract. The variability behaviour of BL Lac objects on time-scales of days in the optical regime has been studied. Since variability in the optical on such time-scales is most likely intrinsic in nature, this study allows to determine the duty cycle, the intrinsic time-scales and amplitudes of the variability.

The 1 Jy sample of radio-selected BL Lac objects as well as the EMSS and EXOSAT samples of x-ray selected BL Lac objects have been observed. The duty cyle in the radio-selected BL Lac objects is very high (i.e. 0.8). By means of structure function and autocorrelation function analysis the typical time-scales and amplitudes of the variability have been determined. The time-scales found were in the range between 0.5 and 3 days, the amplitudes up to 30%/day.

Contradictionary, the duty cycle in x-ray selected BL Lac objects is very low (≤ 0.4). In a few sources a time-scale could be determined, lying in the same range as found for the radio-selected BL Lac objects. The amplitudes of the variability were always below 5%/day.

1. Introduction

Variability measurements of BL Lac objects offer one tool to study the physics and geometry of this class of AGN. For example, strong simultaneous multifrequency campaigns offer the possibility in probing various models (see Maraschi, this volume), extensive global monitoring may test certain predictions of models based on long-term lightcurves (Sillanpää et al., also this volume). Other applications include searching for microlensing events in these objects as has been suggested by Chang & Refsdal (1979). Within this context studying the intraday variability behaviour (IDV) of BL Lac objects can help to discriminate between the different intrinsic variability models, since IDV in BL Lac objects is most likely in nature (Wagner, 1992).

Previous IDV monitoring campagins have shown that BL Lac objects may show a completely different variability behaviour. For example, in S5 0716+714 quasiperiodic oscillations on timescales of days have been found (Quirrenbach et al., 1991), whereas S4 0954+658 displayed strong outbursts lasting 2-3 days (Wagner et al., 1993). Naturally, the question arises, whether this type of IDV behaviour is typical for the BL Lac class itself, typical for some "subclasses" of BL Lac objects or a characteristic property of a few BL Lac objects alone. In order to study these issues in detail, IDV studies of complete samples of BL Lac objects are required.

BL Lac objects can be divided in two classes, the radio-selected (RBL) and the x-ray selected (XBL) ones. They differ by their their α_{r-o} versus α_{o-x} spectral indices (Stocke et al., 1985). Unification of both classes have been proposed e.g. by Ghisellini et al. (1993) or Giommi & Padovani (1994).

In order to test the unification suggestions mentioned above and in order to determine the duty cycle of the BL Lac objects on short time-scales and to search for differences among both classes an extensive monitoring of 3 samples of RBL and XBL have been carried out, whose main results will be presented in this contribution.

2. The samples and observations

The samples chosen for this work were the statistically complete 1 Jy sample of RBL extracted by Stickel et al. (1991) from the 1 Jy catalogue (34 objects) and two flux/area-limited samples of XBL (the EMSS sample compiled by Morris et al. (1991) and the EXOSAT sample given in Giommi et al. (1991)) having 22 and 11 objects, respectively.

The observations have been carried out during 17 observing runs between 1990 and 1994 with telescopes at Calar Alto, Spain, ESO, La Silla, Cananea, Mexico and in Heidelberg. In all cases a CCD was attached to the telescope. The observations were taken through an R filter in order to perform relative photometry. The length of an observing run was typically one week. During a run 4 - 6 BL Lacs have been observed, usually 4 - 6 times per night. The integration times have been chosen such that the integrated fluxes of the comparison stars varied by no more than a factor of 2 during the observations. More details on the observing strategy can be found in Heidt & Wagner (1996). A description of the data reduction and the relative photometry performed can be found in the same paper.

3. Results

In figure 1 two lightcurves of each, RBL and XBL are displayed. The upper panel shows the most extreme case found among both classes, the lower panel a typical example. Note that the most extreme XBL corresponds to a typical RBL, whereas a typical XBL displayed no IDV.

3.1. Intraday variability statistics

In order to determine the duty cycle of the RBL and XBL, a χ^2 test was applied to the lightcurves. A confidence level of 99.5% as was chosen as cutoff. This gave the fraction of variable and non-variable BL Lac objects. Afterwards a linear slope have been removed from the lightcurves of the variable objects and the χ^2 test applied to the residuals. This gave the fraction of the BL Lacs displaying IDV during the observations. The resulting distribution is presented in figure 2. 82% of the RBL displayed IDV during the observations whereas only 27% of the EMMS XBL and 36% of the EXOSAT XBL showed IDV. Including also the

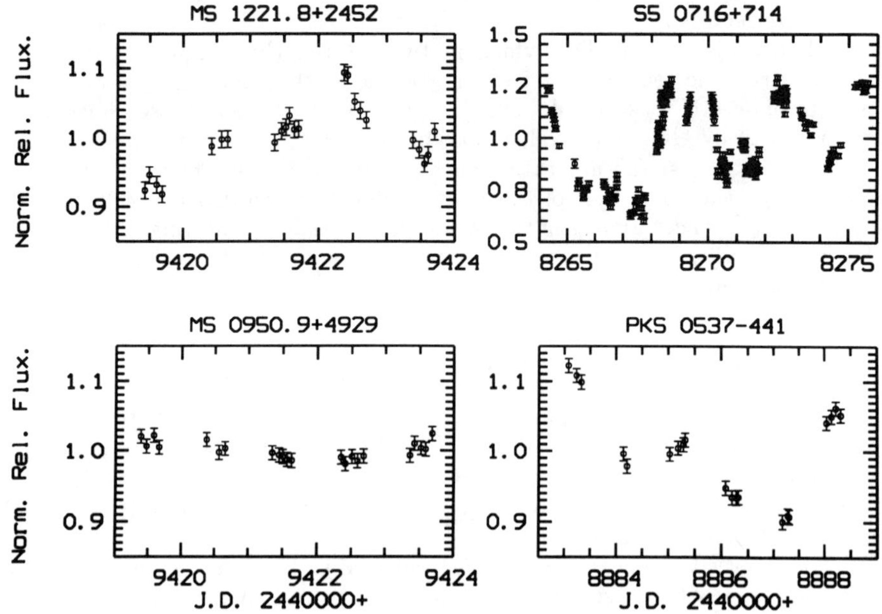

Figure 1. Lightcurves of two XBL (left) and two RBL (right). The upper panel shows the most extreme, the lower panel a typical example.

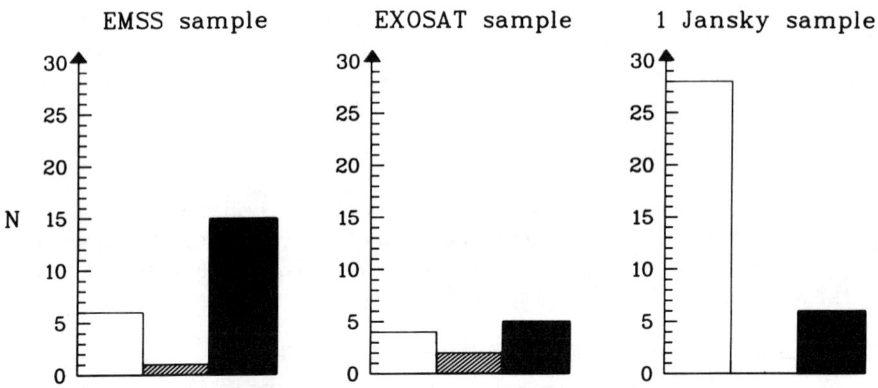

Figure 2. Variability statistics of the 3 BL Lac samples observed. The open area in each diagram gives the fraction of BL Lac objects displaying IDV, the shaded area the variable BL Lac objects, showing no IDV and the black area the number of BL Lac objects showing no variability during the observations.

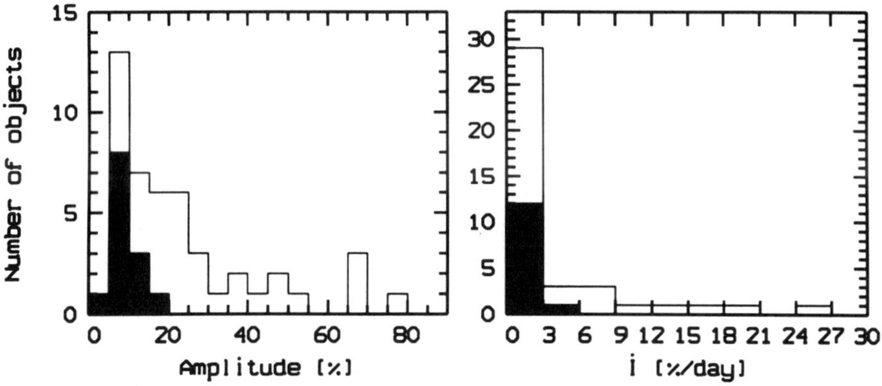

Figure 3. a) Distribution of the variability amplitudes of the variable objects as defined in the text (left). The black area represent the XBL, the open areas the RBL. b) Distribution of $\dot{I}$ for the RBL (open area) and the XBL (black area) (right). See text for details.

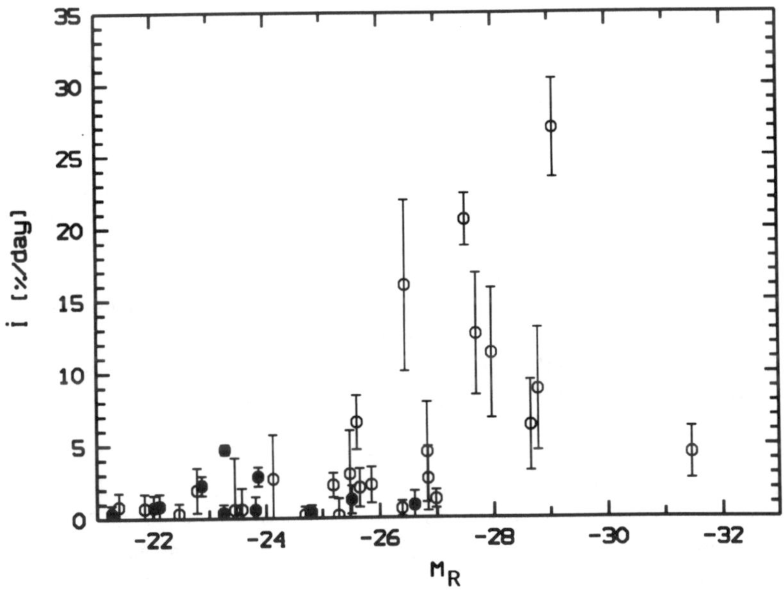

Figure 4. Activity parameter $\dot{I}$ versus absolute magnitude M_R. The dots correspond to the XBL, the open circles to the RBL. The time-scales involved in the derivation of $\dot{I}$ have been corrected for cosmolo/-gical effects.

objects, which showed variability on even longer time-scales, the fraction grows now to 32% in the EMSS XBL and to 55% of the EXOSAT XBL. In summary, the duty cycle in RBL is at least 0.8, whereas in the XBL the duty cycle is 0.4 or less.

Next, the amplitudes defined as $A = \sqrt{(A_{max} - A_{min})^2 - 2\sigma^2}$ where A_{max} and A_{min} are the maximum and minimum values of each lightcurve and σ the measurement errors, respectively have been calculated from the lightcurves of the variable objects. Their distribution is shown in figure 3a. The amplitudes of the RBL differ by a factor of 3 from the XBL. Whereas the RBL showed an average amplitude of 28%, the XBL showed average amplitudes of only 10%.

3.2. Statistical analysis

In order to study the temporal behaviour of the lightcurves a structure function (SF) analysis as described by Simonetti et al. (1985) was carried out with the lightcurves. The slope of the SF characterizes the variability and hence the underlying process, whereas characteristic time-scales can be found by identifying pronounced maxima in the SF (see also Smith, these proceedings). The slopes of the SF of both classes - RBL and XBL - were similar. In both cases the average slope was 0.8 ± 0.6 indicating that the lightcurves are mainly dominated by shot noise. This would imply that in both classes of BL Lac objects the physical mechanism responsible for the observed variability is similar.

In 21 RBL a time-scale could be measured from the SF, which was possible only for 4 XBL. In both cases the time-scales derived are in the range of 0.5..3 days, indicating similar time-scales.

Finally, the discrete autocorrelation function as described by Edelson & Krolik (1988) has been computed from the lightcurves. In order to achieve a typical amplitude per time interval of a BL Lac object the activity parameter $\dot{I} = \sqrt{ACF(0)}/\Delta t$ was measured from the ACF. Therefore we did not normalize the ACF to unity at Δ t = 0. A second-order polynomial was fitted to each ACF. ACF(0) was derived from the maximum of the fit polynomial, for Δt the decorrelation length of the fit polynomial was chosen. The resulting values were corrected for redshift. Their distribution (given in %/day) is shown in figure 3b. The distribution for the RBL is twofolded. 17 RBL displayed an $\dot{I}$ of $\leq 3\%$/day, 10 RBL an $\dot{I}$ of up to 27%/day. With one exception $\dot{I}$ was always $\leq 3\%$/day in the XBL.

The activity parameter $\dot{I}$ was correlated with other intrinsic properties of the BL Lac objects. Whereas no correlation could be found with the redshift (assuming z = 0.3 for those BL Lac objects, where no redshift could be measured so far), a trend for higher $\dot{I}$ towards higher M_R could be found in the RBL. This is shown in figure 4. This may be a hint that relativistic beaming is important in this sources. Note that a similar behaviour can not be seen in the XBL.

4. Summary and concluding remarks

The study of the IDV behaviour of the RBL and XBL samples presented in this article can be summarized as follows:

1. The duty cycle of the RBL and XBL differs ≈ by a factor of 2. Whereas the RBL have a duty cycle of at least 0.8, the duty cycle in XBL is ≤ 0.4.

2. The average amplitudes differ by a factor of 3. They were 30% in RBL and 10% in XBL on average.

3. The SF analysis revealed similar slopes (i.e. shot noise) for RBL and XBL indicating a similar mechanism responsible for the observed variability. The time-scales found were also similar. However, due to the poor statistics for the XBL it is not possible to check, whether XBL have indeed similar time-scales of somewhat longer time-scales than RBL.

4. The activity parameter $\dot{I}$ calculated from the autocorrelation function showed a twofolded distribution for the RBL (17 RBL had a $\dot{I}$ of $\leq 3\%$/day, 10 RBL an $\dot{I}$ of up to 27%/day, whereas in the XBL with one exception $\dot{I}$ was always $\leq 3\%$/day. No correlation between $\dot{I}$ and redshift could be found. There seems to be a correlation between $\dot{I}$ and M_R in the RBL. Such a behaviour was not found in XBL.

The results challenge both unifications proposed by Ghisellini et al. (1993) and Giommi & Padovani (1994). Whereas the former suggestion would imply a similar duty cycle in both classes (which have not been found), the latter allows relativistic beaming also to be present in XBL (this has also not been observed).

Finally, I wish to remark that every model, which is able to explain IDV must take into account the high duty cycle in RBL and the factor of 2 lower duty cycle in XBL.

Acknowledgments. I am deeply indebted to my former supervisor Stefan Wagner. Without his ongoing interest, support and help this work would never had been possible. This work was supported by the Deutsche Forschungsgemeinschaft through SFB 328. A travel grant by the Deutsche Forschungsgemeinschaft under contract AZ 477/288/96 is gratefully acknowledged.

References

Chang, K. & Refsdal, S. 1979, Nature, 282, 561

Edelson R.A. & Krolik J.H. 1988, ApJ, 333, 646

Ghisellini, G., Padovani, P., Celotti, A., & Maraschi, L. 1993, ApJ, 407, 65

Giommi P., Tagliaferri, G., Beuermann, K., et al. 1991, ApJ, 378, 77

Giommi, P., & Padovani, P. 1994, MNRAS, 268, L51

Heidt, J. & Wagner, S.J. 1996, A&A, 305, 42

Morris, S.L., Stocke, J.T., Gioia, I.A., et al. 1991, ApJ, 380, 49

Quirrenbach, A., Witzel, A., Wagner, S.J., et al. 1991, ApJ, 372, L71

Simonetti, J.H., Cordes, J.M., & Heeschen, D.S., 1985, ApJ, 296, 46

Stickel, M., Padovani, P., Urry, C.M., Fried, J., & Kühr H., 1991, ApJ, 374, 431

Stocke T.J., Liebert J., Schmidt G., et al. 1985, ApJ, 298, 619

Wagner, S.J., et al. 1990, A&A, 235, L1

Wagner, S.J. 1992, in Gravitational Lenses, Kayser R., Refsdal S., Schramm T., Heidelberg: Springer, 189

Wagner S.J., Witzel A., Krichbaum T.P., et al. 1993, A&A, 271, 344

Blazar Continuum Variability
ASP Conference Series, Vol. 110, 1996
H. R. Miller, J. R. Webb, and J. C. Noble, eds.

Reflections from the OJ-94 Project

L. O. Takalo

Tuorla Observatory, Tuorla, Fin-21500 Piikkiö, Finland

Abstract. We will describe the large international monitoring project: OJ-94. The aim in this project was to monitor blazar OJ 287 during its predicted outburst fall 1994. This outburst was observed in November 1994, a couple months from the predicted time. OJ-94 project has lasted now over two years, and will continue (at its present format) until next summer. During this time we have collected over 8000 data points on OJ 287. These data constitutes the best ever observed light curves on any extragalactic object. OJ 287 is seen to show variations *always* in timescales from tens of minutes to months. Here we will present some preliminary results from this project. Also we will discuss the impact of this project to the future of blazar research.

1. Introduction

OJ 287 is one of the best observed blazars (e.g., Takalo 1994). It is also one, for which we have observations for over 100 years (Sillanpää et al. 1988, 1996a). This historical light curve shows several large outbursts, that seem to occur every 12 years. Based on the historical light curve Sillanpää et al. (1988) predicted that a new outburst would occur during fall 1994. OJ-94 project was created for verifying this outburst.

2. Project description

Discussions of the project started already during 1989. Then the idea was to apply for the 5% International observing time (ITP) on the Canary Island telescopes for the project. We received this ITP-time for a half a year period, starting fall 1993, when the project observations started. During this ITP time a project gets 5% of the observing time on all the telescopes at Canary Island Observatories. Since we are interested in intensive long-term monitoring on OJ 287, we organized a large international collaboration for this monitoring. At present there are over 50 astronomers from 10 countries involved in the project. Project members use whatever instruments they have at their disposal for this monitoring, and send the reduced data into the data archive. Table 1 lists the telescopes and instruments used.

So far we have had four project meetings. These meetings were attended by a varying number of project members. The first meeting was held in Seili, Finland in July 1993. In this meeting the project strategies and management

Table 1. The observatories and instruments used.

Telescope/Observatory	Size	Instr.	Filters
Jakobus Kapteyn Telescope	1-m	CCD	UBVRIZ
Nordic Optical Telescope	2.5-m	CCD	VRI
		photop	UBVRI
William Herchell Telescope	4.2-m	ISIS	Spectra
Isaac Newton Telescope	2.5-m	IDS	Spectra
RoboScope	60-cm	CCD	V
Boltwood	17-cm	CCD	VRI
Heidelberg	70-cm	CCD	R
Lowell Observatory	1.06-m	CCD	BVRI
Capilla Peak observatory	61-cm	CCD	BVRI
San Pedro Martir	2.1-m	CCD	V
Cananea	2.1-m	CCD	VR
Crimean Observatory	1.25-m	photop	UBVRI
Tuorla Observatory	1.03-m	CCD	V
IAC-80 telescope	82-cm	CCD	BVRI
Calar Alto	1.2/2.2-m	CCD	VR
Perugia Observatory	40-cm	CCD	BVRI
Torino Observatory	1.05-m	CCD	BVR
Telescope Carlos Sanchez	1.5-m	Photom	JHK
Michigan Radio Telescope	26-m	Continuum	4.8, 8, 14 GHz
Metsähovi Radio Station	13.8-m	Continuum	22, 37 GHz

was discussed. The project leadership was divided as follows: Project leader, L. O. Takalo; ITP observers leader, M. R. Kidger; Analysis leader, H. Lehto; Archivist, A. Sillanpää (This task was later given to T. Pursimo). The data archive was decided to be maintained at Tuorla Observatory. The second meeting was in December 1994 at IAC, Tenerife. Here we discussed mainly the two weeks observing run at Canary Island telescopes later the same month. During the third (Orilampi, Finland, summer 1994) and fourth (Oxford, England, autumn 1995) meetings preliminary results were presented and discussed. The talks presented in these last meetings were published in the Tuorla Observatory Report Series. The next project meeting is planned for September 1996 in Girona, Spain.

Blazars 3C66A, S5 0716+714 and AO 0235+164 were selected as comparison objects for OJ 287. We hoped to use the observations of these objects for checking and removing instrumental effects from the OJ 287 observations. But all these objects showed such large intrinsic variations that they could not be used for this original task. Instead the observed light curves give very interesting results on these blazars (see below). The best light curves are on OJ 287 and 3C66A, the other objects have been observed only occasionally.

All the monitoring observations are collected at Tuorla observatory into an archive (Pursimo& Lehto 1996). The archive also contains some other data, e.g., IUE observations, besides the monitoring data. This archive is open to all project members via anonymous ftp. At present this archive contains over 11,000 data points on these blazars. Of these 8000 are on OJ 287, about 2500 on 3C66A and the rest on AO 0235+164 (Pursimo & Lehto 1996). These data sets on OJ 287 and 3C66A provides us with the best ever observed light curves on any extragalactic objects.

3. Results and discussion

The predicted (optical) outburst was observed in OJ 287 during November 1994 (Sillanpää et al. 1996a,b). In the radio frequences OJ 287 was in a very low state during this outburst (Valtaoja et al. 1996). During this time we also got Target of Opportunity observations on OJ 287 with IUE, ROSAT and GRO satellites (Pian et al. 1996; Webb et al. 1996). This was the first time that OJ 287 was detected by the EGRET. Another outburst was observed during December 1995. This double peak outburst behavior is similar to the behavior observed in the previous outburst in 1983/84 (Sillanpää et al. 1988; 1996b). In the optical observations OJ 287 has shown variations ALL the time in time scales from tens of minutes to months (Sillanpää et al. 1996b). This is the first time that such continuous variations has been observed in a blazar. Analysis of this data is in progress. Preliminary results are being discussed in these proceedings.

3C66A has been in outburst throughout the observing period. We have observed it about one magnitude brighter than what it has been seen previously (Takalo et al. 1996). Also it has shown continuous activity in the optical and infrared bands bands, with frequent small flares (Takalo et al. 1996). In the radio bands we have less observations, which do not so any dramatic behavior.

Blazar AO 0235+164 has been very faint during the observing period, with small occasional flares. It has been observed only occasionally and with the larger telescopes. The faintness has allowed us to obtain deep images, looking for the host galaxy. Our images reveal a complex system, with two (perhaps interacting) galaxies at a lower redshift ($z = 0.524$) than the blazar (Nilsson et al 1996). We have also a spectrum of it, observed with the WHT (Table 1). In the spectra we detect, for the first time, broad H-γ and H-δ lines (Nilsson et al. 1996) at the blazar redshift ($z = 0.94$). It seems that BL Lac objects show quasar like emission lines when in faint state (see also Vermeulen et al. 1995).

S5 0716+714 has shown continuous variability with amplitudes up to one magnitude (Sillanpää et al. 1996c).

4. Conclusions

This project has produced several new and very important results on blazars. We have for the first time confirmed a predicted phenomena in an extragalactic object; the outburst in OJ 287. A surprising result is the observed continuous optical variability in OJ 287 and 3C66A. Similar variability characteristics is not seen simultaneously in the radio bands. This is the first time that such variability has been clearly seen in the optical bands. We have also demonstrated that large international, long term multiwavelenght monitoring is essential, if we want to get a complete picture of blazar emission processes. Most of the previous monitoring campaigns have used only one telescope and/or lasted only a few weeks with several telescopes. Thus giving us only a 'snapshot' on the blazar behavior. I would like to see this project continued and expanded to include a few more objects. Possible new objects could be e.g., ON 231 and BL Lac. Of these objects there exists also a long term historical light curves.

Our monitoring observations provide theoreticians with new data for checking their models. It also raises several new questions that a blazar model has to answer. Personally I do not believe that a single emission model (jet or accretion disk model) can be used to model our observations. More likely we need a combination on the jet and accretion disk emissions and propably also each blazar will have it's own combination of these emission processes.

Acknowledgments. This research was supported by the Finnish Academy. Part of the observations used in this work were obtained during the 5% International time on the Canary Island telescopes awarded for the OJ-94 project.

References

Nilsson, K., et al. 1996, A&A, submitted

Pian, E., et al. 1996, in "Workshop on Two Years of Intensive Monitoring of OJ 287 and 3C66A," ed. L. O. Takalo, Tuorla Observatory Reports, 176, 26

Pursimo, T., & Lehto, H. 1996, in "Workshop on Two Years of Intensive Monitoring of OJ 287 and 3C66A," ed. L. O. Takalo, Tuorla Observatory Reports, 176, 2

Sillanpää, A., et al. 1988, ApJ, 325, 628

Sillanpää, A., et al. 1996a, A&A, 305, L17

Sillanpää, A., et al. 1996b, these proceedings

Sillanpää, A., et al. 1996c, in "Workshop on Two Years of Intensive Monitoring of OJ 287 and 3C66A," ed. L. O. Takalo, Tuorla Observatory Reports, 176, 7

Takalo, L. O. 1994, Vistas in Astronomy, 38, 77

Takalo, L. O., et al. 1996, A&A, submitted

Valtaoja, E., et al. 1996, in "Workshop on Two Years of Intensive Monitoring of OJ 287 and 3C66A," ed. L. O. Takalo, Tuorla Observatory Reports, 176, 13

Webb, J., et al. 1996, in "Workshop on Two Years of Intensive Monitoring of OJ 287 and 3C66A," ed. L. O. Takalo, Tuorla Observatory Reports, 176, 20

Vermeulen, R. C., et al. 1995, ApJ, 452, L5

Blazar Continuum Variability
ASP Conference Series, Vol. 110, 1996
H. R. Miller, J. R. Webb, and J. C. Noble, eds.

Optical Variations of OJ 287

A. Sillanpää, L. O. Takalo, T. Pursimo, K. Nilsson, P. Heinämäki, S. Katajainen, H. Pietilä, M. Hanski, and R Rekola

Tuorla Observatory, FIN-21500 Piikkiö, Finland

P. Boltwood

1655, Main St. Stittsville, Ont. K2S 1N6, Canada

G. W. Turner, J. W. Robertson, and R. K. Honeycut

Department of Astronomy, Indiana University, Swain West 319, Bloomington, IN 47405, USA

Yu. S. Efimov and N. Shakhovskoy

Crimean Astrophysical Observatory, P/O Nauchny, 334413 Crimea, Ukraine

M. Fiorucci and G. Tosti

Osservatorio Astronomico, Universita di Perugia, I-601123 Perugia, Italy

G. Ghisellini, C. M. Raiteri, M. Villata, G. De Francesco, S. Bosio, and G. Latini

Osservatorio Astronomico di Torino, Strada Osservatorio 20, I-10025, Pino Torinese, Italy

J. Heidt

Landessternwarte Königstuhl, D-69117 Heidelberg, Germany

Abstract. This short analysis of the different types of optical variations in the BL Lac object OJ 287 is based mainly on the optical data collected in the OJ-94 Project. Besides of these data we have also used the historical data base collected for this object at Tuorla Observatory. We can easily separate at least seven different types of variations. Some examples of all of these different types are presented and some simple possible mechanisms behind them are discussed.

1. Introduction

BL Lac object OJ 287 is one of the most violently variable objects in the whole Universe (e.g., Takalo 1994). We still don't know what really causes these vari-

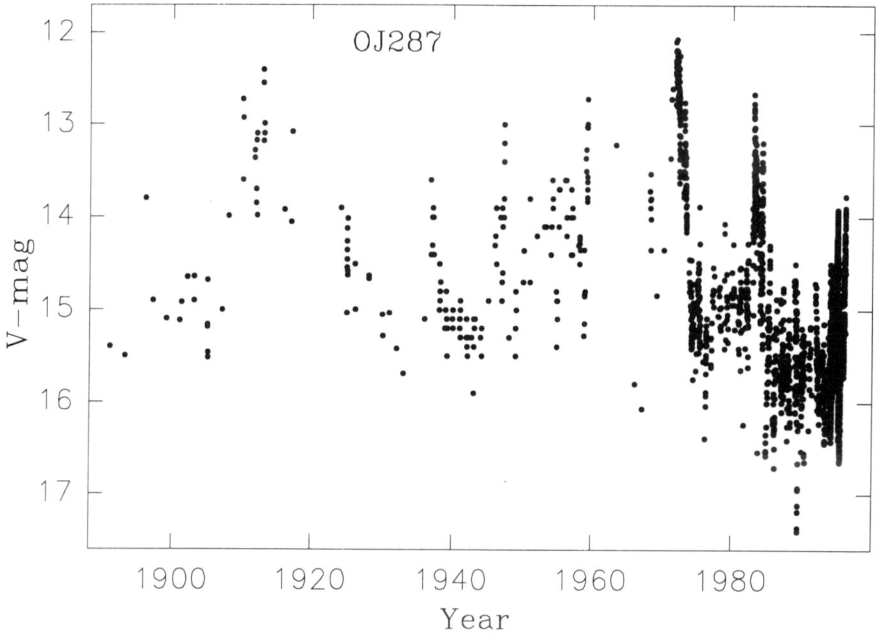

Figure 1. The historical light curve of OJ 287.

ations but it is very obvious that for example in the optical region we can see many different types of variations. It is possible that there are also totally different mechanisms behind some of these variations. From the OJ-94 Project data we can separate possible five different types of optical variations and from the historical behavior two more.

2. Results

We can separate seven different types of optical variations in OJ 287 as follows:

1. From the historical light curve (Fig. 1) we can see very long timescale, slow variation. The amplitude of this variation is very large, more than three magnitudes. The timescale of the oscillation is more than 50 years. A possible mechanism behind this is the slow precession of the jet in OJ 287 (Sillanpää et al. 1996a).

2. The periodicity of about 12 year in the large optical outbursts (Fig. 1) (in more detail, see Sillanpää et al. 1988, 1996a).

3. The double peak structure in the 12 year periodical outbursts. This behavior is easily seen from the OJ-94 Project data (Fig. 2). The reason for these double peaks is uncertain but it can be caused for example by the hits of the secondary black hole through the accretion disk of the main black hole (Lehto & Valtonen, 1996). Another possibility is microlensing

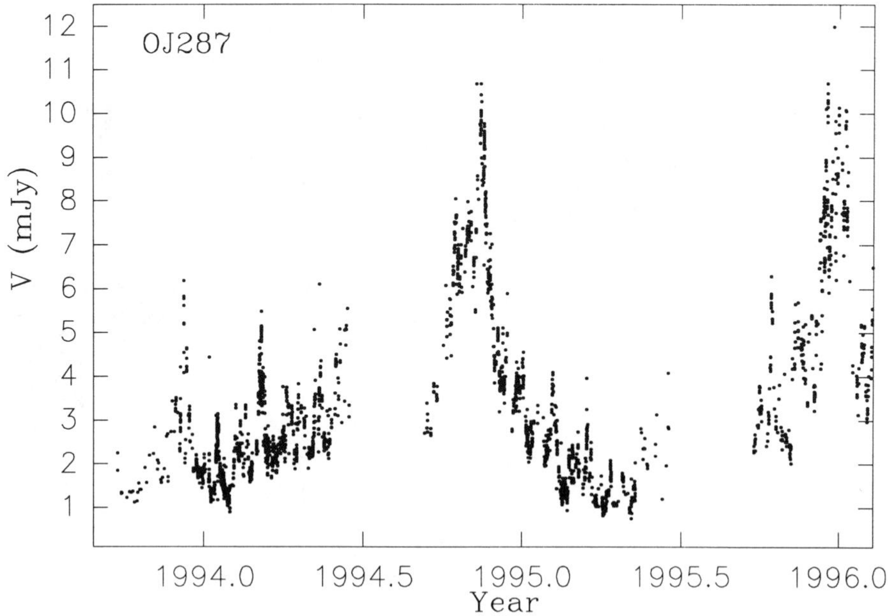

Figure 2. OJ-94 Project data in V-band.

effect. This effect can be caused by foreground star moving near to the line of sight or (still to explain the 12 year periodicity!) by the secondary black hole moving almost exactly between us and the main black hole. There are two strange things in the optical behavior of OJ 287 which can be explained by the lensing effect: We have *not* seen any changes in the optical spectral index during our OJ-94 Project even when the total amplitude of variation has been about three magnitudes (Fig. 3). Another thing is the strange symmetry of two major outbursts observed during the OJ-94 Project (Fig. 2). This can be seen if we put the axis of the symmetry around the year 1995.4. This behavior is discussed in more detail in Sillanpää et al. 1996b.

4. From the OJ-94 Project optical data we can see very many good examples of the "normal" synchrotron flares. One of these examples is shown in Fig. 4. Very fast rising times and much slower declining tails are typical for these flares.

5. So called "lighthouse" behavior is caused by "clumps" of material spiralling in the relativistic helical jet oriented almost exactly towards us (Sillanpää et al. 1991 and Schramm et al. 1993 and referencies therein). One example is seen in Fig. 5.

6. Of course we can also see microvariability in OJ 287. One example of this behavior is seen in Fig. 6. When we are speaking about the microvariability

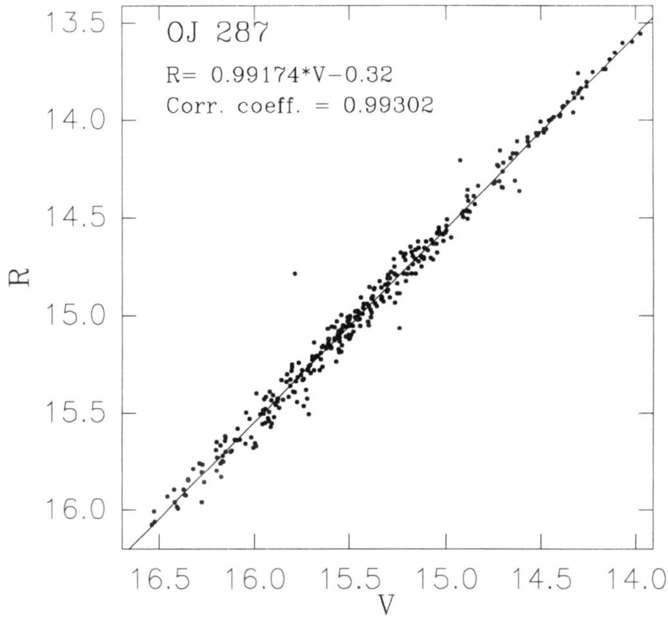

Figure 3. The $(V - R)$ diagram from the OJ-94 Project data. The magnitudes are daily averages.

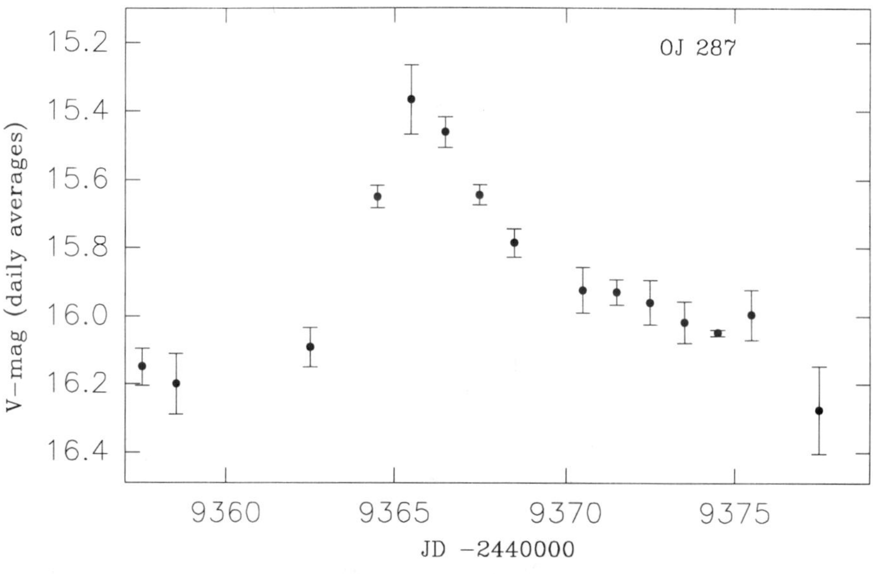

Figure 4. A typical synchrotron flare from the OJ-94 data.

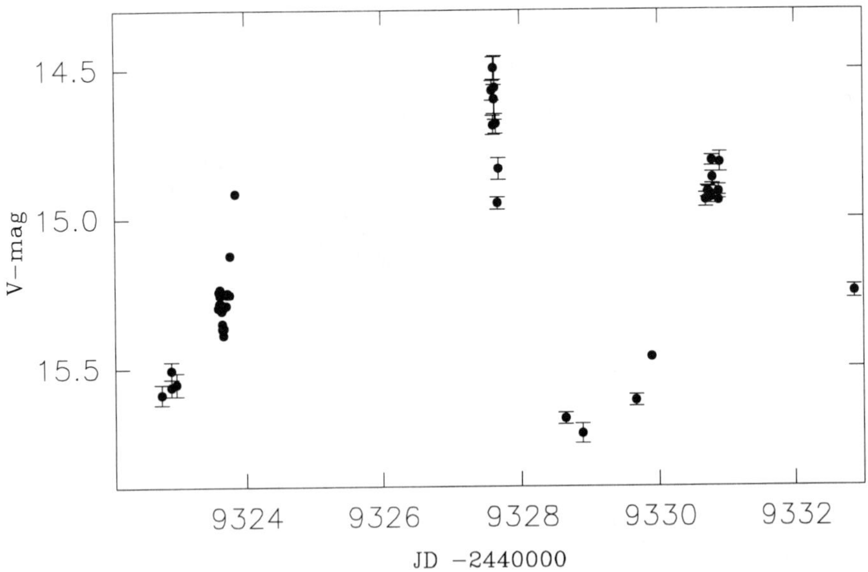

Figure 5. "Lighthouse" swings from the OJ-94 data.

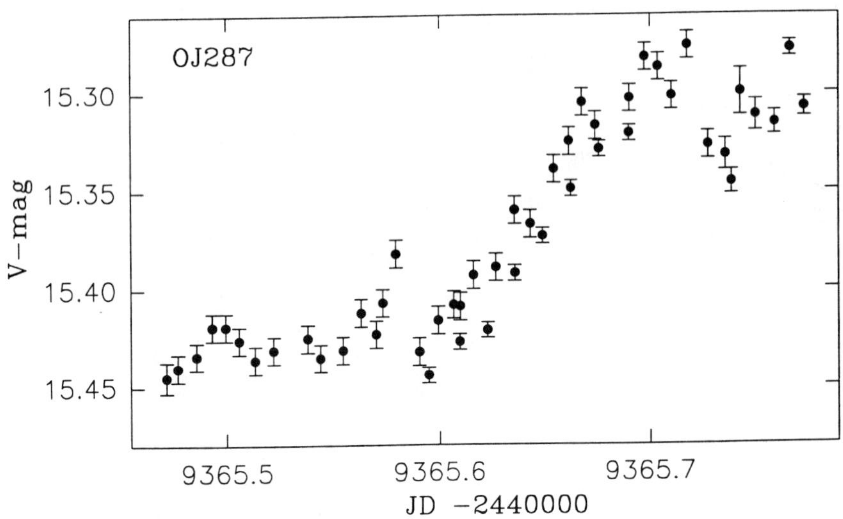

Figure 6. One example of the microvariability taken from the OJ-94 data.

of blazars we must be very careful and separate the real microvariability from the rising and declining tails of the "normal" flares lasting a few days. It is possible that the reason for these two different variations is also totally different.

7. The polarization behavior of OJ 287 is also totally different in different outbursts. Some of them are almost zero polarization flares (for example, the 1983 outburst) and some of them are very highly polarized ones (for example, the outburst in 1984) when the polarization reached almost 40 % (Takalo 1994). It is quite clear that there must also be some differencies behind these flares.

OJ-94 Project has shown that blazar OJ 287 is variable all the time without any quiescent periods. We have also seen that there are many different types of variability. It has also been shown that this kind of projects are really the only way to understand how these different variations are connected and what is the mechanism behind them. We don't see any reason why this is not the situation with all the blazars.

References

Lehto, H. J., & Valtonen, M. J. 1996, ApJ, 460, 207

Schramm, K. J., et al. 1993, A&A, 278, 391

Sillanpää, A., et al. 1988, ApJ, 325, 628

Sillanpää, A., et al. 1991, AJ, 101, 2017

Sillanpää, A., et al. 1996a, A&A, 305, L17

Sillanpää, A., et al. 1996b, in preparation

Takalo, L. O. 1994, Vistas in Astronomy, 38, 77

Blazar Continuum Variability
ASP Conference Series, Vol. 110, 1996
H. R. Miller, J. R. Webb, and J. C. Noble, eds.

Analysis of the Optical Light Curve of OJ 287

Harry Lehto

Tuorla Observatory, Väisäläntie 20, University of Turku, Finland

Abstract. We have analyzed the V-band *flux density* light curve of OJ 287. One main aim is to test whether the quasiregular outbursts appearing in OJ 287 at about 12-year intervals can be thought as a "continuation" of the lower level variability seen during 1993–1996 to higher amplitudes or longer timescales. We will indicate that this appears not to be the case. The cause of the low level variability seen in 1993–1996 is different from the variability seen in the historical light curve. We will also give reasons why we think that many significant outbursts have not been missed in the historical light curve.

1. Introduction

The historical light curve of OJ 287 shows outbursts that appear to be spaced quite regular, at a timescale of about 12 years. Models involving binary black holes (Sillanpää et al. 1988, Lehto & Valtonen 1996) can explain these variations. We will take a sceptical view towards these models. By studying the variability characteristics by different means we will test the working hypothesis that all variability in the V flux density light curve of OJ 287 can be explained by a single process. We will indicate that this working hypothesis appears to be wrong.

2. Wavelets

Wavelet analysis is capable of disentangling variability on various timescales. It can also be used to study local variability. Haar wavelet is the simplest one and has the form

$$\psi(t) = \begin{cases} 1, & 0 \le t < t_0/2; \\ -1, & t_0/2 \le t < t_0; \\ 0, & \text{otherwise.} \end{cases}$$

Wavelet bases are orthogonal and have a scaling property:

$$\psi_{s,l}(t) = \frac{1}{2^{s/2}}\psi(\frac{t}{2^s} - l),$$

where s gives the length of the wavelet and l its location. The wavelet transform of the discrete data train $\{X_i\}$ can be defined as

$$W_{s,l} = \sum_{n=1}^{N} \psi_{s,l} X_i.$$

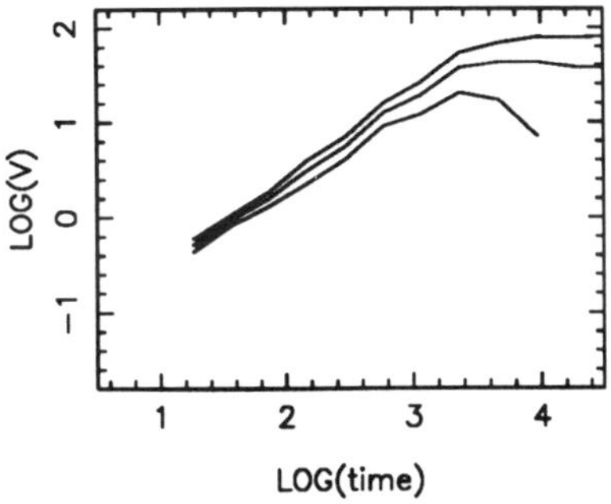

Figure 1. Scalegram of a white noise ($\sigma = 1$) data set with sampling identical to the historical. The unit of time is days. 1σ errors of estimates are shown.

In calculating the wavelet transform it is most convenient to start from the longest timescales and proceed towards shorter ones. The ratio of successive timescales considered is 2, thus if one studies variability, e.g., in 12 dyads, as we choose here, then the ratio of the longest timescale to the shortest one is $2^{12} = 4096$.

We define the scalegram (Scargle 1993) as the sum of the square of the wavelet amplitudes at a given timescale

$$V(s) = \langle \sum_l (W_{s,l})^2 \rangle.$$

To understand scalegrams we applied the method to artificial data which had the time distribution identical the light curves of OJ 287 (pre-1992 light curve, 1993/94 and 1994/95). Light curves of white noise were synthesized with variances equal to 0.01, 0.1, and 1. We also ran simulations with artificial red noise. In addition to the average scalegrams, we also calculated variances of white noise scalegrams giving us estimates of the errors.

The *only* change in the scalegram caused by changing the variance in the simulation was a shift along the ordinate; the shape and the errors remained identical. Without losing generality we now consider only the simulations with variance $\sigma^2 = 1$.

The scalegrams of white noise are not flat as expected for equally spaced data. A cutoff usually takes place at about $\tau = 2000$ days in the century-long light curve and at about $\tau = 4 - 7$ days in the 1993/94 and 1994/95 light curves. The bend is near the timescale below which the time series becomes incomplete. An example of a scalegram of white noise ($\sigma^2 = 1$) is shown in Fig. 1.

3. Noisegrams (continuation of wavelets...)

If we consider two scalegrams of white noise for a fixed time series and divide them by each other we obtain a flat curve, i.e., a constant, which is independent of timescale. This constant has a numerical value equal to the ratio of the variances of the two light curves. If we choose the divider to have unit variance,

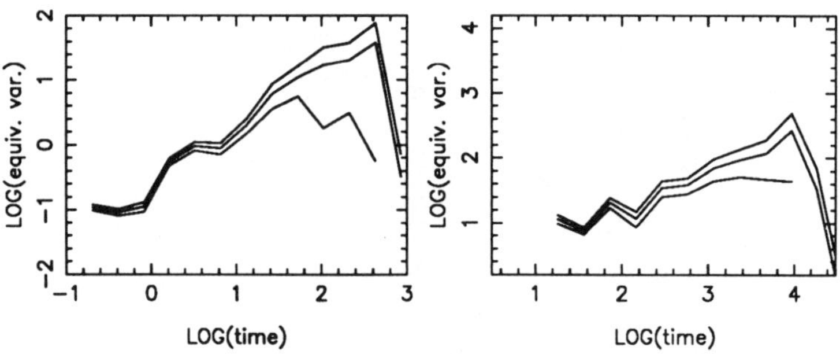

Figure 2. Noisegrams of the 1993–1996 light curve of OJ 287 and of the historical light curve. Note the good connectivity between the two noisegrams. The unit of time is days. 1σ errors of estimates are shown.

then this ratio gives directly the variance of the other light curve. We can define the noisegram of a given light curve as the ratio of scalegram of the light curve to the scalegram of a unit variance white noise light curve with the same time distribution.

The noisegram tells us how much variability is present on the average at a given local timescale.

OJ 287 shows variability on all timescales from 3 hours to decades. The noisegram level we expect from measurement errors is about $\log\langle\sigma^2\rangle = -2.4$ in 1993–96 (Fig. 2a). The noisegrams of the 1993–96 light curve at long time scales connects well with the noisegram from the historical light curve (Fig. 2b). The noisegram of the historical light curve shows a very fast decline at 10^4 days $\sim$ 30 years (half of the long time base line change) (Lehto & Valtonen 1996).

The wavelet analysis indicates that variability occurs in OJ 287 on timescales over 4.5 decades in frequency.

4. Powerspectra

Next we investigated the noise properties of the 1993–1996 flux density light curve of OJ 287. From the appearance of the light curve we anticipated an $(f^\alpha, \alpha < 0)$ -type red noise spectrum. A CLEANed spectrum (Fig. 3) of the 1993–1996 flux density light curve of OJ 287 shows indeed such a spectrum with $\alpha = -1.5$ over about 3 decades from about 0.1 d^{-1} to 10^{-3} d^{-1}.

To avoid (unphysical) divergence of total variance, a spectrum showing slopes $\alpha = -1.5$ must have a low frequency bend in the spectrum. Consider a spectrum for randomly arriving exponentially decaying shots with fixed amplitude, A, and decay rate, $(\frac{1}{\omega_o})$. This process now has a power spectrum of the shape

$$S(\nu) = A^2\lambda\frac{1}{(\omega^2 + \omega_o)},$$

where λ is the average arrival frequency of shots.

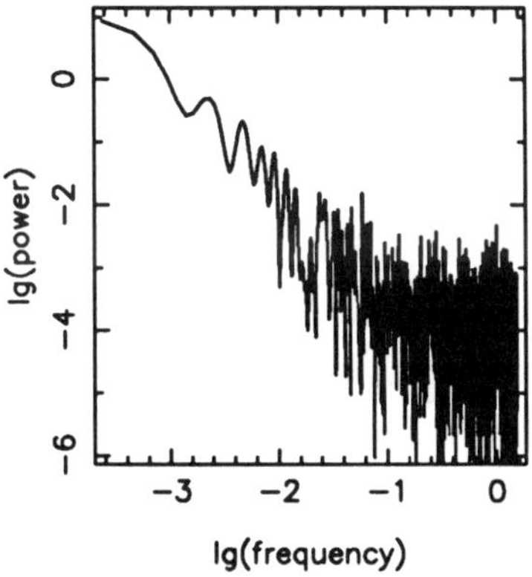

Figure 3. CLEANed powerspectrum of the 1993–1996 light curve of OJ 287. The units of frequency are inverse of days.

A superposition of shots with different decay timescales, which have a larger $A^2\lambda$ for slower shots is capable of producing a spectrum with a slope $-2 < \alpha < 0$. If both amplitude and shot intervals depend on the decay timescales as

$$A \propto t_{\rm d}^{\beta}, \qquad t_{\rm r} \propto t_{\rm d}^{\gamma},$$

then the spectrum will have a slope of about

$$\alpha \approx \gamma - 2\beta, \text{ as long as } -2 < \alpha < 0.$$

From the light curve (Sillanpää 1996, and references therein) it is quite clear that high amplitude shots are less frequent than shots with small amplitudes implying $\beta > 0$. Also it appears that fast shots occur more frequently than slow shots implying $\gamma > 0$.

If shots with different decay timescales have equal amplitudes, then the long timescale shots should be much more frequent. This appears not to be the case.

It is also apparent that the average level of the flux is not considerably higher than the height of the shots. This implies that at any given decay timescale the number of overlapping shots is small, thus, *we are seeing almost individual shots in OJ 287.*

Small amplitude shots ≤ 1 mJy appear to occur at least once per 30 days. High amplitude shots (> 10 mJy) on the other hand take place quite infrequently, once a year at most. Using these rough estimates we can obtain a value for

$$\frac{\beta}{\gamma} = \log(10/1)/\log(360d/30d) \leq 0.92.$$

Combing this with the observed slope of the spectrum yields a requirement $\beta > 2$, and $\gamma > 2$, thus amplitude of shots should be proportional to at least the square of decay times. A shot decaying in 30 days should be about 40 times brighter than a shot decaying in 5 days. This clearly is not the case if one compares, e.g., shots at JD 2449366 and the large shot around JD 2449680. *If the observed spectral slope represents a reasonable realization, then there must be at least two separate ways in forming shots.*

A flatter spectrum would reduce the severity of this contradiction. In that case, qualitatively, *the outbursts of Oct. 1995 and Nov. 1996 should be very infrequent.*

5. Deconvolution to individual shots

We also looked at the distribution of amplitude vs. width of the shot. We fitted gamma-function-shaped shots to the light curve with the height, location, width and symmetry of each shot being free variables. One should be cautious about such fits, but even so, the result was somewhat surprising. If a correlation is present, it is such that slower shots have lower amplitudes quite against our intuition. If this is the case, then the large amplitude shots, such as the 1995 November outburst should consist of a very large number of clustered, very fast shots ($t_d < 4$ d) meaning that the shots would not be random any more. The fitting procedure can be questioned in some cases, e.g., shots around JD 2449415.8 or 9327.5.

6. Lost shots

Because short timescale variability may be distinct in characteristic from longer timescale variability, we will try a more direct and more robust method to estimate the number of large outbursts missed by the sampling of the historical light curve.

The light curves we are using have gaps, which increase as we go into the past. In analyzing such light curves one is bound to ask how can one be sure that you have not missed any important events have not been missed in the past, or alternatively how many events have actually been missed?

We are fortunate to have our light curve sampled in such a way that we can infer its behavior on various timescales from some sections of the light curve.

We calculated the interval of each successive data point. If the separation of the data points was smaller than the timescale, we concluded that a shot with that characteristic timescale cannot occur in the given time interval. If the time interval was longer than the timescale considered, we simply divided the interval by the timescale. Adding up results from different time intervals one can count the total number of shot timescales that are unsampled in the light curve.

If the duty cycle of a certain shot timescale is known, one can estimate how many such events have been missed. This analysis makes no assumption on the shape of the shots or variability events.

First let us consider the lost shots in 1993/94 and 1994/95. Sampling was complete in both cases at timescales longer than 5 days (e.g., Fig. 4a). During 1993–1996 the well-covered section of the light curve is about 600 days in total length. During these 600 days only two (obvious) long timescale ($t > 30$ d) shots are present. This gives us a duty cycle ≤ 0.1 for these kind of events.

The number of lost shots of timescale of 30 days could conceivable be as high as 100 in the 8856 day (1970–1992) light curve. If we take into account the

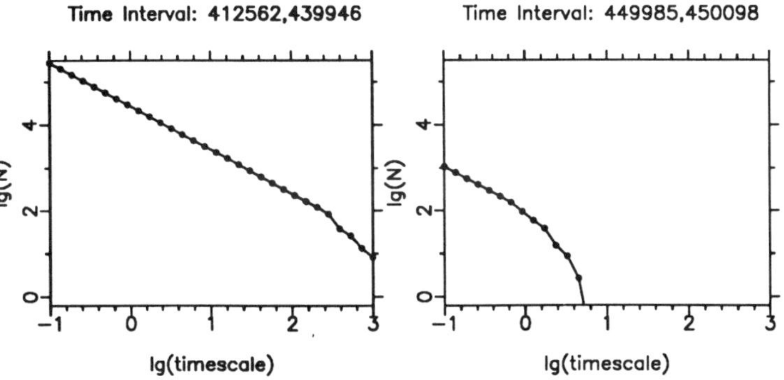

Figure 4. Number of unsampled timebins in the pre 1972 light curve and the 1994/95 light curve of OJ 287. The abscissa is the length of a time bar in days, and the ordinate is the number of time bars that are needed to cover all the gaps in the data set.

duty cycle from above we obtained from the denser light curve, we may set an upper limit to 10 lost shots. However, if these shots are random we should also see 20 such random shots with our 2/3 coverage of the observed light curve. We see only 4 clear cases (or clusters). One of these shots appears to have a long, over 300 day, decay timescale. Based on this light curve it appears that we have at most six 30 day shots and one 300 day event in 8900 days. The respective duty cycles are then 0.02 and 0.03.

Finally, if we look at the pre 1970's section of the light curve, then we can figure out that we may have lost maximally 500 one month long events (Fig. 4b). Taking into account the duty cycle we may claim that we really have lost at most 10 shots with a length of 30 days, and at most one 300 day events. These numbers should be compared with the observed number of 6 quasiperiodic events, which have reoccurred every 10–12 years, and often in pairs.

Acknowledgments. Dr. Markku Lainela made a last minute successful effort in running the Jandel scientific software co. package Peak fit to decompose the 1993/94 light curve into separate shot components.

The OJ-94 project observers have provided wonderful data. Drs. Leo Takalo and Aimo Sillanpää have compiled most of the historical light curve.

References

Scargle, J. 1993, "Wavelet Methods in Astronomical Time Series Analysis" in International Conference on Applications of Time Series Analysis in Astronomy and Meteorology held in Padova, Sept. 1993

Lehto, H. J., & Valtonen, M. J. 1996, ApJ, 460, 207

Sillanpää, A., Haarala, S., Valtonen, M. J., Sundelius, B., & Byrd, G. G. 1988, ApJ, 325, 628

Sillanpää, A., et al. 1996, A&A, 305, L17

Sundelius, B., et al 1996, these proceedings

Blazar Continuum Variability
ASP Conference Series, Vol. 110, 1996
H. R. Miller, J. R. Webb, and J. C. Noble, eds.

Chaotic Properties of the Blazar OJ 287

A. Sadun

Agnes Scott College

Abstract. Large amounts of flux variability data of the blazar OJ 287 are analyzed using nonlinear techniques. Conventional methods include looking at phase space plots for attractors, the autocorrelation function for dependence on neighboring flux values, the correlation dimension as a measure of underlying structure, and using maximum entropy to uncover dominant frequencies. In addition, the twin techniques of singular value decomposition (for noise removal) and model fits to dynamical behavior are employed to great effect on these data. Strange attractors are found and correlations are established. The results thus obtained are discussed.

1. Introduction

An international collaboration of scientists involved in the OJ-94 monitoring project collected enormous amounts of photometric data at many wavelengths of OJ 287. This is the largest and most successful project of its kind. Data has been taken over three separate temporal regimes: a century of archival data, a single observing season, and microvariability. The quality of this large database has afforded us a unique opportunity to study this blazar in terms of chaotic behavior.

Accordingly, data analysis is presented here in an unusual form. Phase space plots of the data are presented, as well as dynamical model equations as derived by the technique of singular value decomposition. Special features in phase space are identified, including strange attractors as well as other areas of attraction and avoidance.

The light curve of OJ 287 is by far the most extensive assembled for any active galactic nucleus (AGN) both in terms of number of data points as well as extent of time monitored. Tens of thousands of data points have been recorded covering over a hundred years of observations, the most recent years being covered the most extensively. This database has been scrutinized very carefully for patterns of any kind. Initially, it was noticed that there appears to be a major outburst every dozen years or so. (This observation is suspect, however, because routine power spectral analysis is difficult to do when the data train is full of gaps, and the preponderance of the data was taken only in the past 30 years or so.) In fact, when looked at more closely, the outbursts do not seem to be strictly periodic at all. This could imply that we are dealing with a chaotic phenomenon, and that we need to treat the data in a correspondingly nonlinear dynamical way. Indeed, based on such characteristics, a binary black hole model has been suggested to explain the outbursts and some of the chaotic features

(Valtonen et al. 1988, Lehto & Valtonen 1996). Fundamentally, chaos means nonlinearity and feedback. It is by looking for evidence of such feedback (by searching for strange attractors and the like in phase space), that even better models can be found to explain the OJ 287 system.

2. Noise removal

Perhaps the greatest challenge of chaos analysis is the attempt to separate the chaotic part of the time series from the true stochastic noise component. Such removal must be done carefully and cautiously. The minimization of noise will be the central feature of the data analysis presented here. To accomplish this, we make use of a recently developed technique referred to as *singular value decomposition* (Broomhead & King 1986, Vautard & Ghil 1989).

The data is linearly interpolated to yield equally spaced points, and a minimal boxcar smoothing is also performed as an initial filter for noise reduction. This prepares the observations for chaos analysis.

The method of singular value decomposition is then employed in which we decompose the original light curve (or any time series) into a set of orthogonal expansion functions. In this respect, it differs little from decomposing a light curve into a Fourier series. However, with a Fourier decomposition, one is imposing a certain mathematical form onto the data. With singular value decomposition, the orthogonal functions are determined from numerical analysis of the data. These functions are the eigenfunctions of a square correlation matrix of dimension M. The correlation matrix is made of values of the correlation function for the time step offset $\tau = 0$ along the diagonal, $\tau = 1$ on either side, etc. The eigenvalues of this correlation matrix correspond to normalized amplitudes of the orthogonal functions, just as one has amplitudes for the normal modes of a Fourier series. The squares of these amplitudes are analogous to the creation of a power spectrum of a Fourier series expansion.

With singular value decomposition, just as with a Fourier series, one can decompose the light curve to a finite number of terms M. After such an expansion, one truncates to a certain number of terms—in this project to $d = 3$ terms. The light curve, now reconstituted from the three largest orthogonal terms only, is in theory capturing most of the signal, and containing little of the noise. The two free independent parameters here are the number of terms of the expansion, M, and the number of terms of the truncated series, d. For the OJ-94 project data, the reconstituted light curve looks almost identical to the original. This is a good sign. It means that there was little noise that had to be eliminated. We will find that there are other tests, such as dimensionality, that show the same thing.

We could stop here, and study the reconstructed light curve in phase space, but that would limit us to past behavior, and say nothing about future behavior. So, at this point, we need to use our newfound three functions to create model equations of a chaotic system. Once we do that, we could in principle get the functions for all time.

Expansion functions for future times are generated by taking linear combinations of products of expansion functions of the data (Sprott & Rowlands 1992). The coefficients are solved for by looking at the last N points and using the method of least squares. Then, the three model functions are plotted against each other in what is mathematically equivalent to a phase space plot.

Now we can display and analyze the data in various ways. We can make phase space plots of the expansion functions, both for times taken during the observations, and for future times. We may look for attractors, usually appearing as loops and representing periodic behavior. We may also look for strange attractors, which often appear as beautifully intricate geometrical patterns, representing areas of phase space especially frequented by the system. We can then measure our confidence that this technique represents reality by looking at how well the model equations fit the data already taken, at least for the last N points. If the least squares fit matches well the data collected for the last N points, we can have a good deal of confidence that the model equations now represented in phase space, for both the present and future, are a reasonable description of the signal without the noise component.

3. Other tests for randomness

There are three other evaluations worth mentioning that can be made of the data, independently of other techniques. The first of these is to investigate the dimensionality of the data. I have chosen to use the *correlation dimension*, which is a measure of how the number of data points in phase space scales within a region of radius R as R approaches zero. The larger the dimension, the greater the degree of randomness of the data. Typical values for this data are between 2 and 3, and a value greater than 5 implies essentially stochastic behavior.

The next test is taking a measurement of the *Lyapunov exponent.* This is a measure of the rate at which nearby trajectories in phase space diverge. All three exponents of periodic orbits would be negative. An exponent of zero would occur near a bifurcation. The larger the exponent, the less periodic the behavior. Table 1 shows only the most positive exponent calculated.

Finally, an *autocorrelation function* (acf) is calculated as well. In this test, the value of each data point is multiplied by the value of a data point τ time steps before, and this is summed over all of the data points. The sum is then plotted as a function of τ. The acf gives a measure of how dependent data points are on their neighbors. Where the function goes to zero gives a measure of the correlation time. Highly random data will have no correlation, so the function will drop abruptly to zero, implying small correlation time. Highly correlated data will have an acf that varies, but will have an amplitude that only slowly decreases with time.

4. Characteristic features

The light curves of OJ 287 show certain interesting features. The long term, century old archival data show what on the surface appears to be periodic outbursts every dozen years or so. This has been difficult to verify because the data initially had been very sparse, and has gotten denser and denser dramatically in more recent years. The gaps in the data make it next to impossible confirm. Indeed, perhaps the periodic behavior is merely conjecture, or perhaps we are

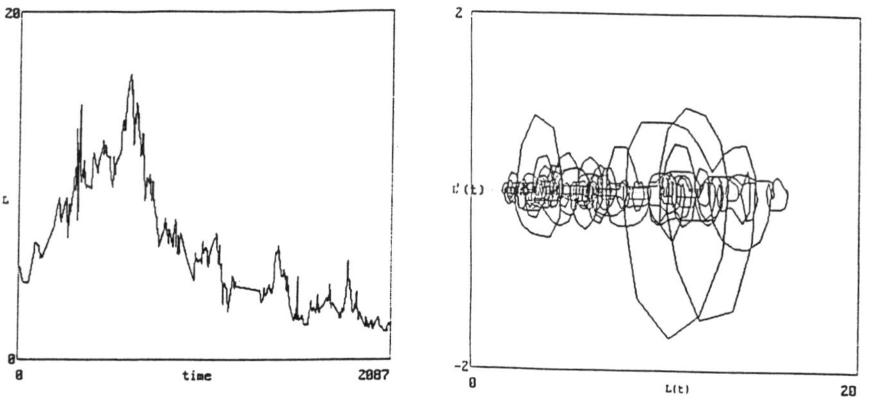

Figure 1. Left: Light curve 1994–5 where each time unit represents 0.1 days. Right: Phase space diagram.

dealing with a system that has slowly or quickly varying periods that only chaos analysis can recover.

The data also show two kinds of luminosity increases, spikes or flares which seem to come quite suddenly in a matter of days with a decay just as fast, and longer outbursts lasting months or even years, and just as great in amplitude as the flares. The suddenness of the flares are seen in considerable detail with some of the recent data taken in 1994–5. [See Fig. 1(Left).]

For investigating intra-night variability, we have included the data of others (Miller, Carini, & Goodrich 1989, Carini, Miller, & Goodrich 1990, Carini, Miller, Noble, & Sadun 1991, Carini, Miller, Noble, & Goodrich 1992) as well as from the OJ-94 project. We found that sometimes the variability shows real structure, and at other times it does not (Sadun 1995). So, when we tried to find a correlation between structure and fractal dimension, we find for 19 data sets that there is indeed a strong inverse correlation between the relative standard deviation and correlation dimension (of 0.82). We also found for a smaller sample of 8 data sets a smaller correlation between average brightness and dimension (of 0.69). Theoretical implications of this follow in the Summary.

Further information can be gleaned from the phase space diagram. [See Fig. 1(Right).] This diagram shows all manner of loops of various sizes (changing periodic behavior), and large excursions (flaring behavior). Taken as a whole, it shows general areas of attraction and avoidance by the system.

Finally, we see model equation phase space diagrams showing attractors and even strange attractors—areas where the system frequents. (See Fig. 2 and 3.) One may also see, by watching these plots form, that there is episodic or transitional behavior in which the system is trying to go towards or away from a strange attractor.

5. Summary

The numerical results of some of the data sets used are shown in Table 1. Here, the dimension, Lyapunov exponent, the first zero of the autocorrelation fuction

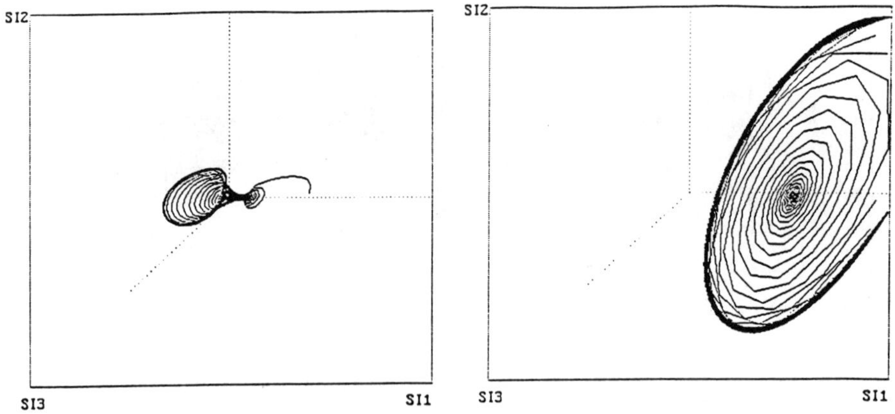

Figure 2. Model fits: (Left) Light curve for 1994–5 with $M = 11$ and $N = 704$; (Right) Microvariability data set Micr1li.dat with $M = 10$ and $N = 16$.

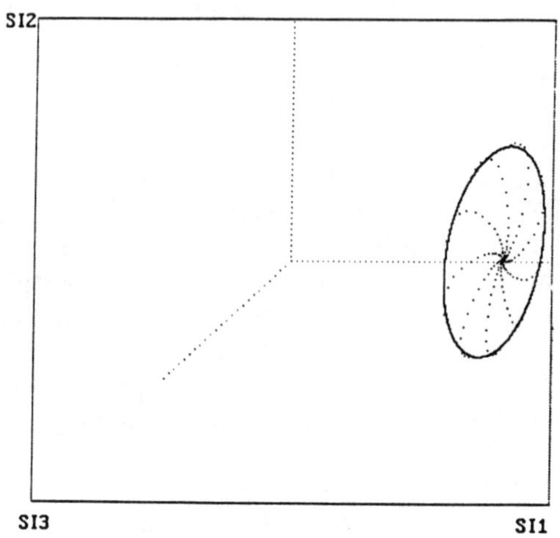

Figure 3. Microvariability model fit (untracked) for data set Micr13.dat with $M = 8$, and $N = 16$.

Table 1. Tabular data for the light curves of OJ 287.

Data Set	Interval (days)	Dim'n	Lya. Exp.	Period (days)	1st Zero of ACF	Comments
rad22li.dat	6	2.447	0.318	1510	1502	Sharp frequency peak
rad37li.dat	5	2.775	0.356	NA	1869	Slow drop in acf, very active
micr1li.dat	0.001	3.884	0.426	0.05	0.048	Large dim. and Lya. exp.
micr2li.dat	0.003	1.958	0.397	NA	0.096	Small dim., sharp drop in acf
micr13.dat	0.0015	2.626	0.363	NA	0.073	No structure in acf
OJ93.lgt	0.1	2.110	0.258	20	33	Wide freq. peak, sharp drop in acf
OJ94.lgt	0.1	2.632	0.439	NA	55	Large Lya. exp.
OJcent	50	2.379	0.350	6467	4428	Long period, sharp drop in acf

are given, among other parameters. Other characteristics noticed by this study are that OJ 287 occasionally goes into episodic (i.e., sudden change) activity, as well as cyclical and pseudo-periodic behavior.

Perhaps the most interesting result noticed by this data is that microvariability can show a great deal of structure which correlates to fractal dimension—that is, the greater the structure, the smaller the dimension. Another correlation for microvariability is between average luminosity and dimension; the greater the average brightness, the smaller the dimension. These two findings taken together indicate that there are two components to microvariable emission. One underlying component is stochastic behavior of constant average luminosity over a time period of several hours. Superimposed on this is the chaotic component, which at times is turned on or off, thus varying the overall structure of the general light curve.

Another interesting result is that many of these data sets show some kind of periodicity, whose value changes by the different types of manipulation imposed (e.g., interpolation schemes, smoothing, truncation, etc.). This indicates to this author that there may not be any single dominant frequency for any particular data set, or if there is, it varies with time. The associated not-too-small Lyapunov values seem to confirm this.

The methods of analysis employed here, of singular value decomposition and model fits, are found to be quite useful in extracting dynamical information. In addition, they allow one to make short term predictions, if the model fits are deemed to be good ones (small standard deviations in the least square fit). If so, one should be able to predict when a major outburst is expected, and should be able to forewarn other observers to concentrate their energies on monitoring intensively OJ 287 during these times of more enhanced activity. Obviously, these nonlinear dynamical techniques can be applied more generally to other AGN as well.

Acknowledgments. I would like to thank Melissa Nysewander for taking the time to help us with some of the data reduction, and Leonid Ozernoy for some very useful discussions.

References

Broomhead, D. S., & King, G. P. 1986, Physica D, 20, 217

Carini, M. T., Miller, H. R., & Goodrich, B. D. 1990, AJ, 100, 347

Carini, M. T., Miller, H. R., Noble, J. C., & Sadun, A. C. 1991, AJ, 101, 1196

Carini, M. T., Miller, H. R., Noble, J. C., & Goodrich, B. D. 1992, AJ, 104, 15

Lehto, H. J., & Valtonen, M. J. 1996, ApJ, 460, 207

Miller, H. R., Carini, M. T., & Goodrich, B. D. 1989, Nature, 337, 627

Sadun, A. C. 1995, BAAS, 27, 1408

Sprott, J. C., & Rowlands, G. 1992, Chaos Data Analyzer, ed. J. S. Risley (New York, AIP)

Valtonen, M., et al. 1988, in Active Galactic Nuclei, ed. H. R. Miller & P. J. Wiita (Berlin, Springer-Verlag), p. 68

Vautard, R., & Ghil, M. 1989, Physica D, 35, 395

Blazar Continuum Variability
ASP Conference Series, Vol. 110, 1996
H. R. Miller, J. R. Webb, and J. C. Noble, eds.

The OJ287 Binary Model and the November 1995 Outburst

Mauri Valtonen, Harry Lehto, Krista Kokkonen, and Seppo Mikkola
Tuorla Observatory, University of Turku, Finland

Abstract. A binary black hole model was proposed for OJ287 on the basis of its historical light curve in 1988 by Sillanpää et al.. The model predicted the next outburst in September 1994 on the basis of the idea that tidal transfer of matter in the accretion disk of the primary is responsible for the increased output of the relativistic jet. Even though the outburst did occur, it was late by about 6 weeks. Similar and even greater discrepancies between the binary model and the timing of the outbursts are obvious in the historical light curve. Therefore the binary model was generalized by Lehto and Valtonen to a companion orbit which is highly inclined to the disk of the primary, instead of the zero inclination orbit of the original model. As a result of the inclined orbit the secondary crosses through the disk of the primary twice per period. A bright thermal flash, seen primarily in ultraviolet and optical, is expected at each disk crossing from the shock heated disk gas. It has been possible to identify many of the disk crossing events in the historical light curve in this way. A very satisfactory explanation of the main features of the light curve, including the apparently early or late arrivals of the outbursts can be given. In fact a unique orbit determination is possible when 5 crossings have been exactly timed. The necessary number of the crossings, "superflares", was known after the 1994 event, and the orbit was solved. The solution predicts the timing of the future plane crossings. The next prediction was a superflare which should begin between November 1 and November 15, 1995. The uncertainty in the prediction was due to uncertainty in the accretion disk model: the exact timing will be important for determining the disk parameters. Once the disk model is well known, the next set of outbursts in 2006 and 2007 can be predicted very accurately, to the extend that one can test the General Theory of Relativity in the strong field limit for the first time.

1. Introduction

OJ287 is one of the very few quasars for which we have a fairly continuous light curve over the period of one hundred years. This is due to the fortuitous situation of OJ287 being near the ecliptic and of its relatively high brightness. This is why it has been accidentally photographed during minor planet searches and other solar system work since late last century. The magnitude measurements from literature were put together by Sillanpää et al. (1988) who found that the object brightens up spectacularly at mean intervals of 11.65 yr. Since the last major flare was observed at 1983.0, the following one was expected at 1994.65. The flare was well observed by the OJ-94 campaign (Sillanpää et al. 1996) as well as others. Its timing was a little late, about 1994.76, but otherwise it lived up to expectations.

Sillanpää et al. (1988) proposed that the basic reason for the rather regularly spaced outbursts is a black hole binary system with a period of 9 yr which translates to the observed outburst interval at the observed redshift of 1.3. The problem with this explanation is the irregularity of the exact times of the major flares: they appear to be sometimes early, sometimes late by as much as one year relative to the mean period. In terms of the binary model this would mean that the outbursts are induced at different phase angles relative to the pericenter, and these phase angles vary as much as $\pm 90°$ on either side of the pericenter. It is difficult to produce such an irregularity in the tidal triggering model of Sillanpää et al. (1988).

In addition, a careful inspection of the light curve during the best observed active phases in 1971 – 1973 and in 1983 – 1985 shows that there are in fact two or even three major peaks in the light curve per active phase. This is also problematic in the tidal triggering model.

Because of these problems Lehto and Valtonen (1996) proposed a new model where the activity is triggered not only by the tides but also by the crossing of the secondary black hole through the accretion disk of the primary. Then two of the major flares per active phase are associated with the impact on the disk and the energy released in the impact while the slower and longer lasting increase in the brightness level is associated with tidal effects.

The attractive feature of this model is the unique solution of the binary black hole orbit and the subsequent possibility of predicting the future light curve of OJ287. The next prediction was the occurrence of another major flare, similar to the 1994 flare, in November 1995. This is associated with the second crossing of the secondary through the disk of the primary. Later also new calculations of the tidal triggering process were carried out and they resulted in the prediction of the third brightening of OJ287 in the early part of 1996 (Sundelius et al. 1996, reported in this conference). Here we concentrate on the former prediction, how well it was fulfilled and what is its significance to the binary black hole model.

2. The binary model

The basic ingredients in the binary black hole model of Lehto and Valtonen (1996) are:

1. the relative timing of five superflares, i.e. outbursts which last 6–8 weeks and reach up to 10 mJy above the base level in the V-band;
2. a model for the time delay between the impact on the disk and the radiation burst in the optical V-band, and
3. a model for the loss of orbital energy.

The binary orbit is fully described by 4 parameters (neglecting the orientation in space) and therefore 4 time intervals, i.e. 5 impact times are necessary for a unique solution of the orbit. In the first approximation, one may assume that time delays are constant irrespective of the radial distance of the point of

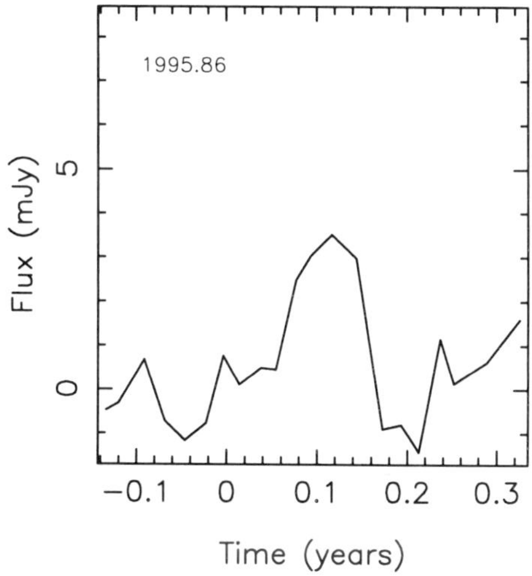

Figure 1. The time averaged light curve of OJ287 around November 1995. The averaging interval is 0.02 yr. The data is courtesy of the OJ-94 campaign (L. Takalo and A. Sillanpää, private communication). The steadily rising brightness has been subtracted.

impact. Also one may neglect the loss of orbital energy. Then one obtains the first model of Lehto and Valtonen (1996), and a prediction that the 1994 and 1995 outbursts should be separated by 1.06 yr.

In the second case of Lehto and Valtonen (1996) a geometrically thin but optically thick α_g-disk model was used (number density $n = 10^{14}\text{cm}^{-3}$, scale height $h = 4 \cdot 10^{14}$cm at the typical impact distance). The collision of a secondary black hole of mass $10^8 M_\odot$ with a speed between 45 000 km s^{-1} (at the pericenter) and 9000 km s^{-1} (at the apocenter) releases typically 10^{55} erg/collision. The energy is transferred by the bow shock of the secondary black hole to the gas of the disk. The shocked, radiation pressure dominated gas is accelerated in the direction of the secondary's orbit. Part of the energy produced in the interaction heats up the unshocked gas of the disk.

The bulk velocity acquired by the shocked gas causes it to exit from the plane of the disk. The gas remaining in the disk will start to close the hole at the local sound speed. Once the shocked gas is outside the disk, it starts to expand. Selfgravity is not very important. The gravity of the secondary is non-negligible when interactions near apocenter are considered.

The gas bubble is initially optically thick. After sufficient adiabatic expansion has taken place, the bubble becomes optically thin. From that moment onwards radiative cooling dominates over adiabatic cooling. This is when an outburst is observed. The model predicts the timelags of the outbursts as compared with the disk-plane crossings. The time delays are small at pericenter and up to two years at apocenter.

The expected amplitudes in a disk interaction model involving a secondary of 100 million solar masses yields amplitudes between about 2 to 8 mJy depend-

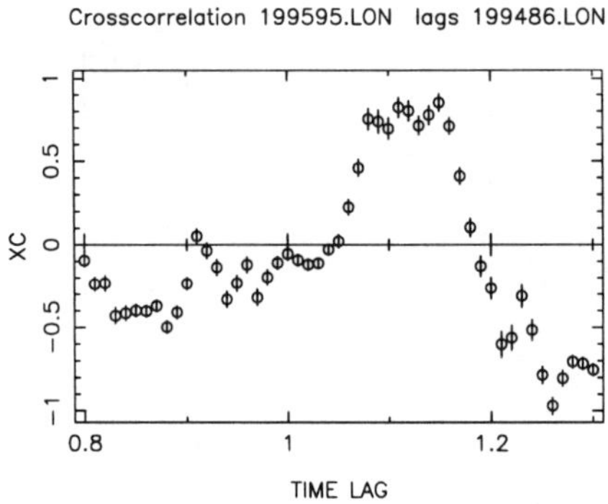

Figure 2. The cross-correlation of the light curves of the 1994 and 1995 superflares.

ing on the orbital radius vector of the secondary on impact. These agree with the observed values.

The prediction of the second model of Lehto and Valtonen (1996) was that the 1994–1995 outburst interval should be 1.10 yr. Orbital energy losses were also neglected in this model.

The 1995 outburst came much as predicted. A time averaged light curve of the outburst event is shown in Fig. 1. Figure 2 shows the cross-correlation of the V-band flux values in the 1994 and 1995 outbursts. It is obvious from Fig. 2 that the outburst interval was greater than 1.06 yr (the prediction of model 1), but that the 1.10 yr interval (model 2) gives an excellent fit to the data. Note that the models were calculated and the paper (Lehto and Valtonen, 1996) was accepted for publication well before there was any indication that a second outburst should occur at all, not to mention the narrow two-week time interval defined by the two models.

In the following we will discuss the details of the model and what one can say about the orbital energy loss due to gravitational radiation.

3. The orbital energy loss

The binary model was described in detail in Lehto and Valtonen (1996). In essence, we have a binary black hole with component masses $17.7 \cdot 10^9 M_\odot$ and about $10^8 M_\odot$ which go around each other in a (redshifted) period of 12.07 yr and in an orbit of eccentricity $e = 0.67$. The major axis of the orbit precesses 33° per period, i.e., the system is extremely relativistic and will collapse due to gravitational radiation within the next $4 \cdot 10^4$ yr even though the shortening of the period was neglected in the model. The semi-major axis of the binary is 0.056 pc which corresponds to about 10^{-5} arc sec in the sky at the distance of OJ287 (for the Hubble constant of $H_0 = 60$ km s^{-1} Mpc^{-1}). It is not possible to resolve the binary with current methods, but it may be within reach in the near future with space VLBI if both black holes are strong radio sources.

How serious is the neglect of the orbital energy loss in the models of Lehto and Valtonen (1996)? We have calculated further models where the gravitational

Table 1. The relative timing of the superflares in different models (m_2 is the mass of the secondary in units of $10^8 M_\odot$) compared to observations.

Model I (no rad'n) $m_2 = 0.98$	Model II $m_2 = 0.98$	Model III $m_2 = 0.72$	Model IV $m_2 = 0.60$	Observed times
1947.30	1947.30	1947.29	1947.29	1947.30
1959.21	1959.22	1959.21	1959.21	1959.22
1972.99	1973.00	1972.98	1972.97	1972.98
1983.00	1983.00	1983.00	1983.00	1983.00
1984.16	1984.16	1984.16	1984.16	1984.16
1994.77	1994.75	1994.75	1994.75	1994.77
1995.86	1995.84	1995.85	1995.85	1995.87
2006.35	2006.30	2006.23	2006.21	

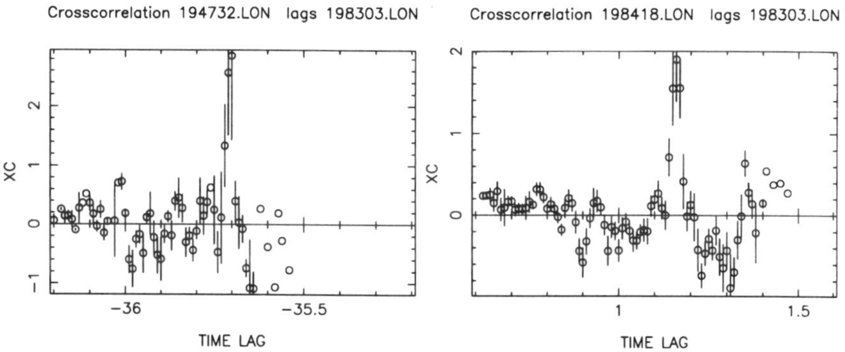

Figure 3. The cross-correlation between the 1983 superflare and some other superflares. Time lag is the time difference between the 1983.00 flare and the 1947.30 flare (left) and 1984.16 flare (right).

radiation terms are included in the equation of motion (see Valtonen et al. 1995 for details). Table 1 shows the timing of the outbursts in a number of the models. Note that the timing is relative to the outburst at the beginning of 1983; it is given the exact time of 1983.0 and it refers to the beginning of the outburst. The brightness maximum for a superflare occurs sometime close to the beginning of the superflare (as in 1983) sometime during the second rise of the superflare about one month later (as in 1994). We do not consider the exact times of the brightness maxima important since they may be related to random processes such as variations in opacity and geometrical configurations of radiating and absorbing gas clouds.

In order to obtain relative timing of the various outbursts, we have carried out cross-correlation studies between the flux values of the 1983 superflare and the other superflares.

Figure 3 shows examples of the correlation diagrams. Using the cross-correlation method as well as the manual superposition of light curves on top of each other we have derived the timings of altogether seven superflares relative to each other. These are also shown in Table 1. As the timing of the superflares is accurate only within ± 0.01 yr (i.e., within one week), it is obvious from Table 1 that currently we cannot claim to have detected the orbital energy losses.

The first opportunity to test the expected radiation loss comes in 2006. Then the superflare may start as early as in March if the radiation is important, but is not expected until May if the radiation is negligible. This will also be a test of the General Theory of Relativity.

References

Lehto, H. J., & Valtonen, M. J. 1996, ApJ, 460, 207

Sillanpää, A., Haarala, S., Valtonen, M. J., Sundelius, B., & Byrd, G. G. 1988, ApJ, 325, 628

Sillanpää, A., et al. 1996, A&A, 305, L17

Sundelius, B., Wahde, M., Lehto, H. J., & Valtonen, M. J. 1996, these proceedings

Blazar Continuum Variability
ASP Conference Series, Vol. 110, 1996
H. R. Miller, J. R. Webb, and J. C. Noble, eds.

Long-time Brightness Variations of OJ287 in the Binary Black Hole Model

Björn Sundelius and Mattias Wahde

Division of Mechanics, Chalmers University of Technology, S-41296 Göteborg, Sweden

Harry J. Lehto and Mauri J. Valtonen

Tuorla Observatory, University of Turku, FIN-21500 Piikkiö, Finland

Abstract. The BL Lac object OJ287 is here modelled as a binary black hole: one supermassive object is orbiting around a second in a precessing orbit that intersects the accretion disc surrounding it. By identifying the observed major outbursts with pericenter passages and superflares with disc crossings, this model has enabled us to predict the latest outburst with high accuracy.

1. Introduction

Sillanpää et al. (1988) proposed that the brightness variations in OJ287 are controlled by tidal transfer of matter in the accretion disc of a black hole when a secondary black hole perturbs it periodically. This model lead to a prediction of the 1994 outburst in OJ287 which was correct within the accuracy of two months. This paper reports an improved model obtained by combining an orbital tracking code with an N-body simulation of the accretion disc. The calculations now include the full set of possible orbit parameters, as well as an attempt to determine the parameters by fitting the observations of the historical light curve to the model. The superflares and their association with the disc crossing events (Lehto & Valtonen 1996) are also included in the model. The model explains the OJ287 light curve very well and predicted the outburst in late autumn 1995 with high accuracy.

2. The model

The model consists of a primary black hole, surrounded by an accretion disc, and a secondary black hole orbiting around the primary in an orbit perpendicular to, and intersecting, the accretion disc. The disc is simulated with 500,000 non-self-gravitating particles moving in a constant potential following the primary. The two black holes move on precalculated orbits, keeping the center of mass for the entire system fixed. The time step used is 0.01 year. Our new model presents "particle outbursts" at pericenter passages and disc crossings. By fitting the times for these outbursts to the observed events we try to determine the parameters of the model: the masses M and m of the black holes and the

Table 1. This table gives the times for different events in the simulation. The distances are in units of 0.001 pc, the delays in years.

Time	Event	Distance	Delay	Expected
1910.65	disc passage	22.1	0.06	1910.71
1910.99	pericenter			1910.99
1911.92	disc passage	38.6	0.29	1912.21
1922.38	disc passage	31.5	0.16	1922.54
1923.06	pericenter			1923.06
1923.52	disc passage	25.6	0.09	1923.61
1933.69	disc passage	50.7	0.60	1934.29
1935.13	pericenter			1935.13
1935.35	disc passage	19.5	0.05	1935.40
1943.94	disc passage	80.0	2.59	1946.53
1947.20	pericenter			1947.20
1947.26	disc passage	17.1	0.04	1947.30
1952.97	disc passage	94.6	4.88	1957.85
1959.19	disc passage	17.2	0.05	1959.23
1959.27	pericenter			1959.27
1962.14	disc passage	75.4	2.09	1964.23
1971.09	disc passage	19.9	0.05	1971.14
1971.34	pericenter			1971.34
1972.61	disc passage	47.0	0.47	1973.08
1982.90	disc passage	26.7	0.10	1983.00
1983.41	pericenter			1983.41
1984.01	disc passage	29.5	0.14	1984.15
1994.43	disc passage	41.4	0.34	1994.77
1995.48	pericenter			1995.48
1995.78	disc passage	21.3	0.06	1995.84
2005.20	disc passage	67.8	1.45	2006.65
2007.55	pericenter			2007.55
2007.67	disc passage	17.7	0.04	2007.71
2014.64	disc passage	92.7	4.50	2019.14
2019.62	pericenter			2019.62
2019.60	disc passage	16.9	0.03	2019.63
2023.54	disc passage	86.4	3.43	2026.97

(redshifted) period P, the eccentricity e, the longitude of the pericenter ω and the precession rate Ψ of the orbit. Because of the varying quality of the observations and gaps in the records, a perfect and unique fit should not be expected. By minimizing the difference between simulations and observations we obtain the following values:

$$
\begin{aligned}
M &= 17.7 \cdot 10^9 \ M_\odot \\
m &= 1.30 \cdot 10^8 \ M_\odot \\
P &= 12.07 \text{ years} \\
e &= 0.698 \\
\omega &= 229.50^\circ \text{ at } 1904.95 \\
\Psi &= 33.3^\circ/\text{period}
\end{aligned}
$$

giving the semi-major axis $a = 0.0558$ pc.

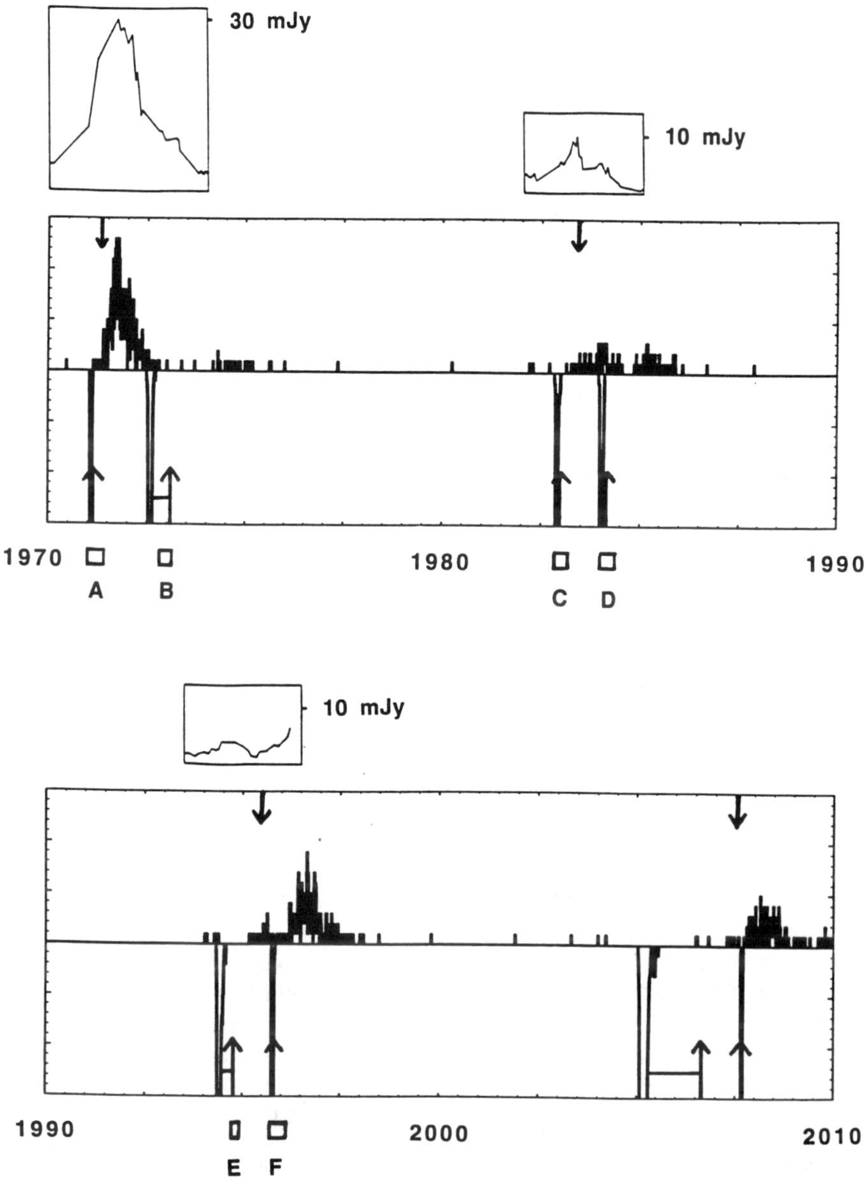

Figure 1. The two large diagrams: the number of particles per time step in the simulation which are captured by the primary (upper panel) and the secondary black hole (lower panel). See the text for the definition of the capture radius. The three small diagrams: the light curves for the 1971, 1983, and 1995 outbursts.

3. The slowly varying major outbursts

The major outbursts, which occur at approximately 12-year intervals, are related in this model to tidally induced mass flows from the accretion disc surrounding the primary towards the black hole. One such outburst is therefore expected to occur for every pericenter passage of the secondary. In the diagrams in Figure 1, the upper panel shows the number of disc particles per time step that cross the border of an accretion area, the radius of which was taken to be 10 Schwarzschild radii, centered on the primary black hole. These particles are considered consumed and are removed from the simulation. The arrows on the upper edge indicate the times for pericenter passages in the numerical simulation. Since the disc is perturbed by an object of relatively small mass on a high inclination orbit, the response of the disc occurs a short time after the pericenter passage, in agreement with Donner et al. (1991). Over the diagrams are, for comparison, the optical light curves for the 1971, 1983, and 1995 outbursts drawn to the same time scale. Although the connection between the material inflow and the optical outburst is complex and not well understood, we let the number of consumed particles represent the light curve and note the striking agreement between simulation and observations. We also note that the 1911 and 1971 outbursts, which were exceptionally bright, are outstanding also in the simulation. This is a consequence of the fact that the major axis of the precessing orbit of the secondary is almost parallel to the disc at these events. The model also predicted the beginning of a major outburst after the 1995 preicenter passage, and this has been observed in recent months. It is most apparent in a time averaged (one-month averages) light curve where the superflares have been taken out (Figure 1). This outburst is expected to last much longer than the typical 0.15 year superflare lifetime.

4. The superflares

The superflares that are superposed on the longer outbursts are related to the secondary crossing the accretion disc of the primary. The lower panel in Figure 1 shows the number of disc particles per time step that enter a circular area, radius 0.001 pc, centered on the orbiting secondary. The arrows connected to the peaks of the curve indicate when the observed outburst should be expected, taking into account the time delay (Lehto & Valtonen 1996). The boxes under the diagram show the time intervals for observed outbursts. These are magnified in Figure 2 to facilitate a detailed study of the best fit between simulation and observations that our model has produced. The analysis of the latest outburst, the start of which was reported at 1995.845 (Lehto & Valtonen 1996), is reported by Valtonen et al. (1996).

5. The time delay

The time delay between the disc crossing and the superflare is an important constituent of our binary black hole model, for which reason it is here given a short description. See Lehto and Valtonen (1996) for a more detailed explanation. The disc around the primary is assumed to be geometrically thin, optically

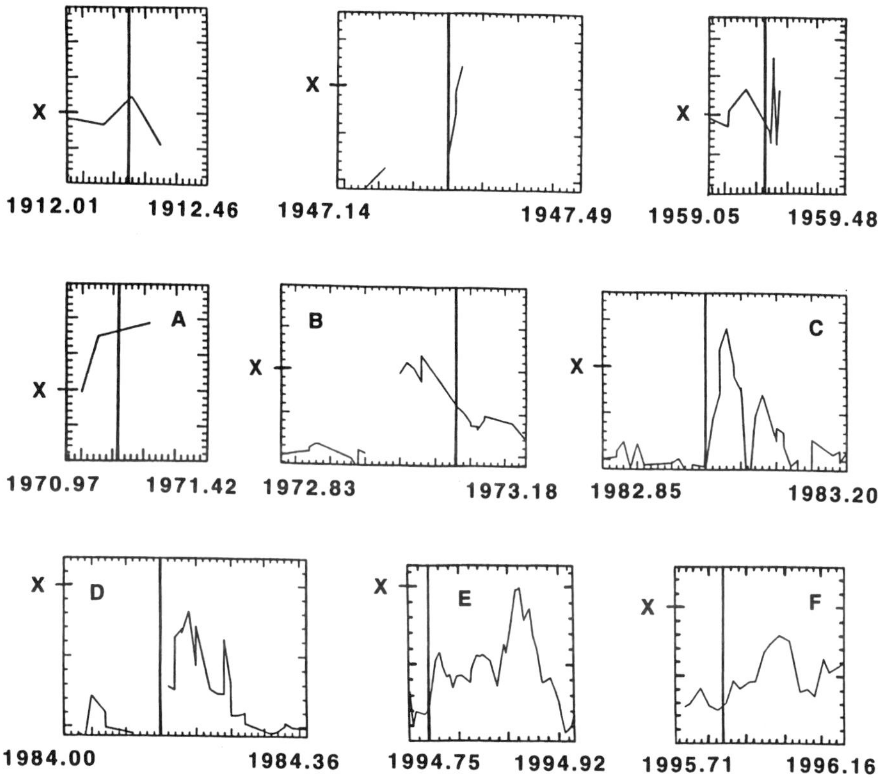

Figure 2. The light curves of the nine observed superflares. The vertical thick lines indicate the times given by the binary black hole model for the expected flare. "X" denotes the flux level 10 mJy.

thick and radiation dominated. The secondary enters the disc with a hypersonic velocity and shocks the gas surrounding the point of impact. This gas obtains a velocity component perpendicular to the disc plane and will emerge from the disc after a time $t_{\rm dyn}$, which depends on the parameters of the disc and on the velocity of the secondary relative to the disc. Since the emerging gas is optically thick it cannot radiate appreciably until it has expanded and cooled sufficiently to become optically thin, after which it will radiate some of its energy by bremsstrahlung resulting in an apparent flaring. The time lag $t_{\rm lag}$ between the emergence of the hot gas and the moment it becomes transparent depends in addition to disc parameters and the relative velocity also on the mass of the secondary black hole. We will keep this mass dependence explicit, and write the total time delay as

$$t_{\rm dyn} + t_{\rm lag} \quad = \quad t_{\rm dyn}(r) + t_0(r)\, m_{\rm sec}^{26/21}$$

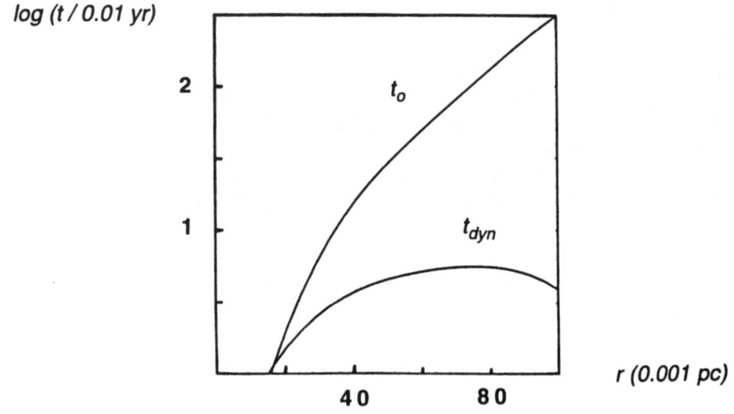

Figure 3. The functions $t_{\rm dyn}$ and t_0.

where $m_{\rm sec}$ is the mass of the secondary expressed in units of $10^8 M_\odot$ and r is the distance between the point of disc crossing and the primary black hole. Adopting an α_g-disc, in which viscosity is proportional to the gas pressure, as a model for the accretion disc (see Sakimoto & Coroniti 1981) $t_{\rm dyn}$ and t_0 can be calculated as functions of r. The result is given in Figure 3. This parameter-dependent and incomplete model for the time delay will of course introduce a degree of uncertainty in our calculations but, fortunately, it is only for disc crossings at large distances from the primary that this uncertainty becomes noticeable. All well observed superflares result from disc crossings very close to the primary black hole.

Acknowledgments. We wish to thank Leo Takalo and Aimo Sillanpää in the OJ94 project for supplying us with the optical light curves for the 1995 outburst and the November 1995 superflare.

References

Donner, Engström, & Sundelius, B. 1991, A&A, 252, 571

Lehto, H. J., & Valtonen, M. J. 1996, ApJ, 460, 207

Sakimoto & Coroniti 1981, ApJ, 247, 19

Sillanpää, A., H., Valtonen, M. J., Sundelius, B., and Byrd 1988, ApJ, 325, 628

Valtonen, M. J., Lehto, H. J., Kokkonen, and Mikkola 1996, these proceedings

Blazar Continuum Variability
ASP Conference Series, Vol. 110, 1996
H. R. Miller, J. R. Webb, and J. C. Noble, eds.

Search for Periodicity in the Brightness Variations of the Active Extragalactic Objects BL Lac, OJ 287, and 3C 120

S. G. Marchenko, V. A. Hagen-Thorn, and V. A. Yakovleva

St.-Petersburg State University. St.-Petersburg, Russia

O. V. Mikolaichuk

Pedagogical Institute, Sterlitamak, Bashkirstan

Abstract. The results of studying the 20-year light curves of BL Lac, OJ 287 and 3C 120 for periodicity are given in the range of $P > 30$ days. For all three objects only long-term components are statistically significant; these must be considered as a trend (maybe with the exception of the 8-year component for BL Lac). No periodic components with $P < 7$ years are found.

1. Introduction

The nature of nuclear activity of extragalactic objects has not been completely elucidated up to now. Among its manifestations the photometric variability is probably one of the most important. The investigation of variability characteristics, in particular, studying the light curves of the objects for periodicity may give a valuable information being useful for constructing an adequate model of AGN. Detailed light curves covering a long time interval are needed for such investigation. Here we summarize some results of our work in this field. For details see our papers in Astronomical Journal of Russia.

2. Observational data

A program of photometric observations of active extragalactic objects was carried out at the Burakan Station of St.-Petersburg University for last 25 years. Two 20-inch reflectors were used for photographic monitoring in the B band and photoelectric brightness estimations. The most detailed data have been obtained by the authors for blazars BL Lac, OJ 287 and Seyfert galaxy 3C 120.

Our data were supplemented with published ones. After examination for homogeneity all the data for each date were averaged, and combined light curves in the B band were obtained. These contain 1418 points for BL Lac in time range 1968–1989, 934 points for OJ 287 (1971–1989) and 548 points for 3C 120 (1967–1989).

The light curves transformed to fluxes using O'Dell et al. (1978) calibration are shown in Fig. 1. The light curves for BL Lac and 3C 120 are more detailed as compared with those investigated for the same time ranges by others. It is these light curves that were analised for periodicity.

Note the different character of variability of three objects. In the case of BL Lac and 3C 120 the long-term variations are smooth while in the light curve of OJ 287 two large outbursts dominate. The activity level may be characterized by the ratio of flux standard σ_F to mean flux $\overline{F}$. These ratios are: $\frac{\sigma_F}{\overline{F}} = 31\%$ for 3C 120, $\frac{\sigma_F}{\overline{F}} = 54\%$ for BL Lac and $\frac{\sigma_F}{\overline{F}} = 117\%$ for OJ 287.

3. The method for a search of periodicity

The method which we used is "whitening" of time series. The basis of this method is the Schuster periodogram:

$$D(\nu_j) = \frac{1}{N^2} \mid \sum_{k=1}^{N} x(t_k) \exp(-i2\pi\nu_j t_k) \mid^2,$$

where $i^2 = -1$, N is the number of data points, $x(t_k)$ is the time series, ν_j is the frequency array of power spectrum.

At first the Schuster periodogram is constructed for initial time series $x(t_k)$. The next step is to find the exact value of frequency ν_1 (and $P_1 = \frac{1}{\nu_1}$), corresponding to $D_{\max}$ of the periodogram.

Then the system of conditional equations is obtained:

$$x(t_k) = A_0 + a_1 \times \cos(2\pi\nu_1 t_k) + b_1 \times \sin(2\pi\nu_1 t_k) = A_0 + A_1 \times \cos(2\pi\nu_1 t_k + \varphi_1).$$

The constants $A_0, A_1 = \sqrt{a_1^2 + b_1^2}$ and $\varphi_1 = \arctan(\frac{b_1}{a_1})$ and their uncertainties are found by the method of least squares (MLS). After eliminating the calculated sinusoid from the initial time series the procedures are repeated and as a result we obtain a number of remaining time series:

$$\xi_m(t_k) = x(t_k) - A_0 - \sum_{i=1}^{m} A_i \times \cos(2\pi\nu_i t_k) + \varphi_i\,.$$

The process is stopped when $D_{\max}$ of current periodogram is no more than the maximum value which may be produced with 1%-probability by time series of the white noise with the same average, dispersion and time distribution as the real time series.

The set of P_i, A_i and φ_i forms the discrete frequency spectrum of the time series under consideration.

This method has a number of merits. It allows one:

- to exclude the maxima in the power spectrum being due to the unequaly spaced data (aliases);
- to regard an elimination of the long-term components as deliverance from a trend when searching high frequency periodic components;
- to use the dependence of remaining time series dispersion on the number of subtracted component as a criterion of significance of periodic components.

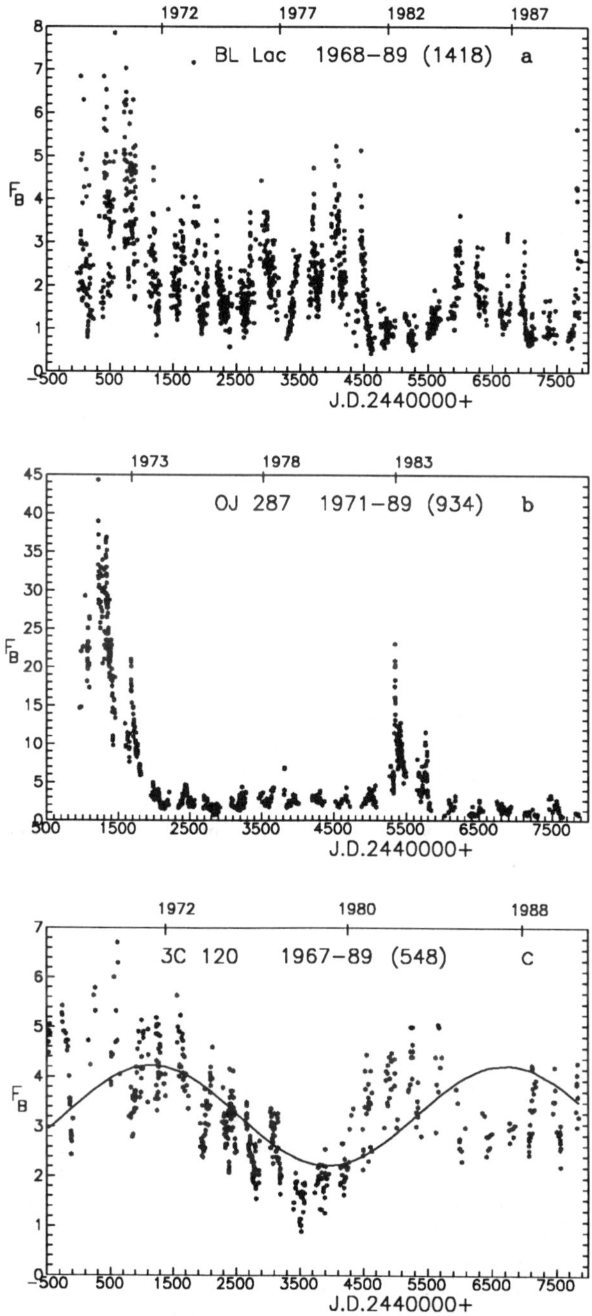

Figure 1. Combined light curves; (a) - for BL Lac; (b) - for OJ 287; (c) - for 3C 120.

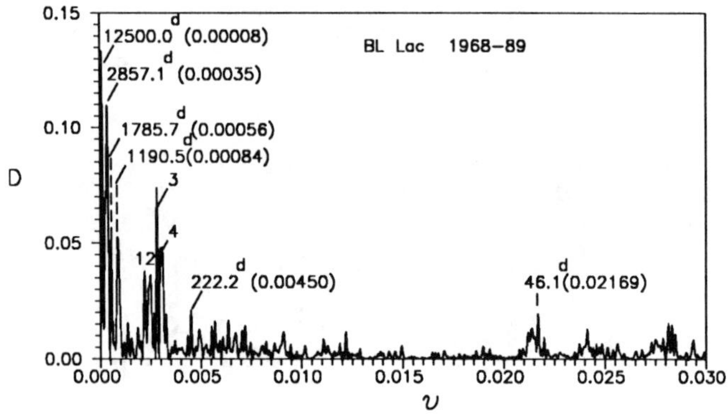

Figure 2. Schuster periodogram for initial light curve of BL Lac.

Indeed, it is easy to calculate the values

$$\Delta_i^2 = \frac{1}{N-1} \sum_{k=1}^{N} \xi_i^2(t_k),$$

where $i = 0, 1, 2, \ldots, m$. The dependence of Δ_i^2 on the number of subtracted component i must be linear for the time series with constant spectral density (Terebizh 1992). Therefore one can suppose that when the dependence Δ_i^2 on i for the real time series becomes nearly linear then remaining time series is close to the white noise. This fact allows to consider that only those periodic components are real for which the deviation of Δ_i^2 from the straight line corresponding to the white noise is statistically significant (say, by Fisher criterion).

4. Results

Because of limited space, we give detailed consideration for only one object—BL Lac. For others, the enumeration of results will be given. Fig. 2 shows the first Schuster periodogram for initial light curve of BL Lac. Six maxima are marked (the aliases due to the unequaly spaced data are indicated by figures).

Applying Δ-criterion is illustrated by Fig. 3 in which remainder dispersions are plotted versus the number of subtracted components. Solid and dashed lines represent the results for the real time series and time series of white noise, correspondingly. One can see that for the first four components values Δ_i^2 don't lie on straight line. But only for the first two components these deviations are significant by Fisher criterion.

It is clear that the first component with $P = 34$ years can be considered with our length of time series as a trend. This component has amplitude of about 0.7 mJy. Fig. 4 shows the light curve of BL Lac after elimination this trend (dots). The solid line is the second significant components with the period $P = 7.8 \pm 0.2$ years and amplitude of 0.60 ± 0.04 mJy. It follows the smooth brightness variations reasonably well.

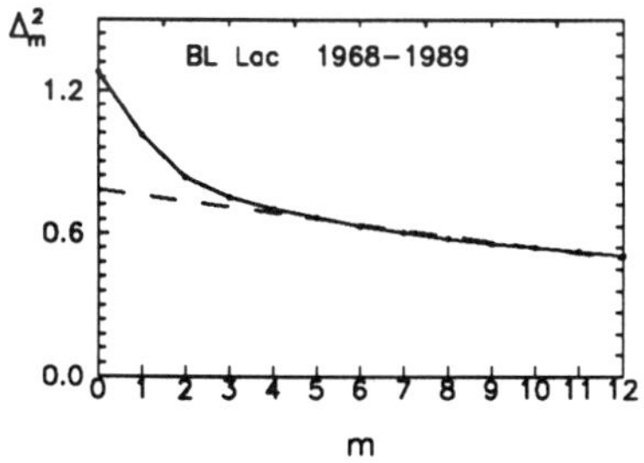

Figure 3. The dependence of remainder dispersions on the number of subtracted components for BL Lac .

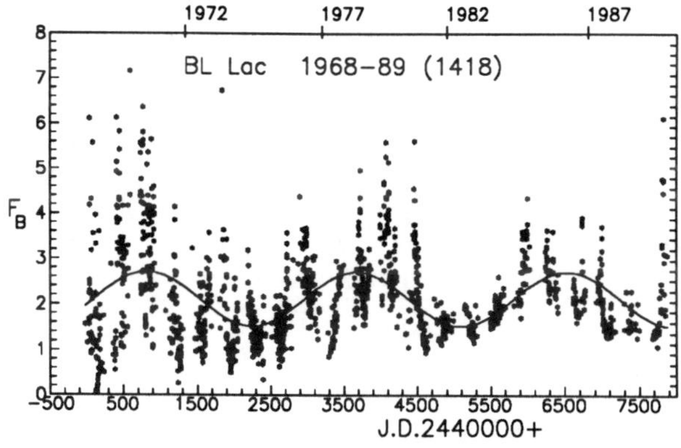

Figure 4. The comparison of observational data of BL Lac (dots) and sinusoid with period of 7.8 years (solid line).

The first periodogram of initial light curve for 3C 120 has one main maximum at the frequency corresponding to the component with period $P = 15.2 \pm 0.4$ years and amplitude $A = 1.01 \pm 0.06$ mJy. Only this component is significant by Δ-ctiterion. Fig. 1c (solid line) demonstrates how good this component follows the observational data. It is seen that there are two time intervals with different agreement level: in the first—1967–1976—agreement is good; in the second—1977–1989—it is not so good.

If these time intervals are investigated separatly then for the first interval the component with period of about 15 years is the most significant one (as for the whole time series) and has the same amplitude. But the second significant component with $P = 480 \pm 5$ days and amplitude $A = 0.53 \pm 0.06$ mJy appears. However, for a number of time intervals this sinusoid contradicts the observational data. Therefore the presence of that kind of periodicity is doubtful.

The Schuster periodogram for the second time interval has two significant maxima at the frequenceis: $\nu_1 = 0.00037 \pm 0.00002$ ($P = 7.4 \pm 0.3$ years), $\nu_2 = 0.00017 \pm 0.00002$ ($P = 16.1 \pm 0.9$ years). The component with period of about 7 years has the maximum amplitude $A = 0.97 \pm 0.04$ mJy while the component with $P \approx 16(15)$ years diminishes its amplitude essentially ($A = 0.48$ mJy instead of $A = 1.0$ mJy). The sum of these two significant sinusoids fits observational data well.

In the case of OJ 287, four maxima of the Schuster periodogram are significant by Δ-criterion:

$$\begin{aligned}
\nu_1 &= 0.00026 \pm 0.00001\ (P = 10.5 \pm 0.2 \text{ years},\ A = 8.1 \pm 0.4 \text{ mJy}),\\
\nu_2 &= 0.00011 \pm 0.00001\ (P = 25 \pm 1 \text{ years},\ A = 5.7 \pm 0.3 \text{ mJy}),\\
\nu_3 &= 0.00038 \pm 0.00001\ (P = 6.2 \pm 0.1 \text{ years},\ A = 5.0 \pm 0.2 \text{ mJy}),\\
\nu_4 &= 0.00071 \pm 0.00001\ (P = 3.9 \pm 0.1 \text{ years},\ A = 2.4 \pm 0.2 \text{ mJy}).
\end{aligned}$$

The first component is connected with two large outbursts in the OJ 287 light curve (1971 and 1983). The component with period of about 25 years describes the decrease of the outburst amplitude in 1983 as compared with that of the outburst in 1971. The components with periods of $\sim$ 6 years and $\sim$ 4 years are most probable due to the method of data processing, namely, to the search for periodicity on the base of sinusoidal functions for light curve with sharp features. Indeed, the analysis of the part of light curve between two outbursts (for 1974–1982) doesn't show significant periodic components.

5. Conclusions

The main results of our work are as follows:

- In the 20-year light curves of BL Lac, OJ 287 and 3C 120 there are no statistically significant periodic components with periods from 30 days to $\sim$ 7 years. The variations at these time-scale are similar to the white noise by their properties.
- For all three objects only long-term components are significant. In the case of 3C 120 and OJ 287 our results don't contradict the existence of periodic components found early from the analysis of the historical light curves

(Webb 1990, Sillanpää et al. 1986). But we stress that the amplitudes of these components may be variable as seen from the analysis of separate fragments of light curves.

- The periodic component with $P \approx 8$ years may exist in the light curve of BL Lac.

Acknowledgments. This work was supported by grants NWN000 of the International Science Foudation and NWN300 of ISF and the Government of the Russian Federation. S. G. Marchenko is grateful to the Russian Fund of Basic Research for the travel grant N96-02-26502 to the workshop.

References

O'Dell, S. L., Pushell, J. J., Stein, W. A., et al. 1978, ApJ, 224, 22

Terebizh, V. Yu 1992, in Time Series Analysis in Astrophysics, Moscow, Nauka

Webb, J. R. 1990, AJ, 99, 494

Sillanpää, A., Haarala, S., & Valtonen, M. J. 1986, Turku University Observatory, Report Series, Inf. No 108, 1

Blazar Continuum Variability
ASP Conference Series, Vol. 110, 1996
H. R. Miller, J. R. Webb, and J. C. Noble, eds.

VRI Microvariability of OJ287 and 3C66A During Single Nights

Paul Boltwood

1655 Main St., Stittsville, Ont., Canada K2S 1N6
email: ae772@freenet.carleton.ca

Abstract. As a part of the OJ-94 Blazar project, the Boltwood Observatory made 7400 differential Johnson-Cousins VRI photometric observations of OJ287 and 3C66A, from September 1993 to January 1996. This large number of homogeneous observations has permitted an analysis to locate nights during which these objects detectably varied.

For OJ287, in the 599 band/nights of observations, the probability that variations were detected exceeded 0.99 for 11% of them (3% with the most deviant point removed). For 3C66A, in the 290 band/nights of observations, the probability that variations were detected exceeded 0.99 for 29% of them (10% with the most deviant point removed).

1. Observations

The Johnson-Cousins V, R, and I band photometric observations made at the Boltwood Observatory cover the period from September 1993 to January 1996. OJ287 was observed all night for about 90% of the time it was visible, starting December 1993. 3C66A was observed when convenient.

There are 5200 OJ287 and 2200 3C66A observations. These observations are in the OJ-94 project archive and will be published at a later date.

Most of the observations are exposure groups of V for 16 minutes, R for 8 minutes, and I for 8 minutes. The methodology used in obtaining these observations is documented in Boltwood 1994 and 1995A.

2. Measuring nightly variability

The VRI photometric data was used to study the variability of OJ287 and 3C66A during single nights.

The nights on which an object varied significantly were located by doing a χ^2 test on each night/band's data. This compared the estimated noise with the actual variation. For this to work it was essential that the noise estimates be accurate. The next section explains how this accuracy was achieved.

The data used was instrumental photometry, not standardized. This was done to remove the complications caused by the color errors that otherwise would have been included. The instrumental bandpasses are close to the Johnson-Cousins standards (Boltwood 1995C).

3. Noise Model

The photometric reduction software used takes into account most or all of the normally recognized error sources. It also is quite effective in filtering out cosmic rays and other outlier-causing events. It is not perfect however.

When the full data base of OJ287 check stars is examined, the $\frac{\sigma_{\text{true}}}{\sigma_{\text{est}}} = 1.12$ due to unknown sources of error.

The OJ287 field check stars were used to produce a model of the known plus unknown errors. This is possible because the data is homogeneous and available in large quantities.

The error estimate model derived for the instrumental linear data (in photons) is:

$$\text{better est} = \sqrt{(1.148 \times est)^2 + 44^2 - (0.008 \times \text{photon brightness})}$$

where *est* is the error estimate in photons provided by the image reduction software.

The reduction in the error of the error estimates due to this equation (in linear terms) is a factor of 2.4 (Boltwood 1995B).

In this preliminary work, the outlier problem was conservatively handled by deleting the most deviant point each night from consideration. Better handling than this is possible, and will be implemented in the future.

4. Nightly variability results

In the tables below are listed the nights where the probability of the variation existing exceeded 0.99, even with the most deviant point removed.

All of the work below is relative to a comparison star. The procedures used in this work only partially detect any variability that might exist in the comparison star. The procedures do not correct for any such variation.

The xxxxx StdDev columns of Tables 1–4 have highly variable reliability. The most truly variable nights could easily have low detection probabilities. Good nights with low noise levels and/or more observations give a higher sensitivity for testing for variation. A better statistical treatment is planned for future work.

Table 2 and Tables 4 & 6 below are the results which include removing the point with the largest contribution to the weighted standard deviation.

5. Control observations

To give confidence in the above results, for each object one of the check stars was subjected to the same procedure.

5.1. OJ287

Star 11 was used as a control. There were no nights where the probability of variation exceeded 0.99 after removing the most deviant point.

Table 1. OJ287

JD	Band[a]	N[b]	Weighted Std. Dev[c]	OJ287 Std. Dev[d]	Reduced Chi-sq.[e]	Proba-bility[f]
2449330	R	8	0.0519	0.0472	5.696	1.00000
2449393	R	19	0.0566	0.0522	6.644	1.00000
2449395	R	25	0.1057	0.0995	8.791	1.00000
2449414	V	15	0.0554	0.0445	2.814	0.99968
2449417	I	12	0.0403	0.0346	3.833	0.99998
2449417	R	12	0.0421	0.0366	4.067	0.99999
2449417	V	11	0.0469	0.0404	3.843	0.99997
2449420	I	12	0.0505	0.0434	3.863	0.99999
2449420	R	11	0.0563	0.0487	3.977	0.99998
2449420	V	13	0.0559	0.0469	3.385	0.99994
2449441	I	11	0.0711	0.0648	5.956	1.00000
2449441	R	11	0.0677	0.0610	5.331	1.00000
2449441	V	11	0.0723	0.0620	3.789	0.99996
2449457	I	9	0.0593	0.0519	4.264	0.99996
2449744	I	15	0.0504	0.0429	3.624	1.00000
2449770	I	14	0.0485	0.0405	3.326	0.99996
2450030	V	8	0.0679	0.0625	6.626	1.00000
2450070	R	128	0.0753	0.0698	7.152	1.00000
2450093	R	143	0.0323	0.0178	1.437	0.99950
2450102	R	6	0.0539	0.0478	4.700	0.99973

[a]instrumental photometric band pass, close to Johnson-Cousins

[b]number of observations

[c]true weighted standard deviation corresponding to Weighted Avg., which is the average of $mag_{(\mathrm{obj})} - mag_{(\mathrm{comp\ star})}$, weighted by $1/(\text{better est})^2$

[d]the standard deviation of the photometry of the listed object after removing the noise

[e]$\sum \frac{mag_{(\mathrm{obj})} - mag_{(\mathrm{comp\ star})} - (\text{Weighted Avg})^2}{(\text{better est})^2} / (N-1)$

[f]the probability that the night's variation is not due to the *better est* noise computed for each observation

N.B. these table notes apply to Tables 2–4 also

Table 2. OJ287 results with the most deviant point removed

JD	Band	N	Weighted Std. Dev	OJ287 Std. Dev	Reduced Chi-sq.	Probability
2449330	R	8	0.0441	0.0384	4.143	0.99964
2449393	R	19	0.0525	0.0472	5.204	1.00000
2449395	R	25	0.0869	0.0778	5.059	1.00000
2449414	V	15	0.0504	0.0374	2.233	0.99352
2449417	I	12	0.0373	0.0312	3.353	0.99978
2449417	R	12	0.0375	0.0315	3.366	0.99979
2449417	V	11	0.0413	0.0335	2.911	0.99811
2449420	I	12	0.0461	0.0383	3.227	0.99964
2449420	R	11	0.0498	0.0411	3.136	0.99912
2449420	V	13	0.0464	0.0357	2.452	0.99537
2449441	I	11	0.0627	0.0560	4.960	1.00000
2449441	R	11	0.0633	0.0560	4.647	1.00000
2449441	V	11	0.0653	0.0537	3.093	0.99898
2449457	I	9	0.0498	0.0414	3.256	0.99815
2449744	I	15	0.0446	0.0361	2.905	0.99969
2449770	I	14	0.0414	0.0317	2.413	0.99600
2450030	V	8	0.0516	0.0450	4.173	0.99966
2450070	R	128	0.0743	0.0688	6.985	1.00000
2450093	R	143	0.0314	0.0162	1.362	0.99721
2450102	R	6	0.0493	0.0413	3.352	0.99056

Table 3. 3C66A

JD	Band	N	Weighted Std. Dev	3C66A Std. Dev	Reduced Chi-sq.	Probability
2449317	V	22	0.0256	0.0220	3.857	1.00000
2449322	V	28	0.0338	0.0298	4.451	1.00000
2449593	I	13	0.0215	0.0179	3.303	0.99992
2449593	R	11	0.0228	0.0199	4.160	0.99999
2449616	R	10	0.0220	0.0188	3.727	0.99989
2449616	V	10	0.0324	0.0297	6.353	1.00000
2449617	I	6	0.0354	0.0331	7.949	1.00000
2449617	R	6	0.0384	0.0362	9.120	1.00000
2449617	V	6	0.0368	0.0339	6.621	1.00000
2449642	R	10	0.0316	0.0285	5.377	1.00000
2449642	V	10	0.0299	0.0258	3.974	0.99996
2449672	V	9	0.0231	0.0196	3.554	0.99960
2449945	I	6	0.0489	0.0476	19.392	1.00000
2449945	R	8	0.0285	0.0266	7.790	1.00000
2449945	V	8	0.0335	0.0318	9.984	1.00000
2449948	R	8	0.0227	0.0208	6.371	1.00000
2449952	I	4	0.0311	0.0296	10.358	1.00000
2449952	V	5	0.0318	0.0301	9.187	1.00000
2449985	R	4	0.0881	0.0843	11.882	1.00000
2449989	I	11	0.0233	0.0210	5.359	1.00000
2449989	R	11	0.0465	0.0454	22.243	1.00000
2449989	V	11	0.0259	0.0239	6.564	1.00000
2450000	I	7	0.0281	0.0263	7.887	1.00000
2450000	R	8	0.0330	0.0315	10.713	1.00000
2450000	V	8	0.0295	0.0272	6.758	1.00000
2450009	I	13	0.0553	0.0539	19.361	1.00000
2450009	V	13	0.0280	0.0255	5.909	1.00000
2450014	R	152	0.0202	0.0120	1.548	0.99998
2450031	I	7	0.0271	0.0247	5.834	1.00000

Table 4. 3C66A results with the most deviant point removed

JD	Band	N	Weighted Std. Dev	3C66A Std. Dev	Reduced Chi-sq.	Probability
2449317	V	22	0.0205	0.0158	2.436	0.99966
2449322	V	28	0.0312	0.0267	3.699	1.00000
2449593	I	13	0.0185	0.0141	2.382	0.99395
2449593	R	11	0.0194	0.0157	2.845	0.99763
2449616	R	10	0.0194	0.0157	2.901	0.99689
2449616	V	10	0.0275	0.0241	4.337	0.99997
2449617	I	6	0.0231	0.0197	3.621	0.99410
2449617	R	6	0.0262	0.0232	4.625	0.99902
2449617	V	6	0.0265	0.0228	3.775	0.99550
2449642	R	10	0.0280	0.0245	4.267	0.99996
2449642	V	10	0.0260	0.0213	3.048	0.99802
2449672	V	9	0.0209	0.0166	2.731	0.99218
2449945	I	6	0.0298	0.0279	8.140	1.00000
2449945	R	8	0.0249	0.0227	6.007	1.00000
2449945	V	8	0.0254	0.0230	5.536	0.99999
2449948	R	8	0.0208	0.0187	5.189	0.99998
2449952	I	4	0.0217	0.0193	4.855	0.99221
2449952	V	5	0.0248	0.0228	6.429	0.99976
2449985	R	4	0.0584	0.0526	5.336	0.99518
2449989	I	11	0.0212	0.0186	4.307	0.99999
2449989	R	11	0.0309	0.0292	9.801	1.00000
2449989	V	11	0.0227	0.0203	4.997	1.00000
2450000	I	7	0.0243	0.0223	6.368	0.99999
2450000	R	8	0.0301	0.0284	9.497	1.00000
2450000	V	8	0.0225	0.0194	3.945	0.99940
2450009	I	13	0.0334	0.0309	7.151	1.00000
2450009	V	13	0.0246	0.0216	4.300	1.00000
2450014	R	152	0.0198	0.0114	1.490	0.99991
2450031	I	7	0.0222	0.0192	4.002	0.99876

5.2. 3C66A

Star B was used as a control. The following are the only nights where the probability of variation exceeded 0.99 after the most deviant point was removed. Tables 5 and 6 below are a subset of those for the tables above.

Table 5. 3C66A

JD	Band	N	Weighted Std. Dev	Reduced Chi-sq.	Probability
2449604	V	6	0.0711	11.755	1.00000
2449678	V	4	0.0553	7.386	0.99994

These controls show much less nightly variation than that seen in OJ287 and 3C66A.

Table 6. 3C66A results with the most deviant point removed

JD	Band	N	Weighted Std. Dev	Reduced Chi-sq.	Probability
2449604	V	6	0.0479	6.334	0.99996
2449678	V	4	0.0409	4.775	0.99156

6. Conclusions

For OJ287, in the 599 band/nights of observations, the probability that variations were detected exceeded 0.99 for 11% of them (3% with the most deviant point removed).

For 3C66A, in the 290 band/nights of observations, the probability that variations were detected exceeded 0.99 for 29% of them (10% with the most deviant point removed).

The nightly variability of OJ287 is not large, and of 3C66A even less so. Care must be taken to separate them from photon arrival and instrument noise.

References

Boltwood, P. 1994, in "Boltwood Observatory and Observing Techniques," Tuorla Observatory Reports, 174, 78

Boltwood, P. 1995A, in "Boltwood Observatory and Observing Techniques (Continued),", Tuorla Observatory Reports, 176, 116

Boltwood, P. 1995B, in "Preliminary Study of the Nightly Variability of the Blazar OJ287," Tuorla Observatory Reports, 176, 50

Boltwood, P. 1995C, in "Photometric Filters For A CCD Camera (addendum)," IAPPP Communications, 62, 45

Blazar Continuum Variability
ASP Conference Series, Vol. 110, 1996
H. R. Miller, J. R. Webb, and J. C. Noble, eds.

The Optical Variability for the Radio-selected Blazar 3C66A

H. R. Miller

Department of Physics and Astronomy, Georgia State University, Atlanta, Georgia 30303

J. C. Noble and M. T. Carini

Center for Automated Space Science, Department of Physics and Astronomy, Western Kentucky University, Bowling Green, Kentucky 42101

M.-H. Jang

Department of Astronomy and Space Science, Kyung Hee University, Yongin-City, Kyungki-Do, 449-701, Korea

Abstract. Rapid optical variations have been detected for the blazar 3C66A with time scales shorter than an hour. These variations are the most rapid which have been reported to date for this blazar. We also report observations investigating the nature of variations on time scales ranging from days to decades. This object clearly exhibits variations on all timescales studied. The variations detected typically are a few tenths of a magnitude on the timescale of days to 1–2 magnitudes on the timescale of decades. However, the long term variations are not always simply an extension of the shorter term trends observed.

1. Introduction

An ongoing program to investigate the question of the existence of variations in the optical flux of Active Galactic Nuclei (AGN) has been undertaken for a large number of objects belonging to the class of AGN known as blazars. Utilizing modern CCD detectors, variations have been detected on timescales significantly less than a day, which we have called microvariability. Initial results reported for a number of blazars (Miller, Carini, & Goodrich 1989; Carini, et al. 1990; Carini & Miller 1992) have demonstrated the reality of microvariability for this class of objects. In the present paper, we report the first results of a similar study of the variability of the blazar 3C66A.

3C66A was first identified by Wills and Wills (1974) as a blue stellar object having a featureless optical spectrum with $(U - B)$ and $(B - V)$ colors similar to those found for quasistellar objects. An early investigation of the optical variability of this object by Folsom et al. (1976) demonstrated that this object was variable; flares displaying amplitudes of approximately 0.5 mag and durations ranging from a few days to a few weeks were detected on several occasions.

These early observations suggested that this object would be a good candidate for inclusion in our current program of studying the microvariability properties of a large sample of blazars.

2. Observations and data reductions

The observations of 3C66A reported here were obtained with the No. 1 0.9-meter telescope at Kitt Peak National Observatory, the 0.6-meter telescope at Capilla Peak Observatory, the 30-inch at University of Florida's Rosemary Hill Observatory, and with the 42-inch telescope at Lowell Observatory, all equipped equipped with similar CCD cameras and autoguiders. The observations were made through B, V, and R filters. Repeated exposures of 180 to 240 seconds were obtained for the star field containing 3C66A and several standard stars. These standard stars, located on the same CCD frame as 3C66A, provided comparisons for use in the data reduction process and come from the comparison sequence of Craine, Johnson, and Tapia (1975). The observations were reduced using the method of Howell and Jacoby (1986) as described in Carini et al. (1990). Each exposure is processed through an aperture photometry routine which reduces the data as if it were produced by a multi-star photometer. Differential magnitudes can then be computed for any pair of stars on the frame. Thus, simultaneous observations of 3C66A, several comparison stars and the sky background will allow one to remove variations which may be due to fluctuations in either atmospheric transparency or extinction. The aperture photometry routine used for these observations is the *apphot* package in IRAF[1].

3. Results

The results of the observations of 3C66A obtained 03–07 November 1989 are displayed in Figure 1. This figure displays the differential magnitude between a comparison star and 3C66A. Significant night-to-night variations are present as well as microvariations with an overall increase in brightness of approximately 0.20 magnitudes. The most significant change found within one night is observed on 07 November 1989 when the source brightened by approximately 0.05 magnitudes in 8 hours. Several small scale excursions were observed superposed on this overall trend.

3C66A was intensively monitored on 05 October 1993 simultaneously utilizing the 42-inch telescope at Lowell Observatory and the 30-inch telescope at Rosemary Hill Observatory (Figure 2). Microvariability was clearly present and confirmed by the independent observations. An overall increase in brightness was observed during the 4 hours that the object was observed yielding an extraordinary rate of brightening of approximately 0.04 magnitudes per hour.

3C66A was observed again 12–13 December 1994 and varied by approximately 0.1 magnitudes during this time.

---

[1]IRAF is distributed by the National Optical Astronomy Observatories, which is operated by the Association of Universities for Research in Astronomy, Inc. under contract to the National Science Foundation.

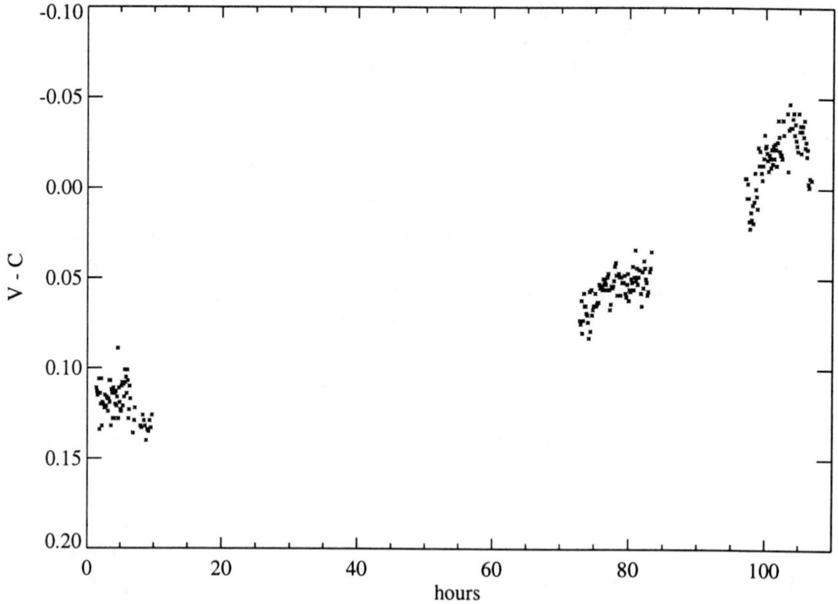

Figure 1. Difference plot of 3C66A and a comparison star for 89 November 03–07.

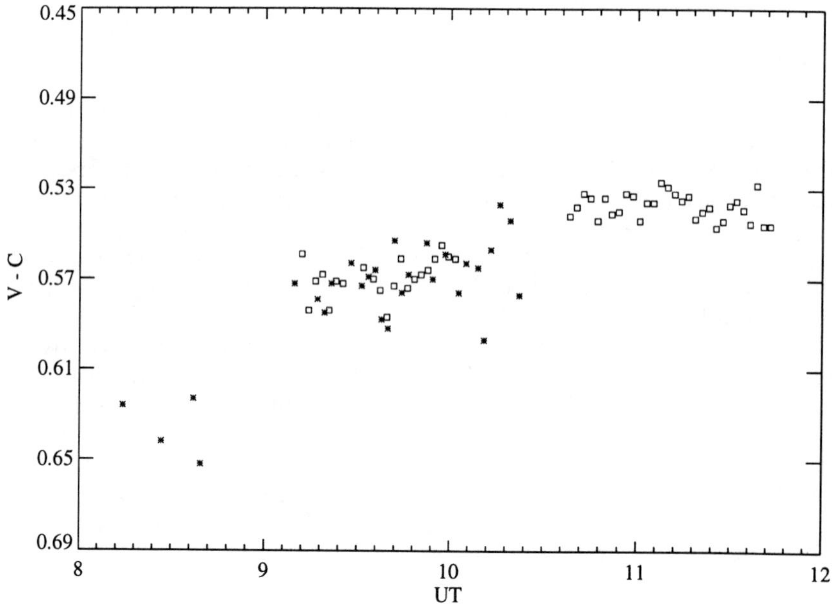

Figure 2. Difference plot of 3C66A and a comparison star for 93 October 05.

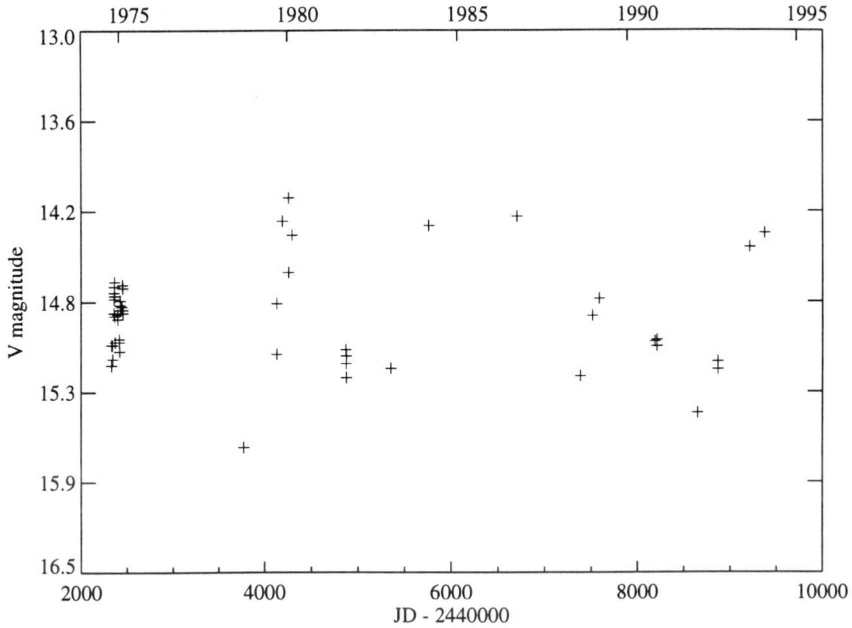

Figure 3. The long-term light curve for 3C66A.

Figure 3 displays the long term variations of 3C66A from 1974 to 1994. A total range of more than 1.0 magnitude is detected, but few well-defined trends can be detected, perhaps due to the intermittent nature of the monitoring.

The question of whether or not these small, several percent, variations observed in blazars are the result of global or local changes in the source region still remains unanswered. Thus, the appropriateness of extrapolating such variations to calculate a doubling time and hence source parameters is also unclear. However, if one observes an event which approaches an observed doubling in the source luminosity, one can be reasonably confident that such an event represents a global, rather than local, change in the source. However, no such changes have been observed for this investigation.

These variations can be modelled using shocked jet based models proposed by Marscher and Gear (1985), Hughes et al. (1985), and Valtaoja et al. (1988) where the rapid variations are due to shocks propagating down the relativistic jet. Alternatively, (Wiita et al. 1991a,b) would explain the observed microvariability as the result of excess emission produced by *flares* or *hot spots* randomly appearing and disappearing on the accretion disks around supermassive black holes. The lack of observed periodicities is in accord with this class of models, which assume contributions typically from 10–100 flaring regions at any given time.

It would be extremely useful to continue the monitoring of 3C66A as well as many other members of our sample. Broadening the temporal coverage in the optical, by involving a string of longitudinally spaced observatories with one-meter class telescopes to provide continuous coverage of individual sources over periods of weeks to months, could really test for the presence of periodicities and provide more well-defined PDS. Theoretical models could be tested and constrained when substantial amounts of multi-color data for a group of sources becomes available.

In summary, the optical variations reported here are similar to those detected for a number of other blazars. These events may be used to place constraints on the size and to test models of these source regions. However, the luminosities observed are sub-Eddington, and therefore, one does not necessarily need to resort to relativistic beaming in order to explain these variations.

Acknowledgments. The authors wish to thank the University of Florida, the University of New Mexico, and Lowell Observatory for providing allocations of observing time to this project. This work has been supported in part by an award from GSU's Chancellor's Initiative Fund to PEGA, and by grants from the Research Corporation and NASA (NAGW-4397). MTC and JCN were supported, in part, by a grant from NASA (NCCW-0085).

References

Carini, M. T., Miller, H. R., & Goodrich, B. D. 1990, AJ, 100, 347

Carini, M. T., Miller, H. R., Noble, J. C., & Goodrich, B. D. 1992, AJ, 104, 15

Craine, E. R., Johnson, K., & Tapia, S. 1975, PASP, 87, 123

Folsom, G. H., Miller, H. R., Wingert, D. W., & Williamon, R. M. 1976, AJ, 81, 145

Howell, S. B., & Jacoby, G. J. 1986, PASP, 98, 802

Hughes, P. A., Aller, H. D., & Aller, M. F. 1985, ApJ, 298, 301

Marscher, A. P., & Gear, W. K. 1985, ApJ, 298, 114

Miller, H. R., Carini, M. T., & Goodrich, B. D. 1989, Nature, 337, 627

Valtaoja, E., et al. 1988, A&A, 203, 1

Wiita, P. J., Miller, H. R., Carini, M. T., & Rosen, A. 1991, in Structure and Emission Properties of Accretion Disks, IAU Colloquium 129, ed. J. P. Lasota et al., p. 557

Wiita, P. J., Miller, H. R., Gupta, N., & Chakrabarti, S. K. 1992, in Variability of Blazars, ed. E. Valtaoja & M. Valtonen (Cambridge, Cambridge University Press), p. 311

Wills, B. J., & Wills, D. 1974, ApJ, 190, L97

Blazar Continuum Variability
ASP Conference Series, Vol. 110, 1996
H. R. Miller, J. R. Webb, and J. C. Noble, eds.

Simultaneous Optical–Near-infrared Microvariability of 3c66a: a Case of Chaos?

Mark R. Kidger and José N. González-Pérez

Instituto de Astrofísica de Canarias, E-38200 La Laguna (Tenerife), Spain

Alberto Sadun

Bradley Observatory, Agnes Scott College, Decatur, Georgia, USA

Abstract. We present simultaneous monitoring of 3c66a, taken on October 5/6th 1995, in the visible (*BVRI*) and near infrared (*JHK*) with two telescopes situated at Teide Observatory, Tenerife, Spain. The observations show a low degree of microvariability (amplitude ~10%) in *BVRI* and *H*; in contrast, the *J* and *K* near-IR bands show a large amplitude of variation, with the former brightening at one point by some 40% whilst the latter fades by ~60%. The most likely explanation is a large synchrotron flare, generated by an intense local magnetic field, but the exact mechinism of variation is very unclear. Chaos analysis of the infrared light curve shows underlying structure; the light curve is chaotic and deterministic, not stochastic, and has a strong memory of its state.

1. Introduction

Regular infrared monitoring of 3c66a began at Teide Observatory in 1990 as part of an extended programme of monitoring of a small sample of blazars. These objects are observed both in the Service Observing Programme of the Carlos Sánchez Telescope (TCS) and in scheduled observing runs. Since late 1993 observing efforts have been intensified as 3c66a was included as the principal target, with OJ287, of the OJ-94 International Time Project, which has been jointly coordinated from Tenerife and from Tuorla (Finland) (see Takalo, Sillanpää, & Kidger 1994). Since Autumn 1993, 3c66a has also been regularly monitored in the visible at Teide Observatory, usually in *BVRI* and, less regularly, in La Palma, in *UBVRIZ*. The long duration of the rise to maximum has been confirmed by the optical data taken since 1988. The observations which are reported here were taken during the night of October 5/6th 1995, with the 1.5-m Carlos Sánchez Telescope (IR) and the 82-cm IAC-80 Telescope. Both are sited in Teide Observatory, Tenerife (Canary Islands), at 2340m altitude. The two telescopes are separated by ~50m distance and thus have, to all intents and purposes, identical observing conditions.

2. Observations and reductions

The TCS uses a InSb detector, cooled to 50K with pumped Nitrogen, within a single channel chopping photometer. The telescope uses a standard ABBA cycle, with a typical chopper throw of 25 arcseconds and an aperture of 15 arcseconds. To compensate for beam-shifting due to flexure, the aperture was recenterd at regular intervals with a nearby bright star. More than 20 calibration stars were observed during the night.

The IAC-80 Telescope has a 1024×1024 Thompson CCD, cooled with liquid nitrogen, with a scale of 0.435 arcseconds per pixel and a field of view of 7 arcminutes. Various nearby stars are observed simultaneously with 3c66a allowing relative photometry to be taken. For these measurements Star A of the standard sequence was used as the principal reference and Star B as the photometric check star. A filter cycle *VRB VRI VRB VRI* was used for the observations, giving a resolution of approximately 15 minutes in V and R and approximately 30 minutes in B and I. There is a long gap in the observations whilst Landolt standards were observed; this gap coincides with one of the IR calibrations.

The infrared magnitudes have been adopted from Kidger (1993). The visible magnitudes for Star A have been carefully calibrated using Landolt standards. A final calibration of the field is in course; we have used the following provisional magnitudes, with their respective errors on the mean, in this work. Adopted magnitudes are: $B = 14.110 \pm 0.006$, $V = 13.620 \pm 0.003$, $R = 13.281 \pm 0.018$, $I = 13.030 \pm 0.011$.

3. Results

The light curves in *BVRIJHK* are shown in Figure 1, each with the same horizontal and vertical scale. The instrumental magnitude of the reference star (Star A) is shown in the top curve; the standard deviation of the instrumental magnitude is 0.010 magnitudes, but part of this is due to the changing airmass of the star (from 1.17 to 1.04). We can thus say that the night was photometric to, at worst, 1%.

3c66a brightened slightly in *BVR* & I during the four and a half hours of observations. Completely different behavior is seen in the IR bands. There is some scatter due to the very short integrations which were used, but we can see that, for example, the magnitude in H brightens approximately 15%, before remaining almost constant for some 2.5 hours to the end of the observations. In J, 3c66a fades by ~20% initially, before rising in approximately 1 hour almost 0.5 magnitudes and then fading again to a level virtually identical to its initial level, whilst in K the magnitude fades from $K = 10.5$ to 11.1, before starting to rise again. The scatter in the IR standard stars over the night is $\lesssim 0.015$ magnitudes, hence the night was photometric and no logical instrumental explanation can be found.

4. Analysis

To investigate the variations in the continuum spectrum, the observations were divided into intervals of 0.02 days (approximately half an hour) and all inte-

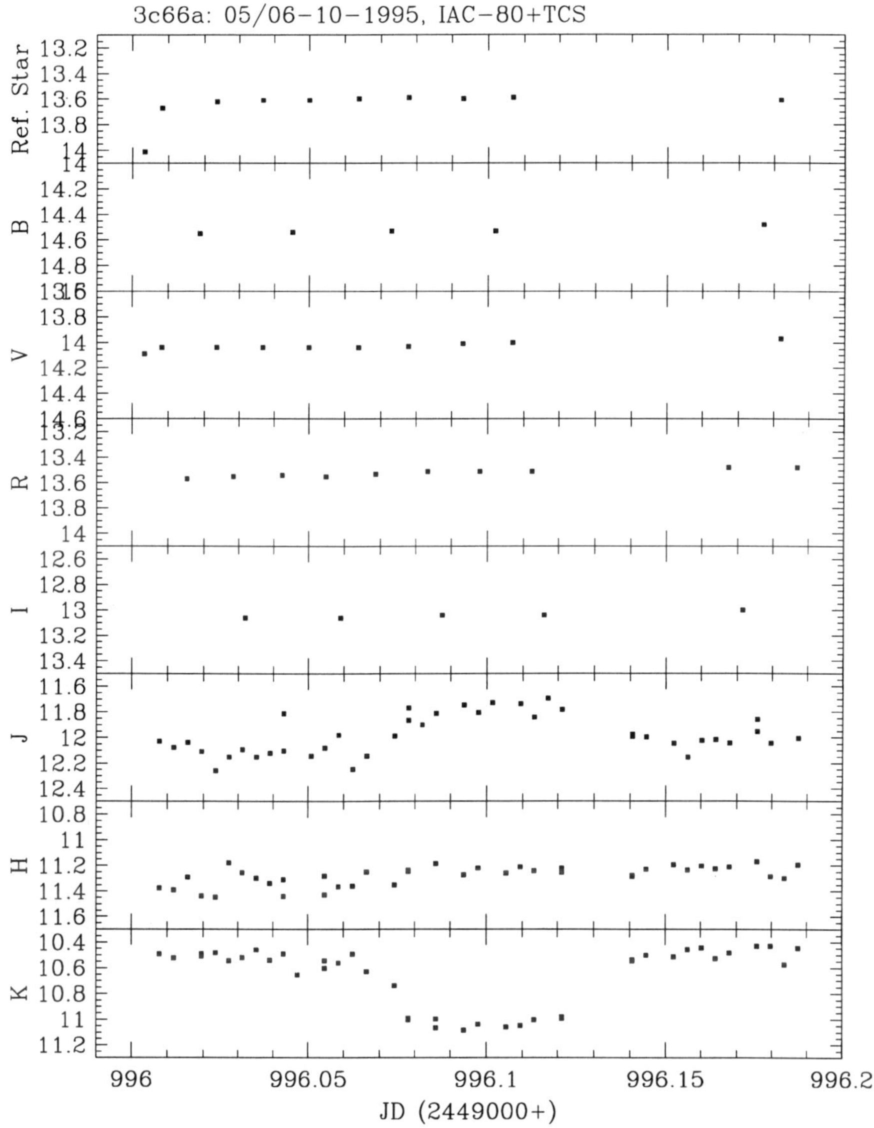

Figure 1. The light curve of 3c66a for the observations reported in the text.

grations in a single filter within that time were averaged. This gave a total of ten "simultaneous" spectra; in two of these time intervals there are no visible data but, in the rest, usually six visible and infrared bands were covered. Figure 2 shows the evolution of the spectrum during the observations. The calibration given by Mead et al. (1990) is used to convert magnitudes to fluxes. The infrared fluxes are found to be consistently slightly below the power law extrapolation of the visible. The spectra at the start of the time series do not show any particularly unusual characteristics apart from this.

The most unusual aspect of the spectra is the progressive flattening of the spectrum in the H and K bands. The spectrum is actually inverted for a time (in other words, the flux, instead of increasing at longer wavelengths, decreases). This flattening and inversion of the spectrum is similar to, but less extreme than, that seen in OJ287 in 1989 (Kidger et al. 1991). From the initially relatively normal spectrum we see how, as the flare at the highest near-IR frequencies proceeds, the flux in J approaches the extrapolation of the visible flux, whilst, in H and K, the spectrum flattens and then turns over. As the flare ends, the turn-over unwinds and disappears and the spectrum returns to its initial state.

4.1. Chaos analysis

The light curves of the three infrared bands have been subjected to chaos analysis using a method described by Sadun and Boltwood (1995). The high Lyapunov exponent shown in all three bands (0.504, 0.639 & 0.681 in J, H & K) suggests that the light curve is strongly aperiodic. The correlation dimension for the light curves are small (1.878 in K, 1.528 in H), showing that the light curve in this band is not stochastic, but rather chaotic and deterministic. The light curve record in all three bands can also be analysed for evidence of Brownian motion, or "random walk." This is measured in terms of the Hurst exponent. If the Hurst exponent is greater than 0.5, this indicates that past trends persist into the future (long term memory), whereas a value less than 0.5 indicates that past trends invert themselves. Values close to 0.5, in contrast, demonstrate random, uncorrelated behavior, with no long term memory of the initial, or any later state. In all three cases $E_{\text{Hurst}} > 0.5$, suggesting some degree of long term memory. The exponent is largest for the K band ($E_{\text{Hurst}} = 0.932$ indicating a strong memory of the past state, in J the value is also strongly indicative of memory $E_{\text{Hurst}} = 0.862$, whilst the H band shows only a very weak effect, consonant with its low degree of variability ($E_{\text{Hurst}} = 0.618$).

5. Discussion

The microvariability event which is observed, if it genuinely pertains to 3c66a and is not a strange instrumental artifact which we cannot identify, appears difficult to explain. The explanation which we preferred to explain the 1989 episode in OJ287 (which was only observed in the visible) was for a large, self-absorbed sychrotron flare on a limited part of the accretion disk, with a 30G field for a 0.1-AU emitting region and 3×10^5G for a 1-AU sized region. Such explanations are to be preferred to models invoking shocks in the jet because of the very rapid time scales of the variations. The variation that we report in 3c66a had a time scale of $\sim$1 hour, implying a maximum size of emitting region

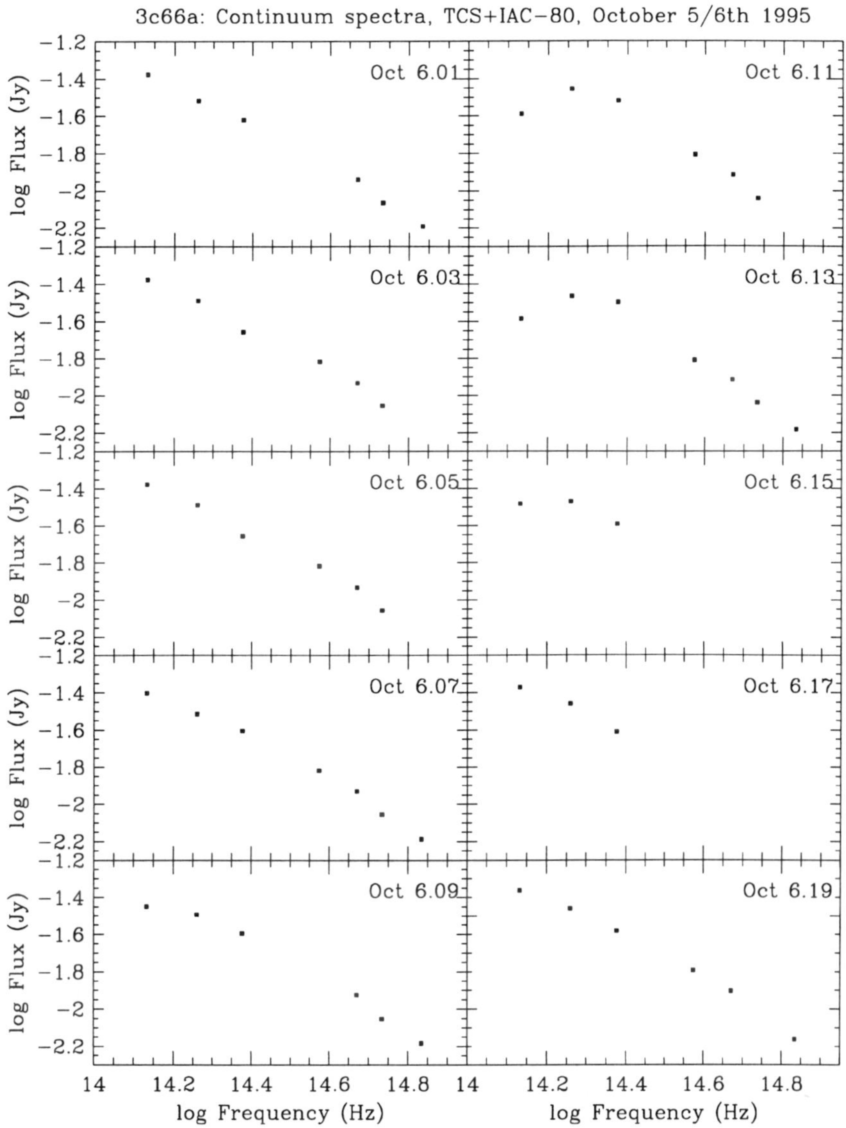

Figure 2. The light curve of 3c66a for the observations reported in the text.

of ~7AU and the fastest variations actually occur on a time scale a factor of ~2 faster. Such emitting region sizes are typical of the inner accretion disk rather than the jet.

To obtain a reasonable fit to the observed spectrum we must assume that the synchrotron spectrum of the flare is steep, with $\alpha_{\rm thin} \sim 2 - 2.5$. Very strong magnetic fields must be supposed, almost certainly $> 10^2$G. Such fields have been postulated recently by Lobanov (1996) and by Unwin et al. (1996) in 3c345. Previously, Blandford and Payne (1982) had considered such a model, with $B \sim 10^2 - 10^4$G as a method of powering jet outflows from accretion disks. Very large fields may only occur in very extreme circumstances during particularly violent, or unusual activity and that the *quiescent* fields are much weaker. Very strong fields will also only occur very close to the central singularity and will be diluted when integrated over the whole emitting region.

6. Conclusions

We believe that 3c66a suffered a massive self-absorbed synchrotron flare implying a very strong magnetic field in an extremely limited region of the accretion disk, with a spectrum such that almost all the energy from the flare was emitted in the near IR. This explanations does not seem completely plausible at this time, so other, more acceptable alternatives may have to be found later. We feel though that instrumental effects are definitely not the cause of the variation which is observed.

Acknowledgments. The Carlos Sánchez Telescope and IAC-80 Telescope are operated in Teide Observatory by the Instituto de Astrofísica de Canarias. We thank Peter Hammersley for providing us with his infrared reduction code.

References

Blandford & Payne 1982 MNRAS, 199, 883

De Diego, J. A., Kidger, M. R., González-Pérez, J. N., & Lehto, H. J. 1996, A&A, in press

Hammersley, P. L. 1995, 1996 , private communications

Kidger, M. R. 1993, Carlos Sánchez Telescope Technical Note No. 16

Kidger, M. R., Takalo, L. O., Valtaoja, E., de Diego, J. A., & Sillapää, A. 1991, A&A, 252, 538

Lobanov, et al. 1996, BAAS, 27, 1445

Mead, A. R. G., Ballard, K. R., Brand, P. W. J. L., Hough, J. H., Brindle, C., & Bailey, J. A. 1990, A&AS, 83, 183

Sadun, A., & Boltwood, P. 1995, BAAS, 27, 1408

Takalo, L. O., Sillanpää, A., & Kidger, M. R. 1994, in Workshop on Intensive Monitoring of OJ287, ed. M. R. Kidger & L. O.Takalo, Tuorla Observatory Reports, Informo No. 174, p.3

Unwin et al. 1996, BAAS, 27, 1409

Blazar Continuum Variability
ASP Conference Series, Vol. 110, 1996
H. R. Miller, J. R. Webb, and J. C. Noble, eds.

Observational Evidence Linking AGNs with Optical Transients

R. Hudec

Astronomical Institute, Academy of Sciences of the Czech Republic, CZ-251 65 Ondřejov, Czech Republic

F. J. Vrba and C. B. Luginbuhl

U.S. Naval Observatory, Flagstaff Station, P.O. Box 1149, Flagstaff, AZ 86002-1149, USA

D. H. Hartmann

Dept. of Physics and Astronomy. Clemson University, Clemson, SC 29634-1911, USA

Abstract. We report on observational evidence linking short-lived Optical Transients (OTs) of astrophysical origin with AGNs. The OTs have been detected during searches for optical counterparts to Gamma Ray Bursts. We have detected three highly certain OT events at mid- to high Galactic latitudes on archival astronomical plates and have investigated their positions for underlying quiescent counterparts. We find one certain QSO with $z = 1.78$, one likely QSO with $z \sim 1$, and one possible QSO/AGN based on broad-band colors. These results provide evidence for the occurrence of brief (< 1 hr), large-amplitude ($\geq$ 10 mag) optical flares from at least some AGNs. We discuss the estimated rate of these events. We suggest wide-field monitoring that may detect such OT events frequently, and in addition a program of automated monitoring of selected AGN targets with improved time resolution. Such observations could significantly increase our understanding of the AGN phenomenon.

1. Introduction

Optical Transients (OTs) are thought to be short-duration (less than hours), large-amplitude (more than several magnitudes) optical flashes from unknown astronomical sources. Historically, OTs have been detected by visual observers and on archival photographic plates. Though the field has suffered from controversy due to false events caused by e.g., plate flaws (e.g., Zytkow 1990), in recent years evidence has begun accumulating that there are genuine OTs of astrophysical origin (see Hudec 1993a,b and references therein for a review). The most reliable OT events have been detected during searches for optical counterparts to Gamma Ray Burst (GRB) sources (Schaefer 1981, Hudec et al. 1990, Hudec et al. 1994a,b, Hudec 1996). These searches have been directed at selected GRB localizations, and have included plate materials covering up to $\sim 20,000$ hours of total monitoring time for the most heavily observed field. Very few astronomical phenomena have been so extensively, if retrospectively, monitored.

Our approach to the problem of OTs has been to obtain deep broad-band CCD imaging of the best-quality OT candidates identified on archival plates (Vrba et al. 1994, Luginbuhl et al. 1996). Since the OT error boxes are rather small (radii of 30 arcsec. or less) the strategy is to search for a common class of underlying object within the OT error boxes. If such a class of object is found with a small probability of random coincidence, this would provide strong evidence concerning both the reality of the OT and the nature of the underlying host for the OT.

In this paper we present evidence that OTs may be related to faint QSOs or AGNs. Of three high Galactic latitude OT fields examined, we detect three confirmed or likely QSOs/AGNs. Below we summarize results obtained for these three OTs and discuss the implications for OT rates and the likehood of OT detection with current and proposed monitoring programs. (In the following we will refer to QSOs and AGNs generally as AGNs for simplicity, except where the context clearly indicates one or the other).

2. The objects

2.1. OT 1905

- OT detected on 1905 May 10
- One OT detected in 18,046 exposure hours examined
- $b^{\rm II} = +56$
- Magnitude B ≤ 5.8
- Duration < 35 min
- Candidate quiescent counterpart V= 20.6, $z = 1.78$, QSO from MMT Spectrum (Vrba et al. 1994)
- 0.5% probability of random coincidence
- Amplitude > 15 mag

2.2. OT 1946, 1954

- OT detected twice on 1946 Aug 31 and on 1954 Apr 27
- Two OTs detected in 7,153 exposure hours examined
- $b^{\rm II} = +21$
- Magnitude ≤ 11.1 and ≤ 9.5
- Candidate quiescent counterpart V = 22.9, B$-$V= 0, U$-$B= -0.9, with QSO/AGN colors (Motch et al. 1990)

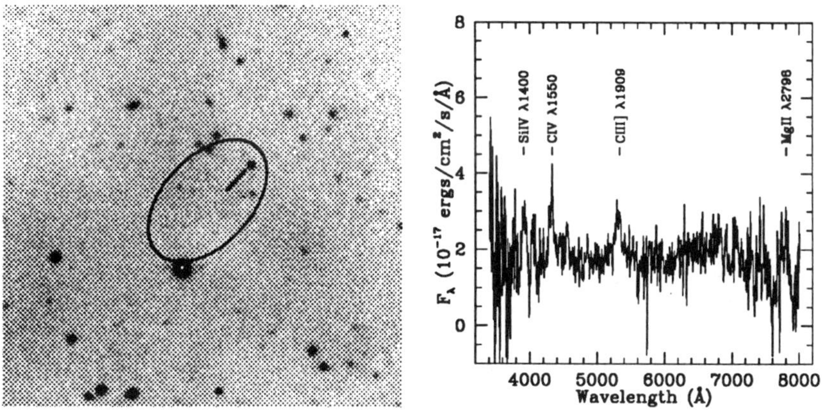

Figure 1. **Left:** V-band CCD image showing the 38×62 arcsec. OT050510 error ellipse and possible quiescent counterpart QSO (north up, east left). **Right:** The MMT spectrum of the QSO.

- MMT/Keck spectra show featureless spectrum; probable $z \sim 1$ QSO
- 1.5% probability of random coincidence (Vrba et al. 1996)
- Amplitude ≥ 10 mag

2.3. OT 1963

- OT detected on 1963 July 24
- One OT detected in ~ 3000 exposure hours examined
- $b^{\rm II} = -26$
- Magnitude B ~ 10.2
- Candidate quiescent counterpart V ~ 20 with (B−V, V−I) colors of the QSO/AGN (Luginbuhl et al. 1996)
- 0.1% probability of random coincidence
- Amplitude ~ 10 mag

3. Conclusions and discussion

The results described above present strong evidence for the first time that AGNs can generate extremely large amplitude, apparently very short duration OT events. The probability of a QSO appearing by chance coincidence at the OT position for each of the objects described here is typically 1×10^{-2}, based on the poorly known areal densities of faint QSOs: for the three such objects (out of three examined) the probability is less than 8×10^{-8}. Nevertheless, we consider it

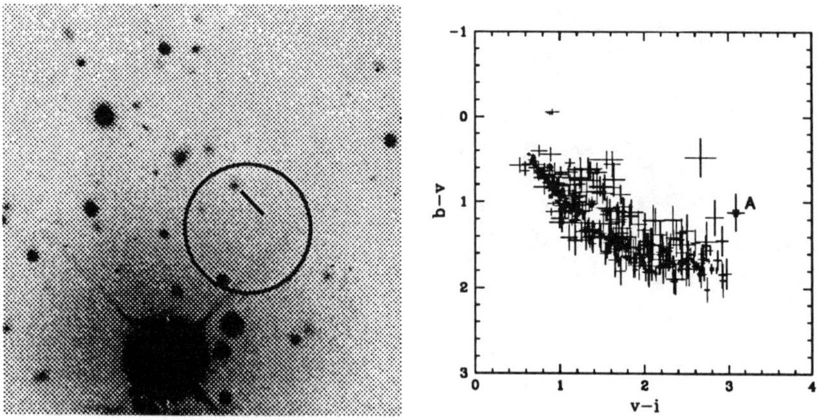

Figure 2. **Left:** I-band CCD image of the region containing the 20 arcsec. diameter OT630724 error circle. The QSO/AGN candidate is indicated (north right, east down). **Right:** Instrumental $v - i$ vs. $b - v$ color-color diagram for objects near the OT location. The object marked in the left figure (indicated with "A" on the graph) has colors consistent with a V= 20 QSO or AGN.

of vital importance to investigate additional high-quality OT candidates to further strengthen the association and to provide firmer estimates of the statistics of such events (discussed below).

A reasonable question that now arises is why such large amplitude variations have not been detected in other AGN observations. The general problem of detecting rare and brief optical flares from AGNs is that their light curves are usually drastically undersampled (e.g., Wagner et al. 1993). This is easily demonstrated by the simple calculations below which, we caution, are very approximate due to inhomogeneous datasets, small number statistics and large background contamination. We approach this calculation from two perspectives. In the first approach we include only the three OT/AGN pairs described above. In the second approach we include the larger dataset comprised of all known (both published and unpublished) OT searches and detections.

OT frequency estimated from likely OT/AGN pairs The three OT/AGN pairs described here have been found in $\sim$ 28,000 hrs of monitoring data. The observed frequency of OT events is thus roughly one OT brighter than mag 11 per area 300 arcmin^2 (this is the average area covered by archival plate searches) per 7,000 hrs (one of the OTs was observed twice); or one OT per 100 arcmin^2 per 21,000 hrs. This approach, however, provides only a strong upper limit to the OT per AGN rate. The actual rate will be lower since (i) searches that did not detect OTs were not taken into account and (ii) the AGNs mentioned here may represent a subclass of AGNs with higher than average OT rates.

OT frequency estimated from all known OT searches The total monitoring time examined by different authors for $\sim$ 30 GRB localizations is $\sim$ 170,000 hrs with each field covering ~ 300 arcmin^2 (Hudec 1993a,b and references therein, Hudec 1995, Hudec 1996). Besides the four OT events described here, three other certain OTs have been found (Schaefer 1981, Hudec et al. 1994b, Greiner & Motch 1994, Luginbuhl et al. 1996), one of unknown but astrophysical origin, and two likely related to flare stars. About 10 more OTs exceeding 11 mag have also been found with less certainty about their astrophysical origin (Hudec 1993a,b, Hudec 1995, Hudec 1996). Since many of these are likely of non-astrophysical origin or due to flare stars, we will assume that 2 of these OTs are of AGN origin. The frequency implied by these OT events is then one OT per 300 arcmin^2 per 24,000 hours, or one OT per 100 arcmin^2 per 72,000 hours. The area of 100 arcmin^2 contains, very approximately, 1 QSO brighter than mag 21 (Boyle, Shanks, & Peterson 1988). Hence, once per 72,000 hours is also the expected approximate rate of large amplitude (> 10 mag) AGN outbursts if we assume that the seven considered OTs are all of AGN origin. The rate will be lower if more than 80% of the unverified OTs are of flare star and background origin. Further, there may be a selection effect since we cannot exclude the possibility that there is a subclass of AGNs responsible for OTs, with lower surface density than AGNs in general. This again makes the one OT event per 72,000 hours per AGN an upper limit.

Whichever method is used to estimate the OT per AGN frequency, it is clear that the currently available realtime AGN monitoring database does not approach the duration necessary to detect OT events.

If the optical transient connection to QSO/AGNs proves solid, it will substantially extend the known amplitude of AGN variability (e.g., Wagner & Witzel 1995). However, since these large-amplitude variations are so rare, verification with real-time photometric monitoring will prove extremely difficult. Nevertheless, there are hints that sudden and rapid flares may occur, though of much smaller amplitude (e.g., von Linde 1993). Most (82%) of blazars studied optically for rapid light changes show variability on time scales of days or less; however the observations available are not dense enough to resolve faster variations (Wagner & Witzel 1995, Heidt & Wagner 1995).

Recognizing the problem of detecting rare events with current monitoring programs, and further that our OT searches are generally only sensitive to extremely bright (< 11th apparent mag) events due to the small apertures of the patrol cameras and difficult verification process for faint triggers, it is possible that AGNs may undergo "moderate" amplitude (5–10 mag) variations that would remain undetected by either type of observational program. Such moderate flares could be observed only by the larger aperture patrol telescopes with smaller fields of view, which provide a much smaller archival database.

If, however, these moderate amplitude variations occur more often than the OT rate discussed above, it may be correspondingly easier to detect them with extended optical monitoring of AGNs. Most of the ongoing AGN monitoring projects are based on relatively infrequent observations (one per day or less) of individual AGN targets. We suggest a new strategy of monitoring selected AGNs continuously with small robotic telescope systems and, in the case of a trigger due to a sudden light increase, then provide rapid photometry with fine time resolution (of order of minutes) to follow the flare. Already a 10-inch aperture robotic telescope based on inexpensive parts (e.g., Soldan & Hudec 1996) is able to provide automated monitoring down to mag. 19.

Additionally, very wide-field monitoring systems originally developed for OT and GRB counterpart detection (e.g., Hudec & Soldan 1994) may be used to monitor sudden bright flares similar to those reported here over larger fields of view (up to 50° diameter and over 30° diameter for detecting ≤ 11 mag triggers). The 30° diameter field includes $\sim 2 \times 10^4$ QSOs brighter than mag 21. Thus we may expect to detect as many as $(2 \times 10^4)/(5.7 \times 10^4) \leq 0.4$ events per hour of observing if all QSOs are of the same nature regarding the production of OT events. Alternatively, assuming that OT-producing QSOs are a fraction F of all QSOs, 1000 hours of observation with such a system has a 50% chance of detecting an OT if $F \geq 0.003$.

Acknowledgments. RH acknowledges the support provided by the Grant Agency of the Czech Republic, grant 205/93/0890, and by the Grant Agency of the Academy of Sciences of the Czech Republic, grant 303103.

References

Boyle, B., Shanks, T., & Peterson, B. 1988, MNRAS, 235, 935

Greiner J., & Motch M. 1994, A&A, 294, 177

Heidt J., & Wagner S. 1995, A&A, 305, 42

Hudec, R. 1993a, Astrophys. Letters and Communications 28, 359

Hudec, R. 1993b, A&A, 270, 151

Hudec R. 1995, Proc. IAU Symposium 151, "Flares and Flashes," in press

Hudec, R. 1996, A&A, in preparation

Hudec R., et al. 1990, A&A, 235, 174

Hudec, R., et al. 1994a, A&A, 284, 839

Hudec, R., et al. 1994b, A&A, 284, 499

Hudec, R., & Soldan, J. 1994, ApJS, 92, 675

Luginbuhl, C., et al. 1996, AIP Conf. Proc., 3rd Huntsville GRB Workshop, in press

Motch, C., et al. 1990, A&A, 235, 185

Schaefer, B. E. 1981, Nature 294, 722

Soldan, J., & Hudec, R. 1996, AIP Conf. Proc., 3rd Huntsville GRB Workshop, in press

von Linde, J., et al. 1993, A&A, 267, L23

Vrba, F. J., et al. 1994, ApJ, 424, 68

Vrba, F. J., et al. 1996, in preparation

Wagner S., et al. 1993, A&A, 271, 344

Wagner S., & Witzel 1995, ARA&A, 33, 163

Zytkow A. 1990, ApJ, 359, 138

Blazar Continuum Variability
ASP Conference Series, Vol. 110, 1996
H. R. Miller, J. R. Webb, and J. C. Noble, eds.

The Optical and Ultraviolet Polarization of BL Lac Objects and OVV Quasars

Paul S. Smith

Steward Observatory, University of Arizona, Tucson, AZ 85721

Abstract. Optical and ultraviolet polarimetry has played an important role in investigating the "blazar" (BL Lacertae objects plus OVV quasars) phenomenon. High and rapidly variable optical polarization are two of the defining characteristics of this class of radio-loud AGNs, and the polarizing mechanism is clearly identified as synchrotron radiation from small regions where the magnetic field is well aligned. Though the polarizing mechanism is easily identified in these objects, the rapid fluctuations are as difficult to interpret and model as the flux variations. This, together with the fact that long-term correlations between optical polarization and flux are rarely observed, has greatly slowed progress in understanding the underlying physics and structure of the continuum emission region at these wavelengths. However, since polarimetry gives us the only direct method of studying the magnetic field in blazars, any serious model of the continuum source must explain the dramatic polarization behavior of these objects.

Although a coherent picture has yet to emerge which explains both polarization and flux variability at high frequencies, observations of how the polarization varies with wavelength are able to constrain or (in some cases) rule out models for the UV/optical/near-IR continuum. I review what has been learned by studying the wavelength dependence of the optical polarization of blazars and summarize the implications of UV spectropolarimetry obtained using *HST*.

1. Introduction: the success of polarimetry in AGN research

Studies of the linear polarization of nearly every class of Active Galactic Nuclei (AGNs) have greatly contributed to our understanding of the inner structure of these enigmatic objects. Indeed, spectropolarimetry of radio-quiet AGNs during the past decade have advanced the view that many of the disparate properties observed in various classifications of objects can be explained simply as the result of our viewing angle, and not necessarily because of large intrinsic differences (see Antonucci 1993 and references therein). In this "unified" model of radio-quiet AGNs it is effectively the inclination of an obscuring torus to our line of sight that determines, for example, whether we perceive a Seyfert 1 or Seyfert 2 nucleus.

Since Antonucci and Miller (1985) pointed out that the nuclear polarized flux spectrum of the Seyfert 2 galaxy NGC 1068 reveals a blue continuum and broad permitted emission lines characteristic of Seyfert 1 nuclei, optical spectropolarimetry has been utilized as a powerful tool to detect hidden broad-line regions (BLRs) and continuum sources in AGNs. Subsequent observations of Seyfert 2s have shown that NGC 1068 is not a unique case, and several (though possibly not all) objects would look like Seyfert 1 nuclei if viewed from a different direction (Miller & Goodrich 1990; Tran, Miller, & Kay 1992; Tran 1995a,b,c).

The basic "unified" Seyfert model has an opaque torus blocking the view of the inner region of the AGN in the equatorial plane. This region of course includes the continuum-emitting accretion disk (whose material is presumably drawn from the torus) and BLR located within several light weeks of the central, massive compact object (see Peterson 1993 and references therein). Electrons and/or dust within or just outside of the narrow-line region (NLR), located above and below the equatorial plane, scatter the nuclear radiation. Therefore, if our line of sight is oriented near the plane of the torus we only see the nuclear continuum and BLR in scattered, polarized light and the NLR dominates the emission-line spectrum as it does in Seyfert 2s. In this model the observed position angle of the linear polarization (θ) would be parallel to the plane of the torus. On the other hand, we directly observe the inner AGN regions if our viewing aspect is nearer the axis of the torus and the spectrum of the total flux exhibits broad permitted emission lines and a bluer continuum.

Recently this basic model of the structure of Seyfert nuclei has been applied to the entire luminosity range of radio-quiet AGNs. For example, Hines et al. (1995) have shown that the "hyperluminous" infrared galaxy F15307+3252 exhibits a polarized spectrum indistinguishable from an optically-selected QSO. At the lower end of the AGN luminosity range, spectropolarimetry and imaging polarimetry of the nearby Seyfert/LINER galaxy NGC 4258 has also provided substantiating evidence for an AGN hidden by an obscuring torus (Wilkes et al. 1995). In this case, θ is found to be exactly parallel with the H_2O maser disk discovered by Miyoshi et al. (1995) in orbit around a central mass of $\sim 3.6 \times 10^7\ M_\odot$. Given that the masing disk is oriented edge-on to our line of sight, and the fact that the X-ray spectrum from the weakly active central engine is heavily absorbed, there is little doubt of the existence of an obscuring torus. Optical emission from the central engine is seen only in scattered light and the close alignment between the orientation of the plane of the maser disk and θ is expected if the scattering regions are centered about the orbital axis of the disk.

Spectropolarimetry has also advanced the notion that broad absorption-line QSOs (BALQSOs) are those QSOs oriented in such a way that the torus partially hides our view of the AGN (Glenn, Schmidt, & Foltz 1994; Cohen et al. 1995; Goodrich & Miller 1995; Hines & Wills 1995). The ablation of material from the torus is possibly responsible for the characteristic broad absorption lines observed. In the "unified" scheme BALQSOs are the transition objects. QSOs completely obscured from direct view have observed properties similar to F15307+3252, whereas objects viewed more "pole-on" are optically-selected QSOs typified by strong UV-excess objects such as the PG sample of QSOs.

The importance of polarimetry to the study of AGNs is by no means limited to the radio-quiet objects. The large polarizations observed for high-z radio galaxies clearly indicates that light from a high-luminosity AGN, hidden from our direct view, is reflected into the line of sight by scattering material distributed within the host galaxy (Cimatti et al. 1993; Jannuzi et al. 1995). These very basic observations greatly weakens arguments that the blue colors of these objects are the result of intensive star formation (see e.g., McCarthy et al. 1987 and Eisenhardt & Dickinson 1993) and the ensuing interpretation that they represent primeval galaxies.

High and rapidly variable linear polarization are unique defining properties of the radio-loud AGNs classified as "blazars." Because of this, polarimetry is a vital tool that must be utilized in attacking the difficult problem that blazars present to astronomy. In this paper I summarize the polarization characteristics of these objects and what has been learned by monitoring the wavelength dependence and variations of their polarization at UV, optical, and near-IR wavelengths. The term blazar will be used for any AGN that shows rapidly variable optical polarization. However, as with emission-line strength, and certain radio properties, it is convenient to separate the blazar class into three types of objects based on their optical polarization properties. These are: (1) radio-selected BL Lacertae objects (RBLs), (2) X-ray-selected BL Lacs (XBLs), and (3) optically violent variable quasars (OVVs).

2. The variability of polarization in Blazars

Roughly three decades ago it was found that RBLs and OVVs exhibit high levels of linear polarization (see Angel & Stockman 1980 for a review of the early work on blazars). Even more remarkable is that the polarization is observed to vary greatly and often over time scales of <1 day. The rapid variability in degree of polarization (P) and θ at optical wavelengths is as stunning and unpredictable as the huge flux variations and argues for a synchrotron origin for the polarization. A scattering origin for the polarization of blazars, as favored for the many types of AGNs mentioned in §1, is highly unlikely for two reasons. First, incoherent, optically-thin synchrotron radiation best explains the featureless power-law continuum of BL Lac objects and is intrinsically highly polarized if the magnetic field is well aligned throughout the emission region (Ginsburg & Syrovatskii 1965). The degree of magnetic field alignment can be very high in the optical-emitting regions of blazars as the present records for both RBLs (AO 0235+164; Impey, Brand, & Tapia 1982) and OVVs (3C 279; Mead et al. 1990) have $P \sim 45\%$.

The essence of the second argument against scattering is that rapid variations in polarization require that the distribution of scattering material (such as dust and/or electrons) around the illuminating source change substantially in less than an hour. Dust is easily ruled out as a source of significant polarized flux from blazars since (1) grains could not possibly survive so close to the nuclear continuum source, and (2) the observed wavelength dependence of P is inconsistent with scattering by small particles (see §3). In any event, it is hard to visualize any cloud of scattering material changing its size and structure so rapidly in an AGN. In fact, until recently, polarized radio-quiet AGNs (objects where scattering is believed to be the polarizing mechanism) have shown no evidence for polarization variability. In the one well-documented case (I Zw 1; Smith et al. 1996), changes in polarization appear to occur over periods of several years to a decade.

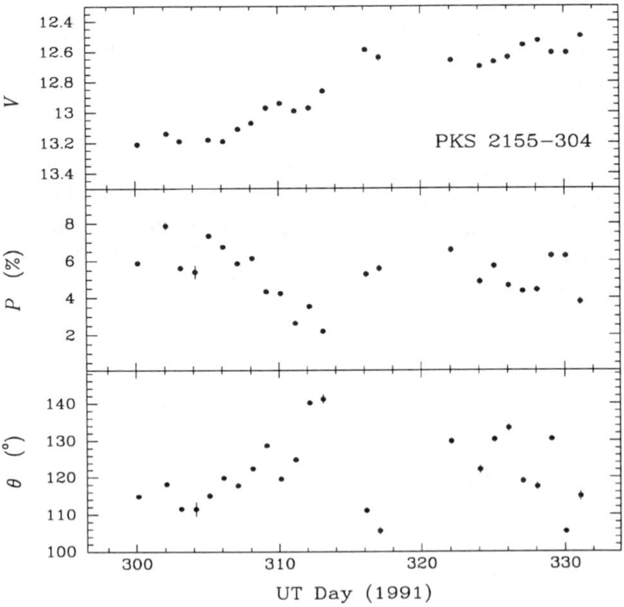

Figure 1. V-band photometry and unfiltered polarimetry of PKS 2155−304 during 1991 October/November (from Smith et al. 1992).

Although the rapid variations observed in blazars unambiguously identify the polarizing mechanism, they greatly hinder interpretation and modeling of the available polarization monitoring data. A relatively recent example of the typical optical polarization behavior of a blazar over several weeks is given by PKS 2155−304 during the intensive, multifrequency monitoring of this XBL in November 1991 (Edelson et al. 1995). Polarization monitoring of this object during this period (Smith et al. 1992) is shown in Figure 1. Note that θ appears to have stayed between 100° and 150°, but that 25° changes can occur in a day. In addition, P varies from 2–8% with daily variation approaching 3%. RBLs have exhibited much larger daily changes in P and comparable position angle fluctuations (Sitko, Schmidt, & Stein 1985; Smith et al. 1987). For example, the polarization of AO 0235+164 increased from ~12% to ~24% in a 24 hour period in 1983 January. OVVs have also shown amazing polarization variability exemplified by B2 1156+295 between 1984 June 9 and 10 when P increased from ~5% to ~20%. It is obvious that daily sampling of the optical polarization of blazars is generally inadequate to follow the fluctuations.

The monitoring of BL Lac by Moore, Schmidt, and West (1987) still represents the densest sampling of the optical polarization of an AGN. The major result was that $\Delta P \sim 1\%$ and $\Delta\theta \sim 5°$ in an hour. Fifteen minutes was about the minimum time interval that showed significant variations given the observational uncertainties (dominated by photon statistics). Fortunately, the current generation of dual-beam CCD polarimeters are fully capable of similar monitoring of many blazars even with moderately-sized telescopes (apertures $\gtrsim$ 1 m). However, the need for continuous coverage, few available instruments, and obvious drawbacks to ground-based observing means that polarization monitoring programs that complement current long-term (several days to a month) flux monitoring are virtually impossible until there is a space-based polarimetric analog to *IUE*.

The top panel of Figure 1 displays the V-band photometry of PKS 2155−304 measured simultaneous with the polarimetry. PKS 2155−304 shows the classic (and irritating) trait exhibited by blazars in general: there are no hints of a direct correlation between polarization and flux. Since polarimetry gives us the only direct information about the magnetic field (in particular, the degree of field alignment and the orientation of the field projected onto the sky), models of the synchrotron-emitting region(s) are incomplete if they do not incorporate polarization information. The importance of including polarization into physical models that attempt to explain detailed variations has been demonstrated at radio wavelengths. Hughes, Aller, and Aller (1989) were able to explain *both* polarization and flux behavior of BL Lac over a period of ~2 years by modeling a shock propagating down a relativistic jet. In this case it is the polarization information that largely constrains the model.

Of course the situation is much more difficult at higher frequencies where the variations in flux and polarization are exceedingly rapid compared to those at centimeter wavelengths. As a result, it remains unclear whether detailed modeling at optical wavelengths is possible at all. Even the limited success at radio wavelengths will likely not be reached at higher frequencies given the usual method of blazar monitoring, which involves measuring only the flux of the object. This means that in many research programs 2/3 of the information the photons can yield is thrown away. Multifrequency flux programs like the 1991 November PKS 2155−304 campaign are well suited for determining if the synchrotron continuum that dominates in the optical also dominates at other wavelengths (though, as will be discussed in §3, polarimetry is also a powerful technique that can be applied to this problem), but are probably of limited use by themselves in deriving robust physical models that explain the dramatic behavior of blazars. On the other hand, polarimetry suffers from not having a space-based platform that can be devoted to blazars for a significant amount of time, a lack of ground-based instrumentation, and the present inability to measure polarization at γ-ray, X-ray, and far-IR wavelengths.

Ideally both flux and polarization must be measured together and sampled sufficiently well to identify and constrain discrete events that can be modeled. Identifying a connection between the flux and polarization variations would then be the first step. This alone appears to be a difficult task for blazars at optical wavelengths. So far there seems to be little evidence for correlations between flux and polarization. However, the OVV 3C 345 gives us a well-documented example of a clear connection between brightness and P as well as θ. The left two panels of Figure 2 plots the data obtained during and after a large flare that peaked in early 1983. As the flare faded, P decreased and θ rotated from ~15° to ~90°. Though some of the decrease in P can be attributed to the presence of an unpolarized continuum source (see §3.3) that remained relatively constant as the synchrotron light diminished, most was due to an overall decrease in the intrinsic nonthermal polarization. In addition, the trend seen between flux and θ

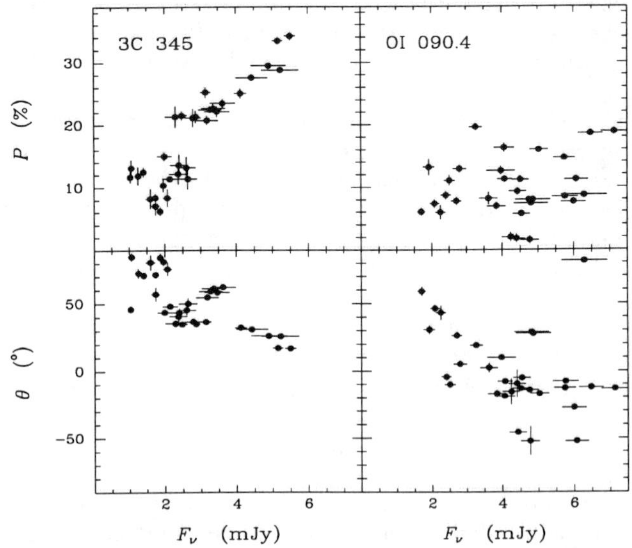

Figure 2. Long-term correlations between optical polarization and flux density in the OVV 3C 345 and the RBL OI 090.4. Data are taken from Sitko et al. (1985) and Smith et al. (1986a & 1987) and show R-band observations from late 1982 to mid-1984.

is independent of any unpolarized light contributing to the total observed flux. Sporadic, unpublished monitoring of 3C 345 by the author indicate that the quasar has typically remained faint over the past decade, and in keeping with the trend described above, P has generally remained $\lesssim$5%.

For comparison, Figure 2 also shows the equivalent plot for an RBL. In this case, OI 090.4 reveals the more typical pattern obtained for blazars and no correlation is seen between optical flux and P. This object does show a trend in θ such that the position angle appears to systematically rotate in the Q–U plane as the object varies in flux. Note, however, that θ varies over the entire 180° range.

Identification of both long- and short-term trends between polarization and flux, must first be made before polarimetry is considered a valuable tool in the analysis of blazar variability and not a complication. It must be remembered, however, that there is little chance of producing compelling models that will be able to follow the variations of *all* the Stokes parameters over an extended length of time if polarization data are not acquired in the first place.

3. Wavelength-dependent polarization

Observations of how the linear polarization varies with wavelength have been more fruitful than monitoring temporal changes largely because of the problems in properly monitoring blazars which result from the unavoidable facts of the fluctuations being so rapid and the lack any clear relation to the flux variations. Spectropolarimetry of strong-lined AGNs has been successful because important information on the structure of an object can be gathered by analyzing how the polarization varies across emission lines. Here we are at a disadvantage with blazars since, by definition, XBLs and RBLs have very weak emission lines. Spectropolarimetry available for OVVs (which have line luminosities comparable with those of normal QSOs) yields no evidence that the BLR or NLR are significantly polarized (de Diego et al. 1994; Smith et al. 1994). Therefore, any current discussion of wavelength-dependent polarization in blazars is limited to analysis of how the continuum polarization varies with wavelength. This analysis has led to some important conclusions about the source of the continuum at IR, optical, and near-UV wavelengths.

The biggest polarimetric difference between BL Lac objects and OVVs is the wavelength-dependent P (WDP) properties they exhibit (WDθ will be used to shorten "wavelength-dependent polarization position angle"), and they are discussed separately. I have also divided the discussion of RBLs and XBLs into two separate sections since they exhibit quite different polarization behavior (though there is as yet no substantial evidence for distinguishing these two types of objects based on WDP or WDθ).

3.1. RBLs

Broad-band optical/near-IR polarimetry by several groups during the 1980's has given us a good picture of the range of polarization behavior presented by RBLs (e.g., Sitko et al. 1985; Brindle et al. 1986; Smith et al. 1987; Mead et al. 1990). Summaries of the wavelength-dependent properties observed can be found in Smith, Elston, and Balonek (1986b) and Ballard et al. (1990). Briefly, nearly every observed RBL has exhibited WDP and WDθ, but dramatic wavelength dependence, such as reported by Holmes et al. (1984) for OJ 287, is very rare. In general, θ is independent of wavelength throughout the optical bandpass. Weak WDP is usually observed in RBLs and appears to be independent of the brightness or polarization level of the object. The most pronounced trend is that when WDP is observed, it is almost invariably in the sense that P *increases* into the blue.

A synchrotron-emitting region with an isotropic distribution of electron velocities and uniform magnetic field results in linear polarization that is independent of wavelength. The fact that wavelength dependence is often observed leads us to the following three possibilities (or combination thereof): (1) the continuum emission region is complex (e.g., the magnetic field might be better aligned in the region radiating the majority of the blue light), (2) there are multiple nonthermal components, each with a different polarization, contributing to the spectrum, and (3) there are sources of continua unrelated to the nonthermal flux that produce the observed wavelength dependence. The latter possibility is sorely tested by the available multi-color polarization data. Indeed, there is a strong argument that the IR–UV continuum of RBLs is dominated by synchrotron emission. Any continuum sources that do not share the polarization and variations of the nonthermal continuum are definitively ruled out. For example, stars in the RBL's host galaxy are a source of unpolarized, non-varying

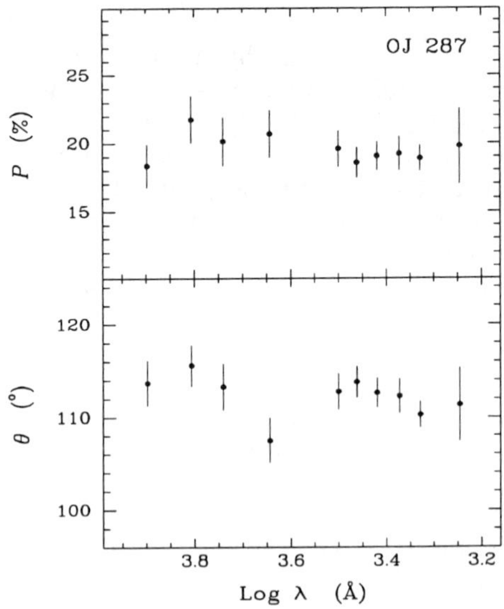

Figure 3. Simultaneous optical and *HST* UV polarimetry of the RBL OJ 287. Data are from Smith et al. (1993) and displayed in the observed reference frame.

light with typically much redder colors than the nonthermal spectrum. Such a light source cannot explain any occurrence of WDθ, but gives the correct sense for WDP (the red stellar continuum contributes more to the total flux in the red than in the blue and therefore dilutes the nonthermal polarization more effectively in the red). However, since WDP is variable, and shows no clear trend with brightness, such a static unpolarized source of flux can be discounted. An excellent example of how wavelength-dependent polarization can be unaffected by large flux changes is discussed in §3.2.

Another strong argument against a significant contribution to the continuum by any sort of non-synchrotron source in an RBL is revealed by simultaneous optical/UV polarimetry of OJ 287 (Smith, Allen, & Angel 1993). Figure 3 shows broad-band optical polarimetry and the binned UV spectropolarimetry obtained with the Faint Object Spectrograph (FOS) aboard *HST*. These observations were made on 1992 May 3 when OJ 287 was about 4 mag fainter ($V \sim 16$) than its historical photometric maximum. Even in such a faint state there is still no hint of wavelength dependence that would suggest the host galaxy is an important component to the observed spectrum. The polarization is constant from 1250 Å to 6100 Å (rest frame; $z = 0.306$) with $P \sim 20\%$ (a wonderful example of how highly polarized a blazar can be even when faint) and $\theta \sim 111°$.

The same general argument can be made against "contamination" of the synchrotron spectrum by thermal emission from an accretion disk [or alternative cause of the typically unpolarized feature known as the "Big Blue Bump" (BBB) in AGNs]. This continuum source would likely be variable in flux over much shorter time scales than the stellar population of the host galaxy, but large variations on the time scales observed for the nonthermal continuum would not be possible for a disk several light days across. In this case, since the BBB is much bluer than the nonthermal power law, P would be expected to decrease into the UV. The wavelength independence of P in OJ 287 during the *HST* polarimetry can place stringent limits on the luminosity of an accretion disk because the continuum of the disk would likely peak in the near-UV. In fact, Smith et al. (1993) place a 2σ-upper limit of only 2% of the total flux received at a rest frame wavelength of $\sim$4200 Å. That is, even with OJ 287 so faint, the synchrotron continuum still completely dominates in the optical and UV. A detailed discussion of how optical polarimetry alone can set hard limits on the near-UV brightness of an accretion disk can be found in Smith and Sitko (1991).

With likely external causes for wavelength-dependent polarization ruled out, we must conclude that the cause is intrinsic to the emitting region. It follows then that observations of WDP and WDθ give us information about the processes and/or structure of the nonthermal source. One possibility is that multiple, independent emission regions result in wavelength-dependent polarization. Multi-component (i.e., two-component) models are certainly tempting for situations when there is strong wavelength dependence accompanied by rapid evolution of WDP and WDθ. Examples of this behavior can be found in Holmes et al. (1984) and Smith et al. (1986b). Two completely independent components should show the entire range of wavelength-dependent polarization possible depending on how disparate their spectral indices can be. We would expect to see as many instances of P increasing with wavelength as decreasing if no free parameters are constrained. Since WDP almost invariably is in the sense of P decreasing with wavelength, the more highly polarized component must almost always have a harder optical spectrum. Also, there are many cases where the polarization and flux vary greatly, but the wavelength dependence of the polarization remains constant (see §3.2) requiring the two components to vary together. Constraints such as these favor inhomogeneities in a single source being the cause of the wavelength dependence (Björnsson 1985). However, reversals in the sense of both WDP and WDθ observed during a four day period in 1983 January for PKS 0735+178 (Smith et al. 1986b) support the view that at least occasionally two fairly distinct sources contribute to the polarization of RBLs. Note that PKS 0735+178 showed one of the rare occurrences of P increasing with wavelength during the 1983 event. As usual for blazars, both multi-component source models and single, inhomogeneous source models (Björnsson & Blumenthal 1982; Björnsson 1985) have enough free parameters to handle the available observations. Distinguishing between the models will most likely rest on their ability to realistically follow variations in WDP and WDθ seen in future well-sampled, high-quality data.

3.2. XBLs

The polarization properties of XBLs are thoroughly described by Jannuzi, Smith, and Elston (1994). These objects differ from RBLs in several major aspects of their polarization behavior. The maximum level of polarization for XBLs is around 10%, as opposed to 30–40% observed for many RBLs. Like their observed flux variations, rapid polarization changes in most XBLs are rare. The polarization position angle in XBLs tends to vary over a much more restricted range than most RBLs. Although several RBLs have preferred polarization position angles, this remains to be demonstrated for the majority of RBLs. In contrast, almost all XBLs exhibit preferred angles based on three years of monitoring of the *Einstein* MSS sample of BL Lacs (Jannuzi, Smith, & Elston 1993).

Keeping in mind that there are relatively few multi-color observations of XBLs, there is no major difference in the wavelength-dependent polarization properties observed between the two types of BL Lacs. Jannuzi et al. (1994) find that WDθ is rare given the typical accuracy of their filtered measurements, and when WDP occurs it is always in the sense of P increasing into the blue. Because wavelength-dependent polarization has only been monitored in one XBL (PKS 2155−304; see below) at a level comparable to RBLs and the fact that the optical nonthermal luminosities of XBLs are generally lower than RBLs (at least in our line of sight), the question is still open as to whether starlight from the host galaxy results in the observed WDP. Most XBLs have a well-determined redshift based on the identification of stellar features in their optical spectra and this indicates a more substantial stellar component to the overall spectrum than in RBLs. The rare instances of WDθ argues that at least some of the wavelength dependence is due to some cause intrinsic to the synchrotron source.

PKS 2155−304 has been the most intensively observed blazar during the 1990's. As a result, we have a fairly complete picture of its optical polarization behavior, and have even glimpsed its UV polarization. The X-ray/optical/radio "colors" of PKS 2155−304 clearly place it among the XBLs, but it seems to be the most RBL-like of its class based on its polarization and flux variability (Figure 1). Indeed, its wavelength-dependent polarization properties are indistinguishable from those of an RBL. For example, during nightly monitoring in 1991 October/November brightness, P, and θ varied greatly but the optical spectral index, WDP, and WDθ remained almost constant (Smith et al. 1992). While the flux increased by about a factor of 2, P always gently increases into the blue. Again, this rules out any significant flux contribution by the host galaxy or an accretion disk. Optical polarimetry alone can set a relatively low limit on the flux from an accretion disk in the UV (see Smith & Sitko 1991). However, PKS 2155−304 is the only BL Lac object where near-simultaneous optical and UV polarimetry shows strong WDP (Allen et al. 1993). These data reveal P smoothly increasing into the ultraviolet and that the cause of the wavelength dependence affects a broad spectral region. The ultraviolet observations confirm that at optical and UV wavelengths the continuum is dominated by the same synchrotron source. Subsequent analysis of the 1991 November multi-wavelength flux monitoring program (Edelson et al. 1995) also provides strong evidence that the UV/optical continuum must be produced by synchrotron emission and extends this conclusion to X-ray wavelengths as well.

3.3. OVVs

Observations of wavelength-dependent polarization in OVVs have revealed a major difference between these objects and the weak-lined, BL Lac-type of blazars. Added to the differences already noted between BL Lacs and OVVs in emission-line strength and VLBI polarization properties (Gabuzda et al. 1992) is the fact that several OVVs exhibit strong WDP in the ***opposite*** sense shown by BL Lacs. Monitoring of 3C 345 during its 1983 outburst showed P consistently decreasing toward the blue with little WDθ (Smith et al. 1986a). In fact, WDP increased in strength and the optical spectrum hardened as the quasar faded during 1983–84. This is exactly the behavior expected if in addition to the flaring nonthermal continuum, there is a relatively static blue continuum source that is essentially unpolarized. The BBB is a likely candidate for the blue continuum, and although monitoring of normal PG QSOs was shown substantial variability in their BBB-dominated optical continua (Maoz et al. 1994), the variability has a much lower amplitude and is not nearly as rapid as seen from the synchrotron source in blazars. Also, the survey of PG QSOs conducted by Berriman et al. (1990) determined that the BBB is almost invariably an unpolarized feature.

Broad-band polarimetry by Smith et al. (1988) and Wills et al. (1989) identified several other OVVs with WDP similar to that displayed by 3C 345. Among the OVVs showing evidence for a diluting, blue continuum component are 3C 454.3 and PKS 1510−089. In all cases the diluting component has to have an optical luminosity comfortably within the range found for normal (unpolarized) PG quasars in order to account for the decrease in P. However, other OVVs have yet to show any sign that there is any non-synchrotron light present in their observed continua. One such example is 3C 279 which is identical to a BL Lac object in its polarization properties, though multi-color polarimetry of this quasar has only been reported for periods when the object has been bright. 3C 279 occasionally can become faint ($V \sim 18$–19) and it may only be during these periods that the effects of any polarization-diluting component become apparent. B2 1156+295 also shows no hint of a BBB (or any other diluting component) while bright. However, P was observed to decrease toward the blue when the quasar was as faint as $V \sim 17.4$ (Smith et al. 1988). There is undoubtedly a range in luminosities for the BBB in blazars and the objects identified so far as harboring this component may simply be the OVVs with high-luminosity "normal" QSO continua. Indeed, BL Lacs may also possess a similar continuum source, but the observations reviewed in §3.1 ensure that the analogous BBB components, if they exists, have luminosities more than an order of magnitude lower than those seen in objects like 3C 345.

Spectropolarimetry of PKS 1546+027 presents an excellent example of the wavelength-dependent polarization properties of OVVs (Smith et al. 1994). The flux and polarization spectra for the object are depicted in Figure 4. Here we see P about a factor of 2 lower at 4000 Å than at 8000 Å (observed frame) and θ is constant with wavelength. The polarization is also depressed at the location of prominent emission lines as expected for the case where synchrotron emission produces the polarized continuum and the emission lines are not highly polarized. Significant line polarization cannot totally be ruled out, but the invariance of θ across the lines and the fact that 3C 345 also exhibits the same traits (de Diego et al. 1994) argue for unpolarized line-emitting regions. The steep, featureless power-law spectrum of the polarized flux confirms the synchrotron origin of the polarization. Reasonable decomposition of the total flux spectrum into nonthermal and unpolarized continuum (plus emission lines) components yields a blue diluting spectrum identical to a normal, low-polarization QSO.

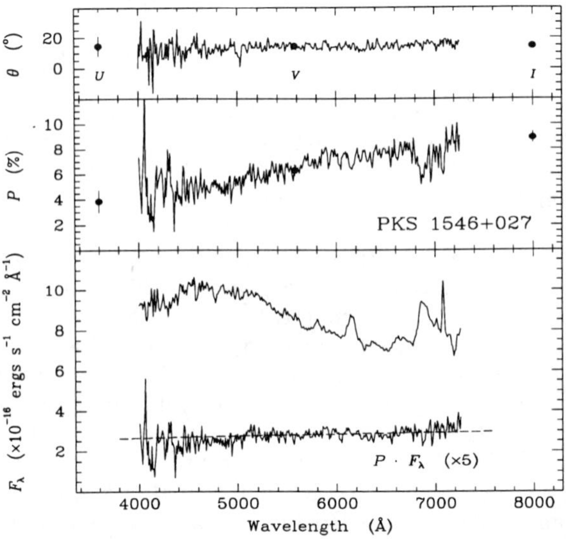

Figure 4. Spectropolarimetry of the OVV PKS 1546+027 (from Smith et al. 1994) obtained in a 40 minute observation at the KPNO 4 m Mayall telescope. Simultaneous broad-band *UVI* measurements made at the Steward Observatory 2.3 m telescope are also shown. The lower panel displays both total and polarized flux ($P \times F_\lambda$) spectra. The polarized flux is multiplied by a factor of 5 and a power-law fit ($F_\nu \propto \nu^\alpha$) to the data is denoted by the dashed curve.

4. Future prospects and directions

With the scheduled decommissioning of the FOS in early 1997, the prospects for UV polarization studies of blazars in the foreseeable future are poor. We will have to be content with the observations already acquired (most of which were made before the COSTAR corrective optics were installed) for many years since there are no current plans for any space missions having UV or optical spectropolarimetric capabilities. This is unfortunate for nearly all AGN research, not just blazars, as many of the recent strides in modeling the structure of AGNs have come from polarimetry.

After the second *HST* refurbishment mission WFPC2, FOC, and NICMOS still give the telescope imaging polarimetry capability. Unfortunately, all of these instruments were not designed with polarimetry as a high priority. As a result, accurate imaging polarimetry with *HST* is a difficult and time-consuming task. Also, because none of the instruments simultaneously samples both senses of linear polarization, measurements are susceptible to many systematic effects. Independent of the lack of efficiently designed space-based polarimeters, the unique and incredible capabilities of *HST* make it unusable for the types of research programs that need to be pursued for blazars. These objects must be intensively monitored in order to make any sense of the dramatic behavior observed at high frequencies. The large amount of time necessary for monitoring an AGN is nearly impossible to allocate for a telescope that is able to make important contributions in almost every field of astronomy.

The FOS polarimetry results show that the UV polarization of BL Lacs can be as high as at optical wavelengths, and that the wavelength dependence (or independence) of the optical polarization optical extends smoothly into the ultraviolet. These results lead to the conclusion that the UV and optical synchrotron radiation arises from the same region and completely dominates the continuum at these wavelengths. Recent multifrequency flux monitoring has independently confirmed this conclusion and the logical next step would be intensive monitoring of both flux and polarization in as many wavebands as possible. Unfortunately the retirement of *IUE* is a heavy blow to the ambitious multifrequency effort and will undoubtedly slow progress in understanding the behavior of blazars. If a replacement for *IUE* is launched, it is imperative that the new facility have polarimetric capability if a large portion of its mission is devoted to blazars. Real progress in understanding these objects is hard to envision if we continue to ignore much of the information carried by the photons entering our telescopes.

The ability to measure polarization at higher frequencies would give us the best way to test for non-synchrotron emission at X-ray and γ-ray wavelengths. For example, the synchrotron continuum could be separated from any synchrotron self-Compton component since some signature would be apparent in the wavelength dependence of the polarization. The spectral decomposition of the two continuum sources would be very similar to the way the synchrotron and BBB continua in OVVs are presently distinguished.

Finally, the development of VLBI polarimetry has recently provided another vital tool by actually measuring the radio polarization in individual radio jet components as well as the unresolved radio core. In addition to clearly showing differences between BL Lacs and OVVs in the magnetic field structure along their jets, VLBI polarization monitoring may yield important clues as to the location of the optical emission region(s). Projects designed to obtain simultaneous optical and VLBI polarimetry of blazars have provided some encouraging preliminary results (see e.g., Gabuzda & Sitko 1994). However, much more work needs to be done in the coming years to link the optical continuum to a particular radio component. Such experiments most likely represent the only method for determining the location of the highly variable optical-emitting region until we are actually able to resolve blazars.

Acknowledgments. Polarimetry of blazars at Steward Observatory and with the *Hubble Space Telescope* has been supported during the past decade by NASA contract NAS 5–29293 and grant NAG 5–1630. I thank Richard Allen, Richard Elston, Dean Hines, Buell Jannuzi, and especially Gary Schmidt for crucial contributions and discussions and marvelous collaboration during my 10 years at Steward Observatory.

References

Allen, R. G., Smith, P. S., Angel, J. R. P., Miller, B. W., Anderson, S. F., & Margon, B. 1993, ApJ, 403, 610

Angel, J. R. P., & Stockman, H. S. 1980, ARA&A, 18, 321

Antonucci, R. 1993, ARA&A, 31, 473

Antonucci, R. R. J., & Miller, J. 1985, ApJ, 297, 621

Ballard, K. R., Mead, A. R. G., Brand, P. W. J. L., & Hough, J. H. 1990, MNRAS, 243, 640

Berriman, G., Schmidt, G. D., West, S. C., & Stockman, H. S. 1990, ApJS, 74, 869

Björnsson, C.-I. 1985, MNRAS, 216, 241

Björnsson, C.-I., & Blumenthal, G. R. 1982, ApJ, 259, 805

Brindle, C., Hough, J. H., Bailey, J. A., Axon, D. J., & Hyland, A. R. 1986, MNRAS, 221, 739

Cimatti, A., de Serego Alighieri, S., Fosbury, R. A. E., Salvati, M., & Taylor, D. 1993, MNRAS, 264, 421

Cohen, M. H., Ogle, P. M., Tran, H. D., Vermeulen, R. C., Miller, J. S., Goodrich, R. W., & Martel, A. R. 1995, ApJ, 448, L77

de Diego, J. A., Kidger, M. R., Pérez, E., & Takalo, L. O. 1994, ApJ, 424, 76

Edelson, R., et al. 1995, ApJ, 438, 120

Eisenhardt, P., & Dickinson, M. 1993, in Proc. 3d Teton Summer School on The Evolution of Galaxies and Their Environment, ed. D. Hollenbach, H. Thronson, & J. M. Shull, NASA CP, No. 3190, 29

Gabuzda, D. C., Cawthorne, T. V., Roberts, D. H., & Wardle, J. F. C. 1992, ApJ, 388, 40

Gabuzda, D. C., & Sitko, M. L. 1994, AJ, 107, 884

Ginzburg, V. L., & Syrovatskii, S. I. 1965, ARA&A, 3, 297

Glenn, J., Schmidt, G. D., & Foltz, C. B. 1994, ApJ, 434, L47

Goodrich, R. W., & Miller, J. S. 1995, ApJ, 448, L73

Hines, D. C., Schmidt, G. D., Smith, P. S., Cutri, R. M., & Low, F. J. 1995, ApJ, 450, L1

Hines, D. C., & Wills, B. J. 1995, ApJ, 448, L69

Holmes, P. A., et al. 1984, MNRAS, 211, 497

Hughes, P. A., Aller, H. D., & Aller, M. F. 1989, ApJ, 431, 68

Impey, C. D., Brand, P. W. J. L., & Tapia, S. 1982, MNRAS, 198, 1

Jannuzi, B. T., Elston, R., Schmidt, G. D., Smith, P. S., & Stockman, H. S. 1995, ApJ, 454, L111

Jannuzi, B. T., Smith, P. S., & Elston, R. 1993, ApJS, 85, 265

Jannuzi, B. T., Smith, P. S., & Elston, R. 1994, ApJ, 428, 130

Maoz, D., Smith, P. S., Jannuzi, B. T., Kaspi, S., & Netzer, H. 1994, ApJ, 421, 34

McCarthy, P. J., van Breugel, W. J. M., Spinrad, H., & Djorgovski, S. G. 1987, ApJ, 321, L29

Mead, A. R. G., Ballard, K. R., Brand, P. W. J. L., Hough, J. H., Brindle, C., & Bailey, J. A. 1990, A&AS, 83, 183

Miller, J. S., & Goodrich, R. W. 1990, ApJ, 355, 456

Miyoshi, M., et al. 1995, Nature, 373, L127

Moore, R. L., Schmidt, G. D., & West, S. C. 1987, ApJ, 314, 176

Peterson, B. M. 1993, PASP, 105, 247

Sitko, M. L., Schmidt, G. D., & Stein, W. A. 1985, ApJS, 59, 323

Smith, P. S., Allen, R. G., & Angel, J. R. P. 1993, ApJ, 415, L83

Smith, P. S., Balonek, T. J., Elston, R., & Heckert, P. A. 1987, ApJS, 64, 459

Smith, P. S., Balonek, T. J., Heckert, P. A., & Elston, R. 1986a, ApJ, 305, 484

Smith, P. S., Elston, R., & Balonek, T. J. 1986b, in Continuum Emission in Active Galactic Nuclei, ed. M. L. Sitko (Tucson, NOAO), p. 171

Smith, P. S., Elston, R., Berriman, G., Allen, R. G., & Balonek, T. J. 1988, ApJ, 326, L39

Smith, P. S., Hall, P. B., Allen, R. G., & Sitko, M. L. 1992, ApJ, 400, 115

Smith, P. S., Schmidt, G. D., Allen, R. G., & Hines, D. C. 1996, ApJ, submitted

Smith, P. S., Schmidt, G. D., Jannuzi, B. T., & Elston, R. 1994, ApJ, 426, 535

Smith, P. S., & Sitko, M. L. 1991, ApJ, 383, 580

Tran, H. D. 1995a, ApJ, 440, 565

Tran, H. D. 1995b, ApJ, 440, 578

Tran, H. D. 1995c, ApJ, 440, 597

Tran, H. D., Miller, J. S., & Kay, L. E. 1992, ApJ, 397, 452

Wilkes, B. J., Schmidt, G. D., Smith, P. S., Mathur, S., & McLeod, K. K. 1995, ApJ, 455, L13

Wills, B. J. 1989, in BL Lac Objects, ed. L. Maraschi, T. Maccacaro, & M.-H. Ulrich (Berlin, Springer-Verlag), p. 109

Blazar Continuum Variability
ASP Conference Series, Vol. 110, 1996
H. R. Miller, J. R. Webb, and J. C. Noble, eds.

Frequency Dependent Polarization Variability of AGN

Gang Bao and Paul J. Wiita

Department of Physics & Astronomy, Georgia State University, Atlanta GA 30303, USA

Petr Hadrava

Astronomical Institute, Academy of Sciences, 251 65 Ondřejov, Czech Republic

Abstract. We report preliminary work on polarization variability due to the orbital motion of spots on accretion disks and general relativistic effects. Both the observed degree of linear polarization and the angle of the plane of the polarization of a single clump or spot changes rapidly within an orbital period, with gravitational bending and special relativistic aberration playing a vital role in these changes. The amplitude of the change depends on the eccentricity, the radius, r, and the inclination of the spot orbit, while in the Newtonian case no significant changes in either the degree or plane of polarization are likely to occur. Since the temperature distribution of spots is likely to be a rapidly varying function of their distances from the central black hole, the polarization variability of a collection of spots should be frequency-dependent or energy-dependent. Thus, detections of frequency-dependent variability in a source might serve as evidence for the existence of a black hole in the system.

1. Introduction

The inner regions of accretion disks around black holes are probably "patchy" and "noisy" as shown by the observed chaotic X-ray and UV lightcurves of both active galactic nuclei and Galactic black hole candidates. Non-uniform features of an accretion disk in both the X-ray and UV/optical bands can be approximated by a group of "spots" orbiting around the central black hole (e.g., Abramowicz et al. 1989; Mangalam & Wiita 1993). The observed anti-correlation of luminosity and variability amplitude in X-ray emission of Seyferts (Lawrence & Papadakis 1993) can also be understood in this disk fluctuation scenario (Bao & Abramowicz 1996).

In both the radio and optical bands temporal and frequency changes in fractional polarization and in polarization angle of AGN have been observed (e.g., Ballard et al. 1990; Hughes, Aller, & Aller 1985). Modeling of polarization variations has concentrated on radio emission using shock-in-jet models (e.g., Hughes et al. 1985; Gopal-Krishna & Wiita 1992). General relativistic modifications to disk polarization were first considered by Stark and Connors (1977).

2. Polarization variability of an orbiting spot in General Relativity

Assume a spot is on an eccentric orbit around a Schwarzschild black hole with a eccentricity of e, semimajor axis a, and the orbit has an inclination i. The mass of the central black hole is taken as the unit of length.

2.1. Equations of geodesic motion

The four-velocities for both the source and the photon are expressed as

$$\begin{aligned} u^t &= \frac{r}{r-2}E, \\ u^r &= \left[E^2 - \frac{r-2}{r}(\mu + r^{-2}b^2)\right]^{1/2}, \\ u^\theta &= r^{-2}\left(b^2 - l^2/\sin^2\theta\right)^{1/2}, \\ u^\phi &= r^{-2}l/\sin^2\theta, \end{aligned} \tag{1}$$

where $\mu = 0$ or 1 correspond to photons and emitting matter respectively, and E, b and l are the standard constants of motion.

Explicitly for photons ($\mu = 0$), one defines $B = b/E$ as the total impact parameter of a photon, while L is the impact parameter around the symmetry axis. Since photon motion is planar, it can be written as $L = -B \sin i \sin\psi'$. Here $\psi' = \sin\phi_s(1 - \cos^2\phi_s \sin^2 i)^{-1}$ is the observed position angle of the orbiting source (Bao, Hadrava, & Østgaard 1994) with ϕ_s the angular location of the spot. Thus photon motion in Schwarzschild space-time is solely described by the impact parameter B.

The non-zero components of the velocity of the spot measured by a static observer are $v_1 = \left[1 - \frac{r-2}{r}\left(\frac{1}{E_s^2} + \frac{1}{r^2}\left(\frac{l_s}{E_s}\right)^2\right)\right]^{1/2}$ and $v_2 = \sqrt{1 - \frac{2}{r}\frac{1}{r}\frac{l_s}{E_s}}$, where $v_1 \equiv v^{(r)}$ and $v_2 \equiv v^{(\phi)}$ in units of c, the speed of light. The energy and the angular momentum of the orbiting spot are E_s and l_s, respectively.

2.2. The polarization angle

For the polarization vector parallel to the spot's orbital plane, we have

$$p^{\hat\mu} = \frac{1}{\sqrt{u^{\hat t 2} - u^{\hat\theta 2}}}(0, \quad -u^{\hat\phi}, \quad 0, \quad u^{\hat r}), \tag{2}$$

where $u^{\hat\mu}(\mu = t,\ r,\ \theta,\ \phi)$ is the four-velocity of the photon measured in the comoving frame. The transformation of any vector $\vec{f}$ between the static frame and comoving frame can be carried out through

$$f^{(\mu)} = \Lambda^{(\mu)}_{\hat\nu} f^{\hat\nu}, \tag{3}$$

where

$$\Lambda^{(\mu)}_{\hat\nu} = \begin{pmatrix} \gamma & \gamma v_1 & 0 & \gamma v_3 \\ \gamma v_1 & (\gamma-1)\frac{v_1^2}{v^2}+1 & 0 & (\gamma-1)\frac{v_3 v_1}{v^2} \\ 0 & 0 & 1 & 0 \\ \gamma v_3 & (\gamma-1)\frac{v_3 v_1}{v^2} & 0 & (\gamma-1)\frac{v_3^2}{v^2}+1 \end{pmatrix} \tag{4}$$

and $\gamma = \frac{1}{\sqrt{1-v^2}}$, $v^2 = v_1^2 + v_3^2$. Since photon motion is planar, any angle measured in the static frame should be the same as that measured by an observer at infinity. Therefore, it is only necessary to derive the polarization angle in the static frame. One can get the polarization vector measured in the static frame through $p^{(\mu)} = \Lambda^{(\mu)}_{\hat{\nu}} p^{\hat{\nu}}$.

The unit normal vector $\vec{n}$ of the plane in which photon moves is,

$$n^{(\nu)} = \frac{1}{\sqrt{u^{(t)2} - u^{(r)2}}}(0,\ 0,\ -u^{(\phi)},\ u^{(\theta)}). \tag{5}$$

Thus, the angle between the polarization vector and normal to the photon-motion plane is

$$\psi = \cos^{-1}[n^{(\theta)} p_{(\theta)} + n^{(\phi)} p_{(\phi)}]. \tag{6}$$

Finally the observed angle of the polarization plane is:

$$\psi_p = \psi' + \psi \pm \frac{\pi}{2}; \tag{7}$$

where "$\pm$" means that ψ_p describes the orientation of the polarization plane. Both the normal vector to photon plane, $n^{(\nu)}$, and the polarization vector, $p^{(\nu)}$, are expressed in terms of the four-velocity of the photon, which is solely determined by the total impact parameter, B. Since B is a function of i, ϕ_s, e, and r, i.e., $B = B(i; \phi_s; e; r)$, we have $\psi_p = \psi_p(i; \phi_s; e; r)$. As there is no analytic expression for B, it must be solved for numerically (Bao et al. 1994). Note that B is a strong function of r.

2.3. The degree of polarization

As an example, assume that the polarization arises from electron scattering. In this case, the polarization vector lies in the disk plane and the degree of polarization δ is a strong function of the angle between the normal of the disk and the direction in which photon is emitted (Chandrasekhar 1960). This δ changes from 11.7% when the disk is edge-on to 0 when the disk is face-on. The angle measured in the comoving frame is found from:

$$\cos\hat{\theta} = \frac{p^{\hat{\theta}}}{p^{\hat{t}}} = \frac{\sqrt{(\frac{B}{r})^2 - (\frac{L}{r})^2}}{\gamma(\frac{1}{\sqrt{1-\frac{2}{r}}} - \frac{v_1^2}{v}\sqrt{(1-\frac{2}{r})^{-1} - (\frac{B}{r})^2} - \frac{v_3^2}{v}\frac{L}{r})}. \tag{8}$$

Again $\hat{\theta} = \hat{\theta}(i; \phi_s; e; r)$ and thus the degree of polarization $\delta = \delta(i; \phi_s; e; r)$. Although the degree of polarization, δ, is a Lorentz invariant (i.e., it does not change along a given ray), it changes during the orbital period in which different beams of photon are emitted. Fig. 1 shows the values of δ and ϕ_p plotted against ϕ_s for several values of r and i.

3. Polarization properties of an orbiting spot in Special Relativity

In flat space time photons go along straight lines, and an analytic expression for the total impact parameter B can be found. For the simplest case of a circular

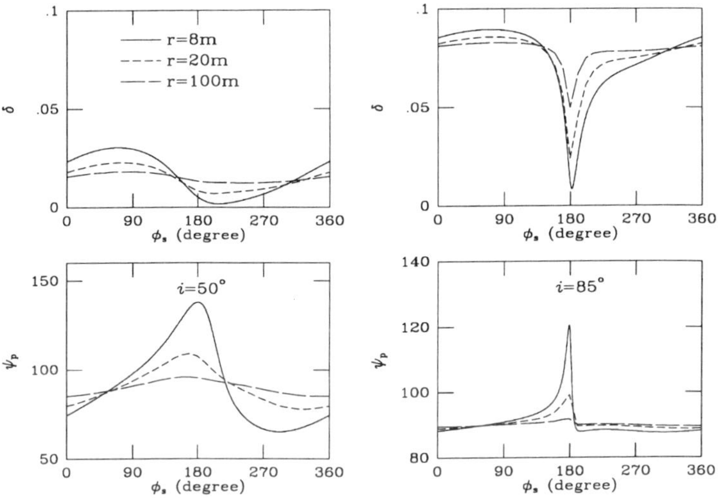

Figure 1. Variation of both polarization and plane of polarization with orbital phase for different inclinations and different radii. The source is on a circular orbit.

orbit ($v_1 = 0,\ v_3 = v = \frac{1}{\sqrt{r}}$) and where the polarization vector lies in the disk plane, we find the angle between polarization vector and the plane of the photon is

$$\psi = \cos^{-1}\left[\cos\psi' \frac{\sin i\gamma}{\sqrt{\gamma^2(1 + v\sin i\sin\phi_s)^2 - \cos^2 i}}\right], \tag{9}$$

where $\gamma = \frac{1}{\sqrt{1-v^2}} = \sqrt{\frac{r}{r-1}}$. We thus see that ϕ_p is a weak function of R, especially when $r \gg 1$. For slow motion of a spot, $v \sim 0,\quad \gamma \sim 1$, so $\psi \approx \psi'$. The physical solution of the polarization angle is

$$\psi_p = \pm\frac{\pi}{2}, \tag{10}$$

implying *that polarization lies in the disk plane* and no rotation occurs during the orbital motion of spot. These conclusions do not depend on the radial location of the spot r at all. For low inclinations, $\psi_p = \phi_s$, i.e., ψ_p changes linearly with the orbital motion of the spot. When the polarization vector is perpendicular to disk's plane, the above basic results still hold. Similarly, we find

$$\hat{\theta} = \cos^{-1}\left[\cos i \frac{1}{\gamma(1 + \frac{v}{r}\sin i\sin\phi_s)}\right]. \tag{11}$$

When $r \gg 1$ we get (unsurprisingly), $\hat{\theta} = i$. This implies that the degree of polarization does not depend on r either.

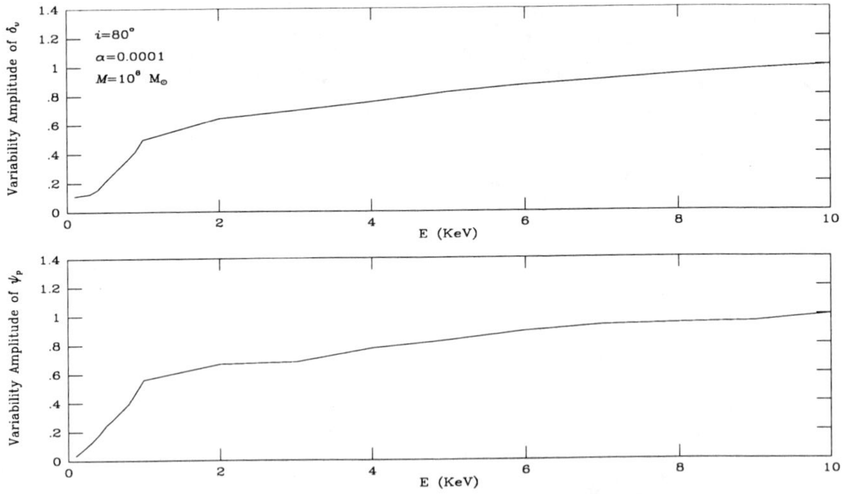

Figure 2. The variability amplitude of δ_ν and variability amplitude of φ_p at 2×10^{-3} Hz plotted as a function of energy. One hundred sources were uniformly distributed on the surface of a Novikov-Thorne disk from $r_{in} = 4r_g$ to $r_{out} = 60r_g$. All sources have equal lifetimes, taken as $4P(r_{out})$.

4. Frequency-dependent variability of polarized spots

Now consider many orbiting spots on the disk surface which have a power law distribution in radius but are distributed randomly in orbital phase. The polarization of radiation from spots can be described by the normalized Stokes parameters, i.e., $u_\nu = U_\nu/I_\nu$, and $q_\nu = Q_\nu/I_\nu$, where I_ν is the intensity at frequency ν. The degree of polarization is $\delta_\nu = \sqrt{u_\nu^2 + q_\nu^2}$, and the angle of the plane of polarization is $\psi_p = \frac{1}{2}\tan(u_\nu/q_\nu)$.

The Stokes parameters (u_ν, q_ν) and the intensity of the spots are calculated through the following formulæ:

$$I_\nu = \sum_{i=1}^{n} \int g_i^3 I_\nu^i d\Pi_i \,, \tag{12}$$

$$u_\nu = \frac{1}{I_\nu} \sum_{i=1}^{n} \int g_i^3 \delta_i \sin(2\psi_p^i) I_\nu^i f(\hat{\theta}) d\Pi_i \,, \tag{13}$$

and

$$q_\nu = \frac{1}{I_\nu} \sum_{i=1}^{n} \int g_i^3 \delta_i \cos(2\psi_p^i) I_\nu^i f(\hat{\theta}) d\Pi_i \,. \tag{14}$$

Here ψ_p^i and δ_i are the polarization angle and the degree of the polarization of the ith spot and Π_i is the solid angle that spot subtends at the observer.

The function $f(\hat{\theta}) = (1 + a\cos\hat{\theta})\cos\hat{\theta}^b$ describes the angular dependence of the radiation, where a and b are constants (both taken as 0 below). The redshift factors, g_i, are the ratios of photon energies measured at infinity to the energies measured in the comoving frame of each spot:

$$g = \frac{p^t}{p^{\hat{t}}} = \left[\gamma\left(\frac{1}{\sqrt{1-\frac{2}{r}}} - \frac{v_1^2}{v}\sqrt{\left(1-\frac{2}{r}\right)^{-1} - \left(\frac{B}{r}\right)^2} - \frac{v_3^2}{v}\frac{L}{r}\right)\right]^{-1}. \tag{15}$$

We suppose the specific intensity on the surface of a blob (in its proper frame) to be given by its LTE thermal radiation, $I_\nu = B_\nu(T)\,[1 - \exp(-\alpha s)]$. Here B_ν is the Planck's function, $\alpha = \alpha_\nu(T, N)$ is the opacity of the blob (with the temperature T and particle density N) and s is the typical size of the blob. This formula leads to the black-body radiation B_ν in the optically thick case $\alpha s \gg 1$ and to $\alpha s B_\nu$ in the optically thin case.

Fig. 2 illustrates the results for an optically thick accretion disk with α viscosity prescription (Novikov & Thorne 1973) around a $10^6 M_\odot$ black hole with $\alpha = 0.0001$, $i = 80°$. We see that significantly more polarization variability occurs at higher frequencies.

Unfortunately, X-ray polarization measurements cannot currently be made, but future instrumentation may be able to detect such fluctuations. We are performing computations for the observationally accessible variations in the UV/optical wavebands and including more detailed physics for polarization production and radiative transfer.

Acknowledgments. This work was supported in part by NASA grant NAG 5-3098 and Chancellor's Initiative funds at Georgia State University.

References

Abramowicz, M. A., Bao, G., Lanza, A., & Zhang, X.-H. 1989, in Proceedings of the 23rd ESLAB Symposium, eds. J. Hunt & B. Battrick (Noordvijk, ESA) p. 871

Ballard K. R., Mead, A. R. G., Brand, P. W. J. L., & Hough, J. H. 1990, MNRAS, 243, 640

Bao, G., & Abramowicz, M. A. 1996, ApJ, 465, in press

Bao, G., Hadrava, P., & Østgaard, E. 1994, ApJ, 425, 63

Chandrasekhar, S. 1960, Radiative Transfer (New York, Dover)

Gopal-Krishna & Wiita, P. J. 1992, A&A, 259, 109

Hughes, P. A., Aller, H. D., & Aller, M. F. 1985, ApJ, 298, 301

Lawrence, A., & Papadakis, I. 1993, ApJ, 414, L85

Mangalam, A. V., & Wiita, P. J. 1993, ApJ, 406, 420

Novikov, I. D., & Thorne, K. S. 1973, in Black Holes, eds. C. DeWitt & B. S. DeWitt (New York, Gordon & Breach) p. 343

Stark, R. F., & Connors, P. A. 1977, Nature, 266, 429

Blazar Continuum Variability
ASP Conference Series, Vol. 110, 1996
H. R. Miller, J. R. Webb, and J. C. Noble, eds.

Liberated Energy and Storage Timescale Comparison of the BL Lac Object AO 0235+164

Pamela Jenkins

Department of Physics and Astronomy, University of Oklahoma; Department of Physics, Florida International University

Abstract. A time series analysis was performed on the optical data of the BL Lac object AO 0235+164. Its long-term light curve exhibited no strict periodic characteristics. However, there did appear to be a similarity in the nearly-regular outbursts on longer timescales. The energy released during outbursts was compared to the storage timescales in order to test AO 0235+164 as a relaxation oscillator. Any correlation between the storage time and the energy liberated is discussed.

1. Introduction

The study of multifrequency quasar light curves can provide astronomers with important information about the processes that drive these non-linear dynamic systems. The energy released and the timescales that follow give an estimate of the time needed to re-fuel the outburst mechanisms. Still, the diverse nature of quasar morphology and optical variability create modeling problems that have yet to solved.

2. Time-series and spectral analysis

The AO 0235 data is unequally spaced; therefore, fast fourier transform (FFT) algorithms which are commonly used on evenly sampled data were not be applied. Deeming (1975) used a discrete fourier transform (DFT) to handle unequally spaced data. The frequency spectrum of a continuous function of time can be measured at a finite number of discrete times. (Roberts et al. 1987). The long-term optical light curve of AO 0235 is shown in Figure 1.

Observable functions, such as the dirty spectrum and the spectral window, are essential for detecting aliases and spurious features due to sampling. The CLEAN algorithm utilizes the dirty spectrum and spectral window to non-linearly deconvolve the power spectrum. Time series analysis of a few quasars have resulted in claims of periodic behavior. However, the clean spectrum of AO 0235 only shows moderately strong components. Although Webb et al. (1988) found no strong features in the power spectrum, several more years of data have been collected. Current attempts to model AO 0235's light curve have not adequately characterized the optical variations observed.

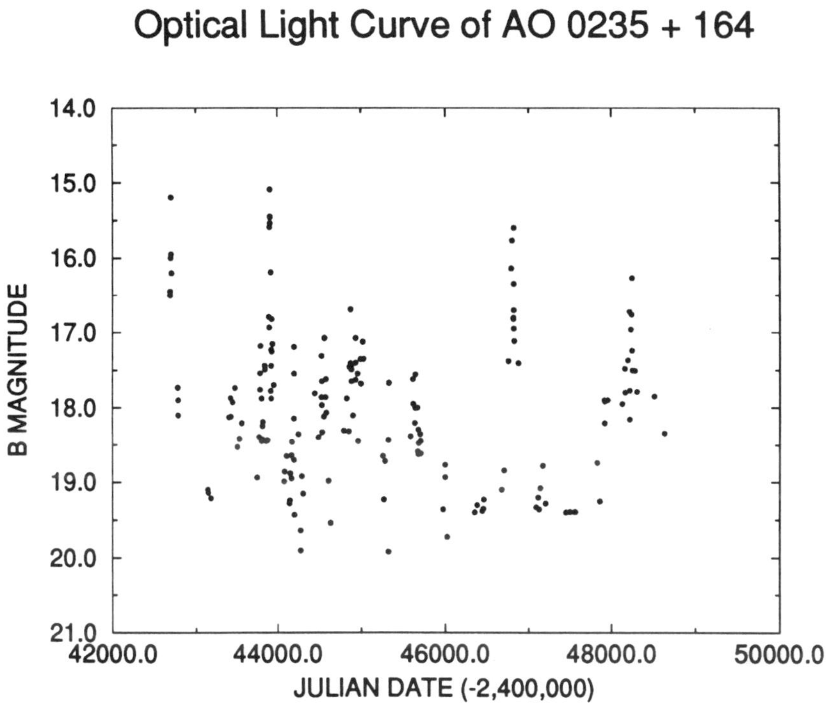

3. Morphology of outbursts

The similar shapes and timescales of the outbursts indicate non-random variations. A thorough investigation of the energy released and the corresponding timescales may shed some light on the seemingly regular occurance of AO 0235's larger outbursts. About four years separate the first two bursts as well as the last two bursts. Approximately eight years lie between the second and third outburst. A closer inspection of the four large outbursts in AO 0235 reveals a double-peak phenomenon that is seen in various other astronomical objects, such as X-ray bursters, dwarf novae, and supernovae. Three of the four bursts exhibit this complex pulse structure. This observation has created an entirely new set of questions. One area of interest is the relationship between the energy of the burst ΔE and the time to the following burst Δt. If ΔE is proportional to Δt, then AO 0235 may be a relaxation oscillator (Brecher et al. 1977). Table 1 lists various parameters, such as log energy, time between bursts, and burst duration. Figure 2 shows the relationship between the time to the following burst and log energy. The data points along the $t = 0$ line correspond to areas of rise and decline that were above selected baselines.

4. Discussion

The timescales for BL Lac objects are much longer than that of X-ray bursters. The longer timescales, combined with the sparceness of observations due to

Table 1. AO 0235 Energy-Time Relations.

Outburst	Time Since Previous Burst (Days)	Burst Duration (Days)	log E (ergs)
1	0	1453.80	54.41
2	690	592.07	54.23
3	2385	1928.63	54.50
4	465	3060.02	54.70

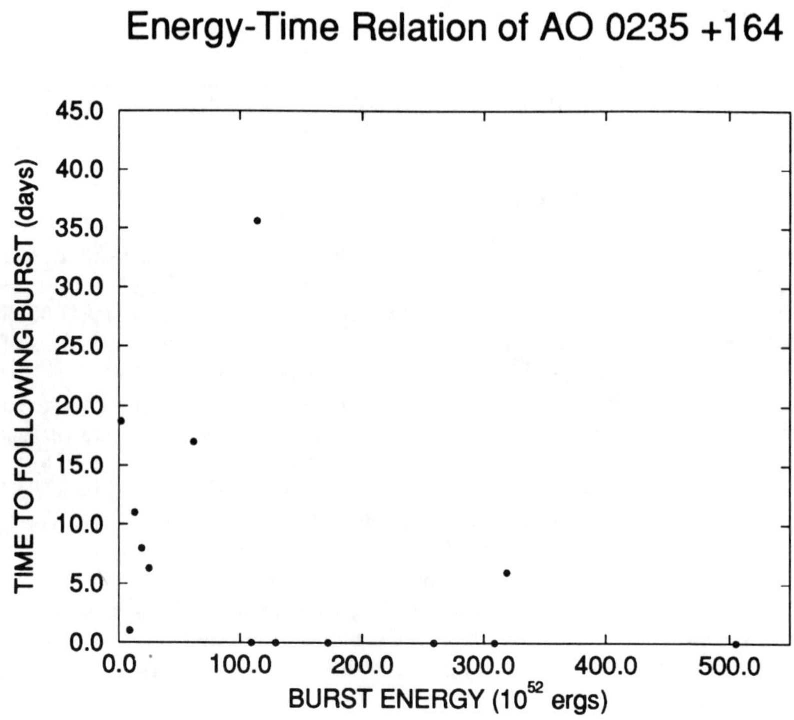

the faintness of the source, make it difficult to accurately test these objects. Small outbursts can be lost due to long gaps in the light curves, while many of the larger bursts are separated by poorly resolved smaller bursts. AO 0235's outbursts exhibit roughly similar ΔE-Δt relations, which suggests that it has a relatively steady mass transfer, storage, and triggering instability (Jenkins 1995).

5. Conclusions

More outburst data is needed to truly test AO 0235 as a relaxation oscillator. Then a more detailed analysis can be performed using the energy of the burst and the time to the following burst. Regardless of the differences between BL Lac objects, X-ray bursters, dwarf novae, or supernovae, they all have accretion in strong magnetospheres. The multiple peaks that AO 0235 and the other various objects exhibit in their light curves may describe the process or structure in which matter accretes into the magnetosphere and forces the release of energy.

Acknowledgments. I am very grateful to Dr. James R. Webb for his guidance, patience, and support.

References

Brecher, K., Morrison, P., & Sadun, A. 1977, ApJ, 217, L139

Deeming, T. J. 1975, Ap&SS, 36, 137

Jenkins, P. J. 1995, Master's thesis, Florida International University

Roberts, D. H., Lehar, J., & Dreher, J. W. 1987, AJ, 93, 968

Webb, J. R., Smith, A. G., Leacock, R. J., Fitzgibbons, G. L., Gombola, P. P., & Shepherd, D. W. 1988, AJ, 95, 374

Blazar Continuum Variability
ASP Conference Series, Vol. 110, 1996
H. R. Miller, J. R. Webb, and J. C. Noble, eds.

Powering the Broad Hα Line in BL Lacertae

Elizabeth Corbett, Andrew Robinson, James Hough, and Stuart Young

Dept. of Physical Sciences, University of Hertfordshire, College Lane, Hatfield, Herts. AL10 9AB, UK

David J. Axon

Affiliated to the Space Sciences Division of ESA, Space Telescope Science Institute, 3700 San Martin Drive, Baltimore, MD 21218 (on leave from National Radio Astronomy Observatory, University of Manchester, Jodrell Bank, Macclesfield, Cheshire SK119DR, UK)

Abstract. We present high quality optical spectropolarimetry of BL Lacertae confirming the presence of a broad Hα emission line which is comparable in width (FWHM$\approx$ 4000 km s^{-1}) and luminosity ($L_{H\alpha} \sim 10^{41}$ erg s^{-1}) to those of Seyfert Type 1 nuclei. We tentatively conclude that this feature must have increased in luminosity by a factor $\sim$ 4 since earlier spectroscopic observations. The beamed synchrotron continuum associated with the radio source can account for the observed equivalent width, but the fact that the line has not previously been detected suggests either a sudden injection of cold gas into the radiation field, or the strengthening of an additional ionizing radiation source. Using a simple model for the illumination of the broad line region (BLR) we show that the Hα emission could be powered by thermal radiation from an accretion disk, without significantly affecting the shape or polarisation of the optical continuum.

1. Introduction

Among blazars a distinction is generally drawn between optically violent variable or high polarization quasars (which for convenience we hereafter refer to collectively as OVV quasars) and BL Lacertae objects (BLOs), which lack the characteristic broad emission lines of quasars and whose optical-IR continuum appears to be pure synchrotron emission. This simple empirical distinction has, however, become somewhat blurred by the detection of broad emission lines in the spectra of several sources which were originally classified as BLOs (e.g., OJ287, Sitko & Junkkarinen 1985; Mkn 501, Moles et al. 1987; 1308+326, Stickel et al. 1993; and PKS 0521–36, Ulrich 1981). While it is possible that some of these cases can be attributed to mis-classifications, the recent discovery of a broad Hα line in the *class prototype*, BL Lacertae itself (Vermeulen et al. 1995), now demands a revision of the conventional picture of these objects and their relationship with quasars. In particular it raises the question of whether the designation “BL Lacertae Object” reflects genuinely different physical properties or, for example, describes an evolutionary state, as might be the case if the broad-line emission is episodic.

Here we present optical spectropolarimetry of BL Lacertae obtained two weeks after Vermeulen et al.'s (1995) observations and discuss the merits of alternative models in which the line emission is powered, respectively, by beamed synchrotron emission from the relativistic jet and by thermal emission from an accretion disk. This work is described in more detail by Corbett et al. (1996).

2. Observations and analysis

Spectropolarimetric observations of the Hα region of BL Lacertae were made with the 4.2-m William Herschel Telescope using the ISIS dual beam spectrograph (giving a dispersion of 3Å/pixel) on the nights of June 4–5 and June 6–7. The total exposure time in each case was 2400 s. The strength of the broad Hα line does not differ significantly between the two data sets and hence for brevity measurements made from the average spectrum are given here. After correcting for atmospheric absorption and line-of-sight reddening (assuming E(B–V)=0.3), the Hα+[NII] feature was decomposed by fitting gaussians (Fig. 1). The broad Hα component has a FWHM $\approx 4000 \pm 200\,\mathrm{km\,s^{-1}}$ and an equivalent width $W_{\mathrm{H}\alpha} \approx 5.5 \pm 0.3$Å. Its integrated flux is $3.0 \pm 0.2 \times 10^{-14}\,\mathrm{erg\,s^{-1}\,cm^{-1}}$, corresponding to a luminosity of $1.5 \times 10^{41} h^{-2}\,\mathrm{erg\,s^{-1}}$ where $h = H_0/100\,\mathrm{km\,s^{-1}\,Mpc^{-1}}$. The Hα line is very similar in luminosity and FWHM to that of the classical Type 1 Seyfert NGC4151. The linear polarization varies only slightly over the spectrum and has an average of 4.39± 0.04%, among the lowest recorded values in this wavelength range (Sitko, Schmidt, & Stein 1985). Simultaneous photometric observations show that during the period 4–9 June BL Lacertae underwent night-to-night variations by 0.1 magnitude within the range $15.8 \leq V \leq 15.5$. These measurements place BL Lacertae at the lower end of its typical variability range ($14 \leq V \leq 16$; e.g., Carini et al. 1992), but also show that it was not exceptionally faint.

2.1. Should the line have been seen before?

One major question which will critically effect how we interpret this result is: has the broad Hα line appeared as a genuine new feature in the spectrum or has it always been present, having previously escaped detection? We estimate that the broad Hα would not have been discernable in our own data if the optical continuum had been significantly brighter ($\sim 1.5 - 2^m$) than it actually was during our observations. Nevertheless, BL Lacertae was not unsually faint and so if the broad line only becomes visible when the beamed optical continuum is weak, we would expect that it should have been detected in earlier spectra if it was indeed present at a luminosity comparable with what we measured in June 1995. In fact, previous spectra of BL Lacertae, (e.g., Miller & Hawley 1977, Stickel et al. 1993) show no easily identifiable broad emission line features, even when the source had a similar V magnitude. We estimate that the broad Hα feature would have been undetectable in our own spectra if reduced to ~ 25% of its observed flux and therefore, taking its absence from earlier spectra at face value, we conclude that it must have increased in luminosity by a factor ~ 4.

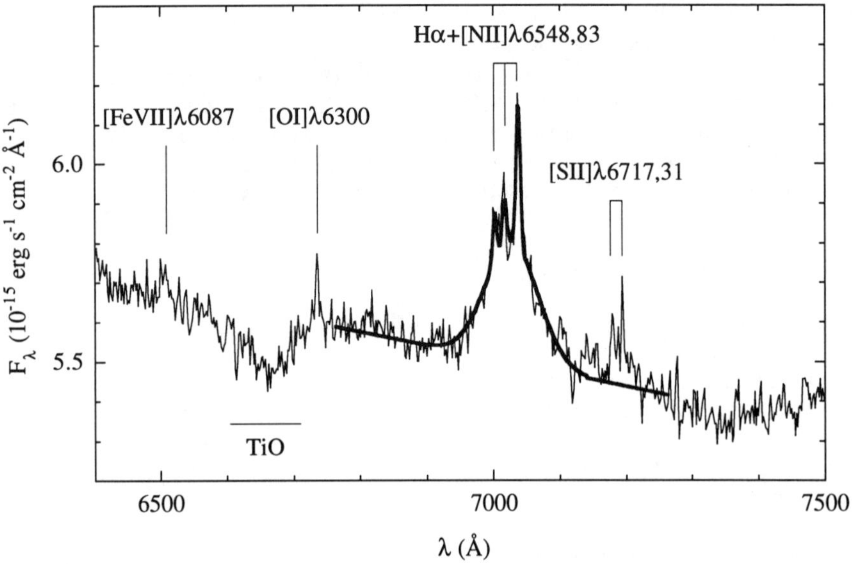

Figure 1. The emission line spectrum of BL Lacertae. Shown here is the average of the spectra obtained on the nights of 1995 June 4–5 and 6–7. The thick lines show the multi-Gaussian fits to the broad Hα+[NII] blend.

2.2. Powering the broad emission line

The broad emission lines exhibited by AGN are believed to arise from photoionization of cool, dense, rapidly moving gas clouds (the broad line region, or BLR) by the central continuum source. The inferred increase in the luminosity of the Hα line in BL Lacertae could therefore be attributed either to an increase in the BLR covering factor, or assuming that enough cool gas is already present, to an increase in the level of illumination by the ionizing continuum. Two main continuum sources are believed to operate in radio-loud quasars: a relativistic jet emitting Doppler-boosted synchrotron radiation, and the hot inner regions of the black-hole accretion disk, emitting thermal radiation in the EUV.

The pure synchrotron illumination model We first consider whether the broad Hα line in BL Lacertae can be explained in terms of illumination of the BLR by the Doppler-boosted synchrotron continuum alone. We model the continuum source as two identical relativistic jets pointing in opposite directions along the same axis. The synchrotron continuum emitted by the jets is represented by the power law $f_\nu \propto \nu^{-\alpha}$ and is Doppler-boosted by the factor $D^{3+\alpha}$, where $D = [\gamma(1-\beta\cos i)]^{-1}$, β is the jet velocity in units of c, i the jet inclination to the line of sight and γ the Lorentz factor. For the spectral index we take $\alpha = 2$ (e.g., Bregman et al. 1990). The observed superluminal motion of the radio source (Mutel et al. 1990) implies that $i \leq 34°$ and $\gamma \geq 3.4$. The BLR is modelled as a uniform distribution of clouds arranged, with a covering factor $C = 0.1$, in a thin spherical shell centerd on the continuum source. The Hα

emissivity is given by $\epsilon_s(\mathrm{H}\alpha) \propto CF(\theta)\,\mathrm{max}[1, \delta]$, where $F(\theta)$ is the photon flux impinging on the shell at an angle θ from the jet axis and $\delta(\propto 1/F(\theta))$ is the ratio of the column density of the clouds to the Strömgren column.

If the BLR is everywhere radiation bounded, i.e., $\delta \geq 1$, for all θ, the measured equivalent width of the broad Hα line can easily be obtained for $\gamma \approx 4$ and $i \approx 20°$. However as both the line luminosity and the underlying continuum are directly proportional to the intrinsic synchrotron power, $W_{\mathrm{H}\alpha}$ would be independent of the continuum luminosity. It is unlikely that a broad Hα line of *constant* equivalent width would have escaped detection in all previous observations so the appearance of the broad Hα line cannot be explained simply by an increase in the beam power with γ fixed.

An alternative possibility is that γ varies, producing changes in D. If the BLR remains radiation bounded, the line luminosity is an increasing function of γ due to enhanced Doppler-boosting of the continuum. On the other hand, the beam becomes narrower so the *observed* continuum flux is initially boosted but then drops rapidly when the value of γ is such that the effective opening angle excludes the line of sight. As a result, relatively small changes in the bulk velocity of material moving along the jet can cause large variations in $W_{\mathrm{H}\alpha}$ with, typically, a decrease in the continuum by $\sim 0.1^m$ corresponding to a doubling of $W_{\mathrm{H}\alpha}$. This is difficult to reconcile with the lack of an earlier detection since we would expect a strong line to appear when BL Lacertae is in a low state.

If the BLR is matter bounded near the beaming axis (i.e., $\delta < 1$ for $\theta < \gamma^{-1}$), some ionizing photons pass through the BLR without being absorbed and somewhat larger values of $\gamma(\approx 4.5)$ and $i(\approx 23°)$ are needed to obtain the observed $W_{\mathrm{H}\alpha}$. For BLR properties typical of Seyferts, these values remain within the limits implied by the superluminal motion but we would still expect to see higher $W_{\mathrm{H}\alpha}$ when the continuum is faint.

Hence although the observed $W_{\mathrm{H}\alpha}$ can be explained by illumination of the BLR by the beamed synchrotron continuum, this model also predicts that it increases or at least remains constant as the observed continuum fades. It follows that the line should have been at least as strong in previous spectra obtained when BL Lacertae was as faint as it was in June 1995. If the broad Hα line is indeed a new feature in the spectrum as we have argued, then its luminosity be cannot be directly related to the intensity of the synchrotron beam and must be attributed either to a sudden injection of cool dense gas into the radiation field or to the strengthening of a second ionizing source which does not contribute substantially to the optical continuum.

An additional continuum source—the accretion disk In Seyfert 1 nuclei and quasars the ubiquitous flattening of the optical continuum shortwards of 4000Å (known as the "big blue bump") is widely attributed to thermal radiation from the accretion disk and is invariably associated with strong emission lines. Although the Doppler-boosted synchrotron emission dominates over much of the IR–optical range in OVV quasars, signatures of the big blue bump are often seen at shorter wavelengths (e.g., 3C345, Smith et al. 1986). An accretion disk is therefore an obvious candidate for an additional ionizing continuum source in BL Lacertae. To investigate this possiblity we have altered our model to include a simple isotropic accretion disk continuum of the form $f_\nu \propto \nu^{1/3}e^{-h\nu/kT}$, where we take $T = 1.5 \times 10^5$, a value comparable to those inferred for Seyfert 1 nuclei.

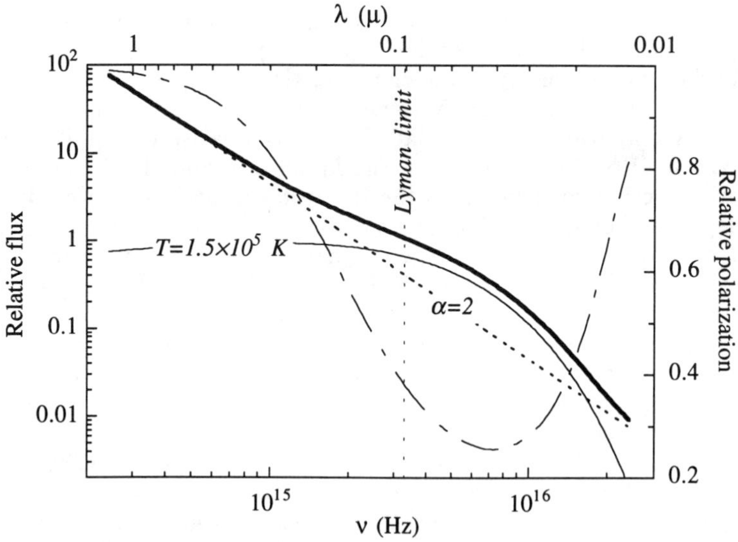

Figure 2. A model in which the continuum spectrum (thick line) has contributions from both a synchrotron beam (dotted line) and a hot accretion disk (thin line), with the beam intensity 25× that of the disk emission at Hα. The dot-dashed line shows the depolarization which would result from dilution of the pure synchrotron continuum (having unit intrinsic polarization) if the disk emission is unpolarized.

The observed continuum flux density is now the sum of the synchrotron and the disk continua but, for simplicity, we assume that BLR is illuminated only by the latter so that for a given covering factor, the Hα line luminosity is determined solely by the power of the thermal source (we might imagine, for instance, that the synchrotron beam shines through a hole in the BLR).

For the combined continuum the observed value of $W_{\mathrm{H}\alpha}$ is obtained when the flux density of the synchrotron continuum $\approx 25\times$ that of the thermal continuum at the wavelength of Hα. This result is effectively independent of γ and i but is sensitive to the disk temperature; as T increases, relatively more ionizing photons are produced and the observed $W_{\mathrm{H}\alpha}$ is achieved with an even smaller thermal contribution to the continuum under Hα.

A thermal continuum sufficiently powerful to produce the observed broad Hα emission by photoionization could therefore be accommodated within the optical spectrum of BL Lacertae without easily detectable modifications to the spectral shape or polarisation, at least for wavelengths > 4000Å (Fig. 2). The line emission would be weakly related to the overall optical continuum brightness which would still be dominated by Doppler-boosted synchrotron emission. Changes in the Hα luminosity could be driven by flares in the accretion disk which are unlikely to be strongly correlated with the variability of the synchrotron beam, so the broad Hα line would not inevitably appear as the optical continuum fades.

3. Conclusion

BL Lacertae is now known to exhibit broad Balmer lines in its spectrum, and our observations show that while the broad Hα line has a relatively small equivalent width, it is comparable in luminosity and width to those of Seyfert 1 nuclei. Given that this feature is apparently absent in earlier spectra obtained when the source had a similar V magnitude, we infer that the line luminosity has increased by a factor $\sim$ 4. The observed equivalent width can be explained if the BLR is illuminated by a Doppler-boosted synchrotron continuum from relativistic jets, but if we take the lack of previous detections of the line at face value, its subsequent appearance is difficult to explain and suggests an injection of gas into the radiation field or the strengthening of an additional ionizing continuum source. A thermal accretion disk can account for the observed equivalent width without easily detectable consequences for either the polarisation properties or the slope of the optical continuum. If the latter explanation is correct, it follows from unified schemes relating BLOs to a parent population of FR I radio galaxies (e.g., Browne 1983) that the latter must also harbour Seyfert-like nuclei including broad emission line regions and accretion disks.

Acknowledgments. We thank R. D. Jeffries and M. R. Thurston for providing us with their photometric data. EAC acknowledges a PPARC reseach studentship and AR thanks the Royal Society for financial support.

References

Bregman, J. D., et al. 1990, ApJ, 352, 574

Browne, I. 1983, MNRAS, 214, 23p

Carini, M., Miller, H., Noble, J., & Goodrich, B. 1992, AJ, 104,15

Corbett, E., Robinson, A., Axon, D., Hough J., Jeffries, R., Thurston, M., & Young, S. 1996, MNRAS, submitted

Miller, J., & Hawley, S. 1977, ApJ, 212, 47

Moles, M., Masegosa, J., & del Olmo, A. 1987, AJ, 94, 1143

Mutel, R. L., Phillips, R. B., Su, B., & Bucciferro, R. R. 1990, ApJ, 352, 81

Sitko, M., & Junkkarinen, V. 1985, PASP, 97, 1158

Sitko, M., Schmidt, G., & Stein, W. 1985, ApJS, 59,323

Smith, P., Balonek, T., Heckert, P., & Elston, R. 1986, ApJ, 305, 484

Stickel, M., Freidel, J., & Kühr, H. 1993, A&AS, 98, 393

Ulrich, M. 1981, A&A, 103, L1

Vermeulen, R. C., Ogle, P. M., Tran, H. D., Browne, I. W. A., Cohen, M. H., Readhead, A. C. S., Taylor, G. B., & Goodrich, R. W. 1995, ApJ, 452, 5

Blazar Continuum Variability
ASP Conference Series, Vol. 110, 1996
H. R. Miller, J. R. Webb, and J. C. Noble, eds.

Quasi-Automated Remote (QuasAR) Monitoring of AGN in the Center for Automated Space Science

Richard Hackney, Karen Hackney, Roger Scott, and Michael Carini

Center for Automated Space Science & Kentucky Space Grant and NASA EPSCoR Programs, Western Kentucky University, Bowling Green, KY 42101

Abstract. Beginning in 1995, NASA provided funding for a five-year study of Active Galactic Nuclei at Western Kentucky University as part of the establishment of the Center for Automated Space Science (CASS), a NASA University Research Center involving a partnership among Tennessee State University, Western Kentucky University, and South Carolina State University. CASS provides a basis for observing AGN using remotely-operated robotic telescopes located in the desert southwest, with control of the observations and transfer of data accomplished using the Internet. The AGN study involves maintaining a program of frequent photometric monitoring of selected active galactic nuclei to provide time series measurements with 0.02 magnitude precision to magnitude 18 with sampling sufficient to define the nature of variations for constraining physical models for the sources. CASS is developing and testing leading-edge artificial-intelligence scheduling algorithms that coordinate the use of the remote observing facilities among several concurrent programs while optimizing the sampling for each program. The CASS AGN program invites collaboration with space-based studies and other ground-based AGN observing programs. Opportunities for funded student participation in the CASS AGN program are available for members of underrepresented groups through NASA center funding and for all students through the Kentucky Space Grant Consortium.

1. Introduction

Beginning in 1995, NASA provided funding to initiate a five-year study of Active Galactic Nuclei (AGN) as part of the establishment of the Center for Automated Space Science (CASS), which is a NASA University Research Center involving a partnership among Tennessee State University (TSU), Western Kentucky University (WKU), and South Carolina State University (SCSU).

Projects in the CASS include studies of chromospherically active stars, studies of solar-type stars, analysis of the chromospheres of cool giant and supergiant stars, studies of the physics of the interstellar medium, studies of the physics of the earth's ionosphere and the AGN monitoring program discussed below.

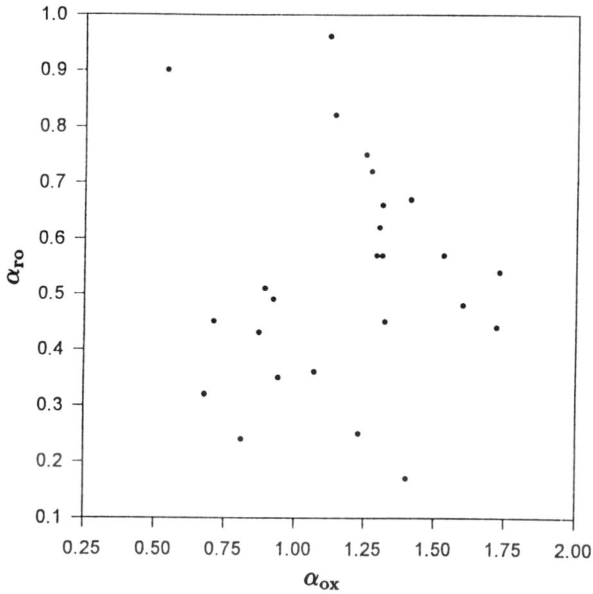

Figure 1. The α_{ro} vs α_{ox} diagram for the CASS sample.

2. The AGN monitoring program

The AGN study is a program of intensive photometric CCD monitoring of a selected set of AGN, with the goal of maintaining a precision of 0.02 magnitude with time-series sampling sufficient to define the nature of variations for constraining source models. One of the particular areas of interest will be to determine if there are differences in the variability characteristics between X-ray strong BL Lacs (X-BLs) and quasar-like BL Lacs (Q-BLs). Our initial target list was compiled with this goal in mind, and is presented in Table 1. Here we list object name, right ascension, declination, type (i.e., RBL or XBL) and redshift. The α_{ro} vs α_{ox} diagram of the majority of our sample is displayed in Figure 1, where α_{ro} and α_{ox} are defined as in Giommi et al. (1990).

This is an initial list. As the telescope comes online, our allocation of time becomes better defined, and our understanding of the system increases, this list will no doubt be modified.

The study will also accommodate cooperative observing to establish optical (*UBVRI*) characteristics during coordinated multi-wavelength studies of particular AGN. We have tried to choose objects which have been the target of previous monitoring campaigns or which the community has indicated are likely candidates for future multiwavelength monitoring efforts. We plan to retain enough flexibility in our observing to support campaigns on particular objects which are not on our primary observing list.

Table 1. Initial CASS AGN Observing List.

Object	RA(1950)	DEC(1950)	TYPE	z
OI 090.4	07 54 22.6	10 04 39.8	R	...
OJ 287	08 51 57.3	20 17 58.2	R	0.306
1011+496	10 11 55.3	49 40 57.2	X	...
MKN 421	11 01 40.6	38 28 43.0	X	0.031
MKN 180	11 33 30.0	70 25 00.0	X	0.046
1156+295	11 56 57.8	29 31 26.0	R	0.729
ON 325	12 15 21.2	30 23 39.8	R	...
ON 231	12 19 01.1	28 30 37.0	R	0.102
3C 273	12 26 33.3	02 19 43.5	R	0.158
3C 279	12 53 35.8	−05 31 08.0	R	0.538
OQ 530	14 18 06.2	54 36 57.1	R	...
1426+428	14 26 35.9	42 53 46.0	X	...
PG 1553+11	15 53 20.8	11 20 30.0	X	0.360
1611+343	16 11 47.9	34 20 20.0	X	...
3C 345	16 41 05.3	39 55 45.3	R	0.595
MKN 501	16 52 11.7	39 50 26.0	X	0.034
I ZW 187	17 27 04.3	50 15 31.2	R	0.055
1ES 1959+650	19 59 30.8	65 08 55.0	X	...
PKS 2155−304	21 55 58.2	−30 27 51.8	X	0.117
BL Lac	22 00 39.4	42 02 08.5	R	0.070

3. Facilities

CASS provides a basis for intensive photometric monitoring of AGN with minimal impact of weather through the use of a 61-cm telescope at Washington Camp in the desert climate of Arizona. The telescope is to be operated remotely, using the Internet to transmit observing scripts and to return CCD images for reduction and analysis. In addition, a 61-cm telescope under dark rural skies near Western Kentucky University will be used for supporting observations for the program and collaborative campaigns, as well as for on-site evaluation of observing and reduction procedures and training of student participants on a "non-virtual" instrument.

4. Input and collaboration

The CASS AGN study is in the organizational phase, with equipment under development and detailed planning of the observing program underway. A major purpose of participating in this conference is to solicit input from the community regarding choices of targets for intensive monitoring that would bear most directly on currently recognized theoretical questions. We seek to coordinate with the entire observational community in optimizing the application of observational resources to get the needed coverage.

It has become very apparent in recent years that no one observatory or observer can obtain enough data to adequately define the optical variations in these objects. The recent monitoring campaigns on PKS 2155−304 and OJ 287 discussed in these proceedings have shown that collaborative efforts are absolutely necessary to define the variability in these objects. To this end, we would like to solicit other observers who have been monitoring these objects to join with us in a collaborative, long term, optical monitoring effort. Data will be compiled and shared in a manner similar to that followed in the multiwavelength monitoring campaigns mentioned above. If you are interested in joining this collaboration, please contact Dr. Michael Carini at carini@wkuvx1.wku.edu.

Acknowledgments. This work has been supported through NASA grants NCCW-0085, NGT-40049 and NCCW-60.

References

Giommi, P., et al. 1990, ApJ, 356, 432

Blazar Continuum Variability
ASP Conference Series, Vol. 110, 1996
H. R. Miller, J. R. Webb, and J. C. Noble, eds.

Optical Monitoring of AGN Using the SARA Telescope: First Results

Rolando Branly

Broward Community College, Davie, FL and Florida International University, Miami, FL

Roy Kilgard

Valdosta State University, Valdosta, GA

Alberto Sadun

Agnes Scott College, Decatur, GA

Alexander Shcherbanovsky

Florida International University, Miami, FL

James Webb

Florida International University, Miami, FL and SARA telescope, Kitt Peak, AZ

Abstract. The SARA Observatory has begun two programs to observe a selected group of Blazars. The first program aims to contribute ground-based optical observations to the general monitoring data bases for these sources. The second is a Target of Opportunity (TOO) program designed to observe Blazars during outbursts or high energy satellite observations. The observatory's automated 0.9-m telescope has already been instrumental in the detection of a dramatic flare in PKS 1156+295, and a subsequent observing run yielded images of BL Lac and 8 other Blazars. The telescope is scheduled to become robotic in late 1996 and all observations will be made robotically after that date.

1. Introduction

In 1992 Florida International University became the fifth member of the Southeastern Association for Research in Astronomy (SARA). The SARA consortium consists of five southeastern Universities: Florida Institute of Technology, University of Georgia, Valdosta State University, East Tennessee State University and FIU. The consortium was formed in 1990 for the sole purpose of acquiring the de-commissioned 0.9-meter NOAO telescope from Kitt Peak National Observatory and to automate and re-install the telescope at Kitt Peak (Oswalt et al. 1995). In 1991 SARA was awarded the telescope over other competitive proposals and a dome was designed by architect Stanley Black and constructed on Mercedes Point near the Case Western Reserve telescope.

As part of the refurbishing, all the drive motors and nearly all the gears were either replaced or reconditioned. SARA contracted with Astronomical Consultants and Equipment Inc. (A.C.E.) to automate the telescope and on January 7, 1995, the "new" SARA 0.9-meter was officially dedicated. The automation system is effectively 95 percent complete at this time.

2. The 0.9-meter telescope

The No. 1-36" telescope, located atop KPNO's south ridge for many years, served as a workhorse instrument at Kitt Peak. It was heavily used until it was chosen to be de-commissioned by NOAO due to budgetary constraints. The Ritchey-Chretien design of the telescope was made specifically for wide-field photography at the Cassegrain focus. The telescope consists of a 0.9-meter aluminized glass mirror and 20-cm secondary which operate at a focal ratio of F/7.5. It is mounted on an English equatorial mount with a "closed tube" design.

The reinstallation of the telescope and the subsequent computer control comprised phase 1 of the automation process. The telescope uses precision stepper motors controlled by a PC running Windows NT and the ACE software for pointing, slewing and tracking. The overall system has a pointing accuracy of 30 arcseconds or less. Phase 2, currently being implemented, will allow observers to operate the telescope remotely while the 3rd and final stage will make the telescope totally robotic. Human intervention will be possible, but not necessary, as the telescope carries out its pre-programmed observational itinerary.

Telescope pointing will be verified with a finder CCD camera on one port of the multi-port instrument selector mounted at the cassegrain focus. A research-grade CCD camera (described below) occupies a second port, and a dual channel photometer is installed on a third port. All the instruments will be operated remotely once the automation system is complete.

3. The CCD camera

The principle detector used in the Blazar programs is the research-grade CCD. The camera is an Axiom AX-4 with a Kodak 4200 chip. This chip is a 2048×2048 array with 9 micron square pixels and a physical size of 18.4×18.4 mm which corresponds to 9.4×9.4 arcminutes on the sky. The grade zero chip is lumigen coated and has no column or row defects, although there are a number of warm pixels when operated at temperatures of −35°C below ambient. The AX-4 camera has a 16-bit A/D converter with MPP electronics with very low noise. The low noise standard is sustained even when operated at higher temperatures. The thermo-electric cooler can reduce the chip temperature up to 50°C below ambient. The camera is controlled by the MIRA software package and camera control is also being adapted for remote use.

4. Blazar research programs

Florida International University is guaranteed roughly 1/5 of the observing time on the SARA telescope and a significant portion of this time will be used observing Blazars. One accepted proposal allocates time for regular monitoring of a list of interesting Blazars, and another proposal is the TOO program which is designed to monitor outbursts or perform satellite backup observations.

Blazars which have a history of high-amplitude outbursts and which are well distributed in RA have been chosen for the monitoring program and are listed in Table 1. Column 1 lists the common name, column 2 the maximum brightness at which the source has been observed in magnitudes, and column 3 the number of years the source has been monitored at the Rosemary Hill Observatory. Column 4 lists the number of times it has been observed at this brightness level during the time RHO observed it, and columns 5 and 6 give the magnitude ranges and object types respectively. The sources must be classified as an OVV quasar or BL Lac object, must have flux variations of at least 2 magnitudes and must be bright enough to be detected with IUE.

Table 1. TARGET OBJECTS

NAME	MAX MAG.	YEARS OBS.	NUMBER of TIMES at MAX	Δ M	TYPE
0048-097	15.3	19	2	2.2	LAC
0215+01	14.5	5	2	4	LAC
0235+164	15.5	12	4	5	LAC
0422+004	15.2	13	2	2.2	LAC
0735+17	14.0	13	2	3.5	LAC
0818-128	15.7	7	?	2.4	LAC
OJ 287	13.0	16	2	5	LAC
1156+295	13.8	7	2	5	OVV
ON 231	15.2	15	4	3	LAC
ON 325	14.6	17	2	2	LAC
3C 279	13.2[a]	13	2	4.5	OVV
1308+32	15.0	11	4	5	LAC
1514-24	14.2	15	1	3	LAC
3C 345	15.0	16	3	2	OVV
NRAO 530	15.5	14	1	4.5	OVV
BL LAC	15.0	15	5	2.5	LAC
3C 446	15.0	15	3	5	LAC

[a]A flare observed in 1938 showed 3C 279 at $B = 10.38$ mag.

5. First results

In May of 1995, a series of preliminary unfiltered test images were made of PKS 1156+295 with the CCD in order to check for comparison star placement. The images revealed a dramatic flare occurring in 1156+295 and we quickly activated an IUE TOO program to obtain UV observations. The results of the optical images were reported in an IAU circular (Webb et al. 1995), and a multifrequency paper on this outburst is currently in preparation.

A subsequent 6 day observing run in early July of 1995 yielded over 90 images of different 9 Blazars, including BL Lac and PKS 1622-29. PKS 1622-29 was observed as part of an EGRET satellite back-up under the TOO program and the others were observed under the general monitoring program. The 1622-29 images will be combined with data acquired by Wagner and published since 1622-29 was undergoing a γ-ray outburst at that time.

The results for the observations of BL Lacertae are shown in Table 2. The images were reduced using the Image Reduction and Analysis Facility (IRAF) from NOAO on a SUN Sparcstation. Bias, Dark, and Flat field frames were handled using the CCD reduction program package. The magnitudes were determined using the aperture photometry package and field comparison stars suggested by Smith et al. (1985). Appropriate aperture sizes were selected by plotting the radial profiles of the objects using the radprof package within IRAF. The broad-band spectrum of BL Lac on July 3, 1995 is shown in Figure 1. The spectrum is not corrected for galactic aborption or reddening. The figure is presented as merely an indication of the data quality we will be able to achieve with the telescope once the telescope automation system is completed.

We are currently finishing the reductions of other Blazars observed at SARA and setting up reduction "pipeline" procedures to be used in the reduction of subsequent CCD observations from the SARA telescope in the future. We hope that observations from remote or robotic facilities such as the SARA observatory can fill in gaps in current monitoring programs and provide critical broadband observations which will result in a better understanding of these enigmatic sources. This telescope will be ideal for TOO programs since we willhave immediate access to the instrument without travel or other complicated arrangements.

Table 2. BL LAC OBSERVATIONS

FILTER	DATE	TIME (UT)	MAG	ERROR
B	July 3, 1995	9:23	16.440	.020
V	July 2, 1995	6:47	15.310	.018
V	July 3, 1995	9:15	15.353	.020
V	July 4, 1995	7:22	15.362	.020
V	July 6, 1995	8:53	15.595	.019
R	July 2, 1995	6:39	14.737	.020
R	July 3, 1995	8:55	14.724	.020
R	July 3, 1995	8:59	14.620	.020
R	July 4, 1995	7:13	14.647	.020
R	July 5, 1995	8:59	14.714	.020
R	July 6, 1995	8:43	14.727	.026
I	July 2, 1995	6:26	13.952	.026
I	July 3, 1995	9:07	13.845	.020
I	July 4, 1995	7:35	13.805	.020
I	July 6, 1995	9:06	14.045	.025

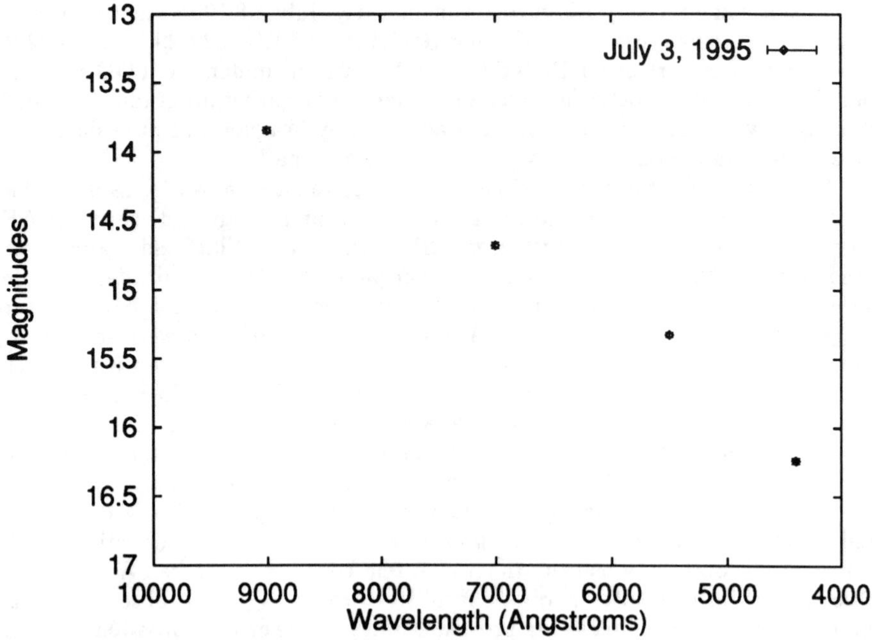

Figure 1. The Broad band Spectrum of BL Lac

References

Oswalt, T., Rafert, J., Smith, J., Wood, M., Castelatz, M., Collins, L., Henson, G., Powell, H., Caillault, J., Magnani, L., Shaw, S., Leake, M., Marks, D., Runstay, K., Webb, J. R., & Van Hamme, W. 1995, B.A.A.S. (185th meeting of the American Astronomical Society)

Oswalt, T. D., et al. 1996, in Progress Report on the SARA 0.9-m Telescope at Kitt Peak, I.A.P.P.P. Communications, in press

Smith et al. 1985, in *UBVRI* Field Comparison Stars for Selected Active Quasars and BL Lacertae Objects, AJ, 90, 1184

Webb J.R., et al. 1995, in Target of Opportunity Multifrequency Observations of BL Lac and OJ287, Proceedings of the Oxford Conference

Blazar Continuum Variability
ASP Conference Series, Vol. 110, 1996
H. R. Miller, J. R. Webb, and J. C. Noble, eds.

Submillimeter and Infrared Variability of Blazars

Ian Robson

Joint Astronomy Centre, 660 N. A'ohoku Place, Hilo, HI 96720

Abstract. Infrared to millimeter studies of Blazars are of particular value as this wavelength regime is almost always dominated by synchrotron radiation from the innermost parts of the jet. Variability of this highest frequency synchrotron component demonstrates the complexity of the jet emission physics and shows that the shocked jet approach is reasonably robust. Early results from near-mm polarization show great promise for this technique in understanding jet physics and the Blazar phenomenon.

1. Introduction

As I was at pains to point out at the Turku Blazar conference, we must always be mindful that the term 'Blazar' refers to a phenomenon, irrespective of how it historically arose, or its use in terms of the sub-classification of the sources. We are talking relativistic jet physics, and there are only a small number of key question in the 'big picture.' Why so few galaxies possess relativistic jets is probably the most important and presumably boils down to why so few of the supposedly common-place supermassive black holes in the centers of AGNs manage to collimate their outflows into powerful jets. Recent investigations of the host galaxies of radio-loud and radio-quiet AGNs (Taylor, Dunlop, Hughes, & Robson 1996) have shed light on this question, finding that the radio-loud phenomenon of relativistic jets seems to be found only in elliptical galaxies. Radio-quiet quasars on the other hand reside in both elliptical and spiral galaxies.

Apart from the question of the central engine, we can ask about the jets themselves. BL Lacs and OVVs are the two most common forms of galaxy exhibiting relativistic jet emission. [Assuming AGN unification to be a valid description on the global scale, radio galaxies can also be added to this list of relativistic jet sources]. There is mounting evidence that there are two basic jet classes. The very powerful FRII jets escape relatively unhindered from the parent galaxy, to produce spectacular and giant radio plumes. The FRI jets on the other hand are less powerful, interacting and slowing down much more rapidly within the parent galaxy, albeit also displaying spectacular plumes that are end-darkened rather than end-brightened.

The investigation of the central engine aspect—the relation between accretion, primary and secondary emissions is intimately tied to what I believe is the most crucial aspect of blazar research today, carefully coordinated multifrequency campaigns. However, although it turns out that the mm–submm–IR are crucial to these studies, I will stick to my brief of variability and not consider these multifrequency topics which will be discussed later in the workshop.

2. The infrared–submillimeter–near-millimeter wavelength regime

The important point to realise is that across this entire wavelength regime, the emission process for the radio-loud blazars (as opposed to the X-ray selected BL Lacs) is dominated by readily-measurable synchrotron emission. In the mm–submm no source shows any contamination by the emission from the extended (pc scale upwards) stellar or dust content of underlying galaxy. Furthermore, very few of the strong radio-loud sources reveal any significant contamination in the near IR (although one that does is 3C 273, e.g., Robson et al. 1986). However, for a number of the weaker radio-loud BL Lac objects, the luminosity of the underlying galaxy is readily apparent in the near-IR, but can be identified and corrected using modern imaging techniques.

Although this paper concerns variability, it is worthwhile re-iterating that inspection of the single-epoch multifrequency continuum of many sources clearly demonstrates that the mm though infrared (and sometimes extending all the way to the X-ray regime—especially for those sources such as the BL Lacs which do not show a blue bump) is a single synchrotron component (e.g., Brown et al. 1989a). Note that at longer radio wavelengths we are presumably seeing the superposition of a number of synchrotron components and their evolution (see the contribution by Margo Aller) and spectral deconvolution can now be undertaken by VLBI studies (see for example Marscher 1988).

3. Origin of the synchrotron emission

We are confident that the synchrotron emission originates deep within the relativistic jet and the infrared–submillimeter–millimeter variability turns out to be crucial in terms of the study of the jet physics. I will give only a brief overview of the modelling work, the intricate details of which are discussed by Alan Marscher later. However, a crucial fact in the observational studies is that the short millimeter and submillimeter regimes sample the optically thin part of the synchrotron emission and therefore monitor those events deep within the jet, inside those regions accessible to the traditional cm-radio wavelengths.

Although we are far from being totally confident of the behavior of the underlying jet (see the summary paper by Roger Blandford), in terms of the flaring behavior—which is what variability mainly addresses—the shocked-jet models as epitomised by Marscher and Gear (1985) (herein MG85) seem to be remarkably successful at explaining much of what is observed. One outcome of these models that is borne out by observation is that the flare emerges at the shortest wavelengths (up to the near-mm domain) almost simultaneously (within timescales of a day). Propagation delays extending up to months then take place as we move up to the cm regime. However, as noted by Esko Valtaoja, even this simplifies the real picture and flare propagation remains an important area for study. Temporal monitoring is therefore crucial, especially in the domain of the 'prompt' emission from the shock, which includes all wavelengths up to the near-mm. Remember, in terms of scale-sizes, monitoring at these wavelengths tells us about parsec scale events in the jet.

4. Facilities

Unfortunately, this domain is not graced by dedicated monitoring facilities, unlike the cm regime with the brilliant work of the Allers with UMRAO. The closest to this is the long-mm and short-cm monitoring from the Metsähovi radio telescope in Finland, which provides excellent coverage. The rest of the wavelengths are sampled very much on an ad-hoc manner. A regular (∼monthly) program of mm–submm monitoring continues at the James Clerk Maxwell Telescope on a sample of ∼20 blazars, while observations at 1.3–3.3 mm are carried out at irregular intervals with SEST and IRAM. In the infrared the situation is even worse. The monitoring program on UKIRT is now almost dead due to the huge oversubscription for time. The 1.5 m Carlos Sanchez Telescope at Izaña is well suited to monitoring studies and is used for campaigns and dedicated monitoring of selected objects. In general, monitoring of a statistically large sample of sources is extremely poor, but across the infrared through millimeter range, much better coverage is obtained for dedicated multifrequency campaigns—as we shall see during this workshop.

5. Results–overview

The simple maxim is that the more you look the more you find. Examination of the lightcurves from eight years of infrared monitoring (Litchfield, Robson, & Stevens 1994) and the mm–submm monitoring (e.g., Stevens et al. 1994) reveals that the picture is indeed complex.

A summary of the results of the long-term monitoring of a sample of 17 blazars presented in 'Blazars V' (Stevens et al. 1994), is that there is no statistically significant difference between the degree of variability of the OVVs and the BL Lacs (supporting the conclusion reached by Lainela & Valtaoja 1993) and there is no evidence to support any particular variability timescale (from years to weeks). Flares tend to attain their maximum amplitude at about 3 mm or longer, with a suggestion that BL Lacs may attain maximum flux at longer wavelengths than OVVs (see later). The trend of a flattening of the submm–near-mm spectral index with flux is well demonstrated by some (7) sources, and generally supported by all—showing a similar physical process for flaring behavior, and the most likely interpretation is a combination of radiation losses, and/or the synchrotron self-absorption turnover moving through the waveband 'window.' However, there is a significant difference between the 0.8–2.0 mm spectral index of the BL Lacs and the OVVs, with the former being flatter (see later).

Within the limitations of the samples, flaring and flickering seem to be relatively common processes at some level. Indeed, this poses a distinct problem because as we shall subsequently see, to investigate the flaring component one would like to be able to remove the 'quiescent' emission of the underlying jet.

The general results of infrared variability were described in Brown et al. (1989a,b) and there have been few changes to their conclusion since. IR monitoring of OJ 287 by Kidger, Takalo, and de Diego (1994) provided evidence for intranight variability, although the uncertainty in the observations needs to be substantially reduced before detailed physics comes out of this, something that will not be an easy task to achieve. Intranight variability has only once been observed at 1 mm wavelength—a crucial observation because this implies either huge boosting factors, or peculiarities in geometry. However, by the nature of things, very few sources are ever measured more than a single occasion in a night! Progress on this front is expected to improve dramatically over the coming year due to new technology (such as SCUBA on the JCMT).

6. Flare development

To investigate flare development one ideally wants nice, big, clearly isolated flares, that are well-sampled as they rise up from a well-defined quiescent jet emission, peak, and decay to the quiescent level before the onset of a second flare. However, reality is rarely so helpful. In the real world lightcurves are far from smooth, showing flickering and often multiple flaring or sub-flaring superimposed on the overall flare. At longer radio wavelengths, this 'noise' is heavily smoothed due to longer time-constants, but it is a real problem in the mm–IR (and optical). Figs. 1 and 2 show nine-year infrared–millimeter lightcurve for 3C 279 and 3C 273, vividly illustrating the problem.

In fact this is probably what we might expect if we approached the subject from the alternate point of view—that of asking what are the likely shock conditions in a relativistic jet—smoothness is probably far from the norm. Also, as Esko pointed out in his presentation, the fact that we do not observe a unique signature, but rather see a wide variety of synchrotron spectral development tells us that there is not a single type of shock. When one thinks about the conditions in a relativistic jet, perhaps this is no longer a surprise, although somewhat messy from both an observational and theoretical point of view.

To determine quantitatively the behavior of one waveband against another—seeing how a flare evolves with time and especially how the peak flux varies with time from the zero point of the flare onset—correlation techniques are essential. A study by Litchfield, Robson and Hughes (1995) employing a large number of Monte Carlo runs using simulated data for submm–long-mm flares (based on a large number of real observations with sampling intervals typical of the wavebands in question) revealed: flickering did not broaden the confidence of the correlation distributions significantly; while the discrete correlation technique had the broadest distribution (which could be interpreted as making it the most optimistic in determining lags) the linear interpolation technique came out of this study better—with a narrower confidence distribution (making it somewhat more pessimistic). However, the peak-to-peak method (the good old 1σ-eyeball) surprisingly (or perhaps not) turns out to be a robust method for determining lags between wavelengths.

Nonetheless, it must be noted that the eyeball can easily pick out false correlations. The superposition of the submm–mm lightcurves of 3C 273 and 3C 279 can easily appear to be frighteningly well correlated—and although these two famous quasars are only a few degrees apart on the sky...! With some relief, a strict correlation test assures us that they are not, in fact, correlated.

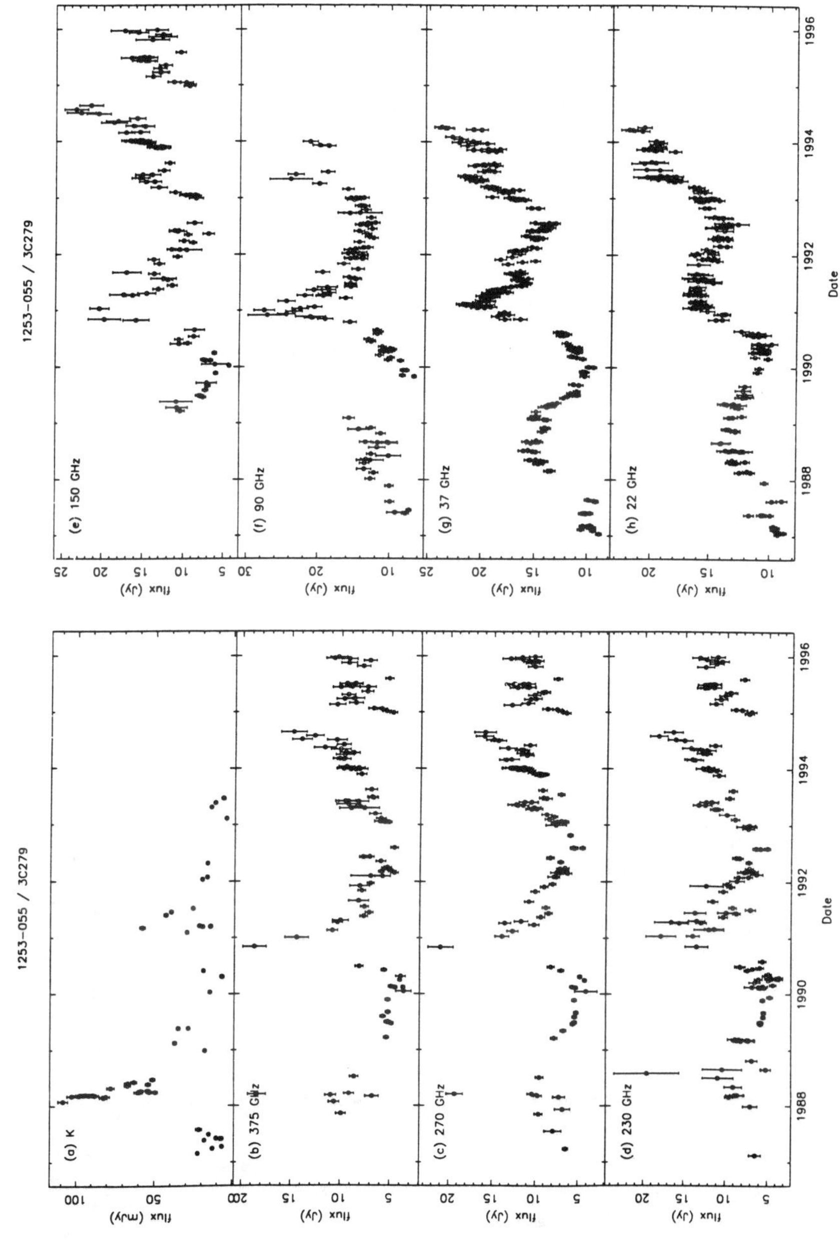

Figure 1. 2.2 μm–1.3 cm lightcurves of 3C 279 from 1987–1996.

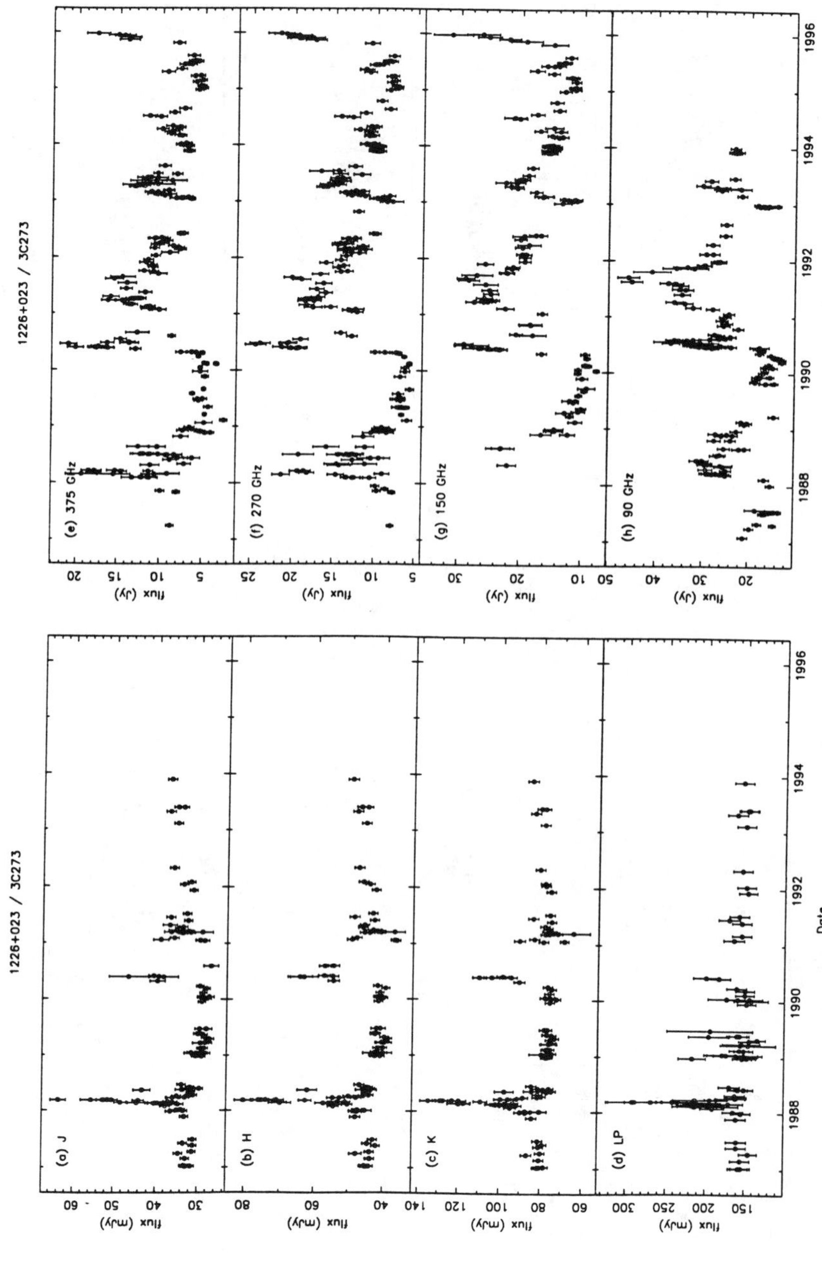

Figure 2. 1 μm–3 mm lightcurves of 3C 273 from 1987–1996.

Overall, the results of a large number of studies on flares from a number of sources confirms the earlier statement that the infrared to short-mm (ranging from 1 mm to 3 mm) is 'prompt' emission, and that beyond these wavelengths the peak of the flare is developed with lags which increase with wavelength, the lag being many months at mid-cm wavelengths.

6.1. Spectral index development

One of the key studies in flare development is to investigate the change of energy spectral index with flux and time—this gives direct information on the radiating electrons and the acceleration/cooling processes. Gear et al. (1986) noted that for OJ 287 there was a clear correlation between the infrared spectral index and flux—the spectrum flattens with higher flux. This was further demonstrated for a number of sources by Brown et al. (1989a). Although Kidger et al. (1994) verified this relationship for OJ 287 they found that it did not hold for a large flare in the same source in 1994 (Kidger et al. 1995). Whether the spectrum evolved at constant spectral index is unclear and much further infrared monitoring is required to pursue the spectral evolution in terms of particle injection and subsequent radiative energy losses. This is an excellent project for an undersubscribed, medium-size infrared facility. Nevertheless, OJ 287 demonstrates that in the infrared at least, different flares behave very differently.

As the flare develops there is undoubtedly a relationship between spectral index and time, (discussed later), but we should note that it is very important to separate the flaring and fading components to determine the precise nature of this spectral index-flux relationship. An example of this is given in Robson et al. (1993) for 3C 273. Of course for OJ 287 with its closely inclined jet and potentially high Doppler factors, relativistic boosting effects cannot be discounted as being intimately linked with the fast variability, which would significantly modify non-boosted deductions about the jet physics (e.g., Marscher, Gear, & Travis 1992). However, whether these can account for variations on the day-to-day scale is unclear, in which case magnetic field variations due to turbulence in the jet provide some ready-made answers (e.g., Ballard et al. 1990).

6.2. Submillimeter flare development

Let us now move on to the general consideration of the submm–mm portion of the synchrotron spectrum. Here the databases are much superior with relatively well sampled light curves. Also, the flickering seems to be less of a hazard than at higher frequencies, so we can investigate and model 'real' flares.

The mm–submm observations by Robson et al. (1983) of the increase of the flare amplitude in 3C 273 with increasing wavelengths compared to the decline in flare amplitude with decreasing wavelengths at much longer radio wavelengths (through expansion losses) led to what is now the 'standard' shock explanation for flare emission. In the first, and most popular of these models, MG85 showed that there are three distinct phases of flare emission, each being dominated by a single emission loss mechanism. In the first stage, inverse-Compton losses dominate and the flare amplitude increases as the turnover moves to longer wavelengths. This is followed by the synchrotron radiation phase in which the turnover continues to evolve to longer wavelengths, albeit it at constant amplitude (hence constant flux). The adiabatic phase then follows in which the

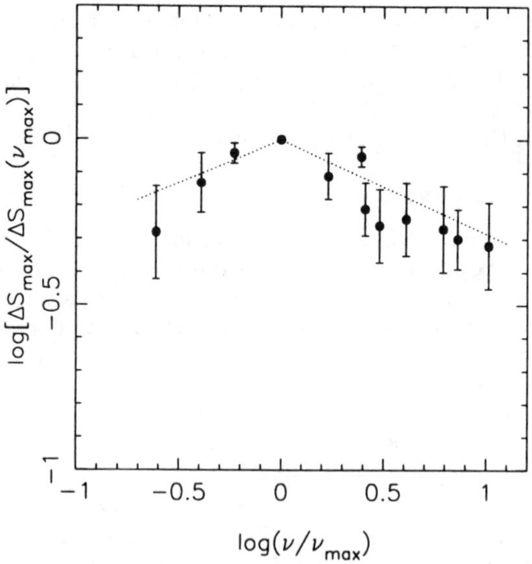

Figure 3. Normalized flare development showing the rapid transition from the Compton-rise phase to the decay phase with little sign of the synchrotron plateau phase (from Stevens et al. 1994).

shocked region expands and the amplitude of emission declines with increasing wavelength.

The models make clear predictions about the behavior in these phases and it is to test and refine the models that the flare development studies are directed. This is not easy due to the complex flaring behavior of most sources, with multiple maxima (or secondary events?). Nevertheless, if we consider the normalised flare amplitude (normalised against its maximum at maximum frequency) against the normalised frequency (normalised against the maximum frequency), then we find there is a trend of the flare development that is in general agreement with the shocked jet models. From 15 flares in 10 sources we see the development shown in Fig. 3, which clearly shows the growth phase (often poorly sampled) the decline phase, but a very short plateau phase, which perhaps has hardly been sampled at all.

For observations lying on the growth phase the maximum amplitude of the flare is attained when the turnover transits to the plateau phase. This suggests two things: there should be no lag between wavelengths on the growth phase; and the flare amplitudes should increase with increasing wavelength that is dependent on the optically thin slope of the synchrotron radiation. The exponent of the growth phase should be equal to the optically thin spectral index. In terms of the lags there is general agreement with the model as far as it can be tested, different flares attaining different maximum amplitudes in the growth phase before the plateau (synchrotron) phase, with almost all still growing at 3.3mm. However, the case is somewhat different for the flare amplitudes.

The slope of the growth phase in Fig. 3 is -0.26 (± 0.08), but, the most straightforward version of the Marscher and Gear model requires the growth exponent to be < -0.5, otherwise the shock becomes radiative, dissipates rapidly and does not propagate appreciably down the jet. Because this is not seen then there may be a problem for this jet model. On the other hand, perhaps the flares are radiative, and unless the shock is very strong, they do not propagate very far down the jet. However, this is unlikely to be the correct answer given the appreciable number of flares (15) and 10 sources. VLBI data on the individual flares will be able to test this hypothesis. Another way out of the dilemma is to note that the radiation is only partially optically thin near the turnover and the resulting deviation from a straight line could easily be masked by the error bars in Fig. 3. However, this seems to be somewhat of a fudge.

In terms of the 'big picture' there is now substantial evidence that the jets of BL Lacs and OVVs are different (e.g., VLBI polarization studies—Cawthorne et al. 1993; FRI/FRII jets and parent populations—e.g., Baum, Zirbel, & O'Dea 1995). Can we see a difference in terms of their synchrotron energy distributions, either for the steady jet, or for flaring components?

The data from Blazars V suggest that there is a difference between the spectral indexes of the BL Lacs and the OVVs, pointing to the BL Lacs having a synchrotron turnover at a longer wavelength than the OVVs. Why might this be? The cm-VLBI polarization measurements have strongly suggested that inferred magnetic fields in the parsec-scale jets are different for BL Lacs and OVVs. In the former, the field is aligned across the jet, whilst in the latter, the field is more aligned along the jet (until a shock develops). This result can be interpreted in a number of ways. The perpendicular field is indicative of a transverse shock, which might mean that the jets in BL Lacs are more often in the shocked state. Alternatively, the BL Lacs could have stronger shocks, or the magnetic field in quasars is usually aligned along the jet, but during a flare a perpendicular component is introduced by the transverse shock and the relative strengths of the two fields then determines the total polarization. For both cases the flares in BL Lacs should reach maximum development at lower frequencies than quasars, and this is supported by the data.

To pursue this topic, Gear et al. (1994) obtained quasi-simultaneous spectra (within ~4 weeks which is adequate for this study) for 22 BL Lacs and 24 OVVs from 6 cm to 0.8 mm. The spectra of all the sources are generally smooth, showing that variability effects are not seriously hampering the sample. When the entire data sets are considered, there is a marginal trend for BL Lacs and OVVs to separate out (at the 94% confidence using a 2D K-S test). What happens when different wavebands are studied? At the 'classical' radio wavelengths of 6 cm to 0.8 cm the samples form a single population to the 99.9% confidence level (from a 1D K-S test), reinforcing the assumption that there is a single physical process (synchrotron radiation and a jet) responsible for the emission from all the sources. However, when considering the shorter wavelength (submm–mm) radio spectral indexes (Fig. 4), the samples clearly separate out into BL Lacs and OVVs at the 98.8% confidence level with, as shown in Blazars V, the BL Lacs having flatter spectra than the OVVs. At the shortest wavelengths of 2.0 mm to 8 mm, the two-point spectral indexes show a separation at the 99.5% level.

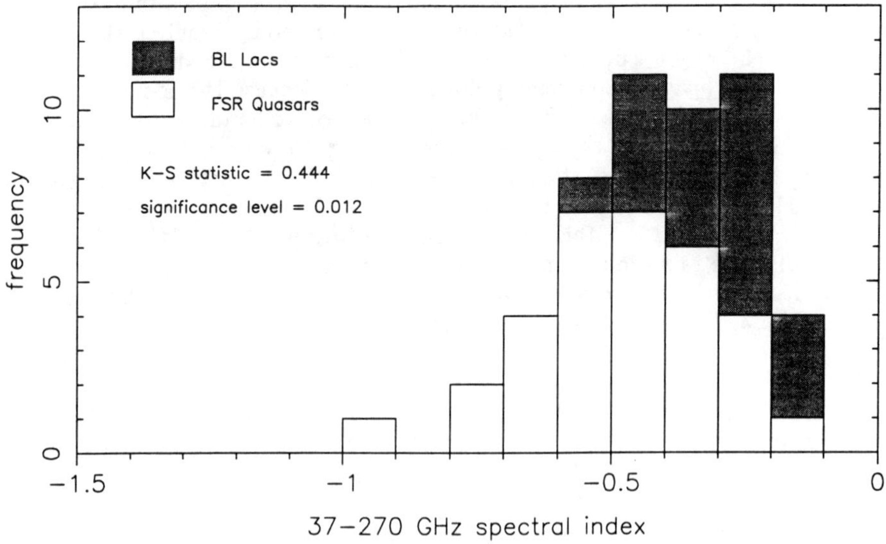

Figure 4. Histogram of the 37–270 GHz (8–1 mm) spectral indices for BL Lacs and quasars. (From Gear et al. 1994).

But in studies such as these one must be ever cautious of possible selection effects. The two most obvious are orientation and redshift. If BL Lacs are more beamed (due to their jets being closer to the line of sight) then their turnovers would be at shorter wavelengths than for quasars and their luminosities correspondingly boosted. We find no such evidence for this. However, there *is* a clear redshift-luminosity correlation in our samples due to their being flux limited. We also find a correlation between the 1–8 mm spectral index and the redshift, for both quasars and BL Lacs. This is believed to be due to the fact that the turnover lies within this window, probably closer to 8 mm, and so this becomes more sensitive to redshift than jet parameters. On the other hand, at 2.0–0.8 mm, the turnover is always outside this window and so is not as sensitive to redshift directly. Therefore, we are left with the strong suspicion that because this difference is only apparent for the low (quiescent) state, the jets of BL Lacs and OVVs are somehow intrinsically different but the flares are basically the same (shocks).

6.3. 3C 273

Study of the development of flares in a number of individual sources serves to emphasise and expand on the general points made above. The best studied source is undoubtedly 3C 273, although we have found over a number of years that it shows very complex behavior in terms of the relationships between the central engine (UV–X-rays–gamma-rays) and the synchrotron components. Nevertheless, it provides a good example of the use of correlation studies to determine the lags between flare development at different wavelengths. The well sampled flares clearly demonstrate that the IR through 1 mm emission is 'prompt' where

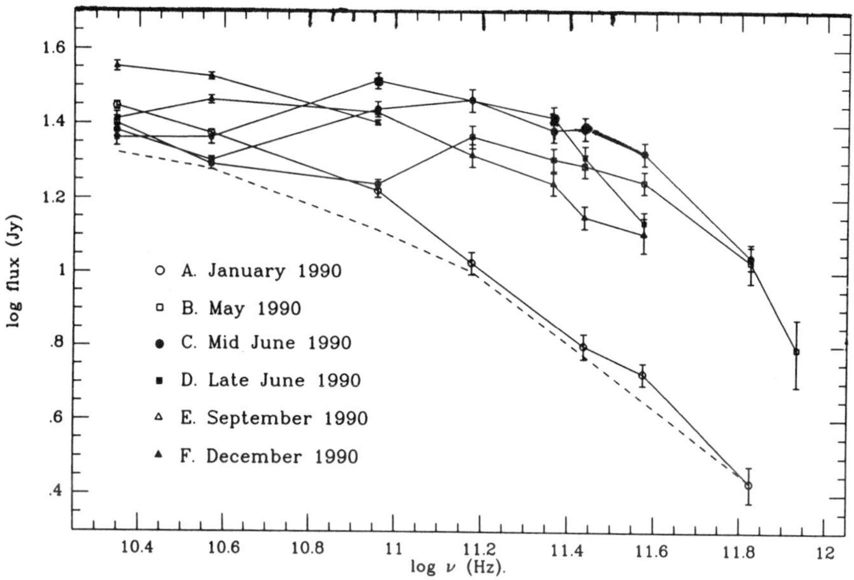

Figure 5. Spectral development of the 1990 flare in 3C 273 from cm–submm wavelengths (From Robson et al. 1993).

this is defined in terms of 1–2 days. Beyond this point, delays of order weeks are seen to long-millimeter, and months to cm wavelengths.

Interestingly enough, the 'non-flare' spectrum of 3C 273 does not seem to be that of pure synchrotron emission in the far infrared, and there is a strong suspicion that cold dust is present and prominent at around 100 μm. ISO should settle this point. (We know that there is a hot dust contribution in the near IR, [Robson et al. 1986] that might be the dust in the torus surrounding the central engine).

3C 273 was the first source for which a detailed description of the spectral evolution of the flaring component was attempted. Fig. 5 shows the cm–submm spectral development of the January 1990 flare. (From Robson et al. 1993.)

6.4. 3C 279

Following the spectacular gamma-ray observations, 3C 279 has become a very popular source for multiwavelength monitoring. The synchrotron component dominates the entire radio through infrared; there is no evidence for cold, warm or hot dust. While the flaring behavior is complex, there is scope for detailed analysis of the flares and the 'quiescent' or 'steady' jet spectrum can be isolated and subtracted to reveal the spectrum of the flaring component. The snapshot flare spectra are reasonably well fitted by a homogeneous synchrotron emission, and while there is good evidence for the flattening of the infrared spectral index with increasing flux, at near-mm wavelengths (1–3 mm) the flux appears to increase at constant spectral index. However, when the 13.6–3.3 mm spectral

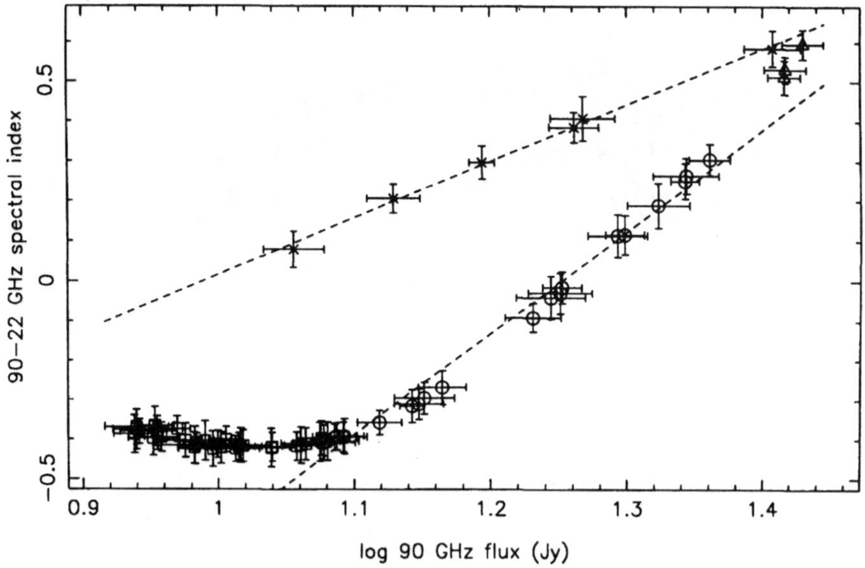

Figure 6. Simulated 22–90 GHz spectral index against log 90 GHz flux for 3C 279 showing the short lived Compton rise-time (crosses), the even shorter-lived synchrotron plateau phase (filled triangles) and the much longer adiabatic decay phase (open circles). (From Litchfield, Stevens, Robson, & Gear 1995).

index versus the 3.3 mm flux is examined, the flattening with increasing flux *is* seen and furthermore, the flaring and fading modes are clearly separated.

This enabled a model to be produced (based on MG85) that describes the evolution of the spectral slope as a function of flux, frequency and time (Fig. 6). The results are in agreement with the general predicted trends of MG85 and again reveal that the plateau phase is very short-lived. The flare moves rapidly from the Compton through the synchrotron (plateau) phase to a general adiabatic decay. For 3C 279, the predicted final phase of the flare decay, when the turnover frequency has evolved to below 13.6 mm (giving almost constant spectral slope with decline in flux) is not observed, perhaps because of the early onset of the next flare. This shows the need for the study of a well-isolated flare but also demonstrates the overall success of the modelling and understanding the emission processes taking place.

6.5. 0420−014

The OVV quasar PKS 0420−014, although having a complex lightcurve, provided good data for further investigating flare evolution. The observed flare development is shown in Fig. 7, from which it is apparent that the synchrotron phase was not observed and must be very short lived. The model simulation of these data is an excellent fit. Stevens et al. (1995) additionally conclude from the studies that: (1) the flare is well approximated by a single homoge-

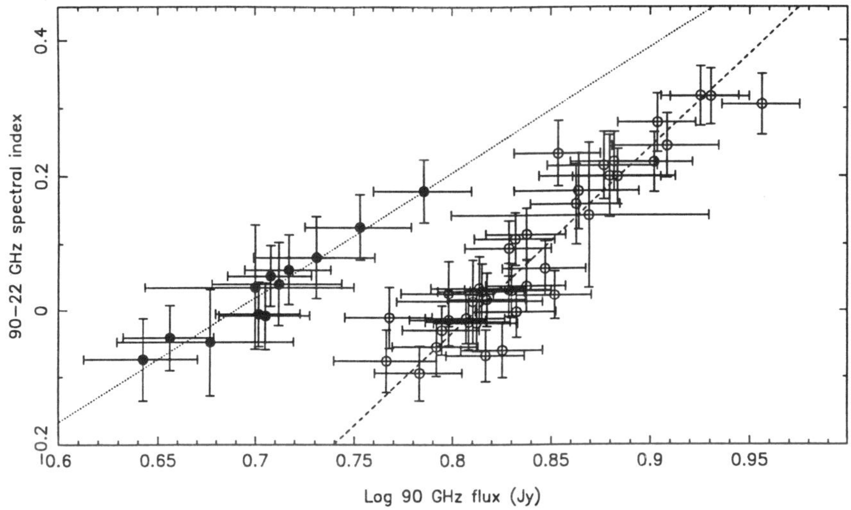

Figure 7. 90–22 GHz two-point spectral index against log 90 GHz flux for 0420−014. The filled circles represent the rising portion of the flare and the open circles the decay phase. There is no indication of the synchrotron phase which must be short-lived. (From Stevens et al. 1995).

neous spectrum, (2) the decay is too rapid for an adiabatic process—possibly suggesting that the jet may bend away from the line-of-sight, (3) there is the possibility of the 1991 flare undergoing brightening farther down the jet and being responsible for the 1992 flare.

6.6. 3C 345

The OVV quasar 3C 345 provided an even better example of a single isolated flare. Using high quality mm–cm data Stevens et al. (1996) showed: (1) that the process of quiescent spectrum subtraction was valid, (2) the flare spectrum could be well-fitted by a homogeneous emission for a period of almost two years—suggesting that the region does not become very inhomogeneous with expansion, (3) the initial optically thin spectral index of the flare was ~ -0.5 (the same as the knots in the jet of M87) which steepened with time to that expected from an instantaneous burst or reacceleration of the electrons that subsequently suffer radiation losses—suggesting that little reacceleration takes place at the shock front, (4) the flare turnover evolved smoothly to lower frequency with time—consistent with expansion which slowed with time, (5) the two flares investigated may be connected (as for 0420−014) or from a change in the jet (or shock) orientation—which could be confirmed by VLBI, (6) the evolution of the peak flux with peak frequency follows a power-law—the initial rise phase is consistent with expansion adiabatically down the jet that bends towards the line-of-sight, (7) the decay phase is too rapid for a simple adiabatic expansion of the shock

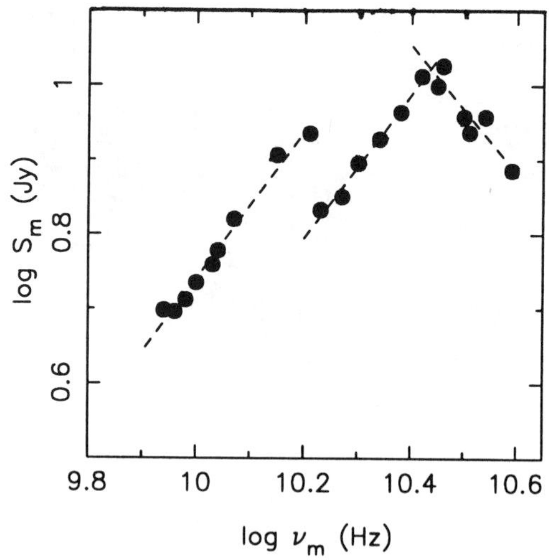

Figure 8. Evolution of the peak flux with frequency for two consecutive flares in 3C 345. (From Stevens et al. 1996).

unless the jet curves away from us; alternatively, the jet is straight but non-adiabatic.

The bending jet models are consistent with the VLBI data which show large bends near the base of the jet (Zensus, Cohen, & Unwin 1995) and gentle bending away from us farther out (Wardle et al. 1994).

7. Millimeter–submillimeter polarimetry

An area of increasing importance is that of mm–submm polarimetry, especially when linked with short-cm VLBI polarization. At a wavelength of 1 mm, Faraday rotation is *never* a problem and the potential ambiguity imposed due to the rotation of the polarization vector by ninety degrees at the synchrotron turnover is also avoided. From a four-year study of the 1.1 and 0.8 mm polarizations of 11 BL Lacs and 16 quasars, Nartallo et al. (1996a,b) have found that the polarization values range from 15% to 1%, with no significant differences between the 1.1 and 0.8 mm values. In terms of the difference between the quasars and BL Lacs, the analysis is still underway (especially determining the VLBI inner jet position angle corresponding to the epoch of the mm measurement) but so far both quasars and BL Lacs tend toward a bimodal distribution, with magnetic fields aligned either along or perpendicular to the inner jet. There is a trend for the low polarization values to be more aligned than perpendicular, but this may be because the quasars are brighter, allowing the polarization to be easier measured. The bottom-line so far seems to be that we do not see the separation observed by VLBI polarization (e.g., Cawthorne et al. 1993).

Another pointer against the strict (quasar : BL Lac) bimodal distribution is that for **3C 279**, even when it is not in a flaring state, the magnetic field is perpendicular to the jet, and the position angle remains almost constant even when undergoing flaring. So perhaps a bimodal distribution of parallel-perpendicular field is correct, but to attribute this to just BL Lac and quasar jet properties might be too simplistic (Stevens, Robson, & Holland 1996). On the other hand, there is evidence that for 3C 454.3, the magnetic field changes direction through ninety degrees during a flare, going from parallel to perpendicular, indicative of a strong shock being formed in the jet (Cawthorne & Gabuzda 1996). This is an area where much more study is required, and will probably form the thrust of future mm work, which will be ideally served by SCUBA and its polarimeter on the JCMT.

Acknowledgments. The James Clerk Maxwell Telescope is operated by the Particle Physics and Astronomy Research Council on behalf of the National Research Council of Canada (NRC), the Netherlands Organisation for Scientific Research (NWO) and the University of Hawaii.

References

Ballard, K. R., et al. 1990, MNRAS, 243, 640

Baum, S. A., Zirbel, E. L., & O'Dea, C. P. 1995. ApJ, 451, 88

Brown, L. M. J., et al. 1989(a), 'Blazars III,' ApJ, 340, 129

Brown, L. M. J., et al. 1989(b), 'Blazars IV,' ApJ, 340, 150

Cawthorne, T. V., et al. 1993, ApJ, 416, 496

Cawthorne, T. V., & Gabuzda, D. A. 1996, MNRAS, 278, 861

Gabuzda, D. C., et al. 1994, ApJ, 435, 140

Gear, W. K., Robson, E.,I. & Brown, L. M. J. 1986, Nature, 324, 546

Gear, W. K., et al. 1994, MNRAS, 267, 167

Kidger, M. R., Takalo, L. O., & Diego, J. A. 1994, A&A, 282, 369

Kidger, M. R., et al. 1995, A&AS, 113, 431

Lainela, M., & Valtaoja, E. 1993, ApJ, 416, 485

Litchfield, S. J., Robson, E. I., & Stevens, J. A. 1994, MNRAS, 270, 341

Litchfield, S. J., Robson, E. I., & Hughes, D. H. 1995, A&A, 300, 385

Litchfield, S. J., Stevens, J. A., Robson, E. I., & Gear, W. K. 1995, MNRAS, 274, 221

Marscher, A. P. 1988, ApJ, 334, 552

Marscher, A. P., & Gear, W. K. 1985, ApJ, 298, 114

Marscher, A. P., Gear, W. K., & Travis, J. P. 1992, in Variability of Blazars (Cambridge, Cambridge University Press), p. 85

Nartallo, R., et al. 1996(a), MNRAS, submitted

Nartallo, R., et al. 1996(b), MNRAS, in preparation

Robson, E. I.,et al. 1983, Nature, 305, 194

Robson, E. I., et al. 1986, Nature, 323, 134

Robson, E. I., et al. 1993, MNRAS, 262, 249

Stevens, J. A., et al. 1994, 'Blazars V,' ApJ, 437, 91

Stevens, J. A., et al. 1995, MNRAS, 275, 1146

Stevens, J. A., et al. 1996, ApJ, 466, in press

Stevens, J. A., Robson, E. I., & Holland, W. S. 1996, ApJ, 462, L23

Taylor, G. L., Dunlop, J. S., Hughes, D. H., & Robson, E. I. 1996, MNRAS, in press

Wardle, J. F. C., Cawthorne, T. V., Roberts, D. H., & Brown, L. F. 1994, ApJ, 437, 122

Zensus, J. A., Cohen, M., & Unwin, S. C. 1995, ApJ, 443, 35

Part II
Radio Variability

Blazar Continuum Variability
ASP Conference Series, Vol. 110, 1996
H. R. Miller, J. R. Webb, and J. C. Noble, eds.

Centimeter-wavelength Flux and Polarization Variability as a Probe of the Physical Conditions in AGN

M. F. Aller, H. D. Aller, and P. A. Hughes

University of Michigan, Ann Arbor, MI 48109-1090

Abstract. Results from a statistical study show that centimeter-λ variability is a nearly universal property of radio-loud sources, and this phenomenon must be accounted for in any theoretical interpretation of extragalactic objects. The characteristic properties of the total flux and linear polarization variability determined from the 30-year University of Michigan monitoring program which is devoted to the study of highly active objects are summarized, and it is demonstrated how such data can be used to constrain shock-in-jet models. The variability data combined with VLBI/P observations provide an effective tool for setting limits on the physical conditions in the radio-emitting parsec-scale jets and for studying structural changes. Examples of such usage are presented. The importance of understanding the effects of jet curvature and orientation when interpreting these observations is emphasized.

1. Introduction

Flux variability of extragalactic objects is now a well-documented phenomenon which occurs from the radio through the γ-ray spectral regions. Variability is important because it allows us to probe the physical conditions in active objects through the temporal and frequency-dependent behavior of the observed emission. We present results from the University of Michigan monitoring program which has been devoted to the study of centimeter-wavelength total flux and linear polarization variations for over three decades. The data obtained from this long-term project provide detailed information on the nature of the variability for over 200 well-observed very active objects in the frequency domain of our program—4.8, 8.0 and 14.5 GHz—and less detailed information for several hundred additional objects. We discuss the characteristic behavior of the flux and polarization variations and illustrate how they may be used to set limits on the physical conditions in the radio-emitting regions of the objects.

2. Characteristic behavior of the flux variability

A very basic question to ask is: how common is flux variability at centimeter-wavelengths? While the majority of our telescope time has been allocated for observations of very active objects, we have also been studying several flux-limited samples with the goal of better defining the statistical radio-wavelength properties of extragalactic objects in general. Among these is the Pearson-

Readhead (1988) sample which has been observed by us systematically since 1984. The sample is based on the NRAO-MPIfR 6cm strong source surveys; objects meeting the criteria: 1) $S \geq 1.3$ Jy; 2) declination 1950.0 in the range $35° < \delta < 82°$; and 3) $|b| > 10°$ are included in our program. Of the 62 objects studied, 28 are QSOs, 27 are galaxies and 7 are BL Lacs; the group is a mix of very well-known AGN such as 3C 345 and BL Lac and objects which would not otherwise have been included in our program.

As a very simple measure of variability, one can define a variability index which measures the peak-to-trough amplitude variation during the time window of the observations:

$$V(\nu) = \frac{(S_{\mathbf{max}} - \sigma_{S_{\mathbf{max}}}) - (S_{\mathbf{min}} + \sigma_{S_{\mathbf{min}}})}{(S_{\mathbf{max}} - \sigma_{S_{\mathbf{max}}}) + (S_{\mathbf{min}} + \sigma_{S_{\mathbf{min}}})}. \qquad (1)$$

The adjustment of the peak and lowest flux densities by the associated measurement error of the flux leads to a slight underestimate of V but greatly reduces the effect of spurious noisy measurements on the determination of this parameter. In Figure 1 we show a histogram of the values of V based on the data at 14.5 GHz.

Inspection of the distribution of V(14.5 GHz) shows that virtually all of the objects in the PR sample are variable and that the majority have variability indices in the range $0.2 < V < 0.9$: 86% of the QSOs, 86% of the BL Lacs, and 33% of the galaxies. This is the same range in V exhibited by our core program of active objects. In the majority of the galaxies we find significant but relatively low-amplitude ($0.1 \leq V \leq 0.2$) variability. As we show in Figure 2, variability is not restricted to flat spectrum sources: several QSOs and galaxies with $|\alpha| \geq 0.5$ also have $V > 0.2$. Thus, as study of the sample shows, variability in our frequency range is a very common feature of extragalactic objects, and it is clearly a phenomenon which must be accounted for in any adequate description of the physical conditions in such objects.

While many objects are variable, all sources do not behave in the same way; the temporal evolution during outbursts shows a wide range of behavior both from source to source and even from event to event in a single source. For example, OJ 287, shown in Figure 3, exhibits nearly continuous activity; many relatively rapid events are superimposed to produce an envelope of long-term activity. As we find for many of the BL Lacs in our program, the outburst shapes in this source are characterized by steep rises and steep declines, and such behavior allows us to, at least partially, resolve the individual frequently-occurring events. However, even in this relatively well-resolved case, the overlapping of events and self-absorption in the emitting region can hamper simple interpretation of the data. The richness of the activity apparent in this light curve highlights the difficulty that one often encounters when attempting to unambiguously associate activity at cm-wavelengths with activity in other spectral regions.

The QSO 1510-089, shown in Figure 4, also exhibits nearly continuous activity. However, while in OJ 287 opacity effects are apparent in the spectral evolution during major events, here the spectral development during outbursts varies from event to event. For example, the spectral index near outburst peak ranges from $\alpha = 0.72$ in mid-1979 to $\alpha = 0.0$ in late 1983 ($S_\nu \propto \nu^{+\alpha}$) indicating that both transparent and moderately opaque events can occur in the same object.

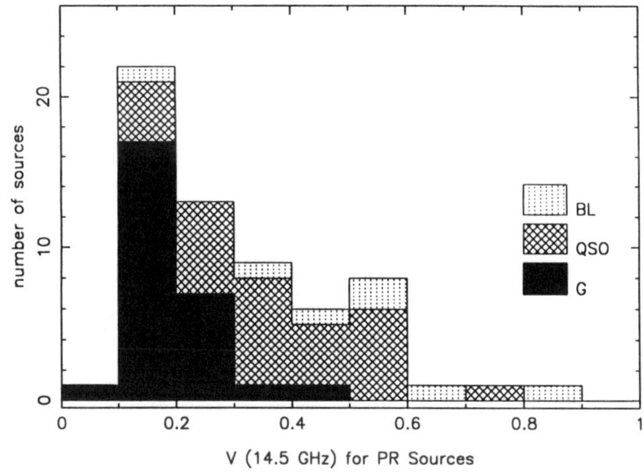

Figure 1. Histogram of V(14.5 GHz) for the Pearson-Readhead sample. The BL Lac object with unusually low-level variability is Mkn 501.

As seen in Figure 5 for the QSO OX 036, the shape of the outburst envelope or active phase can also vary significantly from event to event. During the decay of the 1980 event, the shape is unusually smooth and monotonic. Beginning in mid-1985, we find a longer-term decline on which are superimposed short-term fluctuations, especially evident in the data at 14.5 GHz. The time scales that we find in the Michigan data range from several years, such as we see here, to the kind of intraday variability that has been very well-documented at cm-wavelengths by the Bonn group (e.g., Witzel 1990).

2.1. Structure Function analysis

In an attempt to quantify the broad range of characteristic behavior evident in the light curves, we computed first-order structure functions for our 51 best observed sources (Hughes, Aller, & Aller 1992). The structure function is a useful tool because it gives information about the statistical nature of the process responsible for the variability. The slope of the power law portion provides information on the type of noise process, while the critical time, τ_c, corresponding to the plateauing for time lags greater than the maximum correlation time scale, gives a characteristic time scale for the fluctuations. This work demonstrates that while the characteristic time scales show some dispersion, on the average they are ~ 2 years, and this is, indeed, the kind of number that one would pick from visual inspection of the light curves in well-resolved cases. The slopes, however, show a narrow range and cluster about a value of 1, indicative of shot noise; the narrowness of the range in the slope values suggests that a common mechanism can account for most of the flux variability.

2.2. Evidence for changes in the time-dependent behavior

A question of interest for understanding AGN is: can the flux behavior of a source change over a period of several outbursts. The answer is clearly yes. For

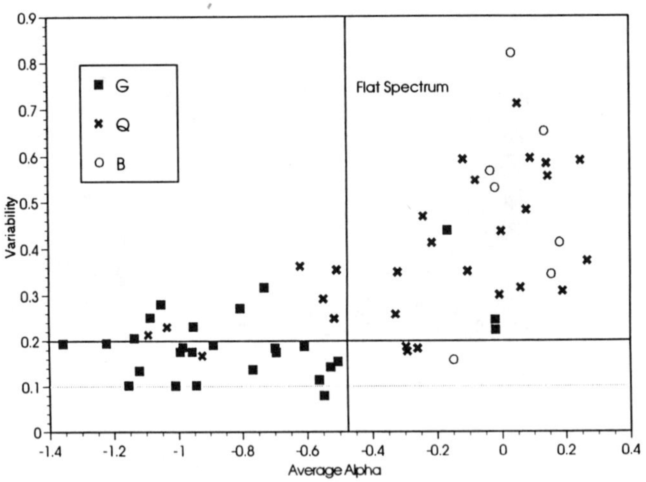

Figure 2. V(14.5 GHz) versus average spectral index between 14.5 and 4.8 GHz for the Pearson-Readhead sample.

example, 3C 120 was a very prominent radio variable in the 1970s and subsequently has exhibited only modest levels of activity. To quantify the impression that temporal changes have occurred in the behavior of several program sources, we computed structure functions based on decade-intervals of data and looked for changes in τ_c. As an example, for OJ 287 based on data obtained during the two periods 1975.0–1985.0 and 1985.0–1995.0 we find that the slopes of the power law portions are comparable while the characteristic time scales decreased from 0.71 to 0.50 years. Thus, while the noise process is unchanged, the more recent events are more rapid, a characteristic that is evident in Figure 3. This change might be explained by reorientation of the jet due to precession of the orbit of a binary companion (Valtonen 1996).

2.3. Evidence for periodicity

We have computed Scargle periodograms for the best observed sources in the Michigan program with the result that there is no strong evidence for periodicity in any of our sources. In OJ 287, however, we find *weak* evidence for periodicity. This result was presented in Aller et al. (1992) based on the data available through 1990; a recent reanalysis including data through 1995 confirms the original result, namely weak evidence from the features in the periodograms for a period of 1.67 years.

3. Importance of linear polarization variability

The study of polarization variability is a major component of the Michigan program and provides significant, unique information about the physical conditions in the sources. Such data are important for several reasons. First, long-term observations provide information about the characteristics of the quiescent flow,

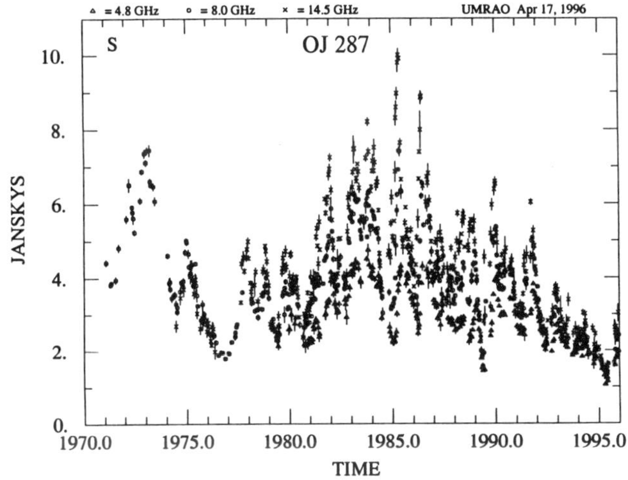

Figure 3. Two-week averages of the total flux density for OJ 287. The crosses, circles and triangles denote the observations at 14.5, 8.0 and 4.8 GHz respectively.

namely the degree of turbulence and the dominant orientation of the magnetic field in the radio-emitting region. This information is especially useful when combined with structural data from VLBI observations. Second, it provides evidence for axial compressions associated with the passage of transverse shocks; signatures of these events appear as temporal swings in polarization position angle through 90° accompanied by large associated changes in percentage polarization (i.e., the ratio of the polarized flux to the total flux). Such behavior provided the impetus for the development of our shock-in-jet model (e.g., Hughes, Aller, & Aller 1989). Some flow parameters determined by the modeling are very sensitive to the time-dependent behavior of the polarization, and our multifrequency measurements of the Stokes parameters I, Q and U are crucial for constraining the models. Thirdly, the polarization data have allowed us to resolve what sometimes appear as hopelessly blended, or even apparently-single, events in the total flux data and hence to separate out the contributions to the integrated total flux from (sometimes) many independently-evolving VLB components. The polarized flux has been an important aid in interpreting the VLBI observations of several sources.

3.1. Characteristic behavior of the polarization

Figure 6 shows both the total flux and polarization observations for BL Lacertae. This source has been a particularly important one in our program because it exhibits unusually well-resolved, relatively transparent outbursts in flux and polarization and hence is ideal for modeling. Additionally, the source has been well-studied using VLB interferometry by Mutel and coworkers (e.g., Mutel et al. 1990) with sufficient time resolution to follow the evolution of the individual components which in this case are relatively short-lived. The arrows in the

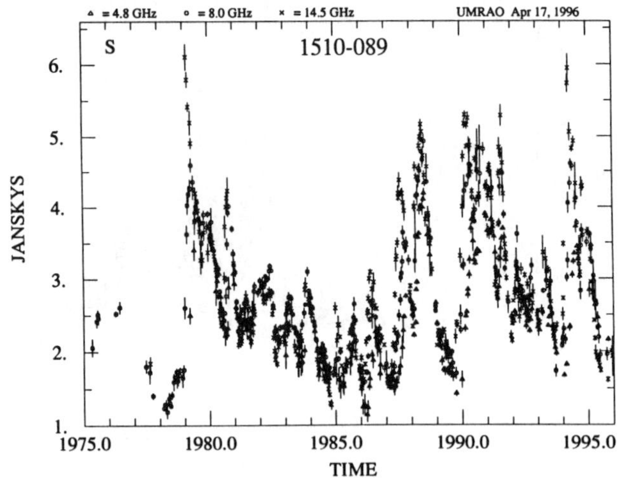

Figure 4. Two-week averages of the total flux density for 1510-089.

middle and lower panels of Figure 6 mark the estimated birth times of these VLB components (S1 through S6), and one can clearly see that these times coincide with the onset of new events in the total and polarized flux light curves. The data for this source represent the single clear example that we find for the direct association of VLBI knot development and outburst onset. The flux and polarization exhibit the signature expected for shocks propagating in the jet. Our shock-in-jet model was first applied to the data for this source, and the horizontal line in the lower panel marks the period modeled. Although there are temporal changes in the polarization position angle, it does exhibit long-term preferred orientations. When compared with the VLBI structural axis one finds that B is oriented perpendicular to the flow direction during the time period when many shock events occurred. Note that the position angle gives the orientation of the electric vector of the polarized emission which is orthogonal to B.

Figure 7 shows the Michigan data for the QSO 3C 273. In this case a long-term preferred orientation of B parallel to the flow is maintained, a characteristic orientation that we often find in the data for QSOs in the program. Note that the magnetic field is perturbed very little by the flow and that the source maintains the preferred position angle orientation in spite of several major outbursts. The percentage polarization remains very low, ranging from about 1% to a maximum of 6.8% in this source. Such low percentage polarizations are characteristic of those that we find in our program sources, and even during outbursts we rarely find $P > 15\%$. To illustrate the range of values for program sources, in Figure 8 we show histograms of the average percentage polarization at 14.5 GHz for two flux-limited samples: 1) the Pearson-Readhead Sample and 2) a flux-limited sample of BL Lac objects meeting the Burbidge and Hewitt (1987) classification criteria that we have systematically been observing since 1979. The values are typically only on the order of a few percent. Note that there is no difference in the distribution of percentage polarization as a function of optical class. BL Lacs do

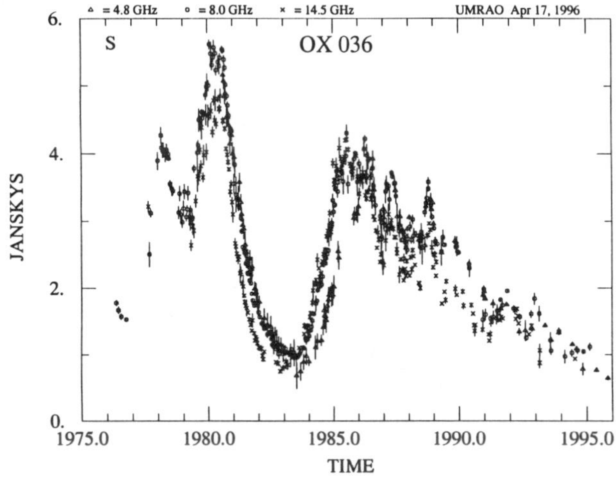

Figure 5. Two-week averages of the total flux density for OX 036.

not exhibit significantly greater degrees of polarization than QSOs and cannot be distinguished from other AGN on the basis of their radio-wavelength percentage polarizations. We have interpreted the low percentage polarizations that we find in the long-term data as evidence supporting the existence of turbulent magnetic fields in the radio-emitting regions (Jones et al. 1985). However, during the times when a shock forms, there is an enhancement of the magnetic field component perpendicular to the flow as the shock compresses the magnetic field in a plane perpendicular to the axis of the flow; this gives rise to an increase in the percentage polarization during outbursts.

Figure 9 shows the data for 3C 279, the second source to which the shock-in-jet model was applied. Inspection of the total flux light curve indicates that the behavior is considerably more complicated than in the case of BL Lac. The effects of overlapping of the individual events, of self-absorption in the emitting region, and of considerable contribution to the integrated flux by the emission from extended structure all complicate the interpretation of single-dish data for this source. To fit the temporal and frequency-dependent behavior of the total flux and polarization during the period modeled requires the propagation of six shocks. Note that the position angle in this source swings from an orientation that is perpendicular to the VLBI structural axis to one that is parallel to it; in this respect the source exhibits behavior sometimes characteristic of a QSO and at other epochs appears to follow the behavior of some BL Lac-type objects. The VLBI observations of this source have not been as well-sampled over the long-term as in the case of BL Lac; the undersampling combined with the fact that the components are long-lived has meant that there is no simple 1:1 correlation between estimated component onset times and the onset of new outbursts in the integrated data such as we found for BL Lac. During the past few years Wehrle, Unwin and collaborators have begun a program to obtain VLBI maps at 22 GHz. Two closely-spaced maps which they obtained in late 1992 and early 1993 (Wehrle et al. 1994) show the rapid development of a new

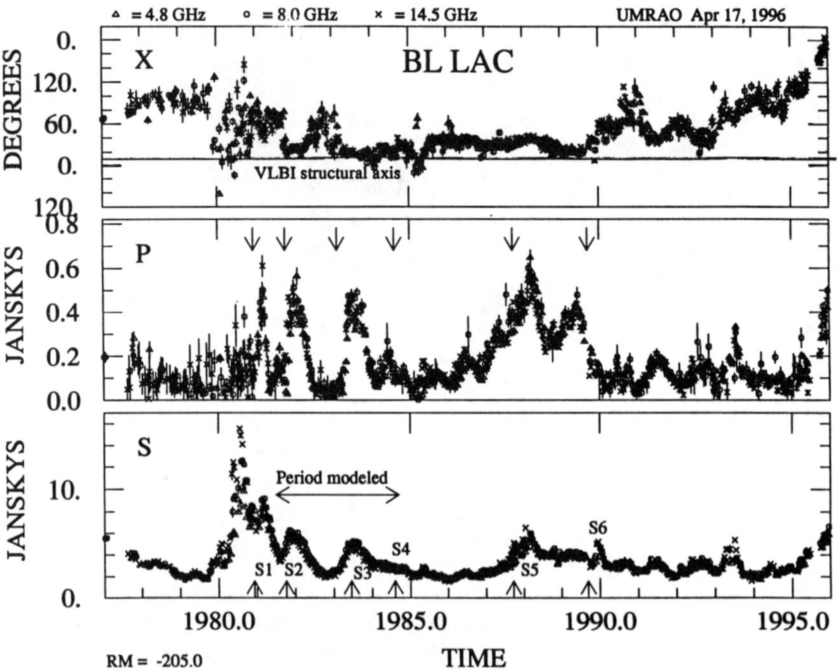

Figure 6. Two-week averages of the total flux density, polarized flux, and position angle of the electric vector of the polarized emission for BL Lac. The polarization has been corrected for Faraday rotation assuming a rotation measure of -205 rad/m^2. The line in the upper panel marks the orientation of the VLBI structural axis in the early 1980s (Mutel et al. 1990).

component between these epochs. Evidence in support of structural changes is also apparent in the UMRAO data at this time, but these changes are masked by the complexity of the light curves. The rapid changes identified by the VLBI data emphasize the importance of closely time-sampled high frequency maps in following the temporal evolution of VLB components. Closely-spaced VLBP maps are also highly desirable for properly interpreting the structural changes that occur, and we have only recently begun a program with the VLBA to obtain this type of data for 3C 279 and other very active objects.

In Figure 10 we show the flux and polarization data for the BL Lac object 0235+164. Here there is no long-term preferred position angle orientation in contrast to the behavior in BL Lac itself already discussed. We find rapid position angle changes in many of the BL Lac type objects in our program, especially at 14.5 GHz, and we believe this behavior arises because the polarization in this class of object is dominated by the contributions from evolving components rather than by an underlying jet.

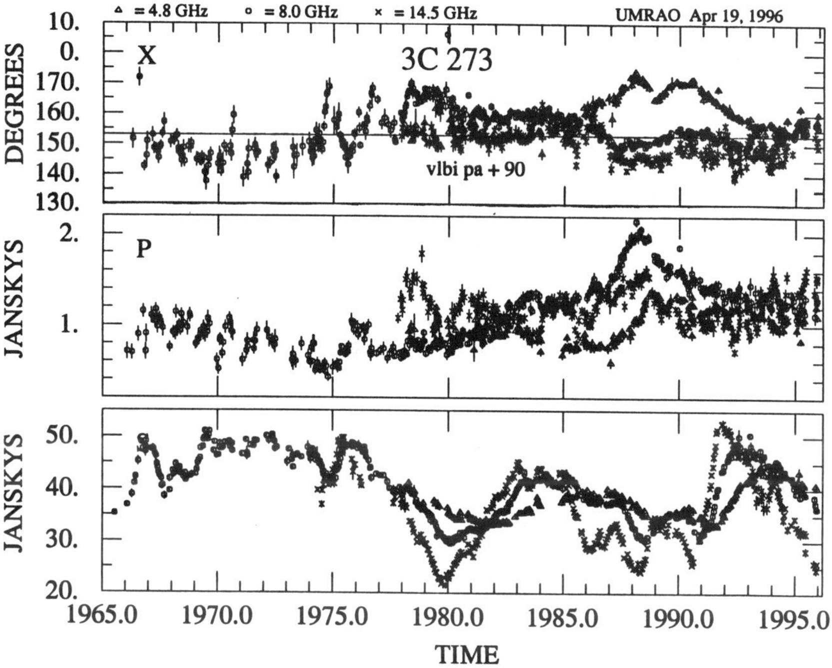

Figure 7. Two-week averages of the total flux density and polarization for 3C 273. The horizontal line marks the orientation *orthogonal* to the VLBI structural axis which is outside the range of the plot.

4. Characteristics of the flows

The data for three sources during active phase—BL Lac, OT 081 and 3C 279—have been modeled very successfully with radiative transfer calculations, to determine the characteristics of the flows in these relativistic parsec-scale jets (Hughes, Aller, & Aller 1989; Hughes, Aller, & Aller 1991). This work demonstrates that weakly-shocked, collimated flows can account for the temporal and frequency-dependent behavior of both the total flux and the polarization in the UMRAO data and the features in VLBI/P maps. The derived Mach numbers are found to be in the range $1.11 \leq M \leq 1.52$, implying that the passage of the weak shocks represent only minor perturbations to the flow. The deduced viewing angles are in the range $29° \leq \Psi \leq 40°$. The fact that the viewing angles are rather large and in a narrow range may represent a selection effect resulting from the need for isolated events for model fits; viewing angles close to the line of sight would produce large Doppler factors and hence compressed time scales as well as a lower source-frame emission frequency and a subsequent reduction in polarized flux by opacity. Viewing angles have been deduced independently from VLBI measurements for BL Lac (Mutel et al. 1990) and for 3C 279 (Carrara et al. 1993) that agree well with the values deduced from our modeling.

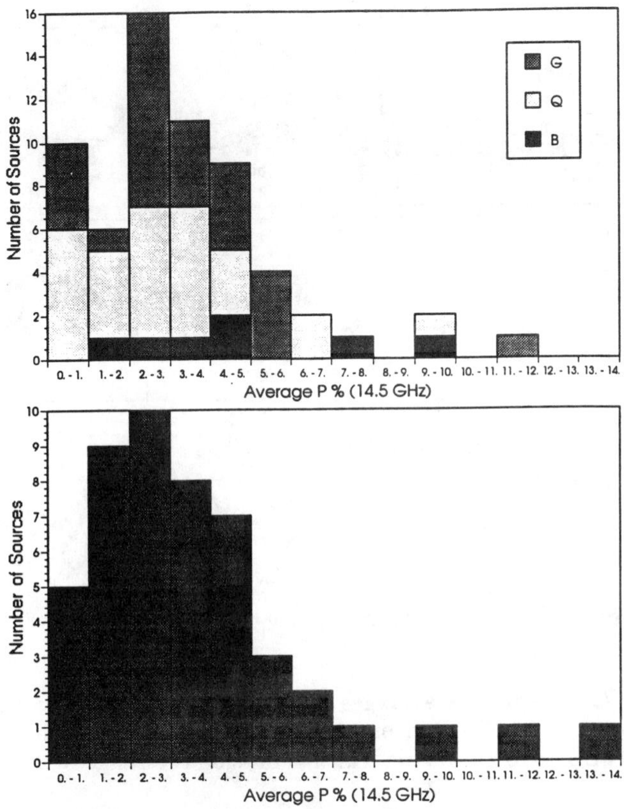

Figure 8. Histograms of the average percentage polarization at 14.5 GHz for two flux-limited samples. Upper panel: the Pearson-Readhead Sample. Lower panel: the BL Lac sample.

4.1. Curvature: complications in interpreting the light curves

Recent VLBI observations indicate that curvature is very common in jets; it can range from small scale kinks to large scale bends. Changes in the flow orientation directly affect the light curves and can complicate the comparison of the single dish data obtained in our program with VLBI maps. As an example, consider 4C 39.25, a source which has been very well-studied by a series of VLBI experiments and is known to consist of both stationary and moving components (Marcaide et al. 1994). The combined single-dish and VLBI data have been interpreted within the framework of a twisted jet (e.g., Marscher et al. 1991). Simulations by us to investigate the effects of curvature on the expected light curves show that modulation of a single outburst by flow curvature can lead to a light curve exhibiting fine-structure, i.e., multiple peaks, which might erroneously be interpreted as indicative of new shocks. Figure 11 contains the results of simulations involving the propagation of a single shock in a jet with modest nonlinearity of the flow (Hughes, Aller, & Aller 1994). Comparison of

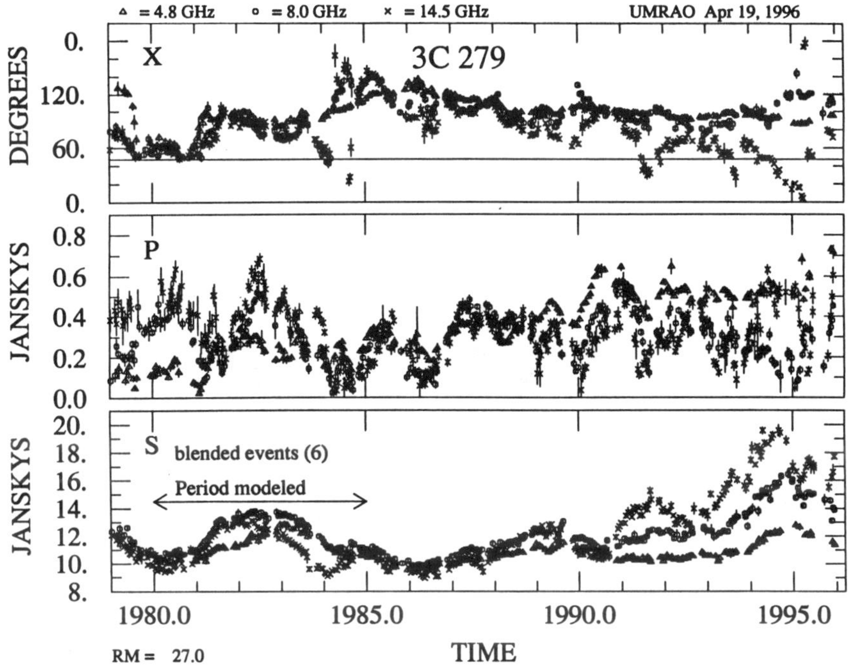

Figure 9. Two-week averages of the flux and polarization for 3C 279. The polarization has been corrected for Faraday rotation assuming a rotation measure of 27 rad/m^2. The line in the top panel marks the orientation of the VLBI jet.

the model predictions and the observed variations shows that the shapes of the light curves can be reproduced, and modeling can also successfully account for the development of both stationary and moving components as seen in the VLBI maps.

4.2. Evidence for temporal changes in the flow

Our recent attempt to model a 1991 event in BL Lac using the parameters adopted in fitting the early-1980 events shows that in order to fit the recent event requires a change in *both* shock strength and flow orientation (Aller et al. 1996), suggesting that significant changes have occurred in the flow. The VLBI monitoring observations indicate that since 1980 the structural position angle has increased by 10° and that the component speed has decreased by $\sim$ 20% (Mutel et al. 1994). The single-dish and VLBI position angle results complement each other in providing evidence for a change in jet orientation during the last decade. This change in orientation combined with flow curvature could account for the apparent slowing and brightness changes of the VLBI components (Mutel & Phillips 1987).

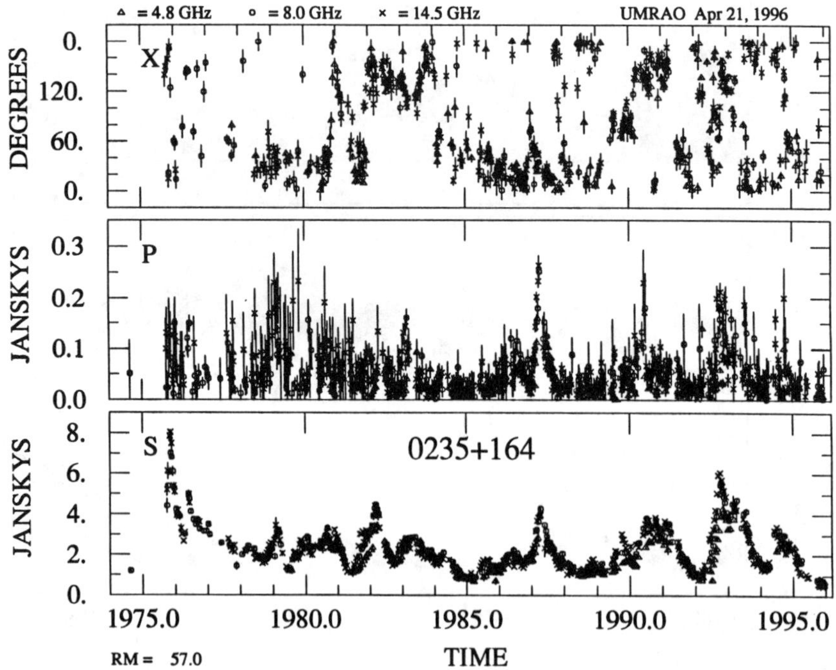

Figure 10. Two-week averages of the total flux density and polarization for 0235+164. The polarization has been corrected for Faraday rotation assuming a rotation measure of 57 rad/m^2. Most VLBI observations indicate that the source is unresolved (e.g., Gabuzda & Cawthorne 1992).

5. Are BL Lac type objects and QSOs intrinsically different?

Several pieces of evidence in the radio regime support the view that QSOs and BL Lacs are intrinsically different. The topological differences in the long-term position angle observations already discussed combined with an analysis of the evolution of the Stokes parameters by Hughes et al. (1994) show clear class differences. The long-term stability that one finds in the position angle orientation in QSOs, where the VLBI hot-spots are typically long-lived, can be explained in terms of a well-ordered quiescent flow (along the jet axis) with shear. In contrast, the rapid position angles changes that we see in many BL Lacs (e.g., 0235+164) where the VLBI components may be shorter-lived is consistent with the interpretation that here the flow is less-ordered and is dominated by the contributions from developing new components. Additionally, the relatively large deduced viewing angles discussed above are inconsistent with an interpretation of class differences in terms of viewing angle differences. The VLBP studies by Gabuzda (1994) also support intrinsic differences. They show that: 1) the distributions of $\beta_{app}h$ for components in BL Lacs and in QSOs are different such that the component speeds are characteristically slower for the BL Lacs, and 2) in VLBP maps

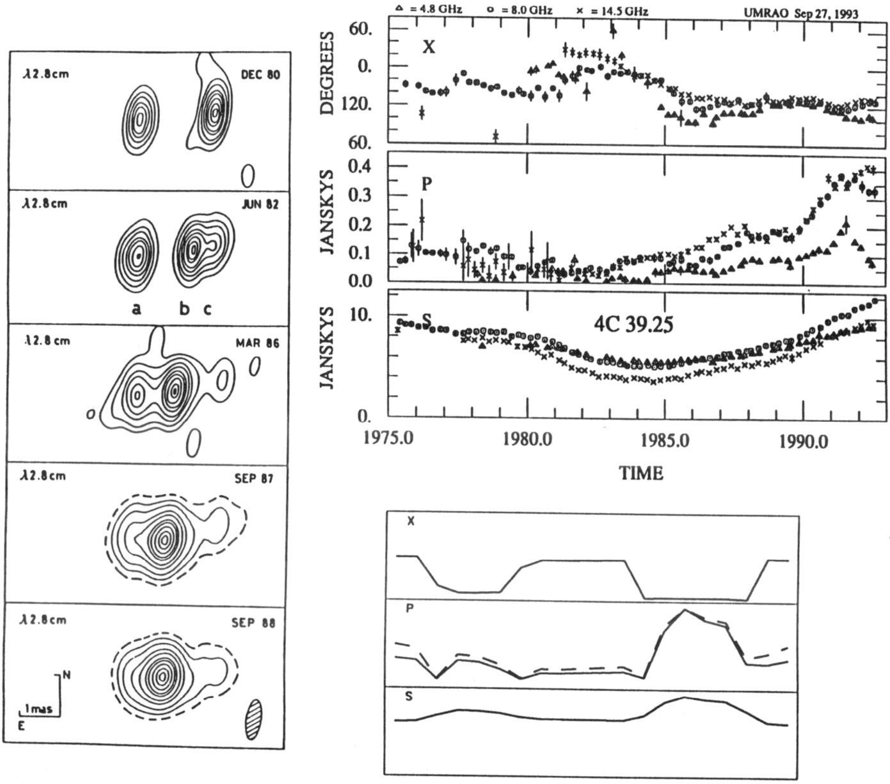

Figure 11. Left: VLBI maps from Marcaide et al. 1990. Right top: light curves based on tri-monthly averages for 4C 39.25. Right bottom: Model light curves.

of BL Lac-type objects the polarization position angles in the jet components are aligned with the structural axis suggesting that shocks are common in BL Lac jets at parsec scales (Gabuzda & Cawthorne 1992). Recent 2-dimensional hydrodynamic simulations by Duncan and Hughes (1994) support the view that flow velocity differences may explain the structural differences apparent in the radio data. In Figure 12 we show results from this work. The density distribution for the nonrelativistic case shows the development of Kelvin-Helmholtz instability, while for the relativistic case this instability is suppressed, and the density distribution is nearly featureless. A possible interpretation for the class differences seen in the single-dish and VLBP data is that the slow BL Lacs are less stable and hence suffer disturbances leading to internal shocks, while for the faster flows associated with QSOs there is little or no structure.

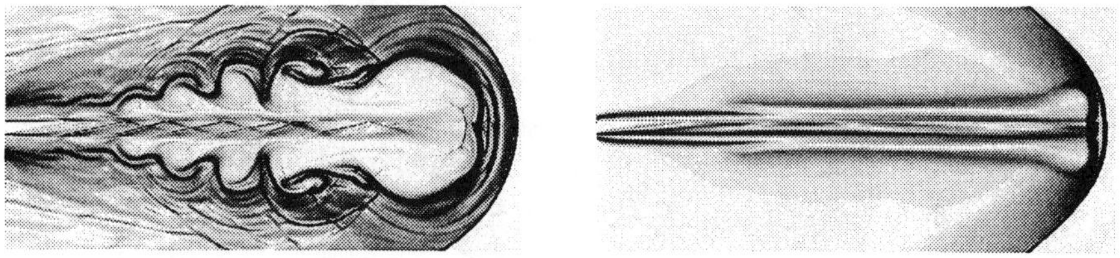

Figure 12. Schlieren-type images of the laboratory frame density gradient. Left: nonrelativistic case ($v = 0.3; \Gamma = 5/3$). Right: relativistic case ($\gamma = 10; \Gamma = 4/3$).

6. Summary

A comprehensive picture of the physical conditions in the radio-emitting regions of AGN has been developed based on the UMRAO cm-wavelength flux and polarization data. The low degrees of polarization that we find suggest that the emission arises from a medium with tangled magnetic field. Typically every two years, axial compressions develop in the flow; model-fits based on the propagation of shocks in the flows of relativistic jets show that they can account for the temporal variations in flux and polarization seen in the Michigan data and reproduce the structures seen in VLBI maps. The shocks are weak and do not greatly perturb the flow. The combined usage of single-dish and VLBI observations is important for fully understanding the physical conditions in the sources; the data support an association of VLB knots with the development of outbursts, but a simple 1:1 correlation for individual events does not exist in most sources due to complications resulting from overlapping events and from self-absorption in the emitting region. Nonlinearities in the flows are well-documented features in VLBI maps, and they must be accounted for when interpreting the light curves. The characteristics of these flows can change significantly over time periods of less than a decade; this is shown quantitatively by the model fits to bursts in BL Lac and by the structure function analyses of data segments. Both the single-dish and VLBP observations support the view that BL Lacs and QSOs are intrinsically different. In QSOs the flows are characteristically along the structural axis, and the integrated fluxes are dominated by contributions from the underlying jet. In BL Lacs, the contributions from possibly shorter-lived, new components dominate the integrated measurements. Relativistic hydrodynamical simulations suggest that the different magnetic field topologies that we see might be explained by speed-dependent stability.

Acknowledgments. We thank M. C. Aller for help in preparation of the figures and gratefully acknowledge partial support from NSF grant AST-9421979.

References

Aller, H. D., Aller, M. F., Hughes, P. A., & Mioduszewski, A. 1996, in Extragalactic Radio Sources, R. Ekers, C. Fanti, & L. Padrielli (Dordrecht, Kluwer), in press

Aller, M. F., Aller, H. D., Hughes, P. A., & Latimer, G. E. 1992, in Variability of Blazars, E. Valtaoja & M. Valtonen (Cambridge, Cambridge University Press), 126

Burbidge, G., & Hewitt, A. 1987, AJ, 92, 1

Carrara, E. A., et al. 1993, A&A, 279, 83

Duncan, G. C., & Hughes, P. A. 1994, ApJ, 436, L119

Gabuzda, D. C. 1994, in Compact Extragalactic Radio Sources, J. A. Zensus & K. I. Kellermann (Socorro, NRAO publication), 211

Gabuzda, D. C. & Cawthorne, T. V. 1992, in Variability of Blazars, E. Valtaoja & M. Valtonen (Cambridge, Cambridge University Press), 238

Hughes, P. A., Aller, H. D., & Aller, M. F. 1989, ApJ, 341, 54

Hughes, P. A., Aller, H. D., & Aller, M. F. 1991, ApJ, 374, 57

Hughes, P. A., Aller, H. D., & Aller, M. F. 1992, ApJ, 396, 469

Hughes, P. A., Aller, H. D., Aller, M. F., & Gabuzda, D. C. 1994, in The Nature of Compact Objects in Active Galactic Nuclei, A. Robinson & R. J. Terlevich (Cambridge, Cambridge University Press), 94

Hughes, P. A., Aller, M. F., & Aller, H. D. 1994, in Multi-Wavelength Continuum Emission of AGN, T. J.-L. Corvoisier and A. Blecha (Dordrecht, Kluwer), 233

Jones, T. W., et al. 1985, ApJ, 290, 627.

Marcaide, J. M., et al. 1990, in Parsec-Scale Radio Jets, J. A. Zensus and T. J. Pearson (Cambridge, Cambridge University Press), 59

Marcaide, J. M., et al. 1994, in Compact Extragalactic Radio Sources, J. A. Zensus & K. I. Kellermann (Socorro, NRAO publication), 141

Marscher, A. P., et al. 1991, ApJ, 371, 491

Mutel, R. L., Denn, G. R., & Dryer, M. J. 1994, in Compact Extragalactic Radio Sources, J. A. Zensus & K. I. Kellermann (Socorro, NRAO publication), 191

Mutel, R. L., & Phillips, R. B. 1987, in Superluminal Radio Sources, J. A. Zensus & T. J. Pearson (Cambridge, Cambridge University Press), 60

Mutel, R. L., et al. 1990, ApJ, 352, 81

Pearson, T. J., & Readhead, A. C. S. 1988, ApJ, 328, 114

Valtonen, M. 1996, these proceedings

Wehrle, A. E., Unwin, S. C., & Zook, A. C. 1994, in Compact Extragalactic Radio Sources, J. A. Zensus & K. I. Kellermann (Socorro, NRAO publication), 197

Witzel, A. 1990, in Parsec-Scale Radio Jets, J. A. Zensus & T. J. Pearson (Cambridge, Cambridge University Press), 206

Blazar Continuum Variability
ASP Conference Series, Vol. 110, 1996
H. R. Miller, J. R. Webb, and J. C. Noble, eds.

Three Decades of Radio Variability: the Michigan Variability Program

H. D. Aller, M. F. Aller, and P. A. Hughes

University of Michigan, Ann Arbor, MI 48109-1090

Abstract. The Michigan program to study the polarization and flux density of extragalactic variable radio sources was started at 8 GHz, 14.5 GHz and 4.8 GHz in 1965, 1974, and 1978 respectively. Since 1977 the 26-meter telescope has operated in an automated mode under computer control, providing an extensive, systematically-obtained data set. A core group of approximately 65 highly active AGN have been observed typically once per week at each of the three frequencies, and less frequent measurements have been made on an additional 200 objects. Example multi-frequency light curves of both flux density and polarization are presented which illustrate the range of observed phenomena. Our data show that flux density and polarization variability occurs on time scales ranging from a day to more than ten years: no periodicities are evident. The typical degrees of polarization of variable sources range from two to five percent, although during bursts a few sources reach polarizations exceeding fifteen percent. For several of the sources we have been able to obtain good fits to our observations with models which invoke shocks propagating along a relativistic flow.

1. The variability program

Active sources are observed weekly at each of our three observing frequencies, 4.8, 8.0, and 14.5 GHz in both linear polarization and total flux density. Our prime focus feed horn arrangement only permits us to observe at a single frequency at any one time; but we change our operating frequency every one to three days. We routinely follow approximately 65 objects intensively and make less frequent observations of about 200 additional sources. Objects which show little variability (and which are not flagged by us or by other researchers as being especially interesting) are only observed occasionally. Quiescent sources which are found to show new activity are added to our intensively monitored group if possible. The 212 sources which have been observed on 100 or more days are listed in Table 1; BL Lac has been observed on more than 2500 days. Examples of the long term behavior of extragalactic variable sources are shown in Figure 1 through Figure 6. The data are presented as monthly averages with the exception of OJ 287 for which two-week averages are shown. The polarization is shown in terms of the percentage of the total flux density. Additional results and a discussion of the variability are presented in Aller, Aller, and Hughes (1996).

Table 1. The 1950 position names of well observed sources.

0003−066	0348+175	0823+033	1148−001	1634+628	2007+777
0007+106	0355+508	0829+046	1156+295	1637+574	2014+370
0016+731	0358+004	0831+557	1157+732	1641+399	2021+614
0022+638	0415+379	0836+710	1215+303	1642+690	2030+407
0026+346	0420−014	0838+133	1219+285	1645+174	2032+107
0038+328	0422+004	0850+581	1225+206	1652+398	2033+187
0040+517	0430+052	0851+202	1226+023	1656+053	2037+511
0048−097	0433+295	0855+143	1253−055	1717+178	2050+363
0056−001	0446+112	0859−140	1254+476	1721+343	2121+053
0106+013	0458−020	0859+470	1307+121	1727−214	2131−021
0108+388	0518+165	0906+430	1308+326	1727+502	2134+004
0109+224	0521−365	0912+297	1328+254	1730−130	2144+092
0116+319	0528−250	0917+458	1335−127	1739+522	2145+067
0127+233	0528+134	0917+624	1345+125	1741−038	2147+145
0133+476	0538+498	0923+392	1358+624	1749+096	2153+377
0134+329	0539−019	0945+408	1400+162	1749+701	2155−152
0153+744	0548+165	0951+699	1404+286	1803+784	2155−304
0202+149	0552+398	0953+254	1409+524	1807+698	2200+420
0212+735	0605−085	0954+556	1413+135	1823+568	2201+315
0215+015	0605+480	0954+658	1418+546	1828+487	2223−052
0219+428	0607−157	0957+227	1422+202	1830−106	2223+210
0220+427	0710+439	1003+351	1458+718	1842+455	2229+391
0224+671	0711+356	1031+567	1504−166	1843−030	2230+114
0234+285	0716+714	1034−293	1510−089	1845+797	2243+394
0235+164	0723−008	1038+528	1514−241	1901+319	2251+158
0256+075	0723+679	1039+029	1514+197	1908−201	2254+074
0300+470	0727−115	1040+123	1524−136	1909+048	2255+416
0306+102	0735+178	1055+018	1538+149	1921−293	2319+272
0315+416	0736+017	1100+772	1543+005	1922+333	2335+031
0316+162	0738+313	1101+384	1548+056	1928+738	2345−167
0316+413	0754+100	1117+146	1606+106	1939+605	2351+456
0323+022	0804+499	1127−145	1607+268	1951+327	2352+495
0333+321	0808+019	1133+704	1609+660	1951+498	
0334+004	0809+483	1137+660	1611+343	1954+513	
0336−019	0814+425	1144+402	1624+416	1958−179	
0340+048	0818−128	1147+245	1633+382	2005+403	

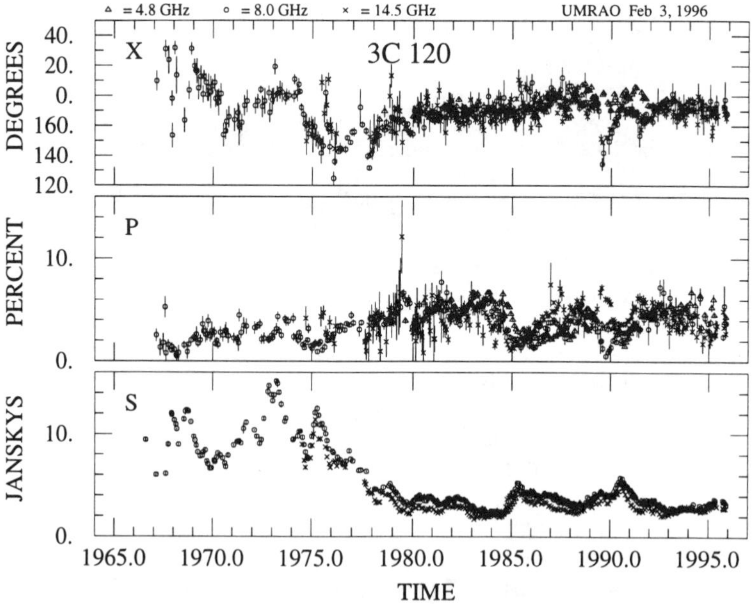

Figure 1. Polarization and flux density of the galaxy 3C 120.

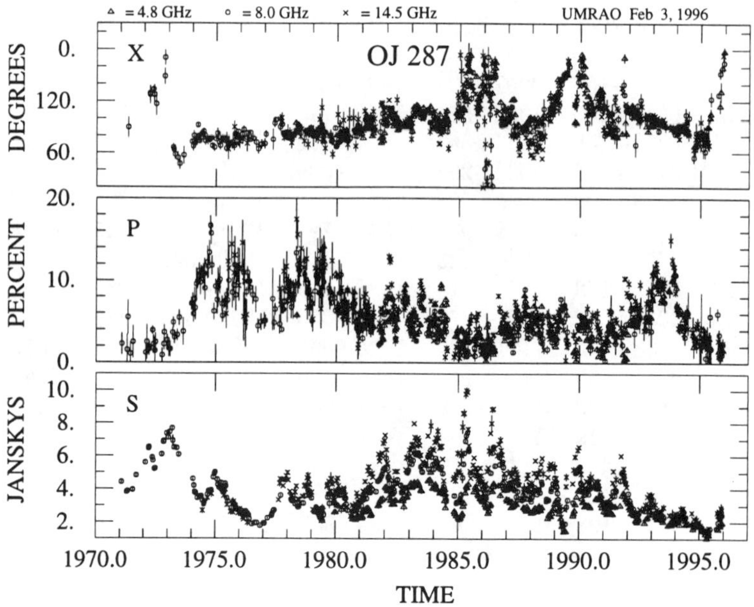

Figure 2. Polarization and flux density of the BL Lac object OJ 287 shown as two-week averages.

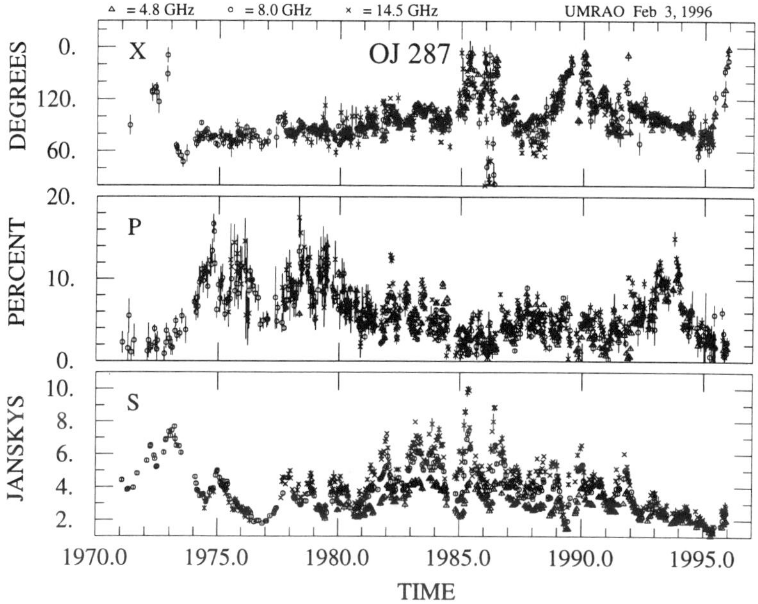

Figure 3. Polarization and flux density of the QSO 3C 279.

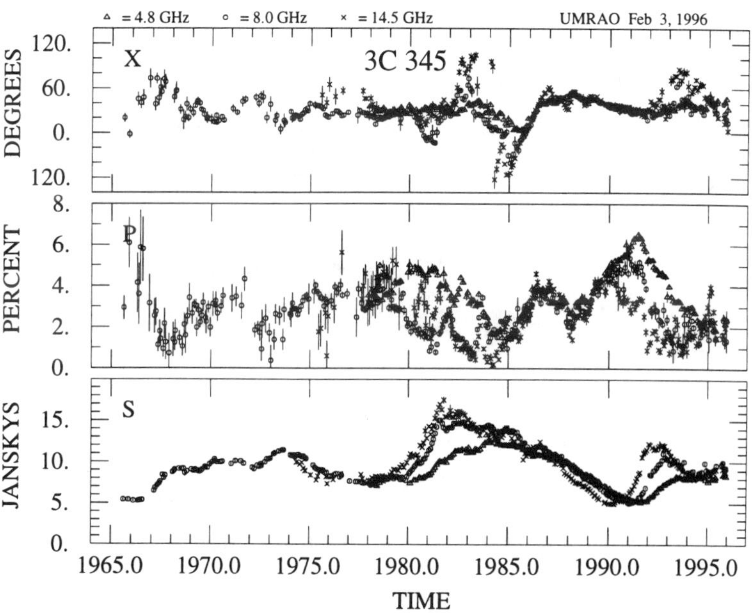

Figure 4. Polarization and flux density of the QSO 3C 345.

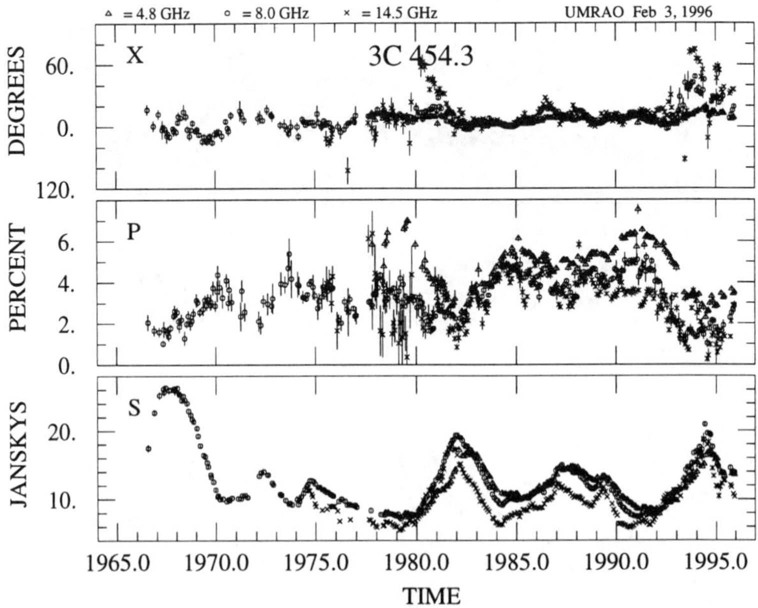

Figure 5. Polarization and flux density of the QSO 3C 454.3.

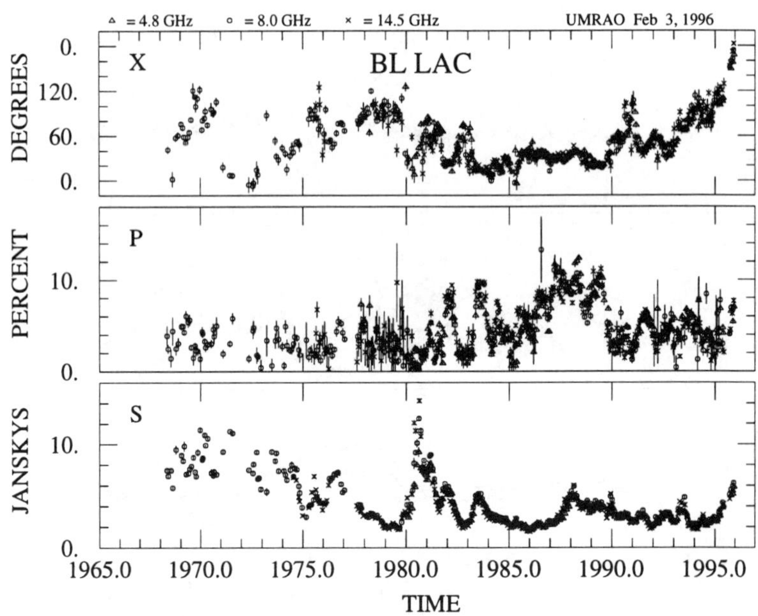

Figure 6. Polarization and flux density of BL Lacertae. The position angles have been corrected for a Rotation Measure of -200 rad. m^{-2}

Starting in 1977, all observing has been done under automatic computer control. We have just completed a rereduction of all our data taken since 1987, and corrected the older data, to take advantage of improved calibration parameters for the telescope. The flux density calibration of our program is based on the Baars et al. (1977) scale which is ultimately based on the source Cas A, with Cygnus A being used as a secondary standard. The instrumental polarization of the telescope is determined from periodic measurements of Galactic HII regions, which are assumed to be unpolarized. The observing and calibration procedures are described in more detail in Aller et al. (1985).

2. Acess to the database

The results of this program through 1984 are presented as monthly averages in Aller et al. 1985. The most recent data are available on the Department of Astronomy World Wide Web page (URL: http://www.astro.lsa.umich.edu) under the section "U of M Radio Astronomy Observatory Database." This is an interactive page which permits the user to select the desired averaging period. We expect in the near future to have all our data available in this manner.

Acknowledgments. This work was supported in part by the National Science Foundation under grant AST-9421979 and preceeding grants.

References

Aller, H. D., Aller, M. F., & Hughes, P. A. 1996, these proceedings

Aller, M. F., Aller, H. D., Latimer, G. E., & Hodge. P. E. 1985, ApJS, 59, 513

Baars, J. W. M., Genzel, R., Pauliny-Toth, I. I. K., & Witzel, A. 1977, A&A, 61, 99

Blazar Continuum Variability
ASP Conference Series, Vol. 110, 1996
H. R. Miller, J. R. Webb, and J. C. Noble, eds.

Rapid Variability in Faint Extragalactic Radio Sources

David J. Helfand and Saswato R. Das

Department of Astronomy, Columbia University, New York, NY 10027

Robert H. Becker

Physics Dept., University of California, Davis, CA 95616

Institute of Geophysics and Planetary Physics, Lawrence Livermore National Laboratory, Livermore, CA 94450

Richard L. White

Space Telescope Science Institute, Baltimore, MD 21218

Richard G. McMahon

Institute of Astronomy, Madingley Road, Cambridge CB3 0HA, United Kingdom

Abstract. In the process of constructing *Faint Images of the Radio Sky at Twenty-cm*, our VLA *FIRST* survey generates a unique database for radio variability studies. To date, we have multiple observations of over 135,000 radio sources in 1550 square degrees of sky complete to 1.0 mJy; thus, we are probing a population over 1000 times fainter and 10,000 times larger than any previous such study. The intervals between observations range from 3 minutes to 3 weeks; most are either a few minutes or 1–3 days, with sources observed from 2–5 times. Of the more than half million flux measurements obtained to date, we have found over two hundred sources which appear to vary by more than 25% between one day and the next; over a dozen objects show fluctuations greater than a factor of two. We have independent evidence that we have selected a special population of objects in searching for such variability, since the rate of matching to optical counterparts on the POSS I plates is greater than that for the catalog as a whole. Following a brief intorduction to the *FIRST* survey, we report here preliminary statistics on the complete variability study, detail our tests to eliminate spurious detections, and present our plans for confirmatory observations.

1. Introduction

Intensity changes in active galactic nuclei (AGN) have been observed on timescales ranging from seconds to decades and in spectral regimes from MHz to TeV. Both extrinsic and intrinsic mechanisms are responsible; interplanetary scintillation measurements were used soon after the discovery of radio sources to place limits on their angular sizes, and light travel time arguments set early constraints on the properties of the central engine in quasars. The principal diagnostic value of AGN variability is that it allows us to measure size scales orders of magnitude below the resolution limit of practical imaging techniques. Rapid variations, then, probe the smallest spatial scales, and the larger the amplitude of the fluctuations, the shorter the timescales which become accessible. It is with this paradigm in mind and largely oblivious to previous negative results that we undertook a search for day-to-day variability in extragalactic 20-cm sources.

We present here a preliminary report of our discovery of large-amplitude variability in these objects. In §2, we introduce the *FIRST* survey from which these data were derived. We then provide a synopsis of our variability search program and present our first results (§3). The concluding discussion (§4) offers support for our contention that we have discovered an unusual class of radio emitters by examining the statistics of their optical counterparts and presents our plans to follow up these sources with additional radio and optical observations.

2. The FIRST Survey

In May of 1993, we began a program at the NRAO[1] Very Large Array (VLA) designed to produce the radio equivalent of the Palomar Observatory Sky Survey (POSS I). Our goal is to cover 10,000 $\deg^2$ of the northern sky to a flux density limit of 1.0 mJy (5σ) with a resolution sufficient to locate each of the 1,000,000 sources detected to an accuracy of better than $1''$. To construct these Faint Images of the Radio Sky at Twenty-cm, we observe in the B-configuration of the VLA, utilizing the bandwidth synthesis mode to obtain complete mapping of the primary beam ($30'$ FWHM) in a series of overlapping snapshot pointings designed to create a uniform and efficient tiling of the sky. A total of over 600 hours were accumulated in the 1993 and 1994 B-configuration sessions, allowing us to complete 1550 $\deg^2$ of the high Galactic latitude ($|b| > 30$ deg) sky between declinations of +28 deg and +41 deg, and it is these data which form the basis of the work reported here. Details of the observing strategy and data analysis procedures may be found in Becker, White, and Helfand (1995).

A typical observing run lasts 10 hours and covers a strip of sky at a constant declination with a small overlap between fields. Since the pointings each last 165 s, this provides two observations of a subsample of sources separated by ~ 3 minutes. While some sources could be variable on such a timescale (e.g., coronally active stars or pulsars suffering interstellar scintillation) this is not the principal matter of concern here. The adjacent declination strips, with considerable overlap, are observed anywhere from 1 to ~ 20 days later, depending on VLA scheduling requirements, the order of sky coverage, etc. Finally, the edges of one year's survey area, as well as areas within that coverage which are reobserved to replace poor-quality or corrupted data, yield multiple observations

[1]The National Radio Astronomy Observatory is operated by Associated Universities, Incorporated under contract with the National Science Foundation

$\sim$ 15 months apart for a smaller set of sources. It is the timescale of days to weeks that is the focus of the results presented below.

In the final step of the reduction process, the overlapping snapshots are merged into a set of "coadded images" which form the principal product of the survey (see Becker et al. 1995 for details). White et al. (1996) describe in detail the procedures we use to extract a catalog of radio sources from these images. In addition, however, we apply the source detection algorithm to the individual snapshots, unconstrained by the catalog produced from the final images. For the current dataset, this yields 11,000 individual source catalogs (one per pointing) containing over 500,000 source fluxes, positions, and extents.

3. The variability search

We begin our search by matching the sources in each of the catalogs to the entire dataset on the basis of positional coincidence. Since *FIRST* positions are accurate to better than 1″ even at the survey threshold and the source-search algorithm forbids the identification of sources less than 3.6″ apart (c.f., the synthesized beam size of 5.4″), this matching is usually straightforward. In the case of complex and/or extended sources and source groups, the algorithm can fit multiple sources to a single region and, since the fitting is done on a field by field basis, the algorithm is not constrained to fit the same source model in each instance, leading to cases of interfield matches with separations

$$1'' < \Delta \text{ pos } < 3.6''.$$

We use our secondary selection criteria to screen out such discrepant fits which could otherwise lead to apparent variability.

Following this procedure, we have a single list with from one to five entries for each source. We then apply the following set of criteria to search for potential variables:

1. *Isolation:* We require that only one source lie within the source-fit region in all maps containing the source.

2. *Unresolved sources:* We require that the major axis of the two-dimensional Gaussian fit to each source be $< 2.0''$; White et al. (1996) demonstrate that sources with larger measured extents are often truly extended. Note that this criterion could exclude variable nuclei embedded in a halo of diffuse emission, but our inability to resolve multiple components in such a source would leave the variability amplitude undetermined.

3. *Peak flux density:* We require that the source appear in at least one map with a peak flux density in excess of 2.0 mJy.

4. *Location in the map:* We require the source to be located $< 24'$ off axis in the field of view, since the primary beam correction beyond this radius is highly uncertain. In addition, we require the source to lie within the inner 95% of the image to avoid spurious edge effects.

We then extract from the remaining list of 75,000 discrete sources those which show variability in any pair of observations that is simultaneously $> 25\%$ and $> 4\sigma$ (using the combined peak flux density uncertainties measured from the Gaussian fits).

The 75,000 sources include a total of 250,000 flux pairs (50% are observed twice, 35% three times, 14% four times, and 1% five times). Over 99.7% of these pairs show no evidence for fluctuations as large as 25%. This provides a stringent test of our reduction and analysis procedures, and acts as the control sample against which we compare the putative variables. Given the number of trials, we would expect 16 sources to show 4σ changes and < 1 to show a 5σ change. We observe 218 which meet our variability criteria.

Of the 218 variable candidates, only 20 show changes on the three-minute timescale. This is close to the expected number of Poisson 4σ results, although at least one such case is a source coincident with a sixth magnitude RS CVn star and probably represents real variability.

As a check on possible systematic errors introduced by the primary beam correction, we determined the ratio of the number of times the source farther from the field center was brighter compared to the cases in which the nearer source was brighter; the result, 0.95 ± 0.06 is consistent with that expected for randomly placed sources exhibiting real variability.

Our next test involves comparison of the flux densities for all sources observed on more than one occassion which fall within the field occupied by a candidate variable; the multiple overlaps for each pointing assure that there are between 45 and 90 such control sources per field. Figure 1 shows an example of the output from this procedure where we plot the flux density of a source in the variable's field vs. the ratio of the flux density measured in an adjacent field to the variable field value, quoted in standard deviations (derived as the local rms in the map at that location—see White et al. 1996 for details). This procedure uncovered a small number of fields in which as-yet-unidentified flux calibration errors generated spurious variability. When these fields were removed, the distribution of variable candidates on the sky was isotropic, as expected for a sample of real variables.

While we continue to devise tests in an attempt to falsify the variability we have measured, we have to date been unable to discredit flux changes in over 175 sources. The variability amplitudes range from the threshold of 25% over one day to a factor of four in a week; the median value is $\sim 50\%$ over a few days. We discuss below further evidence that we have selected through this procedure a special class of sources, and outline our plans for verification and followup.

4. Discussion

In order to take advantage of the astrometric accuracy of the *FIRST* survey, we have undertaken a program to find optical counterparts for as many radio sources as possible using the Automated Plate Machine (APM) scans of the POSS I plates archived at the Institute for Astronomy, Cambridge (McMahon & Irwin 1992). The E (red) and O (blue) plates have limiting magnitudes of 20.0 and ~ 21.5, respectively. In general, we find matches for 17% of all *FIRST* radio sources within $2''$, with a false match rate of 1%. For the sample of point

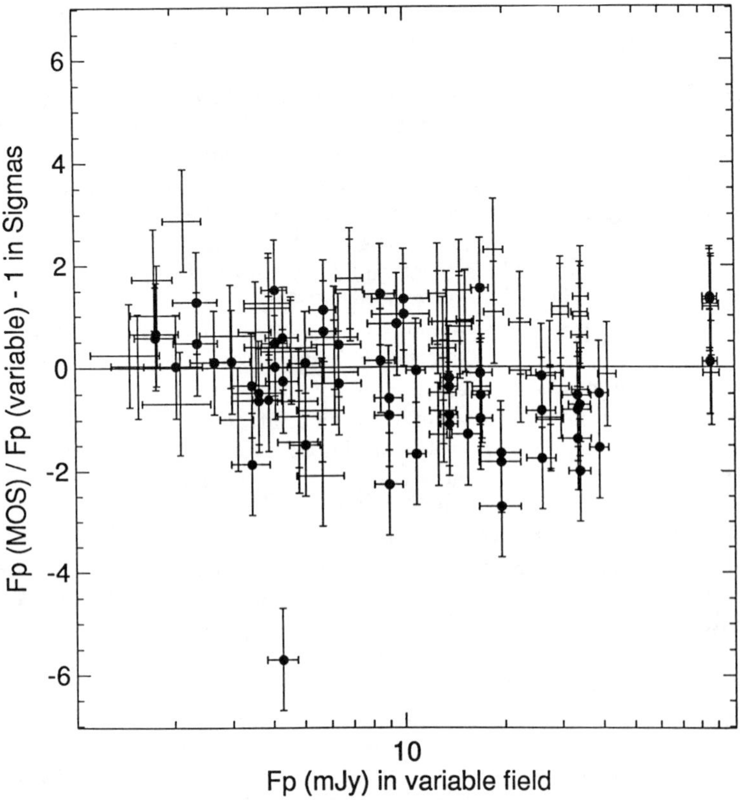

Figure 1. Results of the source variability verification routine. The field centered at RA (2000) = 10 10 30, $\delta(2000) = +36\ 51\ 24$ contains a candidate variable with a flux density of 4.2 mJy. Along with all other sources detected in that field which are also detected in an overlaping field, we compute the ratio of the peak flux density in the overlaping field to that in 10 10 30 + 36 51 24. We plot the results as the ratio of the multiply observed source (MOS) flux density to the flux density in the variable field in units of the standard error (σ) in this ratio, and subtract 1 to make 0 the value expected for a source of constant intensity. Different overlaping fields are represented by different symbols (•, +, etc.). The candidate variable stands out from the $\sim$ 60 other sources in these fields.

sources above 2.0 mJy defined as above, the rate climbs to 24.4%. However, for the sample of candidate variables, the match rate is 40.4%, an excess of 3.8σ over the control sample. Since the entire radio analysis is done independently of the optical database, this apparent excess of optical counterparts to the variable sources provides further evidence that the class of objects we have identified through our procedure is unusual.

The fact that the *FIRST* survey has revealed a new domain of radio variability in extragalactic objects is consistent with the volume of new parameter space the survey has explored: the sources we see varying are $\sim 10^3$ times fainter than those typically monitored for variability on intraday timescales (e.g., see Wagner, this volume) and the number of sources we have observed is $\sim 10^4$ times larger than samples in previous variability studies. The origin of such extreme variability at these relatively low frequencies is, however, a mystery. As Rickett et al. (1995) have pointed out, the short timescale does not distinguish between intrinsic and extrinsic mechanisms: if a source is compact enough to exhibit intrinsic variabilty on intraday timescales, its angular size is small enough that refractive interstellar scintillation must also be taken into account. We regard further discussion of the cause of the variability reported here to be premature pending further investigation of the phenomena.

And such investigation is imminent. Next month (March 1996) we will reobserve ~ 50 of these variable sources at the VLA, recording quasi-simultaneous flux density measurements at 20cm and 3.6cm in order to establish the sources' spectra, gain another light curve datum, and confirm their point-like nature with the higher-resolution 3.6 cm observations. Later in March we have several nights of time scheduled on the 2.1m telecope at Kitt Peak to obtain spectra of the optical counterparts for a subset of these sources. Future observations will include radio monitoring on logarithmically spaced intervals for perhaps a dozen of the best candidates to provide the data necessary to begin to explore the underlying cause of the variations.

Acknowledgments. This work is supported by the National Science Foundation, the National Geographic Society, NATO, the IGPP at Lawrence Livermore National Laboratory, Sun Microsystems, and Columbia University. This is Contribution Number 604 of the Columbia Astrophysics Laboratory.

References

Becker, R. H., White, R. L., & Helfand, D. J. 1995, ApJ, 450, 559

McMahon, R. G., & Irwin, M. J. 1992, in Digitized Optical Sky Surveys, ed. H. T. MacGillivray & E. B. Thompson (Dordrecht, Kluwer), p. 417

Rickett, B. J., Quirrenbach, A., Wagner, R. Krichbaum, T. P., & Witzel, A. 1995, A&A, 293, 479

White, R. L., Becker, R. H., Helfand, D. J., & Gregg, M. D. 1996, ApJ, submitted

Blazar Continuum Variability
ASP Conference Series, Vol. 110, 1996
H. R. Miller, J. R. Webb, and J. C. Noble, eds.

Blazars' Contribution to the Bimodal ΔPA Distribution of Powerful Radio Sources

W. Xu[1]

Infrared Processing and Analysis Center, Jet Propulsion Laboratory, California Institute of Technology, Pasadena, CA 91125

A. C. S. Readhead and T. J. Pearson

Owens Valley Radio Observatory, California Institute of Technology, Pasadena, CA 91125

P. N. Wilkinson

Nuffield Radio Astronomy Laboratories, University of Manchester, Jodrell Bank, Macclesfield, Cheshire SK11 9DL, UK

A. G. Polatidis

Onsala Space Observatory, S-43900 Onsala, Sweden

C. R. Lawrence

Jet Propulsion Laboratory, California Institute of Technology, Pasadena, CA 91125

C. D. Impey and P. S. Smith

Steward Observatory, University of Arizona, Tucson, AZ 85721

Abstract. We present the distribution of position angle differences between radio structures on the milliarcsecond and arcsecond scales (i.e. $|\theta_{\rm VLBI} - \theta_{\rm arcsec}|$ or ΔPA) of the combined first Caltech-Jodrell Bank and Pearson–Readhead sample. The 86 sources with measurable ΔPA show a bimodal distribution, with one peak near 0° and another broad peak near 90°. We confirmed that sources with high optical polarization ($p > 5\%$) show a bimodal distribution with a predominant peak near 90° while sources with low optical polarization peak only near 0°. The excess near 90° is certainly real and entirely due to objects with Blazar characteristics, i.e., high optical polarization, high radio variability, flat radio spectrum and high fractional core flux.

[1]National Research Council Research Associate at Jet Propulsion Laboratory, California Institute of Technology

1. Introduction

The most surprising result from the Pearson-Readhead VLBI survey of a flux density limited sample of 65 powerful radio sources was the discovery of the bimodality of the distribution of position angle differences, ΔPA, between the nuclear radio structure and the extended radio structure (Pearson & Readhead 1988). The ΔPA distribution found in the PR survey showed two distinct peaks, one near 0° (aligned sources) and the other near 90° (misaligned sources). However, the ΔPA distribution of a larger but less well defined sample was *not* bimodal (Rusk 1988). Impey *et al.* (1991) resolved this discrepancy by showing distribution for highly polarized sources ($p \geq 3\%$), which dominate the PR sample, is bimodal, while that for weakly polarized sources ($p < 3\%$), which dominate the larger Rusk sample, is not. The next development came from a semi-independent VLBI survey of a sample of 45 highly variable radio sources (Wehrle et al. 1992). 21 sources in the sample with measurable ΔPA also showed a bimodal distribution, although the peak near 90° was somewhat less pronounced than that in PR. However, the combination of these two samples strengthened the case for bimodality.

Recently we have completed the first Caltech-Jodrell Bank (CJ1 hereafter) VLBI survey (Polatidis et al. 1995; Xu et al. 1995). The combination of the PR and CJ1 sample forms a flux density limited sample of 200 sources. This is the first homogeneous complete sample that is large enough to provide a reliable measure of the bimodality of the ΔPA distribution. In this paper we present new results on the ΔPA distribution, drawn from the combined PR and CJ1 surveys. It is shown that the effect is certainly real, and that the excess of objects in which the nuclear jet is orthogonal to the large-scale jet is due to objects with blazar characteristics, i.e., high optical polarization, high radio variability, and flat radio spectrum, and large fractional core flux.

2. Position Angle Measurements

We took the innermost position angle of the nuclear jet as the position angle on milliarcsecond scales. For the objects in the CJ1 sample, we fitted models of Gaussian components to the self-calibrated 5 GHz VLBI data. The brightest and most compact Gaussian component is taken to be the core. The position angle of the Gaussian component closest to the core is taken as the position angle on milliarcsecond scales. For the objects in the PR sample, we used the models published by Pearson & Readhead (1988) and took the position angle of the component closest to the core as the position angle on milliarcsecond scales.

The position angle on arcsecond scales was determined between the core and brightest feature on either side of the core. In some cases for which there are hot-spots at both sides of the core, the hot-spot on the side that the milliarcsecond jet points to was used to measure the position angle.

3. ΔPA Distribution of the PR+CJ1 Sample

There are 86 sources in the combined CJ1 and PR sample which have measurable position angles on both milliarcsecond and arcsecond scales. The ΔPA of these

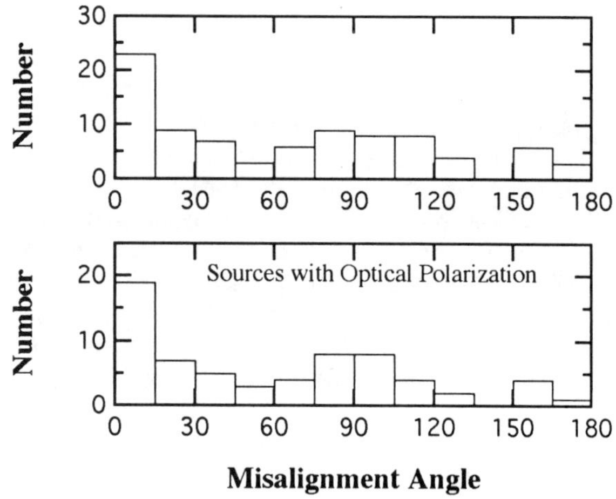

Figure 1. ΔPA distribution of PR+CJ1 sample — (a) upper panel: 86 sources with ΔPA measurement; (b) lower panel: 73 sources with optical polarization measurement.

sources have a weak bimodal distribution (Fig. 1). It shows a broad peak near 90°, contrary to the sharp peak near 90° shown by Pearson & Readhead (1988), Wehrle et al. (1992), and more recently by Conway & Murphy (1993), using both the PR sample and the sample of 45 highly variable sources (Wehrle et al. 1992).

4. Blazars' Contribution to the Bimodal ΔPA Distribution

Impey et al. (1991) found that sources with high optical polarization showed a bimodal ΔPA distribution while sources with low optical polarization showed only one peak near 0°. Similarly, we divided our sample into two groups according to optical polarization.

Amongst the 86 sources with measured ΔPA, 73 have optical polarization measurements. Fig. 2a shows that the 25 sources with high polarization ($p >$ 5%) have a bimodal ΔPA distribution with a prominent peak around 90°, 14 of 25 objects belong to the misaligned population (i.e., $70° < \Delta PA < 110°$), while Fig. 2b shows that the ΔPA distribution of the 48 sources with low polarization has only one peak near 0°. Our results confirm the findings of Impey et al. (1991).

The Kolmogorov-Smirnov test shows that the ΔPA distributions of objects with high and low optical polarizations differ at the 99.8% confidence level.

In addition to optical polarization, we also divided our sample into two groups according to thresholds of spectral indices, radio variability indices, and the fractional core flux density (i.e. the ratio of the flux density of the VLBI

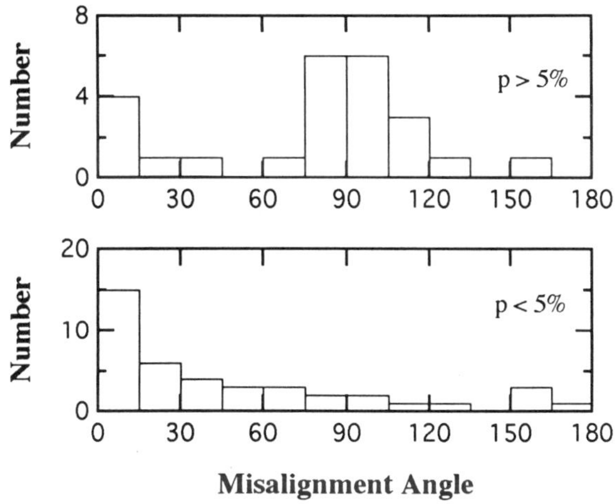

Figure 2. (a) upper panel: ΔPA distribution of sources with optical polarization $p > 5\%$; (b) lower panel: ΔPA distribution of sources with $p \leq 5\%$.

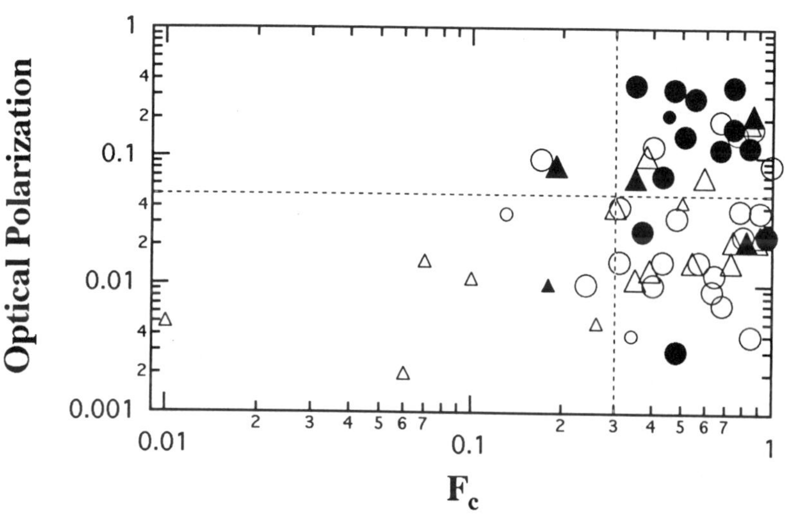

Figure 3. Summary of observation status. Filled symbol — $70° < \Delta PA < 110°$; Open symbol — $\Delta PA < 70°$ or $\Delta PA > 110°$; Large symbol — Flat radio spectrum ($\alpha > -0.5$); Small symbol — Steep radio spectrum ($\alpha \leq -0.5$); Circle — High radio variability ($V > 0.15$); Triangle — Low radio variability ($V \leq 0.15$).

core to the total flux density), F_c. The subgroups with values larger than the threshold of each chosen property shows a bimodal ΔPA distribution while the subgroup with values lower than the threshold shows only one peak near 0°. Detailed results will be published by Xu et al. (1996). In Fig. 3 we plot the optical polarization p vs. the fractional core flux densities F_c with symbols indicating three other properties: ΔPA, radio spectral indices and radio variability. It is clear that the peak near 90° in ΔPA distribution is contributed by a group of objects with some or all of the blazar properties, i.e., high optical polarization, high fractional core flux density, flat radio spectrum, and high radio variability. It is also shown that most sources with blazar properties have ΔPA near 90°.

5. ΔPA Distribution of γ-ray Blazars

There are 7 γ-ray sources (Thompson et al. 1996) amongst the 86 sources (Table 1). All of them have blazar properties, i.e., high optical polarization, flat spectrum, high fractional core flux density, and high radio variability. Six of them have ΔPA near 90°.

Table 1. γ-ray Sources in the PR+CJ1 Sample

Source	ΔPA	Optical polarization	Spectral index	Fractional core flux	Variability index
0716+716	87°	High	Flat	High	High
0804+499	115°	High	Flat	High	High
0836+710	10°	Low	Flat	High	High
0917+449	103°	High	Flat	High	Low
0954+658	104°	High	Flat	High	High
1633+382	74°	Low	Flat	High	High
1739+522	127°	Low	Flat	High	High

6. Possible Models

The physical origin of the bimodal ΔPA distribution is not yet clear. The aligned sources can easily be explained with a simple model in which the jet bends slightly between milliarcsec and arcsec scales (Readhead et al. 1983). However, the misaligned sources must be produced by a different population (Pearson & Readhead 1988).

A physical model has been proposed by Begelman et al. (1980) to account for large misalignments between structures probed by VLBI and the VLA. In their model, jets precess around a cone resulting from a massive binary black hole in orbit about each other. Any extended radio structure associated with jets, built up over many precession periods, will be aligned with the orbital angular momentum, i.e., the cone axis, while the Doppler boosted jet appears orthogonal to the cone axis when observed close to the line of sight. Such a geometry can produce objects with $\sim 90°$ misalignment. Conway & Murphy (1993) used similar geometry and performed detailed numeric simulations. They showed that the peak near 90° could be produced as a result of relativistic beaming with a geometric configuration in which parsec scale jets occur on conical helices with opening angles of 10°–20° and pitch angles of $\sim 1°$. The predicted distribution of misalignment angles is fairly sensitive to the geometry and Lorentz factor of the jet.

It seems that some sort of helical motion of jets may be necessary to produce large misalignments. Since most of blazars in our sample, especially γ-ray blazars, have ΔPA near 90°, it is logical to think that blazars are associated with helical jet motions. The most interesting mechanism to produce a helical motion is precession induced by a binary black hole. This raises the question whether all blazars have a binary black hole in their nuclei. Another mechanism to induce helical motion is Kelvin-Helmholtz instabilities produced by random jitter in the initial jet direction (Hardee et al. 1994). Ram pressure bending and twisted accretion disc models have also been suggested (Appl et al. 1996).

It seems clear that the physical origin of the bimodal ΔPA distribution is closely related to the central engine itself, or the environment near the central engine, or a combination of both. Any finding of the physical origin of the bimodal ΔPA distribution may have a profound impact on our understanding of active galaxies.

Acknowledgments. We thank R. D. Blandford and A. E. Wehrle for useful discussion. Part of this work was performed while W. Xu held a National Research Council-JPL Research Associateship. The work at the California Institute of Technology was supported by the National Science Foundation (grants AST-8814554 and AST-9117100 to the Owens Valley Radio Observatory). The reference search had made use of the NASA/IPAC Extragalactic Database (NED) which is operated by the Jet Propulsion Laboratory, California Institute of Technology, under contract with the National Aeronautics and Space Administration.

References

Appl, S., Sol, H., & Vicente, L. 1996, A&A, in press

Begelman, M. C., Blandford, R. D., & Rees, M. J. 1980, Nature, 287, 307

Conway, J. E., & Murphy, D. W. 1993, ApJ, 411, 89

Hardee, P. E., Cooper, M. A., & Clarke, D. A. 1994, ApJ, 424, 126

Impey, C. D., Lawrence, C. R., & Tapia, S. 1991, ApJ, 375, 46

Pearson, T. J., & Readhead, A. C. S. 1988, ApJ, 328, 114

Polatidis, A.G., Wilkinson, P.N., Xu, W., Readhead, A.C.S., & Pearson, T.J. 1995, ApJS, 98, 1

Readhead, A. C. S., Hough, D. H., Ewing, M. S., Walker, R. C., & Romney, J. D. 1983, ApJ, 265, 107

Rusk, R. E. 1988, Ph.D. thesis, Univerity of Toronto

Thompson, D. J. et al. 1996, ApJS, 101, 159

Wehrle, A. E., Cohen, M. H., Unwin, S. C., Aller, H. D., Aller, M. F., & Nicolson, G. 1992, ApJ, 391, 589

Xu, W., Readhead, A. C. S., Pearson, T. J., Polatidis, A. G., & Wilkinson, P. N. 1995, ApJS, 99, 297

Xu et al. 1996, in preparation

Blazar Continuum Variability
ASP Conference Series, Vol. 110, 1996
H. R. Miller, J. R. Webb, and J. C. Noble, eds.

Blazar Radio Variability

E. Valtaoja

Metsähovi Radio Research Station, Helsinki University of Technology, FIN-02540 Kylmälä, Finland

Abstract. Observations of total flux density variations are in many cases the only way to study the structure and the physics of AGN on the smallest scales. Radio observations can also serve as tracers of other phenomena. The huge amounts of available radio monitoring data can be used in a number of different investigations, among them tests of the shocked jet models, comparisons with VLBI data, connections between radio, optical, and gamma-rays, and the classification of AGN.

1. Introduction

Radio observations of total flux density variations in active galactic nuclei span more than thirty years. By far the largest amount of data we have on AGN consists of these measurements. At Metsähovi alone, over 30 000 measurements have been made during the last 15 years, and UMRAO exceeds this by a considerable factor. With such databases, one does not expect to rush off after an observing run and start writing a Letter to Nature. Although individual observations are often valuable per se—for example, by contributing to a multifrequency spectrum—the real power of radio observations lies in the sheer amount of data, the completeness and the time coverage they provide.

This paper is a complement to the one presented in the IAU Symposium 159 in 1993 (Valtaoja 1994). That review was mainly concerned with the use of radio data in testing the shocked jet models. This time I will concentrate more on the use of radio observations in connection with other data, an area where significant progress has been made during the last few years. Much of this progress rests on the large databases provided by the radio monitoring programs.

1.1. Buried in data

A few dedicated multifrequency monitoring programs have been collecting radio data over the years. At low frequencies, the University of Michigan program has been monitoring quasars at 4.8 GHz, 8.0 GHz and 14.5 GHz since 1965 (Aller, these proceedings). Its higher frequency counterpart since 1980 has been the Metsähovi monitoring program at 22.2 GHz, 36.8 GHz and 90 GHz, complemented since 1987 with SEST observations at 90 GHz and 230 GHz (Teräsranta et al. 1992; Tornikoski et al. 1996). Going still higher, Ian Robson and his collaborators (Robson, these proceedings) are monitoring a small sample of blazars at millimeter and submillimeter wavelengths with JCMT.

High radio frequency monitoring at Metsähovi occupies a special place for two reasons. First (unlike at still higher frequencies), a nearly complete record of flux variations of a large sample of AGN can be obtained. Second (unlike at lower frequencies), the sources are mostly transparent, and connections between radio and other regimes are much easier to study.

1.2. Lies, damn lies and statistics

Only in radio do we have complete records of how AGN behave: we can study a fully sampled UMRAO flux curve or analyze repeated SEST observations of a complete sample of blazars. This is very different from the usual situation in AGN research: a few data points, perhaps obtained in one-of-a-kind, never-to-be-repeated satellite experiment. The data fits our pet theory well, but nagging doubts remain: what about other times? what about other sources? how much could or should we generalize from this one peek through the keyhole? The very uniqueness and spectacular nature of the data is often also working against us.

Even if other data are available, comparisons between theories and observations often remain limited: perhaps a single multifrequency spectrum, or one 'spectacular coincidence' (which, of course, needs not be a coincidence—we just cannot know it!). Considering the number of adjustable parameters in theories and the persuasive appearance of coincidences, it is not surprising that not too many theories are conclusively disproven with such an approach.

Statistical analysis is less eye-catching, but in the final tally the individual, even if spectacular, cases must carry less weight than an average property found in a large sample. To take just one example: the arguably most variable radio source, OJ 287, is a BL Lac object—but when we analyze data from complete samples of BL Lacs, we find that they are rather characterized by a *lack* of rapid or strong radio variations as compared to quasars (Valtaoja 1994). Any serious theory must first and foremost explain this general, statistical result, and not get sidetracked by the extreme radio variability of a few BL Lacs.

1.3. The Apollo 13 approach

In addition to providing large databases from which the characteristic properties of a phenomenon or a class of objects can be found, radio monitoring can also be used in less obvious ways to circumvent our lack of data in other regimes. I will give one such example, the origin of gamma-rays.

Several theories of gamma-ray production can be bent, folded or mutilated to produce an acceptable fit to a single multifrequency spectrum. (It is, of course, no fault of the theorists that there are so few spectra to begin with!) Ideally, we would like to have repeated simultaneous radio-to-gamma spectra of a carefully selected sample of AGN, with a well-sampled record of mid-IR flux density variations also thrown in, thank you, and then we could without doubt find out just what is going on. Obviously we do not have and will not have such data. What then?

We can either sit down in millennial despair and say that we must wait for the next gamma-ray observatory, or we can try to find another way. Radio monitoring may offer such a way . Radio flux is a tracer of the whole infrared-to-UV blazar spectrum, and although the correlation between the various regimes is not in general simple, with enough radio data statistically significant connections may still be recovered and compared to theoretical predictions (see Section 4.2). Radio variations are two steps removed from gamma-ray variations, and so they are not the kind of data we really would like to use, but if your only chance is to fit a circular peg into a square hole, you should at least attempt it.

2. Radio observations and the structure of Active Galactic Nuclei

The innermost cores of active galaxies remain unresolved even with best VLBI. Nevertheless, observations of their continuum spectra, variations, polarizations, correlations and time delays over a wide frequency range have enabled us to study the structure and the physics of these regions.

The 'canonical' shocked jet model (Marscher & Gear 1985) proposes that the quiescent radio flux comes from the unperturbed radio jet powered by the accreting black hole. The radio outbursts are caused by shocks forming and propagating in the jet. The model essentially predicts four things:

1. The only major contributors to the radio flux are the jet and the shocks.
2. The quiescent and the variable part of the radio flux can be separated. The jet flux has long-term (years or more) stability, while the shock flux is the cause of the radio variations in timescales from weeks to years.
3. Coming from a simple shock, the variable flux has the spectrum of a single, homogeneous synchrotron component.
4. The time evolution of the shock spectrum (i.e., the motion of the spectral turnover $S_{\rm m}, \nu_{\rm m}$, since the spectral shape does not change), while model-dependent, reflects the shock's growth and decay stages.

These predictions are, of course, just first approximations to a much more complicated reality. Nevertheless, the simplest theory forms a convenient framework for the analysis of radio data, especially using the concept of 'generalized shock models' (Valtaoja et al. 1992; Stevens et al. 1994).

To compress a decade of painstaking data collection and analysis into a few sentences, the conclusions of my previous review still hold: "The framework of evolving shocks in a relativistic synchrotron jet is sufficient to explain the observations, without the need to invoke alternative explanations for either the spectral shape or the variations. However, at present the observations provide only weak constraints for the shock models (Valtaoja 1994)." The separation of the jet and the shock components, their spectral shapes, the evolution of the shock spectrum, the variability amplitudes and the time delays between different frequencies all agree with the predicitions of the shocked jet model (e.g., Valtaoja et al. 1988; Brown et al. 1989; Valtaoja et al. 1992; Tornikoski et al. 1993; Robson et al. 1993; Lainela & Valtaoja 1993; Lainela 1994; Stevens et al. 1994, 1995; Robson, these proceedings).

Significant advances will require new theoretical modelling, either analytical or numerical simulations. Here, however, the main difficulty may be that anything approaching a realistic description of a relativistic, shocked jet becomes so complicated that comparisons with the observations are not very illuminating. After all, the information content of synchrotron radiation is rather meagre.

Nevertheless, both the quantity and the quality of the radio observations can still be improved, especially regarding millimeter and submillimeter data, and the as yet very scarce high frequency polarization observations. Another approach, which is only now starting to be fully explored, is to connect radio monitoring to VLBI data, and to the rest of the electromagnetic spectrum. The rest of my review will discuss some of these new developments.

3. Total flux density variations and VLBI

Shocked jet models predict that all major flux density variations are related to the formation, growth and decay of new shocks in the jet. VLBI maps show new components ejected from the core; these are assumed to be the same shocks. Why, then, have actual demonstrations of VLBI and continuum connections been mainly limited to vague statements of the kind 'the extrapolated zero epoch of Component C seems to coincide with a radio outburst in early 1990s'?

The main reason is lack of angular and temporal resolution. The vast majority of VLBI is done at 5 GHz or lower, with a global resolution of $\gtrsim 1.3$ mas. Even assuming the rather large expansion speed of ≈ 0.3 mas/yr, the new component would emerge from the core over 4 years after its creation. By this time, the shock has long since peaked, and we are seeing only the last stages of an old, adiabatically expanding remnant. Most likely there are also a number of other detectable remnants; the slow outbursts blend together in the 5 GHz total flux density variations, and comparisons with structural changes are difficult.

At 22 GHz the situation is much improved. The new VLBI components now emerge after a year or so, and in the 22 GHz total flux density monitoring the flares can be separated from each other. The age, the phase, and the contribution of each shock to the flux can now be estimated. Moreover, with typical lifetimes of only a few years there are fewer strong components visible at any given time, making identification easier. Results from 22 GHz VLBI monitoring programs are now becoming available, enabling for the first time proper comparisons between VLBI and radio continuum data.

All radio flares appear to have a similar morphology: an exponential rise followed by a sharp peak and an exponential decrease (Teräsranta et al., these proceedings). Flux variations can therefore be rather well decomposed into a number of self-similar flares. From such decompositions we can follow the flux evolution of a single shock (a single VLBI component) and estimate the number and the fluxes of shocks visible at any epoch, and compare the results with 22 GHz VLBI data.

Figures 1 and 2 show such a decomposition for 3C 345. We find four major components peaking in 1986, 1992, 1993 and 1995. The 'zero epochs' and the flux density evolution of the two first components compare well with those of the the VLBI components C5 and D (Lobanov & Zensus 1996); the two latter correspond to the new components C7 and C8 (Leppänen 1995; A. Lobanov, private communication). The sum of these four 'total flux density' components explains the observed variations in the total flux density, and each of the components evolves as the corresponding VLBI component: the flares seen in flux density monitoring do correspond to the creation of new VLBI components.

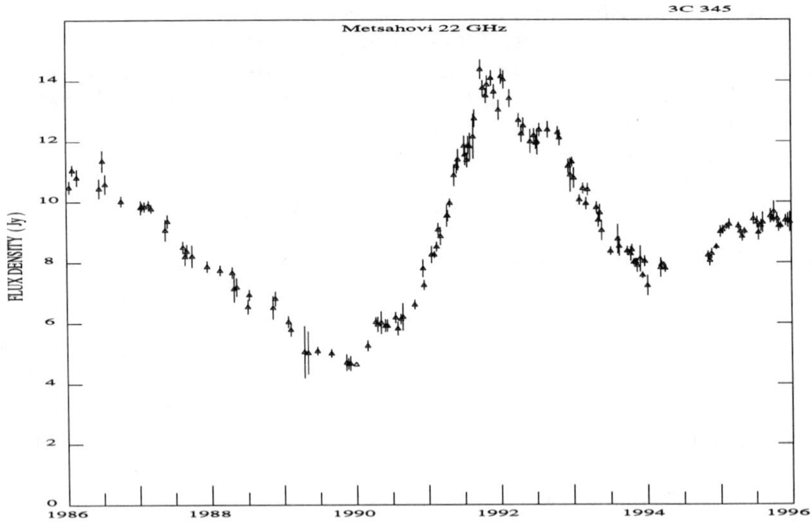

Figure 1. 22 GHz radio variations of 3C 345 from Metsähovi monitoring program.

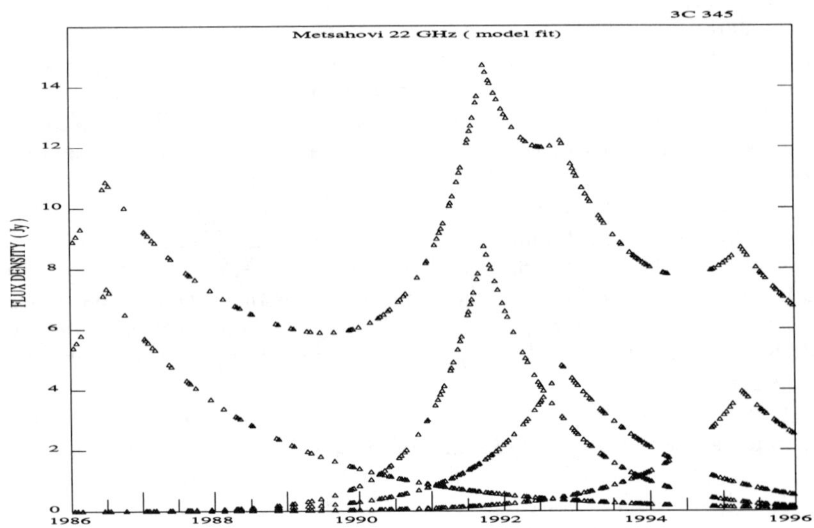

Figure 2. The total flux density variations modelled with four self-similar exponential flares.

There is one snag, however. The VLBI component D is identified with the core. In both the VLBI monitoring and the continuum monitoring decomposition its flux evolved similarly to the shocks C5, C7 and C8; yet no new VLBI component was seen emerging from the core. What was going on? Did the flux of the core itself (i.e., the apex of the underlying jet) vary, closely mimicking the flux variations of shocks, or was there a shock created around 1990, peaking in 1992 and decaying after that without ever moving far enough from the apex to be resolved with 22 GHz VLBI? The second possibility seems more plausible, but must we then postulate the existence of another kind of shocks which evolve—at least in total flux density—as the shocks we are familiar with, but which do not propagate appreciable distances down the jet? The increased resolution provided by space VLBI should provide an answer.

4. Multifrequency connections

A number of different ways may link an event observed at a given frequency to other frequencies. The simplest case is that the same mechanism (e.g., synchrotron radiation) causes the variations. There may be a direct causal connection, for example via the inverse Compton mechanism. Alternatively, there may be an indirect causal connection where some intermediate agent carries the influence from the original to the follow-up event. An example of this could be an increase in the accretion rate to the black hole, which is later reflected in increased jet flow and the formation of a new shock. The connection may also be more of a statistical than of a causal nature, for example with alternating active and inactive periods (of unspecified origin) observable over the whole range of electromagnetic radiation produced by the source. Finally, of course, there may simply be no connection at all, with, say, X-ray and radio variations emanating from completely different parts of the source, separated by light-years and with no knowledge of each other.

All the various possibilities have different observable consequences, and require different observing strategies. However, in all cases we should be wary of 'spectacular coincidences:' for example, blazars are so variable in the optical regime that it would be rare *not* to have something occurring close to the event we are studying—especially if we have no foreknowledge of the reasonable range of time delays or even which event should occur first.

I will discuss two multifrequency connections which have recently been established beyond reasonable doubt. In both cases the key is the large number of data made available by the radio monitoring programs.

4.1. Radio and optical variations

Several mechanisms produce optical radiation in various parts of an active galaxy. The observed spectrum and its variations are a sum of several, often unconnected, mechanisms: lacking spatial information, it is extremely difficult to separate their contributions and to establish connections between radio and optical variations. Any correlation will always be degraded by 'noise' from the other, unconnected, optical variations. One more ambiquity is introduced if low radio frequencies are used in the comparisons, since even between low and high radio frequencies time delays of months or years are common.

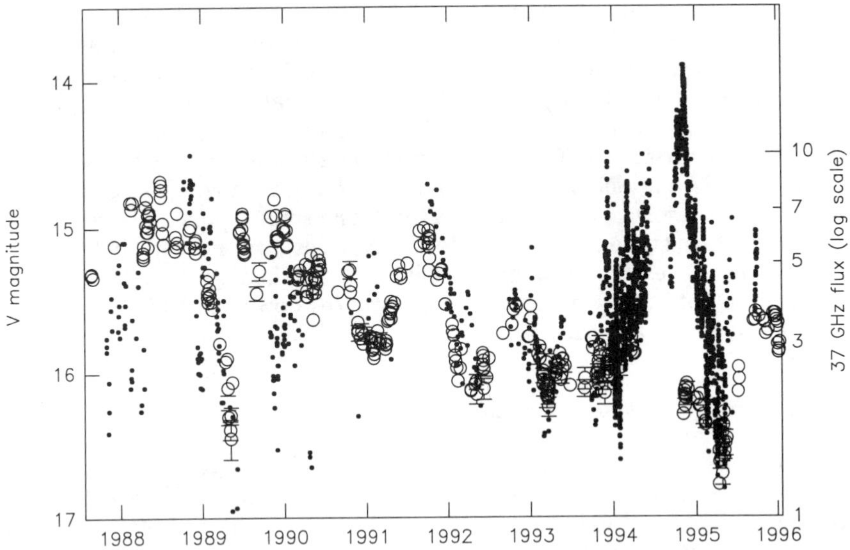

Figure 3. The V band optical (dots) and 22 GHz radio (open circles) variations in OJ 287 between 1988 and 1996. The predicted periodic optical outburst occurred in late 1994.

Some tentative results were found in the 1980s (e.g., Balonek 1982; Valtaoja 1987), but firm statistical evidence has been obtained only recently. Tornikoski et al. (1994) compared the Florida optical monitoring data for 22 AGN with the Metsähovi and SEST 22–230 GHz radio data. Performing a discrete correlation analysis, they found a good correlation in 10 sources, six of them with simultaneous radio and optical variations. Clements et al. (1995) repeated the analysis using Michigan data between 4.8 and 14.5 GHz. The results were very similar: 9 of their 18 sources showed a significant correlation, with time delays ranging from zero to 18 months. The longer time delays can easily be understood as resulting from the lower radio frequencies used in the study.

Tornikoski et al. concluded that many, perhaps all, high frequency radio flares have simultaneous optical counterparts, indicating that in these cases we see optical synchrotron radiation from shocks. However, most optical variations do not have any radio counterparts, and their origin remains unclear. Other questions abound. What is the difference between sources showing correlations and those not showing them? Why are some simultaneous, and some delayed by considerable amounts?

The unique optical flux curves gathered in the large international OJ-94 collaboration (Takalo, these proceedings) are shedding new light on the radio–optical connections. Two of the program objects, OJ 287 and 3C 66A, are both classified as BL Lac objects and have comparable redshifts, optical, and radio luminosities. Nevertheless, their radio–optical connections are quite different (Figures 3 & 4). Both sources are strongly variable in the optical. In the case of OJ 287, radio and optical variations generally track each other quite closely,

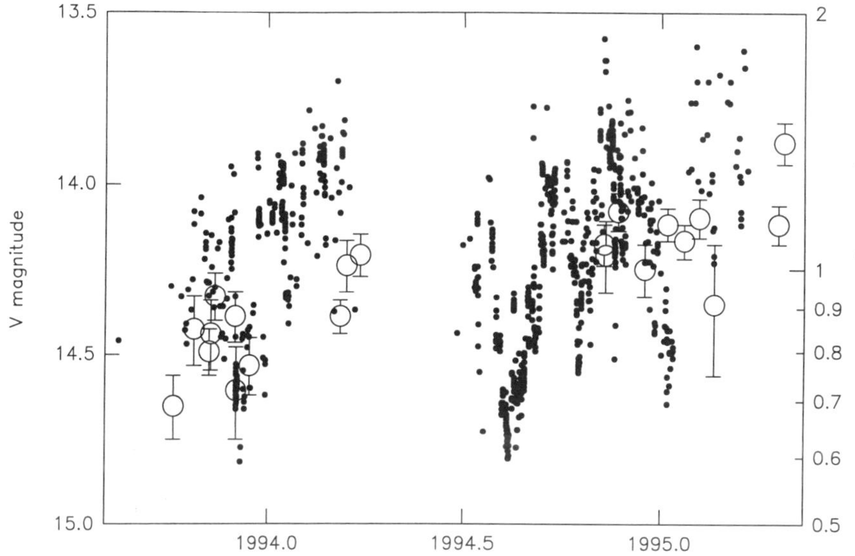

Figure 4. The V band optical (dots) and 22 GHz radio (open circles) variations in 3C 66A between 1993 and 1996 (Takalo et al. 1996).

with little or no delay between the V and the 37 GHz bands. In contrast, the radio variations in 3C 66A appear to be weak and uncorrelated with the strong optical flares (Takalo et al. 1996).

This difference is demonstrated very clearly if we plot the observed maximum amplitude of variations versus frequency (Figure 5). For OJ 287 the amplitude of variations from UV to mm-wavelengths increases smoothly, and then starts to decrease again at centimeter wavelengths. This agrees well with the hypothesis that the synchronous variations come from a single synchrotron component, transparent all the way from UV to mm–radio.

For 3C 66A the UV to near-IR variability is almost identical with that of OJ 287, indicating the same variability mechanism. The radio connection shows that the optical radiation in OJ 287 must be synchrotron radiation from the shocks, and thus the tentative conclusion is that the optical variations in 3C 66A also originate in the shocks. But why, then, do they fail to reach the radio regime, where the variations are an order of magnitude weaker than in OJ 287? Are we here seeing shocks with turnovers always remaining in submm/IR, and never evolving to become visible in the radio regime?

The predicted periodic flaring in OJ 287 in late 1994 (Sillanpää et al. 1996 and these proceedings), adds another mystery. As Figure 4 shows, the radio activity almost ceased during this time. (There is a gap in the Metsähovi data due to re-panelling of the telescope, but SEST and UMRAO data show that there were no strong radio flares during that time.) If optical and radio variations usually track each other, why was the largest optical flaring not visible in the radio? Was it produced by a different mechanism, perhaps thermal radiation

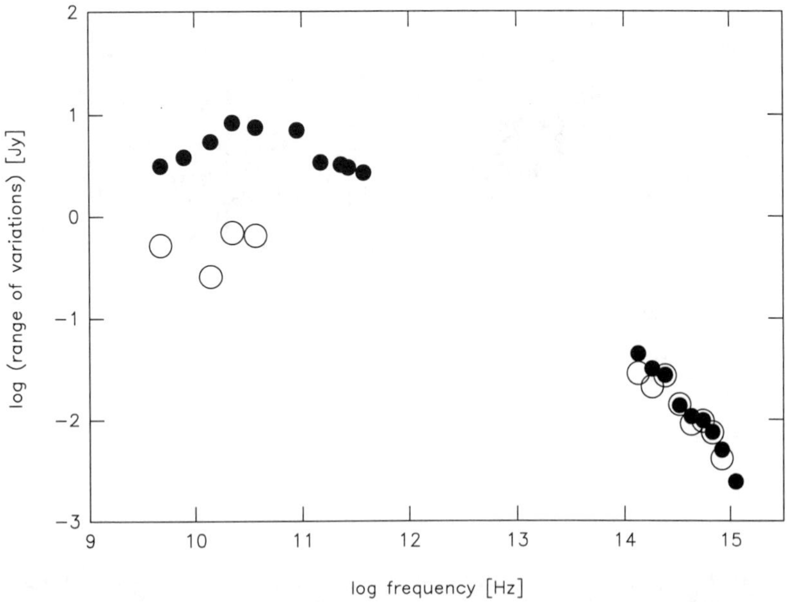

Figure 5. The observed maximum amplitude of flux variations DS at each monitoring frequency for OJ 287 (dots) and 3C 66A (open circles). Note the similar variability amplitudes in the IR-to-UV bands, and the lack of strong radio variability in 3C 66A.

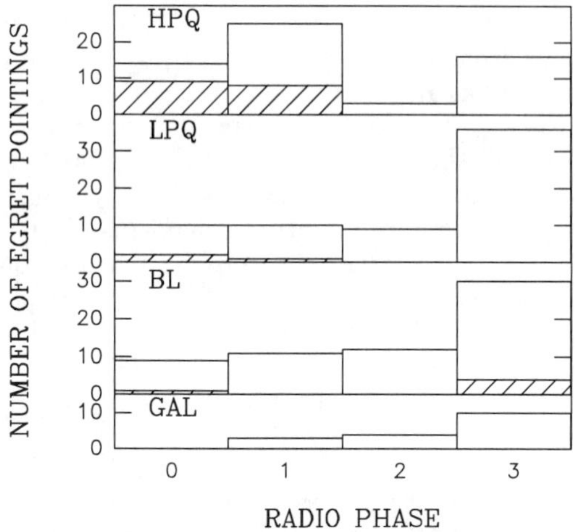

Figure 6. Dependence of EGRET detection probability on the source class and the radio state during the observation (Valtaoja & Teräsranta 1995.) The gamma-ray detections are hatched.

from the accretion disk perturbed by the passage of the postulated secondary black hole with 12-year period (Valtonen et al., these proceedings)?

These questions remain as yet unanswered. However, it is encouraging to see that with sufficient amounts of data we are advancing from anecdotal to statistically significant evidence, are beginning to study the production mechanisms and sites of the optical emission and to separate at least some of the different contributions to the optical flux.

4.2. Radio and gamma-ray variations

As detections by EGRET started to pile up, it become clear that most of the gamma-ray sources were of the 'blazar' type: radio bright, flat-spectrum, highly active sources, often superluminals. However, two major puzzles were why many well-known sources fulfilling these criteria, such as 3C 345 or BL Lac, were not detected, and why others were detected at some epochs only.

A wide variety of gamma-ray production mechanisms have been proposed since the first detections. In most of the viable models the inverse Compton mechanism is assumed to be important. Even within the Compton framework, several different sources for the seed photons and the electrons are in principle possible; as a result, gamma-rays might be produced close to the accretion disk or farther away in the relativistic jet (Marscher & Bloom 1994). The models have mainly been compared against the few available multifrequency spectra, with inconclusive results.

We have tried to take an alternative, statistical approach. As several authors had noted, comparisons with the Metsähovi monitoring data indicated that many of the sources were in an enhanced radio state at the time of the gamma-ray detection. This connection was further investigated by Reich et al. (1993), who obtained radio data for the then known 12 EGRET-detected sources and concluded that for most sources enhanced radio emission occurred with a delay up to several months after the EGRET detection (see also Pohl, these proceedings). Valtaoja and Teräsranta (1995, 1996) used the full Metsähovi radio monitoring data set with the complete Phase 1 EGRET gamma-ray catalog (Fichtel et al. 1994) to investigate all the available relevant data. We used a complete 2 Jy sample of AGN (Valtaoja et al. 1992), enabling us to do unbiased statistics, and included not only detections but also the equally important EGRET nondetections in the analysis. Finally, we used the arguably best available classification (see Section 5), dividing the sources into quasars with high optical polarization (HPQ), quasars with low optical polarization (LPQ), BL Lac objects, and radio galaxies.

For each EGRET pointing we checked which of our radio sources were in the field of view, whether they were detected in gamma-rays, and what was the radio state of the source during the pointing. We characterized the radio states as follows: 0, rising millimeter flux; 1, peaking flux or generally high levels (i.e., several overlapping flares); 2, decreasing flux; 3, flux close to minimum levels.

The main result is shown in Figure 6. A very clear and significant correlation with both the source classification and with the radio state was found. With few exceptions, gamma-rays were detected by EGRET when, and only when, a HPQ was in the initial phases of a high radio frequency outburst. If a HPQ with rising millimeter flux (radio phase 0) happened to be in EGRET's field of view, its gamma-ray detection probability was 0.65; if the HPQ had high flux levels (radio phase 1), the probability was 0.32; if the HPQ had falling or low flux levels (radio phase 2 or 3), the probability was zero. For all other type of sources, the detection probabilities were much smaller: 0.09 in radio phases 0 or 1, and 0.04 in radio phases 2 or 3.

It is specially noteworthy that BL Lac are *not* in general strong gamma-ray emitters. The common statement that 'blazars' are gamma-ray emitters is quite misleading; it is true that 'blazars' as a class are more likely to be detected than 'nonblazars', but this is almost entirely due to the highly polarized quasars, which make up a half of the known 'blazar' population. (The data from later viewing periods is presently being analyzed. However, as the list of presently known gamma-ray emitters shows, the number of detected BL Lacs has not increased significantly although they have been the subject of intense COMPTON investigation—motivated in part, no doubt, by the misapprehension that 'blazars' are strong gamma-ray emitters.)

The relative timing of gamma and radio emission is crucial for theories. The radio flare does not occur after gamma-rays have been emitted—there is no time delay. (A delay *is* found, however, if one compares the gamma-rays with lower radio frequencies, due to opacity effects.) Since the typical timescales of gamma-ray variations are shorter (days to weeks) than those of the millimeter flares (months), the found correlation indicates either repeated gamma-ray flaring during the growing radio flare, or generally enhanced gamma-ray levels making the detection more likely during this time (c.f., Teräsranta & Valtaoja, these proceedings).

The gamma-ray intensity also appears to correlate with the properties of the radio flare, although here the statistical significance is lower. The intensity of the gamma-ray emission correlates at 95 % level both with the amplitude of the radio flare at the time of the EGRET observation, and with the rate of the radio flux increase, i.e., with the associated apparent brightness temperature of the flare (Figures 7 & 8; Valtaoja & Teräsranta 1996).

For any viable model of gamma-ray production it is not enough just to produce an acceptable fit to a single-epoch multifrequency spectrum; the model must above all explain the bulk of the data, the observed statistical properties. The synchrotron-self-Compton framework, unlike some other proposed models, provides a natural explanation for the tight radio–gamma correlations.

Why only some sources are gamma-ray emitters? Since HPQs and some BL Lacs are emitters, while LPQs, most BL Lacs and radio galaxies are not, orientation must be a key factor: only sources with jets making a small angle to the line of sight are detected. This includes HPQs and a small subset of BL Lacs, presumably those with the smallest viewing angles. Large amounts of Doppler boosting are required to make gamma-rays detectable.

Where in the source do the gamma-rays originate? Since gamma-rays are almost exclusively detected only when a millimeter radio flare is rising, they must originate in the growing shocks in the relativistic jets.

What is the gamma-ray emission mechanism? Since gamma-rays are simultaneous with the radio flare, and their intensity depends on the strength and the

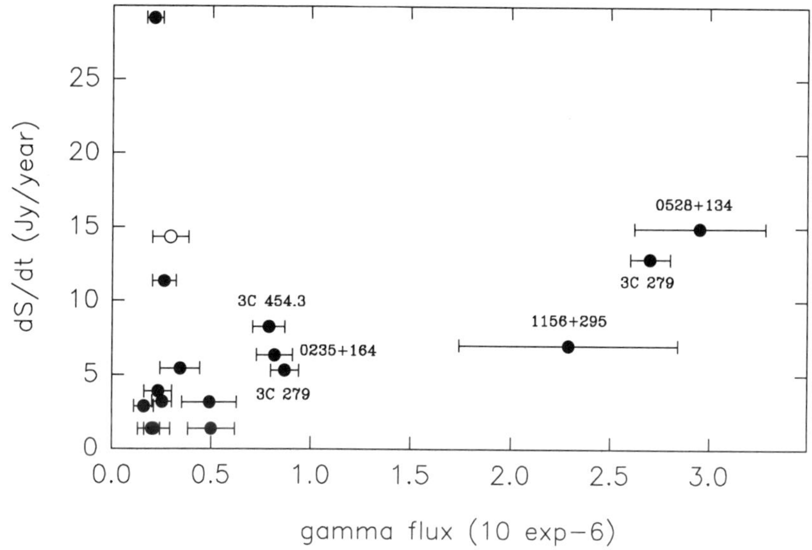

Figure 7. The dependence between the radio flare slope dS/dt and the EGRET gamma-ray flux (Valtaoja & Teräsranta 1996).

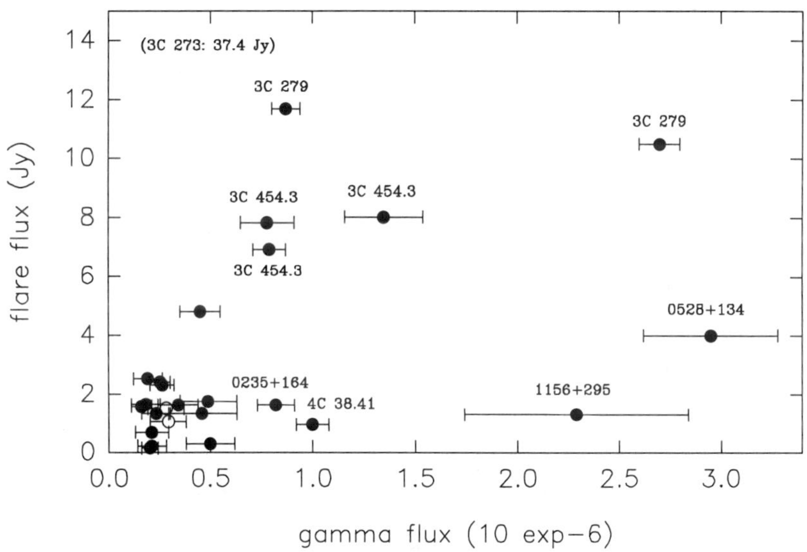

Figure 8. The dependence between the radio flare flux and the EGRET gamma-ray flux at the time of the EGRET pointing (Valtaoja & Teräsranta 1996).

apparent brightness temperature of the radio flare (i.e., the density of the relativistic electrons in the shock and the amount of Doppler boosting), the same shock electrons and photons must be responsible for both the radio and the gamma-ray emission. The gamma-rays are produced by the SSC mechanism in the shocked regions of the jet, with both the relativistic electrons and the seed photons originating in the same shocks.

The SSC mechanism can also explain the observed multifrequency spectra (e.g., Mukherjee et al., these proceedings). The mm-regime photons are not the seed photons of the gamma-ray emission, but the mm-flux is closely connected with the shorter wavelengths; it traces the intensity and the variations of both the electrons and the IR seed photons.

5. Taxonomy

If, as has been claimed, a rose by any other name would still be as sweet, does it matter what name tags we attach to various active galaxies? The classification of AGN is notoriously vague, with names like superluminal sources, blazars, optically violent variables, compact, flat-spectrum and core-dominated sources often used almost as synonyms—but is taxonomy really important?

I would argue that without proper classification there are very few things which you can do properly. Just consider how muddled our understanding of biology and evolution would be if we still continued to clump birds and bats together based on their most apparent, shared observational feature—wings. Yet this is much the way we continue to do in AGN research. Since the whole concept of unified models really is based on taxonomy, it is crucial that we know what are the proper classes of sources whose properties we are trying to compare with each other in our ultimate quest to separate the effects of viewing angles, parent populations, source evolution, and individual differences from each other.

We should do classification without preconceived notions, use such methods as cluster and principal component analysis, and find out what really are the relevant features and the natural groupings. Such an approach has rarely been attempted in AGN research (the PC analysis by Clements, these proceedings, being a step in the right direction). It is clear that large amounts of data are necessary for such a statistical approach to succeed. Here is still another new avenue where the extensive radio monitoring data sets can be used.

In a number of papers analyzing radio monitoring data we have argued that optical polarization is the best single indicator of the viewing angles in quasars. The HPQs, the sources with $P_{\rm opt} \geq 3\%$ at least occasionally, have the smallest viewing angles, while LPQs are the intermediate class between HPQs and FR II radio galaxies with large viewing angles. BL Lacs, coming from a different parent population, form a separate class. There is nothing especially sacred or fundamental in the optical polarization—it simply appears to be the best available single indicator whether the viewing angle is large or small, as demonstrated in a number of statistical analyses of large data sets. I will not repeat the evidence here, but refer instead to Valtaoja (1994) for a concise summary of the original papers.

Using the HPQ/LPQ division, we can use the large radio samples to estimate the relative amounts of Doppler boosting present in the two classes of sources: the spectral shapes (Tornikoski et al. 1993), the characteristic timescales (Lainela 1993), and the brightness temperatures (Teräsranta & Valtaoja 1993) all indicate that the radio flux from HPQs is 2–4 times more Doppler boosted than from LPQs. If, as unification supposes, these sources have similar intrinsic Lorentz factors (say, $\Gamma_{ave} \approx 10$), the average viewing angles of LPQs are about twice those of HPQs (Valtaoja 1994). With the addition of VLBI data, we can even estimate both the viewing angles and the intrinsic Lorentz factors for individual sources, and thus directly see if the sources group themselves in accordance with the unified models . Radio galaxies, LPQs and HPQs appear to form a single set of sources with similar Lorentz factors (10–35) and decreasing viewing angles. BL Lacs form a separate group. with smaller Lorentz factors (4–12) and viewing angles up to 40 degrees (Teräsranta & Valtaoja 1993).

It does matter which name tags we use; not all 'flat-spectrum' quasars are HPQs, and most radio galaxies are expected to be superluminal. For example, several authors have wondered about the lack of strong gamma radiation from 3C 273, one of the brightest, closest, flat-spectrum, superluminal, etc. sources. Nothing to puzzle over: 3C 273 is a firmly established low polarization quasar (c.f., Impey et al. 1989, Valtaoja et al. 1991), should therefore have a moderately large viewing angle, and not be a strong gamma-ray source. Sloppy classification leads to false expectations.

'Blazars' are another example of the confusion resulting from bad taxonomy. 'Blazars' are gamma-ray emitters, but Bl Lacs are not (Section 4.2). By definition, 'blazars' are strongly variable across the electromagnetic spectrum, but BL Lacs are not. The idea of 'blazars' blazing away is so powerful that we tend to forget that the characteristic *lack* of strong or rapid radio variability in BL Lacs was established long ago (Altschuler 1982, 1983) and has since then been confirmed in a number of investigations of large data sets (see Valtaoja 1994 for a summary). 'Blazars' are assumed to have small viewing angles, but most BL Lacs do not (c.f., also Kollgaard et al. 1996). Here, again, we see how bad taxonomy can mislead us. It does not make sense to analyze the properties of 'blazars' as a class, nor is it useful to target 'blazars' as gamma-ray emitters, as demonstrated by the general failure of EGRET to detect BL Lacs.

The blazar classification arose as an after-dinner concept, caught on, and now we still continue to clump together highly polarized quasars and BL Lac objects as 'blazars,' even though they are completely different classes of objects with different parent populations, average luminosities, intrinsic Lorentz factors, etc. BL Lacs and HPQs do have a very apparent, shared observational feature, the 'blazar'-type behavior caused by the dominance of nonthermal radiation, but there is no physically meaningful group of sources that could be classified as 'blazars.' We can reasonably discuss the blazar *phenomenon*, just as we can talk about the phenomenon of flight—but talk about 'blazars' and 'winged creatures' is best restricted to after-dinner speeches.

6. Conclusions

I have tried to argue that the 'statistical' approach made possible by the sheer magnitude of radio data, although less glamorous than the 'striking individual results' approach, is often the only way to firmly settle many outstanding problems in AGN research.

The shocked jet framework has now been established beyond reasonable doubt from a number of analyses of cm-to-submm–radio monitoring data (which does *not* mean that we should stop searching for exceptions!). We are also starting to discern the connections between radio and higher frequency regimes. Here radio observations can be used as tracers of things we would like to observe but for various reasons cannot—such as the emission sites of optical radiation or the gamma-ray seed photon fluxes.

Finally, we are beginning to test the unified models directly by being able to estimate the viewing angles and the Lorentz factors of the sources, and by having enough data for unbiased statistical analysis and proper classification of the active galaxies.

References

Altschuler, D. R. 1982, AJ, 87, 387

Alstchuler, D. R. 1983, AJ, 88, 16

Balonek, T. 1982, Ph.D. thesis

Brown, L. M. J., et al. 1989, ApJ, 340, 129

Clements, S. D., et al. 1995, AJ, 110, 529

Impey, C. D., et al. 1989, ApJ, 347, 96

Kollgaard, R. I., et al. 1996, ApJ, 460, 174

Lainela, M., 1994, A&A, 286, 408

Lainela, M., & Valtaoja, E. 1993, ApJ, 416, 485

Leppänen, K. 1995, Ph.D. thesis

Lobanov, A. P, & Zensus, J. A. 1996 (preprint)

Marscher, A. P., & Bloom, S. D. 1994, AIP Conf. Proc., 304, 572

Marscher, A. P, & Gear, W. K. 1985, ApJ, 298, 114

Reich, W., et al. 1993, A&A, 273, 65

Robson, E. I., et al. 1993, MNRAS, 262, 249

Sillanpää, A., et al. 1996, A&A, 305, L17

Stevens, J. A., et al. 1994, ApJ, 437, 91

Stevens, J. A., et al. 1995, MNRAS, 275, 1146

Takalo, L. O., et al. 1996 (preprint)

Teräsranta, H., et al. 1992, A&AS, 94, 121

Teräsranta, H., & Valtaoja, E. 1993, A&A, 283, 51

Tornikoski, M., et al. 1993, AJ, 105, 1680

Tornikoski, M., et al. 1994, A&A, 289, 673

Tornikoski, M., et al. 1996, A&AS, 116, 157

Valtaoja, E. 1994, in IAU Symp. 159, ed. T. J.-L. Courvoisier & A. Blecha (Dordrecht, Kluwer), 145

Valtaoja, E., et al. 1988, A&A, 203, 1

Valtaoja, E., et al. 1992a, A&A, 254, 71
Valtaoja, E., et al. 1992b, A&A, 254, 80
Valtaoja, E., et al. 1992c, A&AS, 95, 73
Valtaoja, E., & Teräsranta, H. 1995, A&A, 297, L13
Valtaoja, E., & Teräsranta, H., 1996, A&AS, in press
Valtaoja, L., et al. 1987, A&A, 184, 57
Valtaoja, L., et al. 1991, AJ, 102, 1046

Blazar Continuum Variability
ASP Conference Series, Vol. 110, 1996
H. R. Miller, J. R. Webb, and J. C. Noble, eds.

Radio Variability in Simulated Hydrodynamic Relativistic Jets

José-Luis Gómez[1] and Alan P. Marscher
Department of Astronomy, Boston University, Boston, MA 02215, U.S.A.

José Mª. Martí and José Mª. Ibáñez
Departamento de Astronomía y Astrofísica, Universidad de Valencia, Valencia, Spain

Abstract. We study the large scale radio variability in blazars through a set of multi-epoch and multifrequency numerical simulations of the synchrotron emission from hydrodynamic relativistic jets. Variations in the injection velocity of the jet lead to the formation of perturbations travelling along the jet, showing strong interactions with internal recollimation shocks. These perturbations appear as flares in the radio emission, as well as new superluminal components in the radio maps.

1. Introduction

Flares in the radio emission of blazars are often followed by the observation of the birth of new components in very long baseline interferomentric (VLBI) maps. This behavior has been successfully explained in terms of travelling shock waves, approximated by square waves (Marscher & Gear 1985; Hughes, Aller, & Aller 1991; Gómez, Alberdi, & Marcaide 1994). However, in order to obtain a description of the generation, evolution, and influence of these shocks in the radio emission of jets, a more realistic description of the flow hydrodynamics is needed. This has become possible only recently thanks to the development of time-dependent hydrodynamical simulations of relativistic jets (Martí et al. 1994, 1995; Duncan & Hughes 1994), and its implementation in the calculation of radio emission (Gómez et al. 1995, 1996; Hughes, Duncan, & Mioduszewski 1996; Komissarov & Falle 1996).

2. The model

The hydrodynamic code solves the relativistic conservation equations of rest-mass, momentum, and total energy for a perfect fluid. The flow is assumed to be axisymmetric, allowing a two-dimensional representation of the conservation equations in cylindrical coordinates. The code is based on an approximate Rie-

[1]Now at: Instituto de Astrofísica de Andalucía (CSIC), Granada, Spain

mann solver which relies on the spectral decomposition of the Jacobian matrices of the multidimensional system of relativistic equations. The spatial accuracy of the code is set by a conservative monotonic piecewise-parabolic interpolation of pressure, proper rest-mass density, and flow velocity components within the numerical cells. For further details of this code we refer the reader to Martí et al. (1996), where an exhaustive testing of the code can also be found.

To obtain the synchrotron emission from the jets whose relativistic hydrodynamics is calculated as above, we distribute the internal energy among the relativistic electrons following the usual power law $N(E)\,dE = N_o E^{-p} dE$, with $E_{\rm min} \leq E \leq E_{\rm max}$, and spectral index p. Radiative energy losses can be neglected, and therefore the ratio between the maximum and minimum energy can be considered constant all along the jet, and the power law is then fully determined (Gómez et al. 1995). We assume that the magnetic energy density remains a fixed fraction of the particle pressure, and that the magnetic field direction is predominantly randomly oriented.

The absorption and emission coefficients for the synchrotron radiation are computed in the fluid's frame, taking into account relativistic aberration and Doppler transformation of the frequency of observation. To account for time delay effects, the coefficients are calculated using a retarded time that accounts for the delays within the jet. Therefore, time evolution information of the parameters that determine the emission is needed for every cell in the computational grid. The coefficients are then Lorentz transformed into the observer's frame, where the Stokes parameters that determine the emission are calculated by integrating the transfer equations for the synchrotron radiation through all different lines of sight. We refer the reader to Gómez et al. (1993, 1994, 1995) for a detailed discussion of the emission calculations.

3. Outbursts in radio emission

We use as initial condition a jet of constant radius, velocity, and thermodynamic properties, injected at position $z = 0$ and propagating through a specified atmosphere. To allow a radial expansion of the jet, a decreasing pressure gradient in the atmosphere along the axial direction is introduced. The code is then run until a stationary solution is found, with the end of the jet (which contains a bow shock) having propagated off the computation grid.

Once a stationary solution is obtained, the generation of superluminal components, and the outbursts in radio emission associated with them, can be studied by introducing an increase in the velocity of the jet at the injection point. Figure 1 shows five subsequent epochs in the evolution of the pressure and bulk Lorentz factor for a jet in which the injection Lorentz factor has been temporarily increased from the initial value of $\Gamma \sim 3$ to $\Gamma \sim 9$.

Due to an initial overpressure of the jet relative to the external medium, overexpansions and overcontractions caused by inertial overshooting past equilibrium of the jet lead to the formation of standing recollimation shocks and rarefactions. (Because of the negative external pressure gradient, these would occur even if the pressures were matched at the injection point.) The strength and spacing of these shocks is determined by the Mach number, and the gradient in the external pressure. Figure 1 shows a complex, time-variable structure in

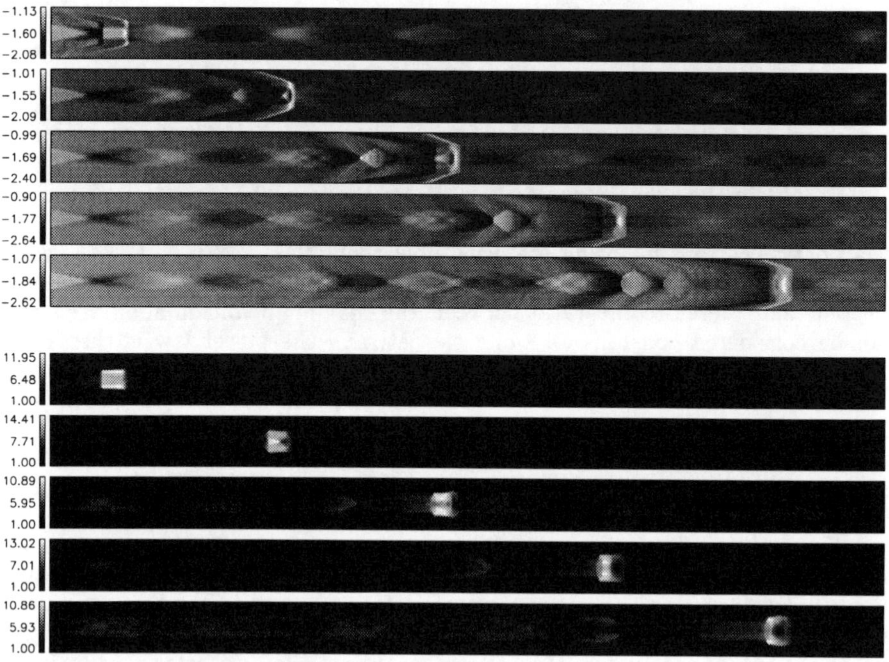

Figure 1. Evolution of the logarithm of pressure (top five panels), and bulk Lorentz factor (bottom five panels) when an increase in the injection velocity is introduced in the jet.

the perturbation, consisting of multiple shocks and rarefactions. Internal and bulk flow kinetic energy are continuously transforming into each other as the perturbation moves along the jet. A region of higher (lower) pressure is found upstream (downstream) due to the higher velocity in the perturbation.

The time evolution of the dynamics in the jet is largely determined by the interactions of the perturbation with the standing recollimation shocks and rarefactions. When the perturbation passes through a standing shock the latter is "dragged" downstream owing to the increased Mach number. The interaction with a rarefaction is similar, with the additional feature that a reverse shock is formed to separate the perturbation from the lower pressure, higher velocity fluid in the rarefaction. These reverse shocks propagate relatively slowly down the jet. The perturbation therefore creates a complex series of flow structures in the wake of the foreshock at the front of the disturbance. Eventually the jet recovers its initial steady state, as long as the jet maintains a constant level of energy and velocity at the input.

Figure 2 shows the total intensity maps corresponding to six subsequent epochs after the ejection of the perturbation for the jet model shown in Fig. 1. The maps have been obtained for a viewing angle of 10° and optically thin frequency of observation, and have been convolved with a circular gaussian beam to allow a better comparison with the observations.

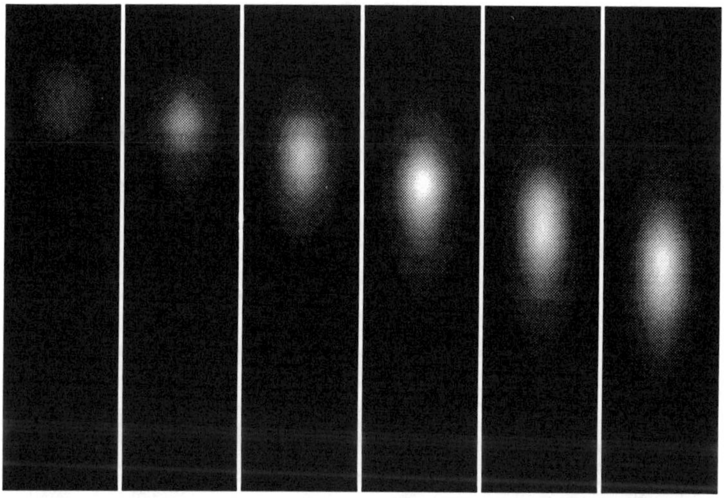

Figure 2. Simulated total intensity images corresponding to six epochs (from left to right) in the evolution of the perturbation. The input point of the jet is at the top of each panel.

The total intensity maps are dominated by the presence of a well-defined component associated with the perturbation. The enhancement of the emission is due to an increase in the Doppler boosting and internal energy density in the perturbation. Part of this flux is also coming from the secondary features, especially the reverse shocks, all blended by the convolving beam. The peak in the intensity maps moves at an apparent velocity of $\sim 7c$, smaller than the value of $\sim 8c$ corresponding to the Lorentz factor of the perturbation, owing to the slower propagation speeds of the reverse shocks.

Time delay effects introduce a significant variation in the emission maps compared with previous calculations neglecting these effects (Gómez et al. 1996). Due to time delays, the main emission feature appears much longer as viewed in the observer's frame, resulting in components with higher emission and an elongated structure for the perturbation considered here. (A weaker perturbation would allow the "core" emission near the input point to be more prominent, while higher resolution would allow the complex structure of the multiple shocks to be revealed.) Observations of circular or transversely elongated components must therefore represent very short-lived variations in the jet inlet.

The light curves covering the time interval of the maps shown in Fig. 2 are plotted in Figure 3 for three different frequencies. The ejection of the component in the emission maps is accompanied by a burst in flux at all three frequencies. However, opacity effects determine the rate of increase as a function of frequency. Initially the jet is optically thin at all three frequencies, but soon after the ejection, the component becomes optically thick at 22 GHz, and the flux at this frequency drops below the emission at higher frequencies. A maximum followed by a slow decrease is expected in the light curves, once the perturbation has travalled farther than shown in our simulations.

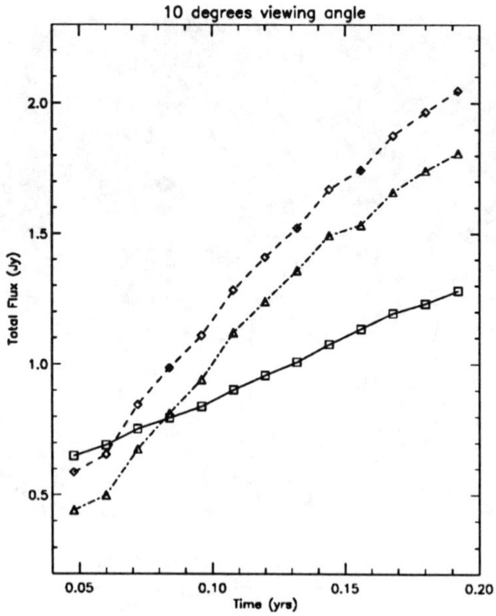

Figure 3. Evolution of the integrated total flux density at 22 (squares), 43 (diamonds), and 86 GHz (triangles). The simulations do not extend to the decay phase of the light curves.

4. Summary and conclusions

Although we are still in the early stages of a long-term project to understand the hydrodynamics of relativistic jets, we can speculate regarding observational predictions that are evident even at this preliminary stage.

Besides revealing the existence of standing shocks and internal structure of moving components, our simulations show that a single perturbation can create several distinct features (corresponding to the formation of reverse shocks) as it propagates downstream. Some indications of this effect are suggested by the C4 component of 3C 345 (see Lobanov & Zensus 1996). In addition, the interaction of moving components with standing shocks may result in a temporary "drag" of the latter. An indication of this behavior has been detected in component K1 of 0735+178 by Gabuzda et al. (1994).

Although we have not yet run a simulation to large enough distances from the injection point to determine the entire light curve of a flare caused by perturbations in the jet flow, it is clear from our work that major outbursts, accompanied by the appearance of new superluminal components, can result from such variations in the jet parameters, as indicated by previous work. The apparent velocity of the superluminal feature is somewhat less than what one would calculate using the Lorentz factor of the foreshock that defines the front of the perturbed region.

Time-dependent hydrodynamical simulations of relativistic jets provide a new, more realistic description of the physics involved in superluminal sources.

We are in the process of exploring a greater range of parameter space so that the rich morphological detail revealed by results of our simulations can be used to interpret the wealth of information provided by high-frequency and space VLBI imaging and multiwaveband monitoring of variations in blazars.

Acknowledgments. This research is supported in part by NASA grant NAG5-2508 and the Spanish DGYCIT (ref. PB94-0973). J. L. Gómez gratefully acknowledges a postdoctoral Fulbright fellowship from the Spanish Ministry of Education and Fulbright Commission.

References

Duncan, G. C., & Hughes, P. A. 1994, ApJ, 436, L119

Gabuzda, D. C., Wardle, J. F. C., Roberts, D. H., Aller, M. F., & Aller M. H., 1994, ApJ, 435, 128

Gómez, J. L., Alberdi, A., & Marcaide, J. M. 1993, A&A, 274, 55

Gómez, J. L., Alberdi, A., Marcaide, J. M., Marscher, A. P., & Travis, J. P. 1994, A&A, 292, 33

Gómez, J. L., Martí, J. Mª., Marscher, A. P., Ibáñez, J. Mª., & Marcaide, J. M. 1995, ApJ, 449, L19

Gómez, J. L., Martí, J. Mª., Marscher, A. P., Ibáñez, J. Mª., & Marcaide, J. M. 1996, in Energy Transport in Radio Galaxies and Quasars, ed. P. E. Hardee, A. H. Bridle, & J. A. Zensus. ASP Conf. Series, Vol. 100, 159

Hughes, P. A., Aller, H. D., & Aller, M. F. 1991, ApJ, 374, 57

Hughes, P. A., Duncan, C., & Mioduszewski, A. 1996, in Energy Transport in Radio Galaxies and Quasars, ed. P. E. Hardee, A. H. Bridle, & J. A. Zensus. ASP Conf. Series, Vol. 100, 137

Komissarov, S. S., & Falle, S. A. E. G. 1996, in Energy Transport in Radio Galaxies and Quasars, ed. P. E. Hardee, A. H. Bridle, & J. A. Zensus. ASP Conf. Series, Vol. 100, 165

Lobanov, A. P., & Zensus, J. A. 1996, ApJ, submitted

Marscher, A. P., & Gear, W. K. 1985, ApJ, 298, 114

Martí, J. Mª., Müller, E., & Ibáñez, J. Mª. 1994, A&A, 281, L9

Martí, J. Mª., Müller, E., Font, J. A., & Ibáñez, J. Mª. 1995, ApJ, 448, L105

Martí, J. Mª., Müller, E., Font, J. A., Ibáñez, J. Mª., & Marquina, A. 1996, ApJ, in press

Blazar Continuum Variability
ASP Conference Series, Vol. 110, 1996
H. R. Miller, J. R. Webb, and J. C. Noble, eds.

Variability of the Non-thermal Emission in the Jets of Blazars

Alan P. Marscher

Department of Astronomy, Boston University, 725 Commonwealth Ave., Boston, MA 02215

Abstract. I discuss mechanisms that can cause variability in the brightness and polarization of the non-thermal emission from the jets of blazars. The complicated light curves can be explained quite satisfactorily by a physical scenario in which a twisted relativistic jet containing a hydromagnetically turbulent plasma undergoes fluctuations in its energy input. This causes shock waves to develop and propagate. The evolution of the emission from such shocks can explain quite well both the long-term and short-term behavior of the observed "messy" light curves of blazars. I offer a guide for interpreting light curves with given characteristics.

1. Introduction

What excites us most about blazars is their rapid variability in brightness, polarization, and structure. The radiation is exotic—synchrotron and inverse Compton emission from ultra-relativistic electrons—and the physics is no less so—relativistically hot, magnetized plasma focussed into jets flowing outward at speeds ranging up to at least $0.995c$. The variability is in many cases rapid enough to observe in one session, even if one is awarded only a single night of observing time, hence blazars provide excitement to one's observing session.

In fact, the observations of blazars have become ever more sophisticated: CCDs on even rather small optical telescopes can be used to follow variations in many blazars night after night, and a number of campaigns have been mounted to observe flares simultaneously across nearly the entire electromagnetic spectrum. As more complex information is obtained, it is important that the theoretical models keep pace, so that the exquisitely detailed light curves can be interpreted and the physics of blazars thereby revealed. With this in mind, here I review the characteristics of variability that one should expect if our current picture of blazars as manifestations of relativistic jets is valid. Many of these ideas are discussed in Marscher, Gear, and Travis (1992; hereafter MGT), but since some of the important points made in that paper seem to have been lost over time (it *has* been four years, after all!), I will repeat and update these to take into account the observational progress that has subsequently been realized.

2. The Relativistic Jet Model

Since I have recently written reviews describing my picture of a relativistic jet (Marscher 1993, 1995, 1996), I will only sketch the basic model here. The central engine produces a jet consisting of relativistic, magnetized plasma (it does not matter much for the purpose of considering the radiation whether the positively charged particles are mainly protons or positrons) that flows outward at speeds very close to that of light. Presumably, an oppositely directed jet also emanates from the nucleus, but for blazars relativistic beaming makes this essentially invisible. Beyond a distance $z_1 \sim 10^{17\pm1}$ cm, the geometry can, for our purposes, be considered that of a cone, although the axis of the cone might be twisted. Between this location and the central engine, the geometry can take on a variety of forms, but given that the multiwaveband spectrum becomes optically thin at $10^{11\pm1}$ Hz, we know that the jet at $z < z_1$ cannot simply be a self-similar continuation of that at $z > z_1$.

To zeroth order, the jet can be thought of as a conical section of a wind, with density falling off as $[r^2\Gamma(z)]^{-1}$, where $r(z)$ is the cross-sectional radius of the jet and $\Gamma(z)$ is the bulk Lorentz factor of the flow. The axial component of the magnetic field should decline as r^{-2}, while that of the component normal to the jet axis should vary as $[r\Gamma(z)]^{-1}$. If there are no energy sources to the flow beyond the injection point, the internal energies (characterized by the internal Lorentz factors γ of the particles) should decay from adiabatic expansion as $[r^2\Gamma(z)]^{-1/3}$.

Jets and winds are not, in general, very steady. Their sources of energy and particles tend to be be unstable, rendering the flow prone to flares and quenchings. Changes in energy flux cause shock waves and rarefactions to propagate downstream. In addition, mismatches in pressure between the jet and its surroundings can cause stationary shocks and rarefactions to form. Even the steady flows (if the solar wind is a typical example) are highly turbulent. A jet is therefore a messy phenomenon. The question is whether we can decipher the observational signatures of the mess and thereby figure out the underlying physics involved. I think that the answer is "yes," but one has first to differentiate the "weather" from the "climate." In the remainder of this section I discuss what I consider to be the main climatic factors that are important in nearly every relativistic jet.

2.1. Propagating shock waves

Major increases in either the bulk velocity or internal energy of the jet flow will cause shock waves to form and propagate down the jet. (See MGT for the criteria under which shocks will occur rather than sound waves.) The function of shock waves is to decelerate supersonic flows to subsonic speeds. Hence, a forward shock decelerates the downstream fluid that it overtakes such that, in the rest frame of the shock front, the fluid velocity becomes subsonic downstream of the shock front (which is upstream in the frame of the ambient jet flow). The shock front therefore propagates supersonically relative to the downstream fluid. A reverse shock, on the other hand, slows high-velocity upstream fluid entering the shocked region from behind down to subsonic speeds, as measured in the frame of the reverse shock front. The reverse shock front propagates

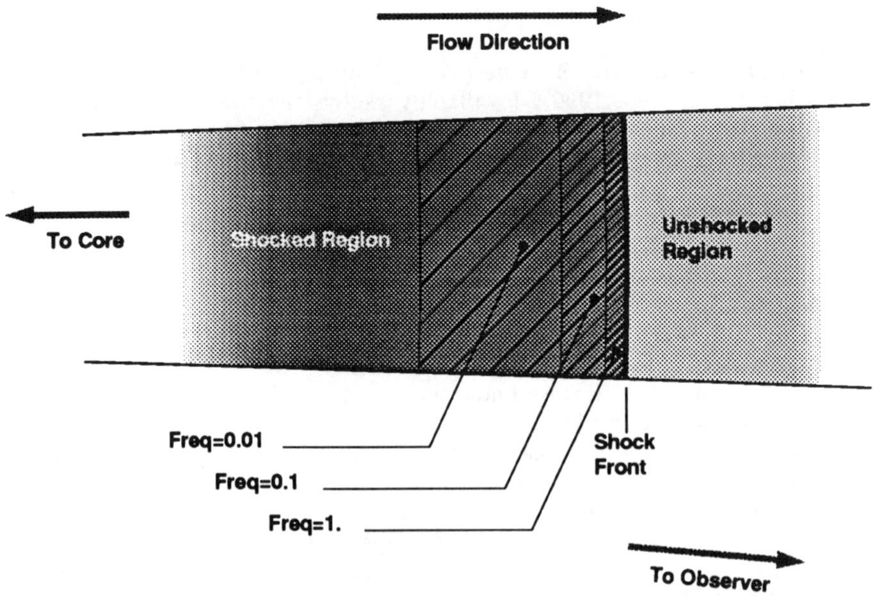

Figure 1. Sketch of a shock in a relativistic jet, with arbitrary scaling. The hatched regions show the sites of emission at different frequencies. The frequencies are scaled such that a value of 1.0 is at the high-frequency end of the synchrotron or inverse Compton spectrum. Adapted from Marscher and Travis (1991).

downstream, but more slowly than does either the forward shock front or the upstream, unshocked gas. A given disturbance can have only a forward shock or both a forward and a reverse shock, with the relative strengths of the shocks depending on the details of the time-variable flow.

A shock front compresses the plasma (which causes an increase in the density as well as an "adiabatic" increase in the internal energy) and enhances the component of the magnetic field that lies parallel to the shock front, both by the inverse of a factor called the compression ratio. (In MGT the compression ratio was defined as the inverse of the standard definition; I will henceforth conform to common practice.) The shock may additionally accelerate relativistic particles through what is usually called the diffusive shock acceleration mechanism. This first-order Fermi mechanism relies on particles crossing the shock front repeatedly, each time turning around through scattering off plasma waves. The acceleration is most efficient when the magnetic field lies perpendicular to the shock front, since the required particle streaming occurs most freely along the field lines. Hence, amplification of the synchrotron emission in shocks results from (1) the increase in density behind the shock front, (2) the increase in the magnetic field strength and adiabatic heating of the electrons for fields parallel to the shock front, and (3) diffusive shock acceleration of the electrons for fields perpendicular to the shock front. If, as is probably the typical case, the magnetic field has a strong stochastic component, diffusive shock acceleration still occurs, and the electron energy distribution in a strong shock (at least in the nonrelativistic case) has a slope of −2.5 (Kirk, Duffy, & Gallant 1996). Deriving this slope has long been the holy grail of particle acceleration theorists.

If the shock accelerates particles only at the shock front, then the highest energy electrons will suffer radiative losses such that they only maintain these high energies over a short distance. The highest frequency radiation can therefore be emitted only within a thin sheet behind the shock front. The thickness x of this sheet increases as the frequency decreases until the frequency is so low that radiative losses are negligible across the entire shocked region. This situation, described in Marscher and Gear (1985; hereafter MG), is shown in Figure 1.

This frequency stratification of the emission in a shock provides a duality that explains very nicely the general characteristics of the multiwaveband light curves of blazars: a flare caused by a shock is spread across many decades of frequency, yet the time-scale of variability can be much shorter at higher frequencies, owing to the thinness of the emission region behind the shock front that results from radiative energy losses of the electrons. MG proposed this model to explain the steepened spectra of blazar flares as well as the general trend of more rapid variations at higher frequencies.

The general evolution of a shock-induced flare in a jet (without bending) is as follows (see MG and MGT for proportionalities giving the quantitative time dependence of the spectrum).

1. The Compton stage, in which the electron energy loss is dominated by synchrotron self-Compton (SSC) scattering. The flux density rises while the self-absorption turnover changes only slightly, perhaps even decreasing with time. The multiwaveband νF_ν spectrum should show the γ-ray portion containing the bulk of the non-thermal apparent luminosity. Not all shocks will go through this phase—it probably requires extreme conditions; on the other hand, many of the blazars we choose to study are extreme objects!

2. The synchrotron stage, in which the energy loss is mainly from synchrotron radiation. The flare spectrum propagates to lower frequency with the peak flux density (at the self absorption turnover) changing slowly or not at all. It is possible that this "sideways" spectral evolution is the main factor responsible for the flat overall spectrum of a typical blazar below the turnover frequency.

3. The adiabatic stage, in which the flare declines as the shock expands down the jet. This phase is also described by Hughes, Aller, and Aller (1985), who show how shocks can explain variations in radio polarization.

The γ-ray and X-ray flux in a shock-induced flare should therefore peak first, followed by the optical and infrared and then radio, as described above. In fact, rise of the inverse Compton emission should reflect the strengthening of the shock as it takes shape. Rapid fluctuations in the γ-ray and high-frequency synchrotron (optical to X-ray, depending on the source) emission can occur if the particle acceleration is unstable during this period, as apparently occurs in Mkn 421 (see § 3 below).

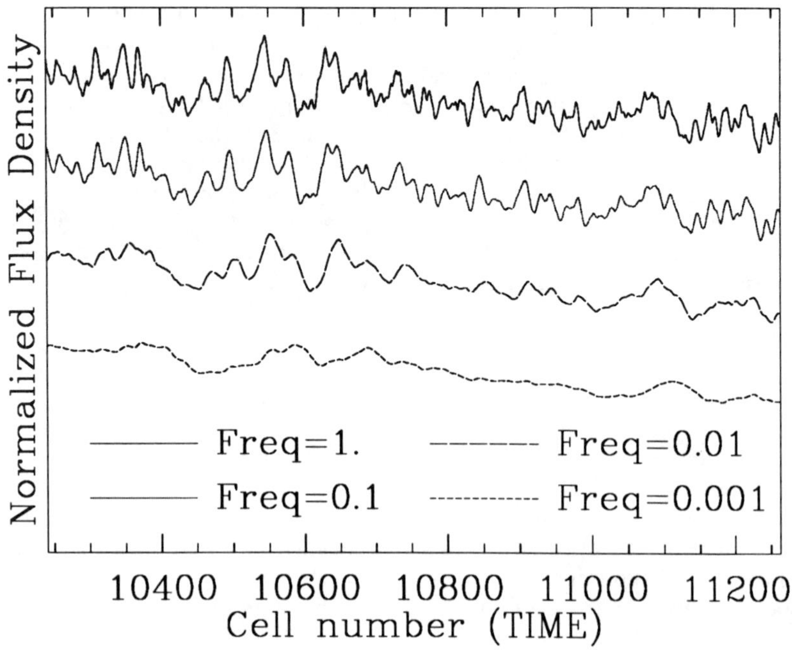

Figure 2. Flickering in the light curve of a blazar as a shock propagates down a hydro-magnetically turbulent jet. The frequencies are scaled such that a value of 1.0 is at the high-frequency end of the synchrotron or inverse Compton spectrum. The slow decline of the flux density is caused by the evolution of the overall emission from the shock, while the rapid variations result from turbulent cells passing into and out of the emission region at that given frequency. From Marscher and Travis (1991).

2.2. Stationary compressions and rarefactions

Not all shocks in jets need to be moving; in fact, standing shocks as well as other stationary compressions and rarefactions are expected in a jet confined by an external medium (Daly & Marscher 1988; Gómez et al. 1995). It is possible that particle acceleration in a standing shock is the mechanism that creates the VLBI core, with the jet being essentially invisible between this point and its base close to the central engine. (One would expect there to be considerable high-frequency emission near the injection point of the jet at the base as well, however.) Furthermore, features in the jet—regions of higher density or magnetic field—could cause flares to occur as they pass through a standing compression. In general, these standing features are oblique to the jet axis.

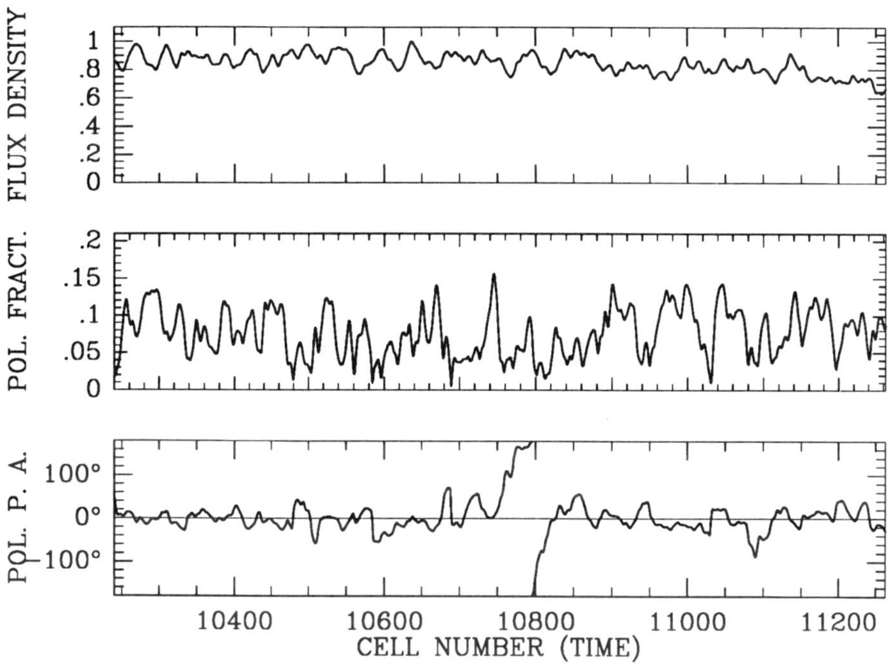

Figure 3. Flickering in the polarization light curve of a blazar as a shock propagates down a hydro-magnetically turbulent jet, as in Figure 2. From Marscher, Gear, and Travis (1992).

2.3. Turbulence

It would be surprising if the plasma inside relativistic jets were not turbulent, given the extremely high flow velocities relative to the external medium, as well as the analogy with the solar wind, which displays a high degree of turbulence. Evidence that the jets are hydro-magnetically turbulent comes from the observation that the linear polarization is considerably less than that expected from standard synchrotron radiation, which indicates that the magnetic field contains a randomly oriented component.

The interaction between shock waves and a turbulent ambient medium in a jet can explain some key characteristics of blazar emission. Since a shock wave amplifies only the component of the field that lies parallel to the shock front, a shock should order the random ambient field such that it lies predominantly in this direction (Jones et al. 1985). This is indeed seen in the bright components (which are here identified as shocks) of BL Lac objects and some quasars (Gabuzda et al. 1989; Kemball, Diamond, & Pauliny-Toth 1996).

I believe that frequency stratified shocks propagating through hydromagnetic turbulence (or, conversely, the turbulent jet medium passing through standing shocks) in a jet are what cause the short time-scale variability in the non-thermal emission of blazars. At low frequencies, the deviations from the mean density and magnetic field are smoothed out statistically because the large emission regions contain a very large number of turbulent cells. At high frequencies, however, the emission regions are very thin and therefore contain a sufficiently small number of cells that fluctuations about the mean emission and polarization are apparent. Simulated light curves computed by J. Travis are shown in Figures 2 and 3. Not only is the flickering seen at high frequencies (and even at lower frequencies, although significantly damped), but even an apparent rotation of the polarization position angle appears in the light curve. This latter feature is most likely when the fractional polarization is low.

The variability in flux density induced by the turbulence arises from fluctuations not only in the density of relativistic electrons and magnitude of the magnetic field, but also in the orientation of the field, since the optically thin flux density varies as $F_\nu \propto N_0(B \sin \phi)^{1+\alpha}\nu^{-\alpha}$, where N_0 is the normalization to the electron energy distribution, B is the magnetic field, ϕ is the angle between the magnetic field and the line of sight (corrected for relativistic aberration), and α is the spectral index. In fact, this latter effect is usually the dominant factor in the variability.

Unfortunately, our understanding of compressible hydro-magnetic turbulence in a relativistic plasma is not very deep. What one would like to do is to be able to predict the power spectrum of variations caused by the interactions of shocks with turbulence and to compare these with observational data. At this stage, however, what one predicts depends too closely on what one assumes for the power spectrum of the turbulence.

Several years ago (Marscher 1990, 1992) I proposed that intraday variability could be the result of shocks moving through turbulence in a relativistic jet. Not much further progress has been made with this model, because standard calculations of shocks indicated that they would be too thick, and the radiative losses too weak to confine the emission to a thin sheet, for such rapid variability to occur at centimeter wavelengths. However, I am now encouraged by the numerical simulations of shocks in jets (Gómez et al. 1996, 1997) which reveal that most of the emission from such shocks occurs over a region comparable to the width of the jet rather than over a sizable fraction of the distance down the jet. Hence, shocks might be thin enough to cause intraday variability even when radiative losses are negligible. This needs to be confirmed with more detailed numerical modeling.

2.4. Multiple shocks from a single event

Progress is finally being made in the area of numerical hydrodynamic modeling of relativistic jets (Gómez et al. 1996, 1997; Komissarov & Falle 1996; Hughes, Duncan, & Mioduszewski 1996). The early results presented in these papers show substantial fine-scale structure both in a steady jet (standing compressions and rarefactions) and in a jet through which a disturbance causes a shock wave to propagate downstream. In the latter case, one or more reverse shocks form to separate very rapidly flowing, rarefied gas from slower but denser and higher pressured material behind the forward shock. The structure is therefore complex, and the light curves should reflect this to some extent. One might therefore expect "bumps" and perhaps even multiple peaks in the light curves (plus flattening of the spectrum near the peak because of the presence of multiple components) even if there has only been a single disturbance in the jet. It will be interesting to compare the light curves of events in numerically simulated jets as these computational studies proceed.

2.5. Bending

A "wild card" in the modeling of flares in blazars is that jets are observed to bend. Theoretically, this can be understood in terms of precession and/or macroscopic instabilities (e.g., Hardee & Clarke 1992). Conway and Murphy (1993) have shown how an otherwise very puzzling result—the tendency of compact jets to be either nearly aligned or at right angles to more extended jets in the same sources—can be explained if jets are helically twisted. The flux density from a feature in a jet depends on the Doppler factor $\delta = [\Gamma(1 - \beta \cos \theta)]^{-1}$ (where Γ and β are the bulk flow Lorentz factor and velocity, respectively, while θ is the angle between the flow direction and the line of sight) as $\delta^{2+\alpha}$ for steady flow (e.g., a stationary shock or the ambient jet) and as $\delta^{3+\alpha}$ for an evolving feature. Therefore, changes in the jet direction can have a major effect on the light curve of a propagating structure such as a shock.

Some crude analytical approximations describing the evolution of the spectrum of a flare caused by a shock moving down a bent jet are given in MGT, while a more detailed analysis is given by Gómez et al. (1993, 1994a, 1994b). A unique property of the light curves is that the variations caused by bending are simultaneous at all frequencies, so that, for example, the flux can increase at both optically thin and thick frequencies with essentially no time delay. Furthermore, the SSC flare has roughly the same fractional amplitude as the synchrotron flare, as opposed to the usual case in which the SSC flare is more pronounced since the flux varies as N_0^2.

The most celebrated case of a source in which bending of the jet appears to dominate the light curve is 4C 39.25 (Marscher et al. 1991). The radio light curves, X-ray variability (the flux increased by the same factor as the radio over a two-year period), evolution of the radio spectrum, and change in apparent superluminal speed are all very well described by emission from a shock propagating down a twisted jet (Alberdi et al. 1993; Zhang & Marscher 1994).

Because of these theoretical successes, it is my opinion that the observed bends of jets are real and not simply illusions, although of course the degree of bending is amplified by projection effects. I caution, however, that there is probably a limit on how short the time-scales of variability resulting from bends can be. (In the case of 4C 39.25 the variability time-scale is about 3 years.) Relativistic jets carry high momentum per unit mass, hence it is difficult to understand how major bends could occur over short enough distances to change the Doppler factor appreciably on, for example, intraday time-scales in the observer's frame.

3. X-ray and γ-ray variability

Roughly 60 blazars have been detected (e.g., von Montigny et al. 1995) at γ-ray frequencies by the high-energy EGRET detector of the *Compton* Gamma Ray Observatory. The γ-rays represent a large fraction (approaching 100% in some cases) of the apparent non-thermal luminosities of these blazars. The most likely emission mechanism is inverse Compton scattering off the relativistic electrons in the jet (see the review by Marscher and Bloom 1994). However, it is not yet clear whether the "seed" photons are SSC radiation from the jet, direct photons from the putative accretion disk (Melia & Königl 1989; Dermer, Schlickeiser, & Mastichiadis 1992), or photons originally from the central engine or even the jet that have been scattered non-relativistically in the ambient medium (Sikora, Begelman, & Rees 1994; Blandford & Levinson 1995; Ghisellini & Madau 1996).

The time variability of the γ-ray emission provides an excellent diagnostic to explore the general nature of variability in blazars, especially in cases in which γ-ray flares are observed simultaneously at other wavebands. A number of such campaigns have been mounted, with some very interesting results. Although most of these studies are still in the analysis stage, the following conclusions can already be drawn:

1. The γ-ray emission is beamed, and therefore is produced in the relativistic jet. This is most graphically demonstrated in the February 1996 γ-ray flare of 3C 279, in which the Rossi X-ray Timing Explorer (RXTE) observed a simultaneous X-ray flare (McHardy et al. 1996). The X-rays and γ-rays are therefore essentially co-spatial. The optical depth to pair production of the X-rays would be too high for the γ-rays to escape unless the γ-rays (and, by implication, the X-rays) are beamed at a Lorentz factor $\gtrsim 4$ (see Mattox et al. 1993), which is similar to the value required to explain the apparent superluminal motion (Wehrle et al. 1996).

2. The γ-ray emission is bright when the overall multiwaveband emission is in a high state (e.g., 3C 279; Maraschi et al. 1994). The ratio of γ-ray to x-ray luminosity is higher than the ratio of X-ray to infrared luminosity, contrary to the predictions of models in which the γ-ray emission is second-order self-Compton scattering (Jones 1979), so that model can be eliminated; i.e., only first-order scattering is viable.

3. Time delays are sometimes seen between γ-ray and lower frequency variations. In 1406–076, an optical flare (Wagner 1995) preceded by one day a γ-ray flare in January 1992. This implies that the γ-rays are produced farther down the jet than is the optical emission. In the accelerating, SSC-emitting jet model (Maraschi, Ghisellini, & Celotti 1992) the γ-rays would then be inverse Compton scattered infrared photons, since these are produced slightly farther down the jet than are the optical photons in inhomogeneous jet models. The internal Lorentz factors of the infrared ($\sim 10^{14}$ Hz) emitting electrons are then constrained to be $\gamma_{IR} \sim 10^4$ and the Doppler factor to be $\delta \sim 20$. The magnetic field must then have a strength $B \sim 0.07$ G in the optically emitting region. (In non-SSC models—except perhaps for that of Ghisellini and Madau (1996)—the one-day delay seems more difficult to explain; see Wagner 1995). In PKS 0528+134 the high-amplitude γ-ray flare of March 1993 occurred near the peak of a months-long millimeter-wave flare (Zhang et al. 1994). This supports the SSC model, which predicts a nonlinear response of the γ-rays to synchrotron variations. This nonlinear response is also seen in the two multiwaveband "snapshots" of 3C 279 in June 1991 and December 1992–January 1993 (Maraschi et al. 1994). Inverse Compton scattering

of seed photons produced outside the jet could also show such a nonlinear response if the Lorentz factor of the jet flow were time-variable. However, in some cases (e.g., 3C 279 and 1406–076) the variability is so rapid that it is difficult to imagine either the Lorentz factor or the direction of motion changing in a relativistic flow on the required time-scale.

4. The γ-ray flare of Mkn 421 observed at TeV photon energies by the Whipple Observatory in May 1994 was simultaneous with an X-ray flare but there was little change in the EGRET, millimeter-wave, infrared, or ultraviolet emission (Macomb et al. 1995). In this case the flare seems to have resulted from an increase only in the efficiency of acceleration of the highest energy electrons (the "tail" of the electron energy spectrum). These observations are not consistent with the pair cascade model (Blandford & Levinson 1995) for producing the γ-rays, since the more rapid GeV flare predicted by the model was not observed. The observations are more consistent with the model in which non-thermal flares are caused by shock waves in the jet if electrons can be accelerated to TeV energies at the shock front and if the efficiency of acceleration fluctuates at the highest energies.
5. In two objects (PKS 0528+134 and 3C 279), new superluminal components observed with VLBI were created near the times of γ-ray flares and at the beginning of mm-wave rises (Pohl et al. 1995; Wehrle et al. 1996). The flares therefore appear to be associated with the production of new knots that flow down the jet.

From the application of the SSC model to 1406–076 (see item 3 above), which indicates a bulk Lorentz factor of 10–20 in the infrared-optical region, it appears that the accelerating jet model would have difficulty explaining the multiwaveband properties of the γ-ray flare in this blazar. The high Lorentz factor is not expected at such a high frequency, since the jet still must accelerate considerably from the near-infrared to the radio emitting section of the jet. The characteristics of this flare are therefore in better agreement with the expectations of the shock-in-jet model, with the shock lying downstream of the portion of the inner jet (which can still be described by the accelerating jet model) that produces infrared-optical photons ambiently. As described above, the frequency stratification of the emission behind the shock front can produce the observed time delays.

4. Summary and conclusions

Given the complex of propagating and stationary hydrodynamical features that must occur inside jets, we should expect that variability in brightness and polarization can have a variety of immediate causes. I have presented a paradigm within which all of these can be understood in terms of a basic phenomenon: shocks in a diverging, twisting, relativistically flowing, turbulent jet of magnetized, relativistically hot plasma. I maintain that, since all these characteristics are prevalent in relativistic jets, they *must* cause variability. This does not preclude other causes, but I know of no mode of variability (other than perhaps cm-wave intraday variability) that appears unable to be explained by these effects.

A guide that matches particular characteristics of light curves to specific physical phenomena can be summarized as follows:

1. A prolonged period of many overlapping outbursts probably indicates that the energy flow into the jet is in a high, unstable state. This has occurred, for example, in 3C 279 from about 1987 and in PKS 0528+134 from 1991 until at least 1996.

2. A major, usually prolonged outburst that propagates from higher to lower frequencies is the signature of the evolution of a shock wave or of a non-shock disturbance that propagates from the base of the jet outwards. The spectrum near the self-absorption turnover could be flatter than for a homogeneous synchrotron source either because (i) the emission is so intense that induced Compton scattering is important (Sincell & Krolik 1994), or (ii) there are multiple shocks caused by a single disturbance, as seen in numerical simulations (see above). The Compton stage should occur for sources that show strong SSC γ-ray emission during the flare. However, for other flares, especially those caused by shocks that form downstream in the radio-emitting section of the jet, the Compton (and possibly the synchrotron) stage may fail to appear. Flares of this type that are apparent at infrared and higher frequencies should show time delays and longer time-scales of variability toward lower frequencies.

3. A major outburst that also propagates from higher to lower frequencies but does not follow the specific behavior predicted for shock models could be caused by a temporary increase in the energy flow at the base of the jet (see Georganopoulos & Marscher 1997).

4. Simultaneous but not extremely rapid variability at optically thick as well as optically thin frequencies is the signature of a changing Doppler factor, probably caused by a shock or other feature propagating down a bent jet. Any associated SSC X-ray or γ-ray emission should vary in step and at the same relative amplitude as the synchrotron emission.

5. Rapid fluctuations (flickering) are likely to be caused by either a shock interacting with an irregular (turbulent) ambient medium or by fine-scale variations in the energy input at the base of the jet. In the former case, the polarization should also vary erratically.

6. Rapid variability of γ rays from blazars can be caused by changes in the density of relativistic electrons. The synchrotron flux density would then increase by a factor of $(1+\epsilon)$, where this is the factor by which the density is increased. In the SSC model, the resulting γ-ray flare should then have a higher fractional amplitude, increasing by a factor $(1+\epsilon)^2$. Scattering of external seed photons by the jet electrons will in this case cause a simultaneous synchrotron/high-energy flare of amplitude $(1+\epsilon)$. If the magnetic field B changes in the same sense as the electron density, the SSC flare will not have as high an amplitude relative to the synchrotron, since it is proportional to $N_0^2 B^{1+\alpha}$ while the synchrotron is proportional to

$N_0 B^{1+\alpha}$. The external seed photon scattering will have an even less than linear response to the synchrotron flare. If the seed photons are primarily jet synchrotron photons from the flare, scattered by an external medium, the amplitudes will mimic the SSC case, but there will be a time delay between the synchrotron flare and the high-energy flare.

Specific versions of these emission models that incorporate particular geometries, etc., have further specific signatures by which they can be tested observationally. I do not go into these here because of space limitations.

The reader should get the impression that the non-thermal emission in relativistic jets is messy because the jets themselves are complex, containing multiple shocks, turbulence, bends, and other substructures. Of course, the light curves are also quite messy. I am often struck by the demands of observers who wish to have simple theories explain their complex data. Actually, the models I have sketched *are* rather simple at least in construct, if not in the light curves they produce, but they can be difficult to invert such that the physical parameters of interest can be determined directly from the data.

I leave observers desperate to interpret their data with reasonably straightforward models the following rule: If you want to explain your messy data with a simple theory, smooth your data!

Acknowledgments. The author gratefully acknowledges partial funding of this research by NASA.

References

Alberdi, A., Marcaide, J. M., Marscher, A. P., Zhang, Y. F., Elósegui, P., Gómez, J. L., & Shaffer, D. B. 1993, ApJ, 402, 160

Blandford, R. D., & Levinson, A. 1995, ApJ, 441, 79

Conway, J. E., & Murphy, D. W. 1993, ApJ, 411, 89

Daly, R. A., Marscher, A. P. 1988, ApJ, 334, 539

Dermer, C. D., Schlickeiser, R., & Mastichiadis, A. 1992, A&A, 256, L27

Gabuzda, D. C., Cawthorne, T. V., Robert, D. H., & Wardle, J. F. C. 1989, ApJ, 347, 701

Georganopoulos, M., & Marscher, A. P. 1997, these proceedings

Ghisellini, G., & Madau, P. 1996, MNRAS, 280, 67

Gómez, J. L., Alberdi, A., & Marcaide, J. M. 1993, A&A, 274, 55

Gómez, J. L., Alberdi, A., & Marcaide, J. M. 1994a, A&A, 284, 51

Gómez, J. L., Alberdi, A., Marcaide, J. M., Marscher, A. P., and Travis, J. P. 1994b, A&A, 292, 33

Gómez, J. L., Martí, J. M., Marscher, A. P., & Ibáñez, J. M. 1997, these proceedings

Gómez, J. L., Martí, J. M., Marscher, A. P., Ibáñez, J. M., Marcaide, J. M. 1995, ApJ, 449, L19

Gómez, J. L., Martí, J. M., Marscher, A. P., Ibáñez, J. M., & Marcaide, J. M. 1996, in Energy Transport in Radio Galaxies and Quasars, ed. P. E. Hardee, A. H. Bridle, & J. A. Zensus, ASP Conf. Ser., 100, 159

Hardee, P. E, & Clarke, D. A. 1992, ApJ, 400, L9

Hughes, P. A., Aller, H. D., & Aller, M. F. 1985, ApJ, 298, 301

Hughes, P. A., Duncan, C., Mioduszewski, A. 1996, in Energy Transport in Radio Galaxies and Quasars, ed. P. E. Hardee, A. H. Bridle, & J. A. Zensus, ASP Conf. Ser., 100, 137

Jones, T. W. 1979, ApJ, 233, 796

Jones, T. W., Rudnick, L., Aller, H. D., Aller, M. F., Hodge, P. E., & Fiedler, R. L. 1985, ApJ, 290, 627

Kemball, A. J., Diamond, P. J., & Pauliny-Toth, I. I. K. 1996, ApJ, in press

Kirk, J. G., Duffy, P., & Gallant, Y. A. 1996, A&A, in press

Komissarov, S. S., & Falle, S. A. E. G. 1996, in Energy Transport in Radio Galaxies and Quasars, ed. P. E. Hardee, A. H. Bridle, & J. A. Zensus, ASP Conf. Ser., 100, 165

Macomb, D. J., et al. 1995, ApJ, 449, L99

Maraschi, L., Ghisellini, G., & Celotti, A. 1992, ApJ, 397, L5

Maraschi, L., et al. 1994, ApJ, 435, L91

Marscher, A. P. 1990, in Parsec-Scale Radio Jets, ed. J. A. Zensus & T. J. Pearson (Cambridge, Cambridge University Press), p. 236

Marscher, A. P. 1992, in Physics of Active Galactic Nuclei, ed. W. J. Duschl & S. J. Wagner (Berlin, Springer-Verlag), p. 510

Marscher, A. P. 1993, in Astrophysical Jets, ed. D. Burgarella, M. Livio, & C. P. O'Dea, STScI Symp. Series (Cambridge, Cambridge University Press), 6, 73

Marscher, A. P. 1995, Proc. Nat. Acad. Sci., 92, 11439

Marscher, A. P. 1996, in Energy Transport in Radio Galaxies and Quasars, ed. P. E. Hardee, A. H. Bridle, & J. A. Zensus, ASP Conf. Ser., 100, 45

Marscher, A. P., & Bloom, S. D. 1994, in The Second Compton Symposium, ed. C. Fichtel, N. Gehrels, & J. P. Norris, AIP Conf. Proc. 304, 572

Marscher, A. P., & Gear, W. K. 1985, ApJ, 298, 114 (MG)

Marscher, A. P., Gear, W. K., & Travis, J. P. 1992, in Variability of Blazars, ed. E. Valtaoja, & M. Valtonen (Cambridge, Cambridge University Press), p. 85 (MGT)

Marscher, A. P., & Travis, J. P. 1991, in Variability of Active Galactic Nuclei, ed. H. R. Miller & P. J. Wiita (Cambridge, Cambridge University Press), p. 153

Marscher, A. P., Zhang, Y. F., Shaffer, D. B., Aller, H. D., & Aller, M. F. 1991, ApJ, 371, 491

Mattox, J. R., et al. 1993, ApJ, 410, 609

McHardy, I. M., et al. 1996, in preparation

Melia, F., & Königl, A. 1989, ApJ, 340, 162

Pohl, M., et al. 1995, A&A, 303, 383

Sikora, M., Begelman, M. C., & Rees, M. J. 1994, ApJ, 421, 153

Sincell, M. W., & Krolik, J. H. 1994, ApJ, 430, 550

von Montigny, C., et al. 1995, ApJ, 440, 525

Wagner, S. J. 1995, in 17th Texas Symp. on Relativistic Astrophysics, ed. H. Böhringer, G.E. Morfill, & J. Trümper, Ann. NY Acad. Sci., 759, 526

Wehrle, A. E., et al. 1996, A&A, in press

Zhang, Y. F., & Marscher, A. P. 1994, in The Soft X-ray Cosmos, ed. E. Schlegl & R. Petre, AIP Conf. Proc., 313, 406

Zhang, Y. F., Marscher, A. P., Aller, H. D., Aller, M. F., Teräsranta, H., & Valtaoja, E. 1994, ApJ, 432, 91

Blazar Continuum Variability
ASP Conference Series, Vol. 110, 1996
H. R. Miller, J. R. Webb, and J. C. Noble, eds.

The Accelerating Jet Model: Steady State and Variability Properties of Blazars

Markos Georganopoulos and Alan P. Marscher

Department of Astronomy, Boston University, 725 Commonwealth Avenue, Boston, MA 02215

Abstract. The accelerating inner jet model is examined, using a numerical simulation, that couples a simple hydrodynamical scheme with optically thin synchrotron relativistic radiative transfer. Adiabatic and synchrotron losses are considered. The model is compared to observations and the hypothesis that different viewing angles explain the dichotomy of radio and X-ray selected BL Lac objects is tested. The emerging picture suggests that the observed properties of a blazar are defined by both the jet orientation and the intrinsic physical properties of the object. Simulations of variability are also compared to observational results.

1. Introduction

The family of blazars includes those AGN that are characterized by compact radio morphology, and a variable and polarized nonthermal continuum. Superluminal motion and gamma-ray emission are also common properties among blazars. When the spectrum exibits the usual quasar-like broad emission lines the object is a Flat Spectrum Radio Quasar (FSRQ). On the other hand the term BL Lacertae (BL Lac) object is reserved for the lineless or almost lineless blazars. The observational taxonomy of BL Lacs, according to the method of discovery, in radio selected (RBLs) and X-ray selected (XBLs) sources corresponds (Padovani & Giommi 1995) to a more physical one, based on the frequency where the peak energy occurs: Low-frequency peaked BL Lacs (LBLs), which are mostly RBLs, and high frequency peaked BL Lacs (HBLs), which are mostly XBLs. The XBLs are less variable and polarized than RBLs (Jannuzi et al. 1994) and have flatter optical spectra (Ledden & O' Dell 1985).

The standard interpretation of the nonthermal continuum radiation of blazars from radio to γ-rays is thought to be synchrotron and inverse Compton radiation from a collimated relativistic plasma jet (Blandford & Königl 1979). The inner jet of a blazar is the section of the jet that connects the central engine with the VLBI core of the radio jet. This is a small ($\lesssim$ 1 pc) region where the jet is formed, collimated and accelerated to speeds close to that of light. In the accelerating inner jet model (Marscher 1980) ultrarelativistic plasma is generated continuously near the central engine of the AGN and is accelerated hydrodynamically (Blandford & Rees 1974). An external pressure (hydrostatic and/or magnetohydrodynamic) that decreases along the jet axis acts as a collimating agent. The internal energy of the plasma is converted to bulk kinetic energy and the jet is accelerated and focused. The electrons interact with a predominately random magnetic field and cool through synchrotron radiation and adiabatic expansion. At the same time inverse Compton losses may become important if the synchrotron photon energy density is comparable to the magnetic field energy density. Higher frequencies are emerging from smaller distances from the central engine, implying shorter flux variation timescales at higher frequencies, as observed (Brown et al. 1989). The velocity of the jet increases with distance; this implies larger Doppler boosting for greater distances down the jet up to the point where the Lorentz factor $\Gamma \lesssim \theta^{-1}$, where θ is the angle between the velocity vector and the line of sight, and therefore at lower frequencies.

Previous jet models used scaling laws to describe the synchrotron and Compton emissivity and the electron energy distribution as well as simplified flow geometries in which either the flow velocity accelerates but does not converge (Maraschi et al. 1992) or the flow velocity converges but does not accelerate (Celotti et al. 1993). This work is a step towards a more constrained and testable model with fewer free parameters. A simple relativistic hydrodynamic scheme that naturally produces a simultaneously accelerating and converging flow is coupled with a detailed calculation of the evolution of the electron energy distribution and synchrotron emissivity due to relativistic electrons radiating in a mostly random magnetic field. The effects of relativistic aberration are included in the calculation. A square density wave is propagated with the jet velocity and the variability thus induced is studied, taking into account time delay effects (Gómez et al. 1994).

2. The model

In the case of quasi-one-dimensional ultra-relativistic (mean total to rest mass ratio per particle $\gamma \gg 1$), stationary adiabatic flow, confined by the pressure of the external ambient medium , with pressure $p \propto z^{-a}$, where z is the axial coordinate of the system, a jet is formed with cross-sectional radius (Blandford & Rees 1974)

$$r(z) = r_\star \left(\frac{z}{z_\star}\right)^{\frac{3\epsilon}{2}} \left[3\left(\frac{z}{z_\star}\right)^{2\epsilon} - 2\right]^{-\frac{1}{4}}. \tag{1}$$

For $z \gg z_\star$, $r \propto z^\epsilon$. The bulk motion Lorentz factor Γ is given by

$$\Gamma(z) = \Gamma_\star \left(\frac{z}{z_\star}\right)^{\epsilon}, \tag{2}$$

where "$\star$" refers to the point where the jet becomes supersonic. The exponent ϵ determines how fast the jet opens and accelerates and is determined solely by the steepness of the external pressure gradient, $\epsilon = a/4$.

The low observed polarization of the optically thin regions implies a random magnetic field, and we assume that there is also a small ordered component parallel to the jet axis. If the magnetic field decays as $B \propto r^{-m}$, we have:

$$B(z) = B_\star \left(\frac{z}{z_\star}\right)^{\frac{-3m\epsilon}{2}} \left[3\left(\frac{z}{z_\star}\right)^{2\epsilon} - 2\right]^{\frac{m}{4}}, \qquad 1 < m < 2. \tag{3}$$

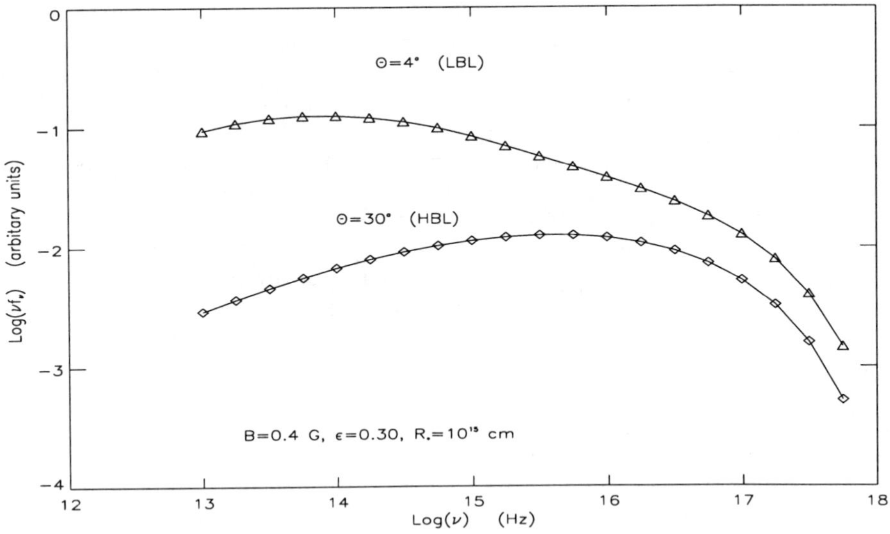

Figure 1. The peak luminosity increases, ν_p decreases and the spectrum becomes flatter as the viewing angle Θ increases from 4° to 30°.

The combined adiabatic and synchroton energy losses give for the electron energy along the jet

$$\gamma(z) = \left(\frac{z}{z_\star}\right)^{-\epsilon} \left[\frac{1}{\gamma(z_\star)} + C_1 I(z)\right]^{-1}, \tag{4}$$

where

$$C_1 = \sqrt{2}\left(\frac{2}{3}\right)^2 \frac{e^4}{m^3 c^6} B_\star^2, \tag{5}$$

and

$$I(z) = \int_{z_\star}^{z} \left(\frac{z}{z_\star}\right)^{-(3m+1)\epsilon} \left(3\left(\frac{z}{z_\star}\right)^{2\epsilon} - 2\right)^{\frac{m-1}{2}} dz. \tag{6}$$

An initial power law electron energy distribution will evolve to a gradually steepening profile towards the higher energies:

$$N(\gamma, z) = N_\star \gamma^{-s} \left(\frac{z}{z_\star}\right)^{-\epsilon(s+2)} \left(1 - \left(\frac{z}{z_\star}\right)^{\epsilon} C_1 I(z)\right)^{s-2}. \tag{7}$$

Given the magnetic field and the electron energy distribution we can calculate the synchrotron emission coefficient in the fluid frame using standard synchrotron formulae (Pacholczyk 1970). To transform the emission coefficient to the observer's frame we need to take into account the relativistic transformations between the observer and the fluid proper frames as well as the aberration of the line of sight (Gómez et al. 1994) . When a disturbance is traveling with a speed βc in a direction that forms an angle θ with the line of sight, then the observed transverse velocity is given by:

$$v = \frac{\beta c \sin\theta}{(1 - \beta\cos\theta)(1 + z)} \tag{8}$$

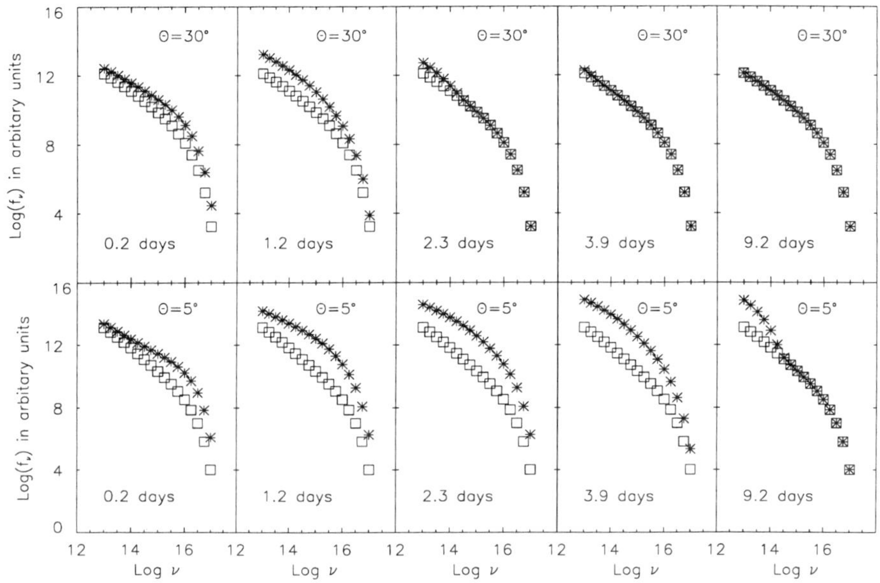

Figure 2. Change of (a) α_{IR-O}, (b) peak (νL_ν), (c) ν_p versus Θ, and (d) α_{O-X}, (e) α_{IR-O}, (f) peak (νL_ν) versus ν_p (see text).

This creates an apparent rotation of the disturbance front that we take into account when calculating the temporal behavior of the disturbance.

3. Results and comparison to observations

The simulations show that the most important free parameters that describe the jet are ϵ, $R_\star$, $B_\star$, and Θ, the angle between the jet axis and the line of sight. The spectrum of a BL Lac object is known to be a featureless curved continuum with no more than two degrees of freedom (Landau et al. 1986). The overall shape is reproduced by the model and is quite insensitive to the choice of the free parameters.

This allows the possibility that BL Lacs in nature cover a quite wide range of parameter space. The observed differences between RBLs and XBLs were initially explained as an orientation effect, with the RBL jets pointing towards the observer and the XBL being misaligned ($\Theta \approx 30°$) (Maraschi et al. 1986). Recently Padovani and Giommi (1995) proposed that the different observed properties of RBLs and XBLs reflect different physical properties. Sambruna et al. (1996) found, using complete samples of blazars, an average peak frequency in the $\log(\nu L_\nu)$ spectrum at 15.93 ± 1.45 for the XBLs and 13.91 ± 0.93 for the RBLs. In our simulation we obtain a range of ν_p from 13.9 to 15.7 by keeping all the parameters constant and varying the angle Θ from 1° to 33° (fig. 1, 2c). Although this range fits nicely to the average values reported by Sabruna et al., it fails to explain the whole range of ν_p under the different angle hypothesis. The infrared to optical spectral index (fig. 2a) in our simulation ranges from

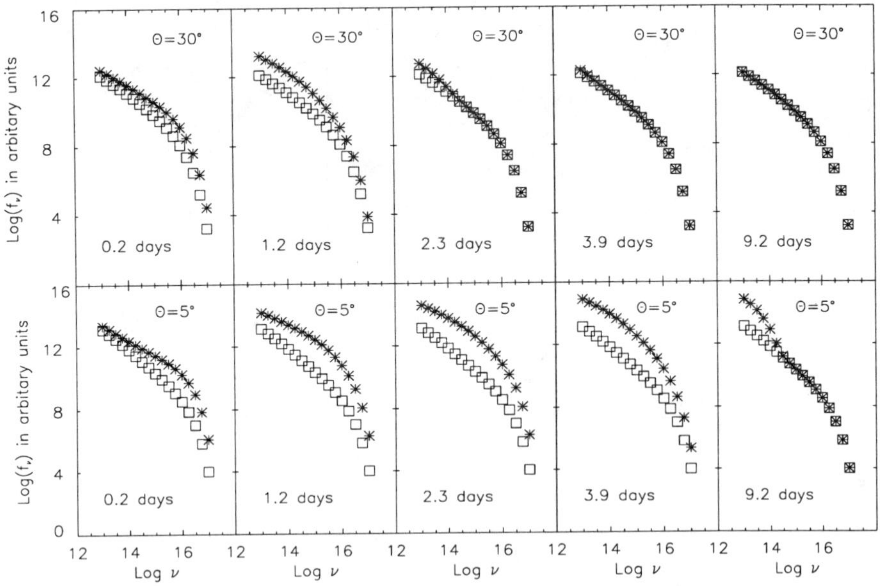

Figure 3. Evolution of a flare as it is viewed at a wide angle (top panel) and a small angle (bottom panel). Squares indicate the steady state spectrum and asterisks the flaring one.

$\alpha_{IR-O} = 0.77$ to $\alpha_{IR-O} = 1.17$ as Θ changes from 33° to 1°. Falomo et al. (1993) reported an average spectral index $\alpha_{IR-O} = 0.68 \pm 0.28$ for 8 XBLs and $\alpha_{IR-O} = 1.20 \pm 0.30$ for 25 RBLs. The peak luminosity in our simulations decreases by 1.5 orders of magnitude as we go from from $\Theta = 1°$ to $\Theta = 33°$, in agreement with Sabruna et al. who find that the integrated luminosity ranges from $L_B = 45.57 \pm 0.37$ for the XBLs to $L_B = 46.49 \pm 0.80$ for the RBLs.

The above results can be viewed in terms of observables only. The optical to 0.4 keV X-ray spectral index in our simulations (fig. 2d) ranges from $\alpha_{O-X} = 1.40$ for objects that peak at $\nu_p \approx 10^{13.9}$ (RBLs) to the flatter $\alpha_{O-X} = 1.11$ for objects that peak at $\nu_p \approx 10^{15.7}$ (XBLs). Sabruna et al. find, using a spectral index defined by $\lambda_o = 5500$Å and 1 keV an average spectral index of $\alpha_{O-X} = 1.49 \pm 0.23$ for the RBLs to $\alpha_{O-X} = 0.96 \pm 0.15$ for the XBLs. A similar picture that agrees with the data of Falomo et al. emerges when we plot (fig. 2e) α_{IR-O} versus ν_p . Finally we find (fig. 2f) that the peak luminostity of a source νL_ν falls as the peak frequency ν_p increases. A similar result is found by Sambruna et al. regarding the relation between the frequency integrated luminosity L_B and the peak frequency ν_p.

The above results indicate that although the viewing angle hypothesis cannot account for all the observations, it reproduces not only the qualitative behavior of the data, but also the mean values of the observed quantities. The significant scattering around these mean values can be interpreted not as an indication of a failure of the viewing angle hypothesis, but rather as a hint that the physical conditions in these sources are not identical (Padovani & Giommi

1995, Sambruna et al. 1996), but vary in ways that have not yet been clearly established. For example, in our simulation, a change from $B_* = 0.4$ G and $\epsilon = 0.30$ to $B_* = 1.0$ G and $\epsilon = 0.40$ gives $\nu_p = 10^{13}$ Hz for $\Theta = 1°$.

The temporal behavior of the model, when a square wave density enhancement is propagated through the jet, agrees qualitatively with the observational picture. In the simulation (fig. 3) the flux starts rising simultaneously in all frequencies, but the rate of ascent and the peak amplitude are higher and the duration of the flare shorter at higher frequencies. This agrees with Impey and Neugebauer (1988) observational results that the amplitude and rapidity of the variability increases with frequency. The model spectrum gets flatter, particularly at high frequencies, when the X-ray flux peaks, in agreement with the results of Sambruna et al. (1994) who found that the spectra of four XBLs flatten with increasing intensity. The simulation also produces small time delays (a few hours) between the x-ray and the optical peak. A 3-hour delay between X-ray and UV-optical peak has been observed for PKS 2155-304 during a month of intense multiwavelength monitoring (Edelson et al. 1995), although in other cases time delays are not observed (perhaps due to a coarse sampling rate). The XBLs have been observed to be less variable than the RBLs (Jannuzi et al. 1994). This is reproduced in the model calculations (Fig. 3), under the hypothesis that the main difference between RBLs and XBLs is the viewing angle.

References

Blandford, R. D., & Königl, A. 1979, ApJ, 232, 34

Blandford, R. D., & Rees, M. J. 1974, MNRAS, 169, 395

Brown, L. M. J., et al. 1989, ApJ, 340, 129

Celotti, A., et al. 1993, ApJ, 416, 118

Edelson, R., et al. 1995, ApJ, 438, 120

Falomo, R., et al. 1993, AJ, 106, 11

Gómez, J. L., et al. 1994, A&A, 292, 33–44

Impey, C. D., & Neugebauer, G. 1988, AJ, 95, 307

Januzzi, B. T., Smith, S. P., & Elston, R. 1994, ApJ, 428, 130

Landau, R., et al. 1986, ApJ, 308, 78

Ledden, J. E., & O'Dell, S. L. 1985, ApJ, 298, 630

Maraschi, L., et al. 1986 ApJ, 310, 325

Maraschi, L., Ghisellini, G., & Celotti, A. 1992, ApJ, 397, L5

Marscher, A. P. 1980, ApJ, 235, 386

Padovani, P., & Giommi, P. 1995, ApJ, 444, 567–581

Pacholczyk, A. G. 1970, in Radio Astrophysics, (San Francisco, W. H. Freeman)

Pian, E., et al. 1994, ApJ, 432, 547

Sambruna, R., et al. 1994, ApJ, 434, 468

Sambruna, R., et al. 1996, ApJ, 463, 444

Blazar Continuum Variability
ASP Conference Series, Vol. 110, 1996
H. R. Miller, J. R. Webb, and J. C. Noble, eds.

Radio Observations of Gamma-ray Blazars

M. Pohl and A. Mücke

MPE, Postfach 1603, 85740 Garching, Germany

W. Reich, P. Reich, and R. Schlickeiser

MPIfR, Postfach 2024, 53010 Bonn, Germany

H. Ungerechts

IRAM, Avenida Divina Pastora 7, 18012 Granada, Spain

Abstract. Starting in 1991 we have performed a regular monitoring of EGRET sources with the Effelsberg 100-m telescope and the IRAM 30-m at Pico Veleta. In comparison with data on a sample of flat-spectrum quasars which have not been seen by EGRET we search for correlations of any kind in the behavior of FSRQ's in the radio and gamma-ray regime. While there is no radio-to-gamma-ray luminosity correlation for these sources, it seems that gamma-ray high states coincide with increased activity in the radio regime with a strong tendency that gamma-ray outbursts precede radio outbursts. The gamma-ray spectra seem to harden with increasing flux level.

1. An example of an individual source: 0528+134

We will first discuss PKS 0528+134, the most luminous γ-ray source known besides γ-ray bursts. 0528+134 exhibits superluminal motion with $\beta_{app} \simeq 4.4$ and indications for even higher values (Pohl et al. 1995). Superluminal motion was expected since given the variability time scales of a day, the redshift of z=2.07 and the γ-ray flux in the OSSE range, strong Doppler boosting is required to satisfy the compactness limit and the Elliot-Shapiro relation (McNaron-Brown et al. 1995). It has been noted that the expulsion of new VLBI components may coincide with γ-ray outbursts (Pohl et al. 1996).

In Fig. 1 we show the radio light curve of 0528+134 in comparison to its γ-ray state for the period 1991 to 1995. The source was quiet in the radio regime between 1985 and 1991 (Zhang et al. 1994). The following conclusion can be drawn: 0528+134 has been very bright in γ-rays either when it was weak in radio or a few months before a mm outburst. It was at medium γ-ray level at the time of the brightest ever-recorded radio outburst, at the end of 1995. Though the γ-ray high states seem to precede the radio outbursts, which is also supported by back-extrapolation of the position of VLBI knots, there is no simple one-to-one relation. The analysis of individual sources is mainly hampered by the limited coverage of the γ-ray light curve and the problem that many sources,

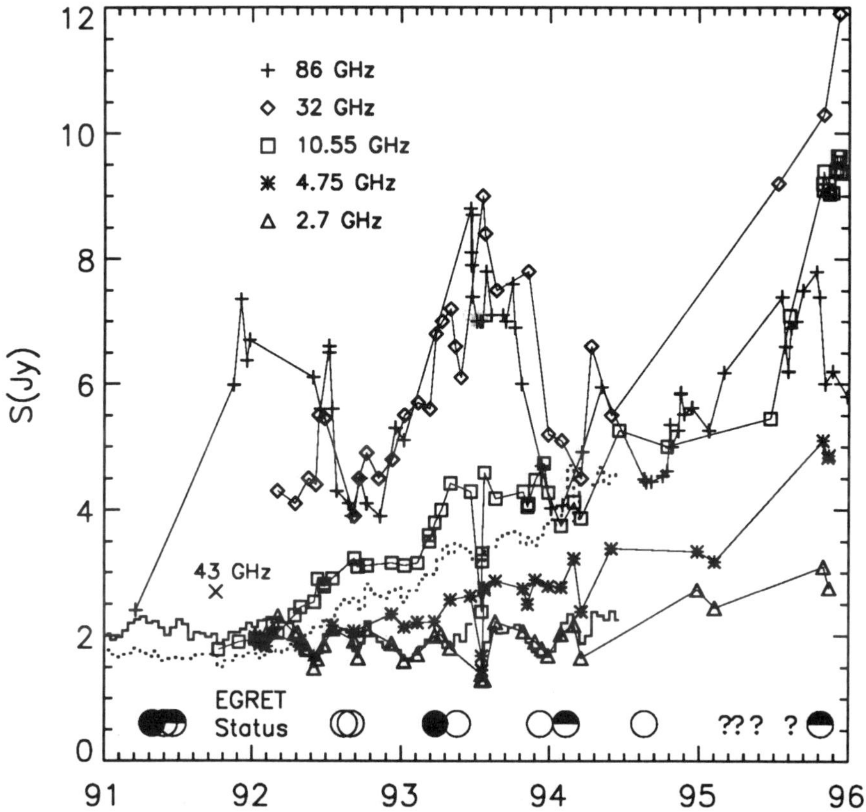

Figure 1. Here we show the radio light curve of 0528+134 based on Effelsberg data and observations at 86 GHz taken with the IRAM 30-m. The histograms give the two-week-averages of the NRL-GBI data at 2.25 GHz (solid line) and 8.3 GHz (dotted line). Observational uncertainties are below 5% except for the data at 32 GHz and 86 GHz which have uncertainty levels around 10%. The unusual depression in July 1993 is most likely an extreme scattering event and not intrinsic to the source. For comparison the state in the EGRET range is indicated by empty circles for low state ($S < 5 \cdot 10^{-7}$ ph. cm^{-2} sec^{-1} above 100 MeV), medium level ($S < 10^{-6}$), and high state ($S > 10^{-6}$). A question mark indicates protected data. It is tempting to relate the γ-ray outbursts in 1991 and 1993 to the mm outbursts a few month later. However, no strong γ-ray outburst has been reported yet for 1995, a few months before the brightest ever-recorded mm outburst. Or is there a time lag of two-and-a-half years, relating the 1991 γ-ray flare to the 1993 radio outburst, respectively the 1993 γ-ray flare to the 1995 radio outburst? Given the superluminal motion we would expect a corresponding VLBI knot to have a core separation of 0.3 mas after that time. In 1992.85 more than 80% of the 22 GHz flux was confined within 0.1 mas to the core (Pohl et al. 1995).

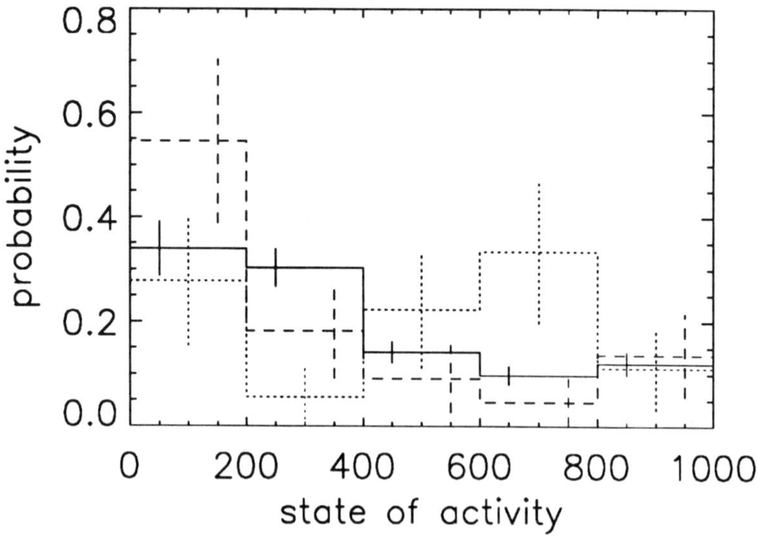

Figure 2. Activity state at 10 GHz simultaneous to EGRET detections (dotted line), averaged over the last three years for EGRET sources (solid line), and for a sample of FSRQ which have not been seen by EGRET. Activity in high-energy γ-rays coincides with activity in the radio regime.

like 0528+134, exhibit γ-ray variability on time scales of days—less than the standard integration time in EGRET observations.

2. The average source

We have seen that for an individual source it is difficult to relate γ-ray and radio outbursts to each other. This is especially true if the source is so active that the radio outbursts blend together.

To circumvent this problem we may also consider the average source. As the absolute flux levels of sources can be different and we are mainly interested in the variability behavior we scale all flux values linearly between historical minimum and historical maximum. In this analysis we accept only sources which have been proven to be variable by at least a factor of two.

During the EGRET sky survey we have observed all blazars with cataloged flux level >1 Jy simultaneously to the γ-ray observations. This allows us to compare the activity distribution in the radio regime of sources which have been seen by EGRET to those sources which have been γ-ray quiet. Here the activity of a source is defined on a scale of zero to thousand between historical minimum and maximum. The resulting activity distributions are shown in Fig. 2 where in case of EGRET blazars we give both the activity at the time of γ-ray detection and the average activity over a three years time period. The average source has

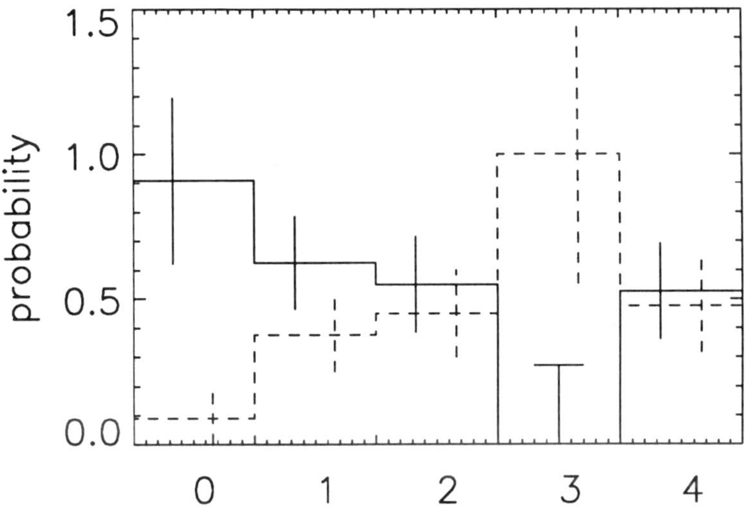

Figure 3. The probability level of detection (solid line) and non-detection (dashed line) of EGRET sources in different radio states: 0: increasing radio flux, 1: medium level without clear trend, 2: high state, 3: decreasing flux, 4: low state. There is a high probability level for detection at increasing radio flux, i.e., in the rising phase of a radio outburst, and a correspondingly small detection probability in the declining phase of radio outbursts.

an enhanced activity level at radio wavelengths when it is bright in high energy γ-rays (Mücke et al. 1996a)

Most EGRET blazars are variable at γ-ray energies. It may be instructive to see whether there is a relation between the radio state and the γ-ray brightness. The γ-ray data for this analysis are taken from the second EGRET catalog (Thompson et al. 1995) where we have excluded observations with large off-axis angles which often yield only poor upper limits. Then the significance of detection is used as discriminator between γ-ray bright and γ-ray quiet. As can be seen in Fig. 3, there seems to be a clear trend, indicating that γ-ray outbursts precede radio outbursts: when the radio flux is increasing, the average source is preferentially bright in γ-rays, and when the radio flux is decreasing, the average source is preferentially weak in γ-rays (Mücke et al. 1996a).

We have also searched for correlations between radio and γ-ray quantities as well as between flux and spectral indices. There is no simple relation between the scaled γ-ray flux and the scaled radio flux or scaled radio spectral index. However, we found evidence for a relation between the γ-ray flux and the γ-ray spectral index. This can be seen in Fig. 4 where both the γ-ray flux and the power-law spectral index have been scaled between historical minimum and maximum. The average source appears to have a harder spectrum at γ-ray high states (Mücke et al. 1996b). The chance probability is of order 10^{-5} and the result is also stable against the omission of individual sources from the sample.

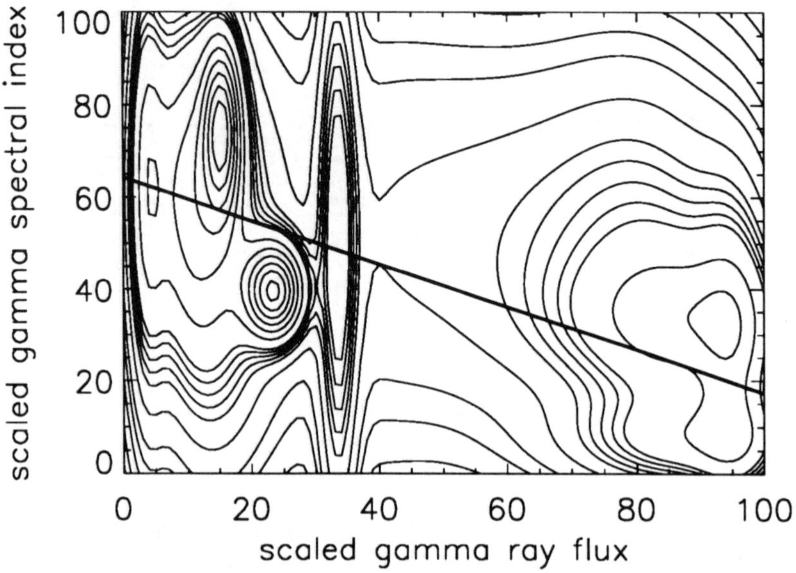

Figure 4. Scaled γ-ray spectral index (0=hardest occurance minus 1σ, 100=softest occurance plus 1σ) in comparison to the γ-ray flux. Here the data are represented by ellipsoidal Gaussians and the fit is based on χ^2 statistic.

3. Luminosity correlations

There is no luminosity correlation between radio and γ-ray emission of blazars.

We have analysed our data with Spearman's correlation method and, also to include the upper limits, with Kendall's τ. These partial correlation coefficients take care of the strong redshift dependence of luminosity, which may mimic a correlation for flux-limited data. In case of the radio-to-γ-ray relation of blazars the apparent alignment of data in luminosity-luminosity diagrams is indeed an artefact arising from sensitivity limits in the sample (Mücke et al. 1996c). For EGRET observations the sensitivity limit is mainly given by count statistic, while in the radio range mainly a credibility limit applies, that in view of their abundance one is reluctant to identify a γ-ray point source with a 100 mJy radio source, apart from the fact that catalogs including spectral information are not complete at these flux levels.

The results of our analysis are summarized in Table 1. We have also compared peak luminosities, here based on radio data taken from the literature. The marginal correlation signal for the 4.8 GHz data is not supported by a similar result for the 8 GHz data and thus probably a statistical fluke.

When averaging over the light curves, i.e., using mean values, the dynamical range gets even more compressed so that a positive correlation signal is artificially produced, though the original data are uncorrelated. Our simulations

Table 1. Results of the correlation analysis of N simultaneously observed data points: partial correlation analysis using the Spearman rank order coefficient R_s or Kendall's τ for the case of censored data (i.e., including upper limits: UL). The third column gives the resultant correlation coefficient and the fourth column the chance probability of erroneously rejecting the null hypothesis. An asterisk indicates that the chance probability is determined on the basis of simulations. The analysis was carried out between the logarithms of the luminosities and K-corrected flux densities, respectively.

data	N	CC	prob.	correlation
10 GHz - γ-ray (simult.)	25	0.158	0.465	NO
10 GHz - γ-ray (simult.+UL)	42	0.079	0.288	NO
2.7 GHz - γ-ray (simult.)	22	0.046	0.751	NO
2.7 GHz - γ-ray (simult.+UL)	41	0.063	0.426	NO
max. 4.8 GHz - γ-ray	12	0.594	0.053	marginal
max. 8 GHz - γ-ray	11	0.363	0.314	NO
4.8 GHz - γ-ray obs. (mean)	38	0.347	0.12 *	NO
8 GHz - γ-ray obs. (mean)	28	0.405	0.11 *	NO

show that the chance probability for a high correlation coefficient is increased roughly by a factor 4 compared to the standard R_s statistic. This may explain earlier claims for a correlation (Stecker et al. 1993; Padovani et al. 1993).

We also searched for linear correlations by a χ^2 test on the flux-flux relation. In no case we obtained acceptable fits of linear relations to the observed simultaneous data.

References

McNaron-Brown K., Johnson W. N., Jung G. V., et al. 1995, ApJ, 451, 575

Mücke A., Pohl M., Reich P., et al. 1996a, A&AS, submitted

Mücke A., Pohl M., Kanbach G., et al. 1996b, in Extragalactic Radio Sources, IAU Symposium 175 (Dordrecht, Kluwer), in press

Mücke A., Pohl M., Reich P., et al. 1996c, A&A, submitted

Padovani P., Ghisellini G., Fabian A. C., & Celotti A. 1993, MNRAS, 260, L21

Pohl M., Reich W., Krichbaum T. P., et al. 1995, A&A, 303, 383

Pohl M., Reich W., & Schlickeiser R. 1996, A&AS, submitted

Stecker F. W., Salamon M. H., & Malkan M. 1993, ApJ, 410, L17

Thompson D. J., Bertsch D. L., Dingus B. L., et al. 1995, ApJS,101, 259

Zhang Y. F., Marscher A. P., Aller H. D., et al. 1994, ApJ, 432, 91

Blazar Continuum Variability
ASP Conference Series, Vol. 110, 1996
H. R. Miller, J. R. Webb, and J. C. Noble, eds.

The Time Profiles of Millimeter-wavelength Outbursts During the Most Intense Gamma-ray Flares in AGN

H. Teräsranta and E. Valtaoja

Metsähovi Radio Research Station, FIN-02540 Kylmälä, Finland

Abstract. We have selected the most powerful gamma-ray outbursts of AGN from the second EGRET catalog and compared them with our 22 and 37 GHz monitoring data. We have determined the phase of the mm-outburst at the time of the EGRET pointings. It seems that the gamma flux is greater when the delay to the mm-maximum is shorter.

1. Introduction

Observations at high energies (E $\geq$ 30 MeV) with the EGRET instrument on board the Compton Gamma Ray Observatory have opened a new window on the energy production mechanisms of AGN. Strong outbursts at gamma wavelengths appear to be rare and of short duration. In order to find if these gamma flares correlate with lower frequency events, we must have well-sampled flux curves. Recent studies (Valtaoja & Teräsranta 1995, 1996) show that the majority of gamma-ray detections have occurred during the rising part of a millimeter outburst in the source (i.e., during the growth stages of a new shock in the jet). A connection between radio and gamma-ray activity was also found in Reich et al. (1993), and in Pohl (these proceedings). However, although the radio-gamma connection is by now well established, and the intensity of the gamma-ray emission seems to correlate with the strength and the variability timescale of the growing radio flare (Valtaoja, these proceedings), it is not known at which phase of the shock growth the gamma-ray emission is the most intense. In addition to the paucity of the gamma-ray data, the study is further complicated by the difficulties in determining the exact phase of the often complex radio flares.

2. Defining the sample

We have used our millimeter monitoring data of AGN at 22 and 37 GHz to determine during which phase of the radio outburst the gamma burst is most likely occur. The gamma-ray data is from the second EGRET catalog of Thompson et al. (1995). We chose only the strongest outbursts, putting the lower flux limit to 80×10^{-8} photons cm^2 and S/N ≥ 4.0. This limits the number of sources to 25. Of these many are still unidentified and thus excluded from this analysis. Some of the sources are too far south for our telescope, and some were not in our sample before the EGRET detection. The data for the events are presented in Table 1.

Taking all into account, only 6 sources are left for further analysis. Of those sources, 4C 38.41 does not show any outburst in radio during the EGRET detection, even though we have covered the observing window quite well. Of the other sources, 3C 279 was undergoing a complex series of outburst with several rising and decaying flares overlapping at the time of the EGRET observations. It is difficult to separate these flares unambigiously, and one cannot know which flare the gamma-rays should be connected to. 3C 454.3 was also undergoing a series of outbursts at our radio frequencies.

Table 1. Sources from the second EGRET catalog with flux $\geq 80 \times 10^{-8}$ photons cm^2 and S/N ≥ 4.0. Notes: *a*—source too far south; *b*—not identified yet; *c*—no Metsähovi data prior to the outburst; *d*—radio outburst complex; *e*—no outburst seen in the radio data.

Source	Flux	Error	vp.	Notes
0208−512	109.7	8.1	10.0	a
	94.7	21.6	220.0	
0235+164	82.5	9.1	21.0	
J0241+6119	121.1	15.7	15.0	b
	113.1	27.5	34.0	
	118.8	13.9	211.0	
0446+11	113.5	20.6	36.+	c
0528+134	128.6	8.6	0.2	c
	83.9	8.1	1.0	c
	295.0	33.3	213.0	
J1021−5835	144.3	29.0	6.0	b
	103.8	20.7	8.0	
1127−145	93.2	22.9	206.0	a
4C 29.45	228.6	54.8	206.0	
1222+216	82.9	20.2	204.0	c
3C 279	287.0	10.9	3.0	d
	92.8	7.7	11.0	d
1406−076	144.0	26.4	205.0	c
	116.3	15.1	207.0	c
J1412−6211	113.9	28.3	23.0	b
J1443−6040	115.1	26.7	12.0	b
4C 38.41	105.4	9.4	9.2	e
J1746−2852	96.6	15.7	5.0	b
	133.1	23.3	16.0	
	131.2	27.6	232.0	
J1747−3039	86.4	21.6	16.0	b
J1811−2339	137.5	31.9	7.2	b
J1813−1229	180.0	30.3	13.1	b
J1825−1307	236.2	48.2	229.5	b
J1828+0145	126.1	23.8	13.1	b
J1835+5919	82.4	15.4	2.0	b
	93.7	21.6	202.0	
	89.7	13.6	203.0	
J2019+3719	87.2	20.2	7.1	b
	99.7	13.7	203.0	
J2020+4026	131.1	15.7	2.0	b
	150.4	14.7	203.0	
	104.8	17.1	212.0	
J2033+4112	115.2	18.8	212.0	b
3C 454.3	81.7	8.8	19.0	d
	131.7	20.7	37.0	d

This leaves only three events for the present analysis: 0235+164 from February 1992, 0528+134 from March 1993 and 4C 29.45 from January 1993. In all of these cases the source had a clear, isolated outburst at both 22 GHz and

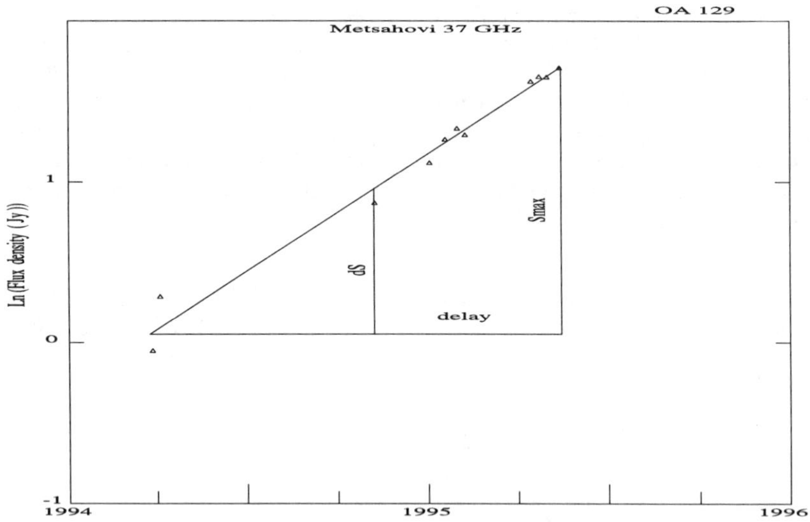

Figure 1. An example of how to determine parameters of the millimeter outbursts in Table 1. A linear fit through $\ln(S - S_{\mathbf{base}})$ for the rising part of the millimeter flare gives dS, $S_{\mathbf{max}}$ and the delay between the gamma-ray and the radio maximum.

37 GHz. In each case the gamma-ray event occured during the rising phase of the outburst, and the relative timing of the gamma-ray and the radio flares is possible.

3. Finding the phase of the mm-outburst

We model the outbursts by first subtracting the quiescent flux density, approximated by the lowest ever recorded flux level at that frequency. After the subtraction of the baseline we plot the flux in logarithmic scale. During an outburst the $\ln S$ seems to both rise and fall linearly. (In other words, the millimeter flares both rise and decay exponentially.) From linear fits to the data we can obtain dS, the flux density of the flare at the time of the EGRET pointing. We can also estimate the time delay between the gamma-ray outburst and the radio maximum, if we assume that the gamma-ray event was of short duration and was observed during its maximum emission. The procedure is clarified in Figure 1, where a linear fit is made to an outburst observed in OA 129 during 1994.

Figure 2 shows one of the simultaneous events, the 37 GHz radio flares in AO 0235+164 observed at Metsähovi. The time of the gamma outburst is marked with a vertical line. The determined parameters for this and the other events can be found in Table 2.

Table 2. Parameters for the linear fits.

Source	f(GHz)	S_{max}(Jy)	dS(Jy)	base(Jy)	delay(days)
0235+164	22	3.70	1.07	0.4	120
0235+164	37	3.79	1.63	0.3	95
0528+134	22	5.91	4.85	1.0	34
0528+134	37	6.50	5.71	1.0	24
4C 29.45	22	1.35	1.02	0.6	47
4C 29.45	37	1.50	1.20	0.5	44

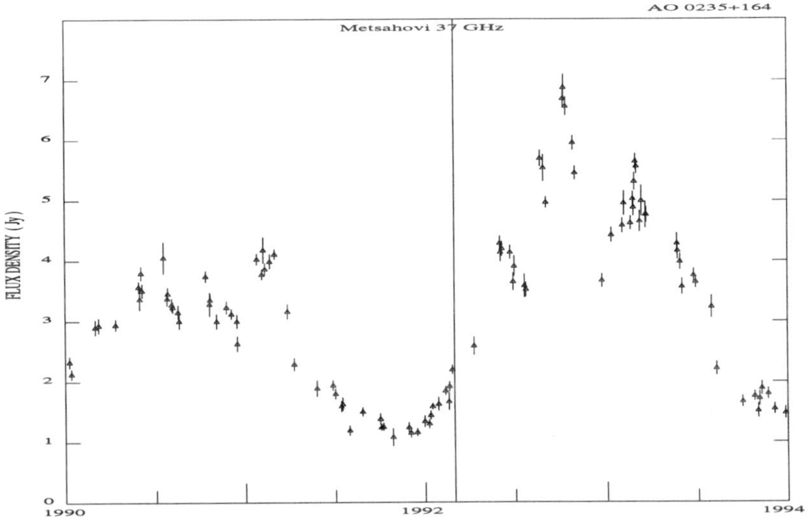

Figure 2. The 37 GHz flux density of AO 0235+164 from 1990 to the end of 1993. The time of the gamma-ray outburst is marked with a vertical line.

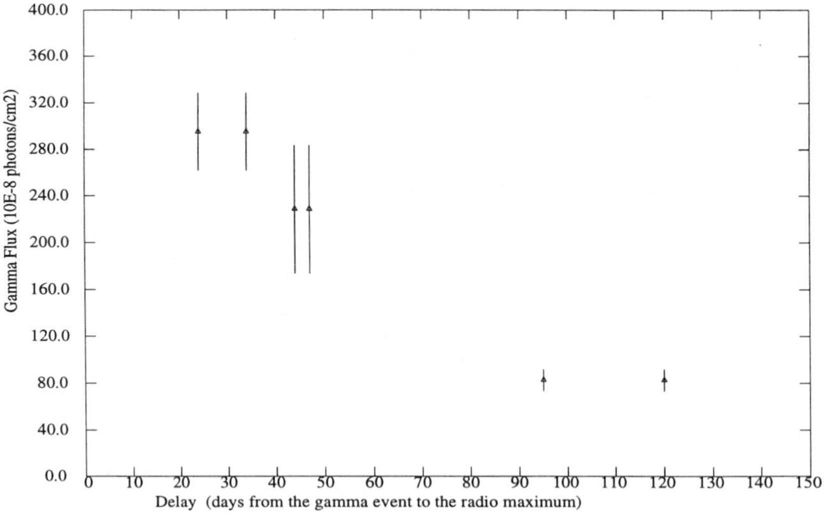

Figure 3. The delay of the mm-outburst maximum from the gamma-ray outburst.

4. Discussion

Figure 3 shows the gamma-ray intensity versus the delay from the gamma outbursts to the radio flare maximum. The stronger gamma events seem to be closer to the maxima of the radio flares. With only three separate events at two frequencies, no very firm conclusions can at present be drawn. This is somewhat disappointing, since knowledge of the temporal evolution of the gamma-ray intensity during shock growth and decay is crucial in testing detailed models for gamma-ray production.

To get better evidence of the relationship we would need well sampled flux curves also at higher frequencies like 90 GHz and 230 GHz, where the individual componets of the flares are more easily separated as they are shorter in duration. New EGRET data, and careful reanalysis of our radio monitoring data using improved flare modelling will also hopefully increase the number of data points.

The amplitudes of the millimeter outbursts associated with gamma flares seem often to be quite small. This makes it difficult to identify new gamma-ray emitting AGN solely on the basis of their radio variability.

References

Reich W., et al. 1993, A&A, 273, 65

Thompson D. L., et al. 1995, ApJS, 101, 259

Valtaoja, E., & Teräsranta, H. 1995, A&A, 297, L13

Valtaoja, E., & Teräsranta, H. 1996, A&AS, in press

Blazar Continuum Variability
ASP Conference Series, Vol. 110, 1996
H. R. Miller, J. R. Webb, and J. C. Noble, eds.

Superluminal Motion in the Gamma Ray Blazar 1611+343

B. Glenn Piner

Astronomy Dept, University of Maryland, College Park MD 20742

NASA/Goddard Space Flight Center, Code 661, Greenbelt, MD 20771

Kerry A. Kingham

U.S. Naval Observatory, 3450 Massachussets Ave N.W., Washington D.C. 20392

Abstract. As part of an effort to observe high energy γ-ray blazars with VLBI, we have produced 8 and 2 GHz VLBI images, at eight epochs, of the quasar 1611+343, which has twice been detected by the EGRET telescope on board CGRO. The VLBI data have been taken from the Washington VLBI correlator's geodetic database. As is expected for a γ-ray quasar, we find that 1611+343 possesses a jet with several compact components moving at apparent superluminal speeds. We have detected five components and have measured speeds for the outer four of 6.7 ± 3.8, 4.0 ± 3.1, 7.4 ± 3.0, and $11.6 \pm 5.9\ h^{-1}c$ from the outermost component inwards. ($H_0 = 100h$ km s^{-1} Mpc^{-1}, q_0=0.5). We have mapped the two-dimensional projected structure of the VLBI jet and clearly see different trajectories for different components at the same radial position, which rules out a continuous jet model for this source. Possible support is found for the idea that a γ-ray flare signals the ejection of a new VLBI component from the core.

1. Previous Observations of 1611+343

The source 1611+343 is an OVV quasar with a redshift of z=1.4 and a spectral index of $\alpha = -0.2$ between 1.5 and 8.1 GHz (Kuehr et al. 1981b), where α is defined by $S \propto \nu^{\alpha}$. A VLA image presented by Murphy, Browne, & Perley (1993) shows that 1611+343 exhibits a triple morphology where the position angle of the brighter southern component is consistent with the position angle of 197° given by Rusk (1988). The quasar 1611+343 was observed with VLBI at 5 GHz in a three antenna experiment by Zensus, Porcas, & Pauliny-Toth (1984). It was well fit by a circular Gaussian of width 0.6 mas. This quasar has also been imaged with lower frequency VLBI at three epochs as part of a campaign to study the milliarcsecond scale structure of low frequency variable sources (Romney et al. 1984; Padrielli et al. 1986, 1991). Observations at the first two epochs at 1.67 GHz were well fit by an elliptical Gaussian elongated along a position angle of 150°. Other VLBI observations are given by Baath (1987) at 932 MHz and by Waak et al. (1988) at 1.66 GHz, with resolutions of about 5 and 10 mas respectively. The source structure is unresolved in both cases. The high energy γ-ray telescope EGRET has observed 1611+343 on five separate occasions (Thompson et al, 1995), and found that like other γ-ray blazars, it is highly variable. It appears to have flared twice in γ-rays; if we use the upper limits prior to each positive detection we can set the times of the flares as between 1992.28-1992.90 and 1993.76-1994.85 respectively.

2. VLBI Observations

The observations used in this paper are from archived geodetic Mark III VLBI observations processed at the Washington VLBI Correlator Facility located at the Naval Observatory. Observations are done at eight separate intermediate frequencies between 8 and 9 GHz and six separate intermediate frequencies between 2 and 3 GHz, and these intermediate frequencies are retained as separate UV plane points. An integration time of 4 seconds is used for each UV plane point. The poor UV plane coverage of many of the observations can be compensated for by adding together observations which were close together in time; this can be done as long as separate flux density data shows that the source was approximately constant over the time span of the summed observations. For this paper we have summed observations with time differences of no more than 42 days; we have verified that the radio flux density of 1611+343 was constant over these times (Waltman, 1995 priv.comm.), and we have performed self-calibration separately on each observation to be summed. The images were produced using standard self-calibration procedures from the Caltech Difmap software package.

3. Results

We have imaged 1611+343 at eight different epochs: 1990.56, 1992.55, 1993.52, 1994.05, 1994.95, 1995.07, 1995.79, and 1996.24. Images for the first three epochs were produced by summing separate observations, with four, five, and four separate observations being summed for these first three epochs, respectively. Five of the 8 GHz radio images are shown in Figure 1. Three images are not displayed due to size constraints, but all images are shown in Piner & Kingham (1996). The typical resolution of the images shown in Figure 1 is about 0.5 mas. In order to quantify the positions and motions of the jet components, we fit Gaussian components to the observed visibilities for each epoch. The locations of these components are also shown in Figure 1. The outermost component, C1, is visible only in the first three epochs and then again in the image from 1995.79 (not shown). There are two jet components which are present in every image; we refer to the outer as C2 and the inner as C3. New components appear in 1993.52 (C4), and in 1996.24 (C5). The motions of the components over time are shown by the dotted lines in Figure 1, which are least-squares fits to the component positions, with an assumed error in the positions of one quarter of the beam FWHM. These fits give component velocities of 6.7 $\pm$ 3.8, 4.0 $\pm$ 3.1, 7.4 $\pm$ 3.0, and 11.6 $\pm$ 5.9 $h^{-1}c$ for C1 to C4 respectively (H_0 = $100h$ km s^{-1} Mpc^{-1}, q_0=0.5). These measurements are consistent with all of the components moving at the average velocity of 6.9 $h^{-1}c$. We have calculated a lower limit to the Doppler beaming factor of this source using the method of Ghisellini et al.

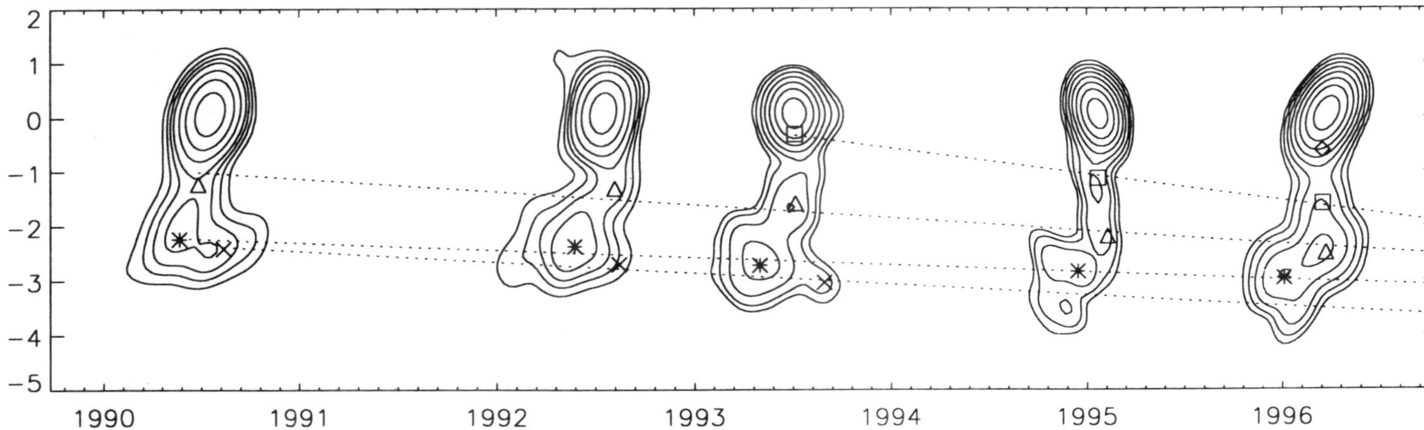

Figure 1: Mosaic plot of images and component motions. Time is plotted along the horizontal axis and relative declination is plotted along the vertical axis. In order to produce a cleaner plot, noise contours are not shown. The lowest detectable flux in the displayed images ranges from 5 mJy for 1995.07 to 15 mJy for 1992.55; the corresponding dynamic ranges range from about 300:1 to 150:1. The lowest contours shown are about twice the level of the lowest detectable flux. Component positions are plotted on top of the images, and the dotted lines show the best fit to the component motions. The Xs represent component C1, the asterisks, C2, the triangles, C3, the squares, C4, and the diamond, C5.

(1993). We find that $\delta \geq 8.6$, which implies an upper limit of the angle of the jet to the line of sight of $\theta \leq 6.5°$ and a lower limit to the Lorentz factor of $\Gamma \geq 7.1$, assuming equal bulk and pattern speeds.

The recent ejection of components C4 and C5 offers the interesting interpretation that VLBI component ejection may be related to flares in the γ-ray regime. The one sigma errors on the slope of the line fitted to the motion of component C4 give a time of ejection from the core between 1988.96 and 1992.94, and if we assume C5 is moving with the average velocity of the other components, then it was ejected between 1992.22 and 1994.13. These times are consistent with the times of the two EGRET flares mentioned in §1. Similar correlations between component ejections and γ-ray flares have been observed for 0528+134 and 3C454.3 (Kirchbaum et al. 1995), and 3C279 (Wehrle et al. 1994), and since correlations between component ejections and enhanced activity at other wavelengths have been observed in many other sources (e.g. Mutel et al. 1990; Zensus et al. 1990), the extension of this correlation to γ-rays is not surprising.

An important question to answer about blazar jets in general deals with the motion of components in bent or curving jets and whether or not there is a fixed track in each source which successive components follow (Padrielli 1989). In order for this question to be answered, sources with bent jets must be monitored over a long enough time for new components to reach the same radial position formerly held by the older components. Figure 2 shows the position on the plane of the sky of every component position at every epoch. Components C1 and C2 have been mapped at approximately the same distances from the core, yet they have very different positions and trajectories. This rules out any kind of continuous jet model where each component follows its precursor. The next important question is whether the components follow a fixed straight path out to a certain distance from the core and then bend, or whether the components start out at the core with their own straight, independent trajectories. Components C3 and C4 define a common path almost due south from the core, and C3 is now far enough out that it is obvious that it has not turned to follow C2, and if it is going to turn to follow C1 it must display a rapid shift towards higher position angles within the next year. All indications show that it is continuing straight along its previous path. We suggest that the strongest interpretation of Figure 2 is that each component is ejected from the core with its own independent position angle, and that it then moves straight out from the core along this trajectory.

The mechanism behind these different position angles remains to be explained. The fact that C3, C4, and C5 have not been ejected at lower position angles than C2 or higher position angles than C1 seems to rule out precession of the jet for this mechanism. An alternative is that the components are being ejected inside a cone with a sizable opening angle, but this possibility has problems as well. Since the minimum opening half angle of the cone defined by the position angle difference between C1 and C2 is greater than the maximum angle from the line of sight defined by the Doppler beaming factor, our line of sight must lie inside the cone. Then the fact that we do not see ejections in all directions must mean that either the azimuth of the ejections inside the cone favors position angles near 180° or our line of sight must lie barely inside the cone, with the majority of the cone lying to the south. Continued observation over the next

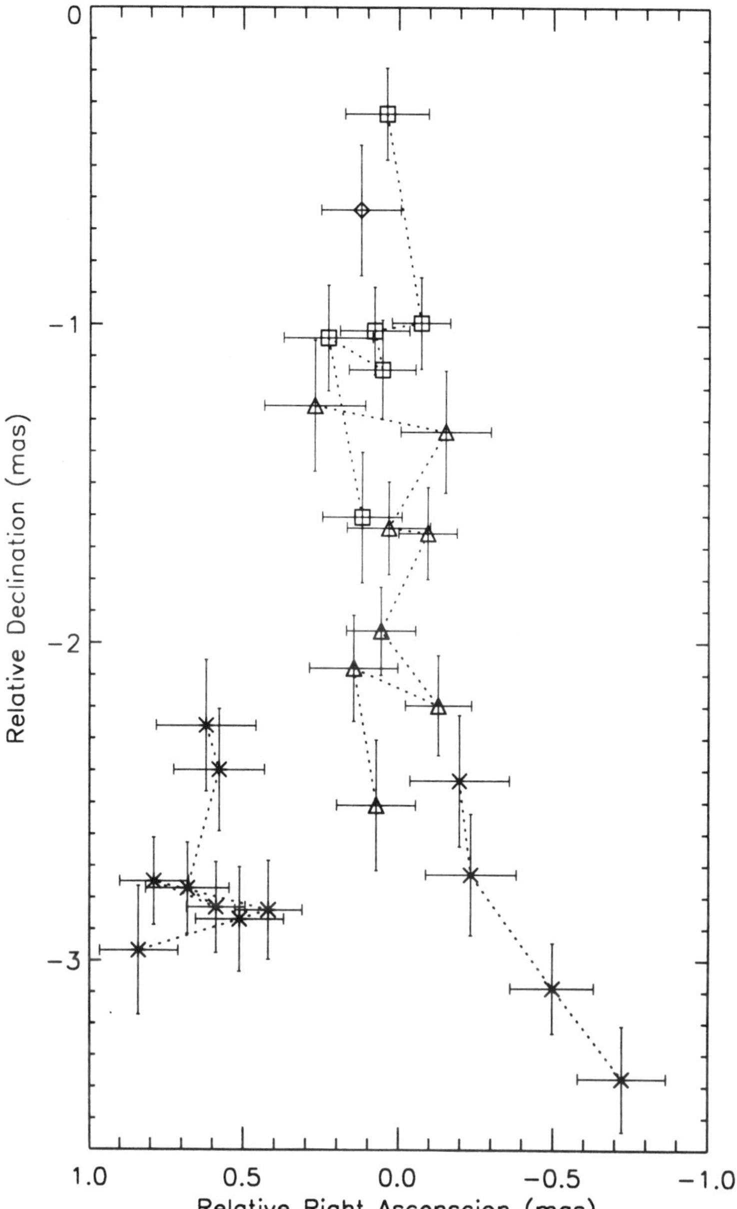

Figure 2: Plot showing the two dimensional position of every measured component position from every epoch. Dotted lines connect the measurements of each individual component in time sequence. The error bars are one quarter of the beam FWHM. The Xs represent component C1 (outer component on right), the asterisks, C2 (outer component on left), the triangles, C3, the squares, C4, and the diamond, C5.

year should be able to settle whether C3 will bend away from its present course to follow C1, and within two years we should have some indication of what will happen to component C4.

Superluminal motion has now been confirmed in fifteen of the forty AGN listed in the second EGRET catalog, and since it is generally agreed that the γ-rays are produced in a beamed relativistic jet (e.g. von Montigny et al. 1995), the other EGRET sources should display this property as well. This paper demonstrates that the pool of EGRET sources which have been relatively sparsely observed with VLBI provides a sample worthy of further study, and it shows that archived geodetic VLBI observations are capable of doing this. Work on imaging more of the EGRET sample with VLBI data from the Washington VLBI correlator's database is currently underway.

Acknowledgments. We acknowledge the support of the VLBI staff at the Naval Observatory and the EGRET team at NASA/GSFC. We also acknowledge helpful discussions and communication of data in advance of publication from Marshall Eubanks, Bob Hartman, Beth Waltman, and Ann Wehrle.

References

Baath, L.B. 1987, in Superluminal Radio Sources, ed. J.A. Zensus & T.J. Pearson (Cambridge:Cambridge Univ. Press), 206

Ghisellini, G., Padovani, P., Celotti, A., & Maraschi, L. 1993, ApJ, 407, 65

Kirchbaum, T.P., Britzen, S., Standke, K.J., Witzel, A., Schalinski, C.J., & Zensus, J.A. 1995, preprint

Kuehr, H., Witzel, A., Pauliny-Toth, I.I.K., & Nauber, U. 1981, A&AS, 45, 367

Murphy, D.W., Browne, I.W.A., & Perley, R.A. 1993, MNRAS, 264, 298

Mutel, R.L., Phillips, R.B., Su, B., & Bucciferro, R.R. 1990, ApJ, 352, 81

Padrielli, L., et al. 1986, A&A, 165, 53

Padrielli, L. 1989, in Very Long Baseline Interferometry: Techniques and Applications, ed. M. Felli & R.E. Spencer (Dordrecht:Kulwer), 319

Padrielli, L., Eastman, W., Gregorini, L., Mantovani, F., & Spangler, S. 1991, A&A, 249, 351

Piner, B.G. & Kingham, K.A. 1996, in preparation

Romney, J.D., et al. 1984, A&A, 135, 289

Rusk, R.E. 1988, Ph.D. thesis, Univ. Toronto

Thompson, D.J., et al. 1995, ApJS, 101, 259

von Montigny, C., et al. 1995, ApJ, 440, 525

Waak, J.A., Spencer, J.H., Simon, R.S., & Johnston, K.J. 1988, in IAU Symp. 129, The Impact of VLBI on Astrophysics and Geophysics, ed. M.J. Reid & J.M. Moran (Dordrecht:Kulwer), 95

Wehrle, A.E., et al. 1994, in The Multimission Perspective (Eureka Scientific:Napa Valley, CA), 27

Zensus, J.A., Porcas, R.W., & Pauliny-Toth, I.I.K. 1984, A&A, 133, 27

Zensus, J.A., Unwin, S.C., Cohen, M.H., & Biretta, J.A. 1990, AJ, 100, 1777

Blazar Continuum Variability
ASP Conference Series, Vol. 110, 1996
H. R. Miller, J. R. Webb, and J. C. Noble, eds.

Misalignments between the Parsec-scale and Kiloparsec-scale Emission of Radio Sources

A. E. Wehrle
Infrared Processing and Analysis Center, Jet Propulsion Laboratory and California Institute of Technology

S. C. Unwin
Owens Valley Radio Observatory, California Institute of Technology

H. D. Aller and M. F. Aller
Department of Astronomy, University of Michigan

G. Nicolson
Hartebeesthoek Radio Astronomy Observatory, South Africa

Abstract. The difference between the position angles of parsec-scale jets and those of kiloparsec-scale emission has been measured for 48 sources in the Variable Source Sample (Wehrle et al. 1992) and Pearson-Readhead sample (Pearson & Readhead 1988). The apparent "bend" was measured by comparing VLBI with VLA images. We confirm the effect found by Pearson and Readhead, namely, that the distribution of bends peaks at $\Delta PA \sim 0°$ and $90°$. A striking characteristic of the BL Lac objects in the sample is that their large- and small-scale structures are all substantially misaligned. Five γ-ray blazars in the Variable Source Sample have highly curved parsec-scale jets; at least one (3C273) is quite straight.

1. Introduction

One of the most interesting peculiarities about the radio structures of some highly polarized, flat-spectrum quasars is the appearance of a 90° bend between their parsec scale jets and kiloparsec-scale hotspots. Early VLBI surveys (Pearson & Readhead 1988, Wehrle et al. 1992) found that flux-limited radio source samples chosen at high radio frequencies with flat spectra, or by strong variability criteria, contained significant numbers of "misaligned" sources as well as "aligned" sources (an "aligned" source has a parsec-scale jet that points directly to the kiloparsec-scale jet or hotspot; a "misaligned" source points roughly 90° away). Xu and co-workers (1995; and these proceedings) have recently confirmed the bimodal distribution of position angle differences in a large VLBI survey.

This paper first presents position angle data for the combined Pearson-Readhead and Wehrle et al. samples. Second, we examine the jet bending

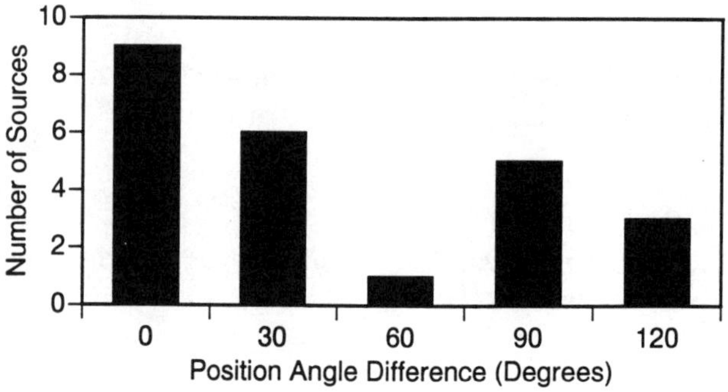

Figure 1. Histogram of position angle differences (ΔPA) for the Variable Source Sample.

properties of several γ-ray blazars in the Wehrle et al. sample and compare our findings to those of von Montigny et al. (1995). Throughout this paper we use a Hubble constant $H_0 = 100$ km s^{-1} Mpc^{-1} and $q_0 = 0.5$.

2. Bend angle data

We have presented elsewhere (Wehrle & Cohen 1989; Wehrle, Cohen, & Unwin 1990; Wehrle et al. 1992) images of 15 sources made with Global VLBI observations at 5 GHz from a flux-limited sample. The sample included all objects north of declination $-18°$ which exceeded 4.5 Jy at 8 GHz at some epoch. It contained 41 sources, including 24 objects for which ΔPA can be measured, with eight that overlap the Pearson-Readhead sample. We measured the parsec-scale position angle from the *overall* structure in 5 GHz VLBI images, assuming that the peak was the core (except for 0923+392=4C39.25 and 2145+067). The measurements of these 24 sources are presented in Table 1. We also remeasured ΔPA for Pearson-Readhead sources for consistency (Table 2).

A histogram of ΔPA for the Variable Source Sample is shown in Figure 1. The distribution showed two peaks near 0° and 90°. The distribution is very unlikely to arise by chance if the parent population has a uniform distribution of ΔPA between 0° and 180° (using K-S test with 5 bins, less than one chance in a thousand).

If we distinguish between quasars, BL Lac objects, and galaxies, as classified by Hewitt and Burbidge (1987), interesting trends emerge. For quasars alone, the distribution is still double-peaked, but less pronounced: 20 quasars have ΔPA near 0° and 11 objects have ΔPA near 90°. Galaxies are clustered near $\Delta PA \sim 0°$. All seven BL Lac objects have $\Delta PA \sim 90°$. All objects with arcsecond "halos" have curved jets, and most are substantially misaligned.

Impey, Lawrence, and Tapia (1991) have studied the optical polarization characteristics of the Pearson-Readhead sample: aligned objects have low polarization while misaligned objects have high polarization, providing direct evidence of beaming in the misaligned population. Since γ-ray blazars are usually highly

Table 1. Characteristics of the Variable Source Sample

Source	Optical Type	Redshift	Small-scale PA (degrees)	VLBI Ref.	Large-scale PA (degrees)	VLA Ref.	Difference |PA| (degrees)	Scale Length (parsecs)	Radio Size (kpc)	150 MHz Radio Power (W/Hz)	Notes
0106+013	Q	2.017	75	W92	185	P82	110	7.3	40.5	---	
0133+476	Q	0.859	317	PR88	---	---	---	10.4	---	9.31E+26	5% level
0234+285	Q	1.213	345	W92	335	M93	10	20.7	43.1	---	
0235+164	B	0.94	unresolved	G89,G92	---	---	---	4.2	---	---	
0316+413	G	0.017	190	PR88	145	P82	45	1.8	---	2.49E+25	5% level
0336-019	Q	0.852	70	W92	345	P82	85	12.1	21.8	---	
0355+508	EF	---	---	---	110	P82	---	---	---	---	
0420-014	Q	0.915	diffuse	W92	---	---	---	5.9	---	---	
0430+052	G	0.033	260	W87	265	P82	5	23.6	378	---	
0458-020	Q	2.286	305	W92	235	B89	70	16.3	19.9	---	
0528+134		---	---	---	---	---	---	---	---	---	
0538+498	Q	0.545	225	PR88	225	R88	0	---	3.7	1.57E+28	15 GHz VLA map
0552+398	Q	2.365	320	W92	---	---	---	3.1	---	---	
0605-085	Q	0.87	---	---	96	P82	---	---	---	---	
0607-157	Q	0.324	---	---	---	---	---	---	---	---	
0727-115		---	---	---	---	---	---	---	---	---	
0735+178	B	>0.424	67	G89	170	P82	103	---	46.2	4.35E+26	
0851+202	B	0.306	260	G89	---	---	---	---	---	---	
0923+392	Q	0.699	277	PR88	259	P82	18	6.7	28	4.00E+27	5% level
1127-145	Q	1.187	65	W92	41	R88	24	67.1	163	---	
1226+023	Q	0.158	230	Z88	222	P82	8	47.7	37.8	---	
1253-055	Q	0.538	230	U89	202	P82	28	34.4	121	---	
1328+307	Q	0.846	260	K88	245	P82	15	---	25.2	1.96E+28	
1335-127	Q	0.541	diffuse	W92	curved	P82	---	5.9	---	---	
1510-089	Q	0.361	---	---	160	P82	---	---	---	---	
1633+382	Q	1.814	115	PR88	147	M93	32	2.1	58.8	4.29E+27	5% level
1641+399	Q	0.595	230	B86	330	P82	100	34.1	103	4.70E+27	cf. U92
1730-130	Q	0.902	180	R84	273	P82	93	---	46.2	---	
1749+096	B	0.321	35	W92	halo	P82	---	6.3	---	---	
1828+487	Q	0.692	328	W87	316	P85,R88	12	64.6	80	4.61E+28	5% level
1928+738	Q	0.302	166	PR88	189	PR88,R88	23	22.5	188	2.24E+26	5% level
2005+403	Q	1.736	diffuse	W92	---	---	---	4.2	---	---	
2121+053	Q	1.178	diffuse	W92	---	---	---	3.9	---	---	
2134+004	Q	1.936	complex	PT90	---	---	---	8.2	---	---	
2145+067	Q	0.99	140	W92	305	P82	15	34.4	29.8	---	
2200+420	B	0.07	182	PR88	305	A86	123	4.3	40.5	7.14E+24	5% level
2201+315	Q	0.297	---	---	---	---	---	---	---	---	
2223-052	B	1.404	90	C91	341	R88	109	38.6	10.7	---	
2230+114	Q	1.037	140	W92	140	P82	0	87.1	15.8	8.64E+27	
2251+158	Q	0.859	310	P87	318	P82	8	40.9	41.7	---	
2345-167	Q	0.6	110	W92	355	P82	115	9.9	7.6	---	

Notes: VLBI and VLA references are given in the text references. "Scale length" refers to the overall extent of the VLBI jet at 1% of peak brightness unless otherwise indicated in the "Notes" column.

Table 2. Characteristics of the Pearson-Readhead Sample

Source	Optical Type	Redshift	Small-scale PA (degrees)	VLBI Ref.	Large-scale PA (degrees)	VLA Ref.	Difference |PA| (degrees)	Scale Length (parsecs)	Radio Size (kpc)	150 MHz Radio Power (W/Hz)	Optical Polarization (percent)	Notes
0016+731	Q	1.781	155	PR88	---	R88	---	2.1	4.2	1.45E+27	0.9	5% level
0108+388	G	---	240	PR88	90	M93	150	---	---	---	---	
0133+476	Q	0.859	317	PR88	---	---	---	10.4	---	9.31E+26	20.8	
0153+744	Q	2.338	155	PR88	---	---	---	39.5	---	4.86E+27	1.1	
0212+735	Q	2.367	90	PR88	---	---	---	48.1	---	4.27E+27	0.7	
0316+413	G	0.017	190	PR88	145	P82	45	1.8	---	2.49E+25	0.6	
0404+768	G	0.598	---	---	---	---	---	---	---	3.72E+27	---	
0454+844	B	---	150	PR88	---	M93	---	---	---	---	14.2	
0538+498	Q	0.545	225	R84	225	R88	0	23.2	3.7	1.57E+28	2.3	15 GHz VLA map
0710+439	G	0.518	0	PR88	---	---	---	82	---	1.80E+26	---	
0711+356	Q	1.62	337	PR88	297	M93	40	22.5	67.2	4.88E+26	1	
0723+679	Q	0.846	272	P87	260	R88	12	---	84	7.64E+27	1.5	
0804+499	Q	1.43	135	PR88	200	M93	65	1	34.4	1.37E+27	8.6	
0814+425	B	---	149	PR88	45	P82	104	---	34	---	8.7	
0831+557	G	0.242	complex	PR88	175	P82	---	9.5	24	6.34E+26	3	
0836+710	Q	2.17	214	PR88	204	R88	10	11.6	20	1.92E+28	1	
0850+581	Q	1.322	156	PR88	139	R88	17	17.2	108	2.29E+27	0.4	
0859+470	Q	1.462	0	PR88	335	P82	25	5.8	25.8	7.37E+27	0.7	
0906+430	Q	0.67	155	PR88	241	P85	86	2.2	13.6	1.47E+28	3.8	
0923+392	Q	0.699	277	PR88	259	P82	18	6.7	28	4.00E+27	0.4	5% level
0945+408	Q	1.252	127	PR88	32	M93	95	22.8	34.4	1.63E+28	0.3	
0954+556	Q	0.909	---	P92	300	P82	---	---	37.8	5.22E+27	6.4	
0954+658	B	0.368	302	G92	205	P82	97	3.1	21.7	1.78E+26	19.1	
1003+351	G	0.0989	292	B85	304	B85	12	20.4	2948	1.69E+26	0.2	
1031+567	G	0.45	---	P92	---	---	---	---	---	6.24E+26	---	
1358+624	G	0.431	---	P92	---	---	---	---	---	1.41E+27	---	
1458+718	Q	0.905	164	PR88	73	PR88	91	110.9	14.7	2.10E+28	1	
1624+416	Q	2.55	239	PR88	352	P82	113	11.3	7.6	1.92E+28	---	
1633+382	Q	1.814	115	PR88	147	M93	32	2.1	58.8	4.29E+27	1.1	5% level
1634+628	Q	0.988	complex	F85	---	---	---	---	---	1.48E+28	---	
1637+574	Q	0.745	unresolved	PR88	305	M93	---	4.1	57.4	1.59E+26	2.4	
1641+399	Q	0.595	230	B86	330	P82	100	34.1	103	4.70E+27	11.2	cf. U92
1642+690	Q	0.751	195	PR88	168	M93	27	14.3	61.5	1.24E+27	16.6	
1652+398	G	0.034	130	PR88	45	U83	85	0.7	13.8	2.47E+24	3	
1739+522	Q	1.375	3	PR88	260	P82	97	4.3	25.8	1.28E+27	3.7	
1749+701	B	---	296	PR88	209	PR88, R88	87	---	---	---	10.5	
1803+784	Q	0.68	268	PR88	195	A86,M93	73	4.5	200	4.90E+26	7	
1807+698	G	0.05	263	PR88	241	W90	22	9.1	20.1	1.55E+25	8	
1823+568	B	0.664	197	PR88	98	R88, M93	99	2.1	97.5	2.44E+27	16.8	
1828+487	Q	0.692	328	PR88	316	P85, R88	12	64.6	80	4.61E+28	0.2	5% level
1845+797	G	0.057	322	PR88	323	R88	1	3.1	158	2.07E+26	1.3	
1928+738	Q	0.302	166	PR88	189	PR88, R88	23	22.5	188	2.24E+26	1.2	5% level
1954+513	Q	1.22	8	PR88	350	P82	18	3.2	73.1	6.94E+27	---	
2021+614	G	0.227	32	PR88	---	---	---	17.9	---	7.32E+25	0.3	
2200+420	B	0.07	182	PR88	305	A86	123	4.3	40.5	7.14E+24	13.7	5% level
2342+821	Q	0.735	---	---	---	---	---	---	---	2.06E+28	---	
2351+456	Q	2	296	PR88	23	R88	87	8.7	9.8	6.91E+27	---	15% level
2352+495	G	0.237	complex	PR88	---	---	---	41.3	---	1.76E+26	1.2	

Notes: VLBI and VLA references are given in the text. "Scale length" refers to the overall extent of the VLBI jet at 1% of the peak brightness unless otherwise indicated in the "Notes" column. Optical polarization data is from Impey, Lawrence and Tapia 1991.

polarized quasars, we examined the structures of γ-ray blazars in our sample in more detail.

3. The γ-ray blazars

The ten γ-ray blazars in Table 1 had ΔPA ranging from 0° to 103°. Some appeared to be "aligned" (c.f., von Montigny et al. 1995) because we measured the position angle of the blob *farthest* from the VLBI core; this means that curvature nearer to the core was ignored. New VLBI images made by various groups reveal complex jet structures in several sources. Five γ-ray blazars have highly curved VLBI jets: 0528+134 (Krichbaum et al. 1995), 0735+178, 3C279=1253–055, 1633+382 (Barthel et al. 1995), CTA102=2230+114 (Rantakyro et al. 1996). One source, 3C454.3=2251+158, has a 40° bend about one milliarcsecond from the core. 3C273 has an "aligned" straight jet. Two sources still have no well-defined VLBI position angle (0235+164 and 0420–014). One source, 1127–145, has a bend of 24° between parsec and kiloparsec scales. It will be interesting to see if the parsec-scale jets of most γ-ray quasars are strongly curved.

4. The Conway and Murphy Model

There is no known physical process which would give rise to the right-angled population, consequently efforts to explain the distribution rely on selection effects. Conway and Murphy (1993) fitted the observed distribution of position angle differences (ΔPA) in the two samples with two populations of sources: those with quite straight jets, and those with a parsec-scale helical jet plus a straight kiloparsec-scale jet. Relativistic beaming with strong Doppler boosting on parsec scales, and orientation near our line of sight were required in the "misaligned" population. The most important result was that the characteristics of the helix were tightly constrained: the parsec-scale helical jet wraps around the surface of a cone, then straightens to follow the side of a kiloparsec-scale cylindrical surface. Possible mechanisms for initial inducement of a helix include Kelvin-Helmholtz instabilities and binary black holes.

Acknowledgments. We are grateful to John Conway and David Murphy for illuminating discussions and to Marshall Cohen for helpful comments on the manuscript. This work was begun while one of the authors (AEW) was at the Owens Valley Radio Observatory. AEW and SCU gratefully acknowledge support from the NASA Long Term Space Astrophysics Program.

References

Antonucci, R. 1986 ApJ, 304, 634 (A86)

Barthel, P. D., Schilizzi, R. T., Miley, G. K., Jagers, W. J., & Strom, R.G. 1985 A&A, 148, 243 (B85)

Barthel, P. D., Conway, J. E., Myers, S. T., Pearson, T. J., & Readhead, A. C. S. 1995, ApJ, 444, 21

Biretta, J. A., Moore, R. L., & Cohen, M. H. 1986 ApJ, 308, 93 (B86)

Cohen, M. H. 1991 private communication (C91)

Conway, J. E., & Murphy, D. M. 1993 ApJ, 411, 89

Fanti, C., Fanti, R., Parma, P., Schilizzi, R. T., & van Breugel, W. J. M. 1985, A&A, 143, 292 (F85)

Gabuzda, D. C., Cawthorne, T. V., Roberts, D. H., & Wardle, J. F. C. 1989 ApJ, 347, 701 (G89)

Gabuzda, D. C., Cawthorne, T. V., Roberts, D. H., & Wardle, J. F. C. 1992 ApJ, 388, 40 (G92)

Impey, C. D., Lawrence, C. R., & Tapia, S. 1991 ApJ, 375, 46

Krichbaum, T. P., Britzen, S., Standke, K. J., Witzel, A., Schalinski, C. J., & Zensus, J. A. 1995 Proc. National Academy of Sciences, 92, 11377

Kus, A. J., Marecki, A., Neff, S., van Ardenne, A., & Wilkinson, P. N. 1988 in IAU Symposium 129, 129 (K88)

Murphy, D. M., Browne, I. A., & Perley, R. A. 1993 MNRAS, 264, 298

Pearson, T. J. 1992 private communication (P92)

Pearson, T. J., & Readhead, A. C. S. 1988 ApJ, 328,114 (PR88)

Pearson, T. J., Perley, R. A., & Readhead, A. C. S. 1985, AJ, 90, 738 (P85)

Perley, R. A. 1982 AJ, 87, 859 (P82)

Pauliny-Toth, I. I. K. 1987 in IAU Symposium 121, 295 (P87)

Pauliny-Toth, I. I. K., Zensus, J. A., Cohen, M. H., Alberdi, A., & Schaal, R. 1990 in Parsec-Scale Radio Jets, ed. J. A. Zensus & T. J. Pearson (Cambridge, Cambridge University Press), 55 (P90)

Rantakyro, F., Baath, L. B., Dallacasa, D., Jones, D. L., & Wehrle, A. E. 1996 A&A, in press.

Romney, J., Padrielli, L., Bartel, N., Weiler, K. W., Ficarra, A., Mantovani, F., Baath, L. B., Kogan, L., Matveenko, L., Moiseev, I. G., & Nicolson, G. 1984 A&A, 135, 289 (R84)

Rusk, R. E. 1988, Ph.D. thesis, University of Toronto (R88)

Ulvestad, J. S., Johston, K. J., & Weiler, K. W. 1983 ApJ, 266, 18 (U83)

Unwin, S. C., Cohen, M. H., Biretta, J. A., Hodges, M. W., & Zensus, J.A. 1989 ApJ, 340, 117 (U89)

Unwin, S. C., & Wehrle, A. E. 1992, ApJ, 398, 74 (U92)

von Montigny, C., et al. 1995 A&A, 299, 680

Walker, R. C., Benson, J. M., & Unwin, S.C. 1987 ApJ, 316, 546 (W87)

Wehrle, A. E., & Cohen, M. H. 1989 ApJ, 346, L69

Wehrle, A. E., Cohen, M. H., & Unwin, S. C. 1990 ApJ, 351, 1

Wehrle, A. E., Cohen, M. H., Unwin, S. C., Aller, H. D., Aller, M. F., & Nicolson, G. 1992 ApJ, 391, 589 (W92)

Wrobel, J. M., & Lind, K. R. 1990 ApJ, 348, 135 (W90)

Xu, W. 1995, Ph.D. thesis, California Institute of Technology

Zensus, J. A., Baath, L. B., Cohen, M. H., & Nicolson, G. D. 1988 Nature, 334, 410 (Z88)

Part III
High-energy Variability

Blazar Continuum Variability
ASP Conference Series, Vol. 110, 1996
H. R. Miller, J. R. Webb, and J. C. Noble, eds.

X-ray Variability of Blazars

Ian McHardy

Department of Physics, The University Southampton SO17 1BJ

Abstract. The current observations of X-ray emission from BL Lacs and OVV quasars are summarised. Both long timescale (weeks–years) and short timescale (seconds–days) flux and spectral variability is reported. BL Lacs have very steep powerspectra, with very little short timescale variability, but a great deal of variability on medium scale (few hourly) and longer timescales. Blazar X-ray variability is somewhat more non-linear than for Seyfert galaxies which is consistent with the interpretation of flares as shocks in the jet. The relationship between X-ray emission and emission in other bands is discussed. For BL Lacs there is a tight correlation between the variability in X-rays and in nearby bands, e.g., UV/optical, and so it is reasonably clear that the X-ray emission is simply the high frequency tail of the synchrotron component. OVV X-ray emission is almost certainly the result of Compton scattering of soft seed photons. However attempts to distinguish between the the external or internal (synchrotron self-Compton) origin for the soft seed photons on the basis of X-ray/other waveband lags have not been overwhelmingly successful. We briefly mention the implications of these results for theoretical models and point the way for future observational programmes.

1. Introduction

The radio through UV emission from blazars is generally agreed to be synchrotron emission from a relativisitic jet oriented towards the observer and I will not argue with that assumption here. However the origin of the X- (and γ)-ray emission, which is crucial to our full understanding of blazars, is far from clear. In this review I will try to provide a summary of the current X-ray observations of blazars and will consider how they help us answer the fundamental questions, namely: 'What is the X-ray emission mechanism?' and 'What does the X-ray emission tell us about the overall structure of the jet?'

I will consider first the X-ray emission from BL Lacs, which is relatively straightforward, and secondly consider the emission from OVV quasars, which is not. There are presently insufficient data to permit separate detailed examination of the variability properties of X-ray selected and radio selected BL Lacs and so, although this is something of an oversimplification, they will usually be discussed together. For organisational purposes I will consider separately flux variability, on both short and long timescales, and will then consider spectral variability and related variability in other bands. I will conclude by mentioning briefly some implications for models of blazars and by suggesting future observations.

In BL Lacs the smooth continuation of the X-ray spectrum, at least in the 0.1–10 keV band, from the UV–optical–IR–radio spectrum is a very strong indication that, as in the UV–radio regime, the dominant emission mechanism is synchrotron radiation. Thus the most important question concerns the structure of the jet. For the OVVs which, in the not too distant past, we were often lucky to detect at all in X-rays, it is generally agreed that the X-ray emission arises as a result of Compton scattering of soft seed photons by the relativistic electrons in the jet and the main debate at present is where these seed photons come from. The two most popular possibilities are that the seed photons are the millimeter/IR synchrotron photons from the jet (the synchrotron self-Compton, or SSC model), or that the seed photons arise externally to the jet, e.g., as optical/UV photons from the accretion disc (the external Compton scattering or ECS model). The reader is referred to Marscher (these proceedings) and Maraschi (these proceedings) for a discussion of the possible theoretical models.

2. BL Lacs

2.1. Short-timescale flux variability

For those BL Lacs which are bright enough to be detected at high significance on short timescales continuous variability seems to be the order of the day. For example, in Figure 1 typical EXOSAT medium energy (ME ~2–10 keV) observation of PKS 2155–304 (e.g., Tagliaferri et al. 1991) is shown. Large amplitude (factor 2) variability occurs on timescales of 5–10 hours but very little shorter timescale variability is seen. PKS 2155–304 is brighter in X-rays than many of the famous Seyfert galaxies such as NGC4051 or NGC5506 (Figure 1) in which scale invariant variability including high frequency flickering is detected (e.g., McHardy 1989) and such variability would certainly have been detected if it had been present.

Powerspectral analysis of the EXOSAT observations of PKS 2155–304 have been performed by Tagliaferri et al. (1991). The powerspectra are dominated by large amounts of power at low frequencies ($< 10^{-4}$ Hz) and consequently all have rather steep slopes (~ -2.5), much steeper than most Seyfert galaxies (~ -1.5; McHardy 1989; Green et al. 1993). They confirm the visual impression that there is very little power at high frequencies ($> 10^{-3}$ Hz). If linear trends are removed from the lightcurves, most of the power disappears from the powerspectra. The ROSAT observations of PKS 2155–304 (Brinkman et al. 1994) look broadly the same as the EXOSAT ones, although the amplitudes of variability are somewhat lower (Section 2.3.1). Mkn 421, another X-ray bright BL Lac object, shows a broadly similar pattern of variation but the observations are not many enough or long enough to permit sensible powerspectral analysis. A further X-ray bright BL Lac, Mkn 501, hardly ever varies at all.

[We note that PKS 2155–304 has an apparent luminosity $L_{2-10\mathrm{keV}} \sim 10^{45}$ ergs s^{-1}, ~ 30 times greater than that of NGC5506 and that, for Seyfert galaxies, more luminous objects are less variable (McHardy 1989). Thus a Seyfert with the same observed luminosity as PKS 2155–304 would also display no high frequency variability, but such a Seyfert would also display very little low frequency

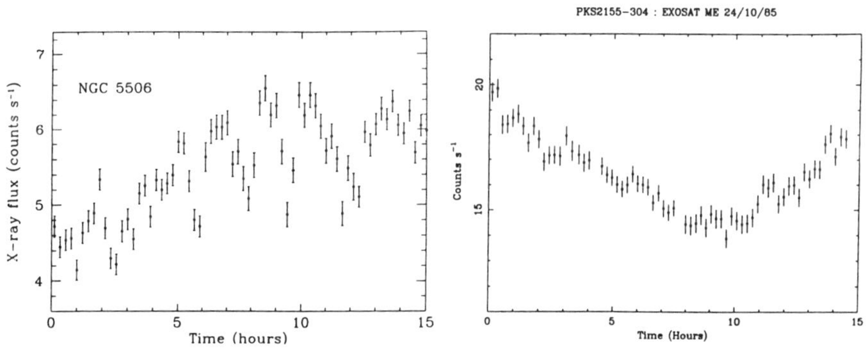

Figure 1. EXOSAT 2–10 keV observations of the BL Lac PKS 2155–304 and of the Seyfert galaxy NGC5506 binned on the same timescale (800s). The same relative flux range is shown. The different character of variability is evident.

variability, unlike PKS 2155–304. However we should caution slightly against comparing apples (relativistically beamed BL Lacs) with oranges (non-beamed Seyferts). The rest frame luminosity of PKS 2155–304 will be considerably less than the observed luminosity—taking it close to the luminosity of NGC5506—and the rest frame variability timescale will be greater, still further reducing the amount of power on short timescales.]

In the only claim for periodicity in X-ray observations of blazars Brodie et al. (1987) point out the possibility of a 17,000s oscillation in EXOSAT observations of Mkn 421 However as only 4 cyles are involved and the fits are at the 'consistent with' rather than 'good' level, this possibility should be classed as 'worth checking further' rather than 'confirmed.' Edelson et al. (1995) searched extensively for periodicities in long ROSAT observations of PKS 2155–304 but also found nothing.

We shouldn't leave short timescale variability without noting the case of H0323+02 (Doxsey et al. 1983) which underwent a dramatic drop in flux to a level consistent with zero for a period of 60s before returning to its previous level. The lightcurve is not of high significance overall and the detector (HEAO-A1) is non-imaging but the drop is quite prominent and, although hard to explain physically, the authors could find no non-astronomical reason to explain the observations.

2.2. Long-timescale flux variability

On longer timescales (days–years) BL Lacs are consistently variable with large amplitude changes ($\sim$ 50%) occuring on timescales of days (e.g., Snyder et al. 1979; Pollock et al. 1985) superimposed on which are occasional very large flares (factors $\sim$ a few variability) on timescales of days/week (see Figure 2). A quantitative study of BL Lac variability as observed by EXOSAT has been performed by Giommi et al. (1991). They describe variability in terms of the amplitude of deviation from the mean level. They find that for 76% of the time BL Lacs are found in a state which is within 50% of the mean level and

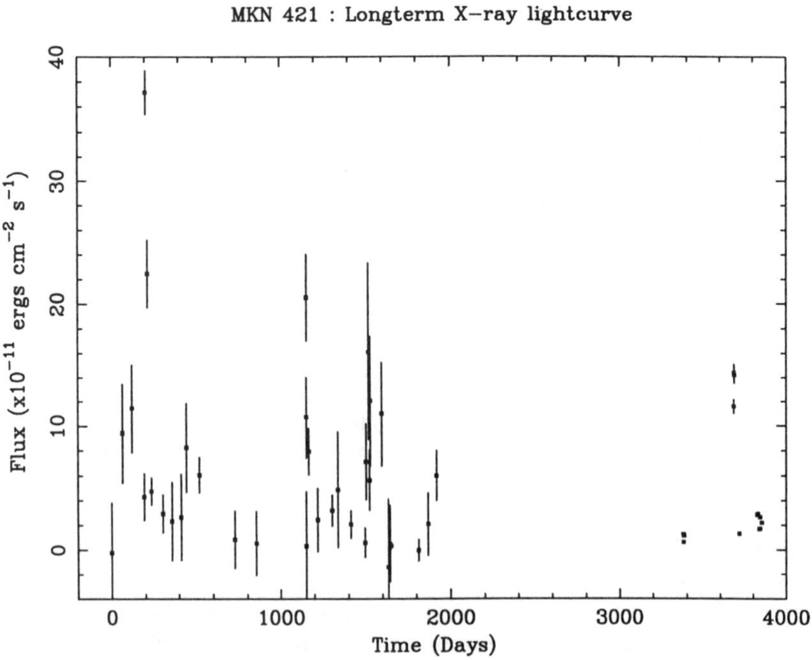

Figure 2. Long term 2–10 keV lightcurve of Mkn421. Day 0 is 27 October 1974. The data from day 0 to 2000 come from the Ariel V Sky Survey Instrument (SSI). Data between days 3000 and 4000 are from the EXOSAT archive.

that variations by factors larger than 2 occur only 7% of the time. Our longer timescale observations (e.g., Mkn421 in Figure 2; PKS 2155–304 looks much the same) agree with this conclusion and we find that the sources are in a flaring state $\sim$ 5% of the time. Unfortunately our data do not have sufficient time resolution to determine the shape of the flares.

A large flare with a rise time of $\sim$ 3 days and decay timescale of $\sim$ 2 weeks, a shape typical of a sudden injection of energetic radiating particles followed by cooling, was found in ROSAT observations of AO 0235+164 in July 1993 (Madejski et al. 1996). The present author ruefully notes that even at its lowest PSPC level the source is about a factor of 5 brighter than it was when he monitored it with EXOSAT. With regard to the γ-ray flares in blazars (e.g., Hartman, these proceedings) it is sometimes suggested that large flares only occur when the sources are already in a high state. We may be seeing a similar situation in AO 0235+164 and perhaps in Mkn421 where Brodie et al. (1987) noted rapid large amplitude variability (25% in 2 hours) when the source was in a particularly bright state.

The tendency of the long term lightcurves to stay close to the mean level for most of the time and then depart from the mean by a large amount, always in the same direction (i.e., flaring) is non-linear behavior. In contrast the X-ray lightcurves of Seyfert galaxies generally have symetric deviations to either side of the mean level and so are linear (e.g., see Vio et al. 1992 who address the optical variability of OVVs, concluding that it is non-linear). Flaring activity in blazars has long been ascribed to shocks—non-linear processes—in the jet (e.g., Marscher & Gear 1985) and the long term lightcurves presented here are consistent with that view.

2.3. Spectral variability

The synchrotron component and hysteresis There have been many observations of the X-ray spectra of the bright BL Lacs (e.g., Mushotzky et al. 1979; Morini et al. 1986; George et al. 1988a and 1988b). As the flux level increases, the medium energy spectra harden, with much greater percentage variability at 10 keV than at 1 keV. In some cases (e.g., Mkn421, George et al. 1988a) simultaneous UV and X-ray observations were arranged which showed that the 2000ÅUV flux varies noticeably less than the 2–10 keV flux and that there is no variation of UV spectral index with flux. Therefore, although there is some variation in the normalisation of the UV/X-ray spectrum, the main change is a pivoting of the X-ray spectrum about ~ 0.1 keV. Although these results require some variation of the normalisation of the electron energy distribution, the most critical variation required is in the upper cut-off of the electron energy distribution.

Later observations (e.g., Sembay et al. 1993; Tashiro 1994; Takahashi, these proceedings) show not just a simple monatonic relationship between spectral index and flux, but indicate some hysteresis in the systems. For example, in Figure 3 we see that as the 2–10 keV flux drops, the photon number index (= spectral index +1) is steeper than at the corresponding rising flux state. This effect is probably the result of synchrotron radiation losses which preferentially remove energy from the highest energy electrons so the higher energy X-ray bands lead the lower energy bands.

Flux-related spectral index variations were not found in ROSAT PSPC observations of PKS 2155–304 (e.g., Brinkman et al. 1994), or in the simultaneous UV and optical observations (Edelson et al. 1995). The PSPC band (0.1–2.4 keV) is very near the 0.1 keV pivot point that George et al. found. Thus flux-related spectral index changes should be small, as was found. Small rises and falls of synchrotron flux with no change of spectral index can be simply explained by mild compression and rarefaction (e.g., Scheuer & Williams 1968) as long as the electron energy distribution has a powerlaw form. Shocks are not required. Lags between various bands require variations in the electron distribution and magnetic field along the jet such that different regions emit preferentially in different bands. As long as, say, the UV and soft X-ray emitting region are close together, then compression/rarefaction waves moving along the jet could successively affect the various regions and so the time delays between the bands give us some idea of the physical separations of the various emitting regions (in PKS 2155–304 the PSPC X-ray flux leads the UV flux by about 2.8 hours and any lag between the UV and optical bands is less than one hour, Edelson et al. 1995). The fact that the lag time is less than the variability timescale can be used to argue for a flattened emission geometry (Edelson et al. 1995).

The model whereby the underlying process is a variation in the upper cut-off of the electron energy distribution of the synchrotron component was dramati-

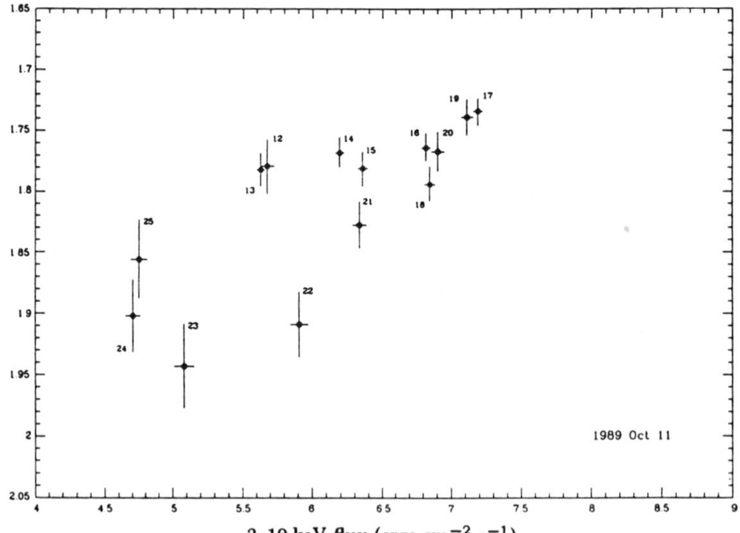

Figure 3. 3.8–17.9 keV photon number index of PKS 2155–304 as a function of 2–10 keV flux (ergs cm^{-2} s^{-1}) during one GINGA observation (from Sembay et al. 1993). The points are numbered sequentially.

cally confirmed by Macomb et al. (1995) who observed quasi-simultaneous X-ray and TeV flaring of Mkn421. The TeV emission, as observed by the Whipple Cerenkov detector, is most likely first order synchrotron self-Compton radiation with the 'low energy' seed photons being the X-ray photons.

The Compton Component In the classic X-ray selected BL Lacs a flattening of the spectrum is sometimes seen at energies above $\sim$ 10 keV. The flattening is usually referred to as a 'hard tail' (e.g., Mkn421 Mushotzky et al. 1979, Hall et al. 1981; PKS 2155–304 Morini et al. 1986). Typically the spectra have energy indices, α, of ~ -1.5 in the 1–10 keV band and $\alpha\sim$ 0 above 10 keV although it is difficult to determine accurately the spectral index of the hard tail. Almost certainly the hard tails are the first order Compton scattered component and may extend all the way to TeV energies (e.g., Macomb et al. 1995). The tails are usually more evident when the sources are brighter (e.g., Mushotzky et al. 1979) but not always. For example in AO 0235+164 Madejski et al. (1996) find that the spectral index determined from ROSAT observations (0.1–2.4 keV) is 2 but that the spectral index as measured later by ASCA (0.5–10 keV) when the source was in a fainter state is 1. They interpret these observations as the synchrotron component having gone away, leaving only the Compton-scattered component. Thus the data are not yet sufficient to tell how the hard tails vary with respect to the synchrotron component.

In the radio-selected BL Lacs the 1–10 keV spectra are significantly flatter ($\alpha\sim$ 1 e.g., Kawai et al. 1991; Tashiro 1994) than in the X-ray selected BL Lacs and do not join smoothly onto the UV spectrum and so we are probably looking at the Compton scattered component. The X-ray spectral indices in radio selected BL Lacs do vary but as yet there is insufficient data to draw any strong conclusions.

3. OVV Quasars

3.1. Flux variability

3C273 With the exceptions of 3C273 and 3C279, most OVVs are faint X-ray sources and so the data on variability—particularly short timescale variability—are sparse. We therefore begin with those two sources. 3C273 is one of the less violently variable of the accepted OVVs but it is brighter in X-rays, by a factor of 5–10, than any other OVV and so has been extensively observed. The first report of rapid variability was of an increase in flux of 50% in 12 hours in observations with the Ariel V SSI (Marshall et al. 1981). However the SSI had a field of view of $\sim 1 \times 14°$ and so we cannot be absolutely sure that the rise in flux came from 3C273. In a two day observation with EXOSAT, 3C273 exhibited almost no variability in the 2–10 keV band (and in the 0.1–2.0 keV band but at a lower significance level) beyond a $\sim 4\%$ variation over a few hours which might be attributed to background subtraction uncertainties. Similarly, in GINGA observations (Lawson 1995; Turner et al. 1990) no variability was detected on minutes/hours timescales. Thus the general belief has grown up that 3C273 does not vary. This belief was finally dispelled with ROSAT observations.

In winter 1992/93 (McHardy et al. 1993) and winter 1994/95 observations were carried out every other day for a month with the PSPC and HRI respectively. Within each short (1–2 ksec) observation there was only marginal evidence for variability but in both winters variations with amplitude of up to 25% on a timescale of ~2–3 days were found. The variability observed with ROSAT is reconcileable with the EXOSAT observations if the latter had occured at a period of low variablity. In sources where it is assumed that beaming is not important, e.g., Seyfert galaxies, rates of change of luminosity are often related to the efficiency of mass to energy conversion in the emitting region (e.g., Fabian 1979) with efficiencies $> 1\%$ being equated to the presence of massive black holes and efficiencies $> 40\%$ requiring exotic explanations such as relativistic beaming. The variability detected in 3C273 does not even require efficiencies $>1\%$. This is perhaps consistent with the relatively low transverse superluminal velocity measured from components in the radio jet of 3C273 ($v = 4.7c$ for $H_0 = 100$ km s^{-1} Mpc^{-1}; Baath 1992) although much of the X-ray emission may still arise in the quiescent jet.

As with short timescale variability, data on longer timescale variability, i.e., months–years, is not extensive. In Figure 4 we show the long term 2–10 keV lightcurve of 3C273. We see continual variation about the mean level. The lightcurve has some similarities to the better sampled lightcurve of the X-ray bright Seyfert galaxy NGC4151 (Papadakis & McHardy 1995) but the sampling are insufficient to draw more detailed conclusions about the character of the variability. The most we can say from these data is that the overall properties of the emitting region(s) do not change greatly on ~ 20 year timescales. There is insufficient data to say whether this conclusion is true of all OVVs although we note that 3C390.3, which is admittedly an even more marginal member of

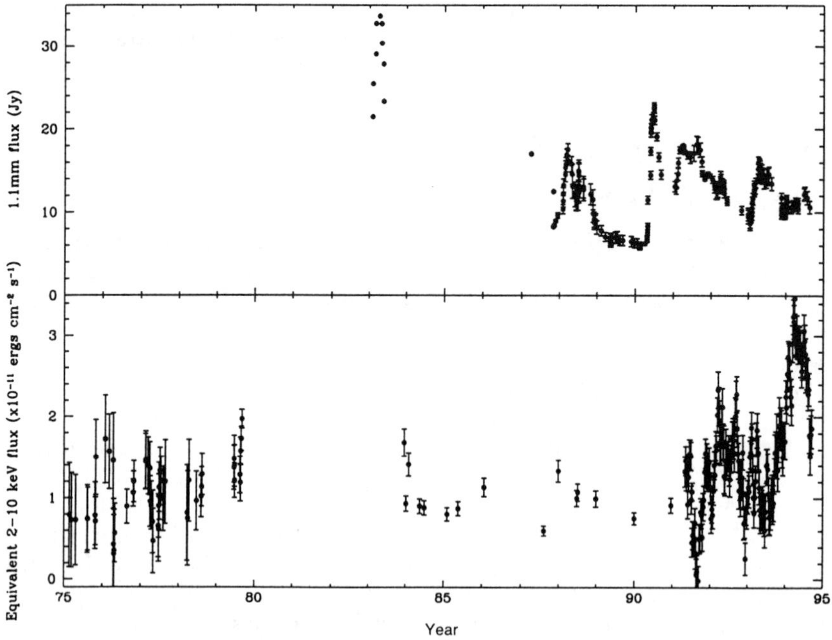

Figure 4. (a) Long term 2–10 keV lightcurve of 3C273 comprising data from the Ariel V SSI, EXOSAT, GINGA and BATSE. Note that the BATSE data, which have been scaled down to the 2–10 keV range cover a rather harder energy band (50–300 keV) than the other instruments and the character of variability may be different at higher energies. (b) Long-term 1.1-mm lightcurve (Robson, private communication), discussed in Section 4. The SSI and BATSE data have been very lightly smoothed.

the OVV class than 3C273, shows an almost monatonic decline throughout the period 1970–1985 (Shafer et al. 1985).

3C279 Another extensively observed OVV is 3C279. Its luminosity ($L_{2-10\mathrm{keV}} \sim$ 2–10 $\times 10^{46}$ ergs s^{-1}) is a factor of a few greater than that of 3C273 but it is a factor 5–10× fainter in X-rays. However, at ∼1–2 mCrab, it is still one of the brighter OVVs and is also very variable. As in other wavebands, it has therefore become everyone's favorite OVV. Let us hope that it does not later turn out to be unrepresentative of the class. In GINGA observations (2–20 keV) covering an 8 hour period, Makino et al. (1989) detect a 20% change in under an hour but otherwise see no rapid variations. The efficiency required by Makino's observation is >100% which is unphysical and so beaming must be invoked to restore physical reality.

On slightly longer timescales (days–weeks) variations of a similar amplitude (∼ 20%) on a timescale of ∼ a week were detected in ROSAT (0.1–2.4 keV) observations (Leach 1995 and Figure 5). During these observations the source

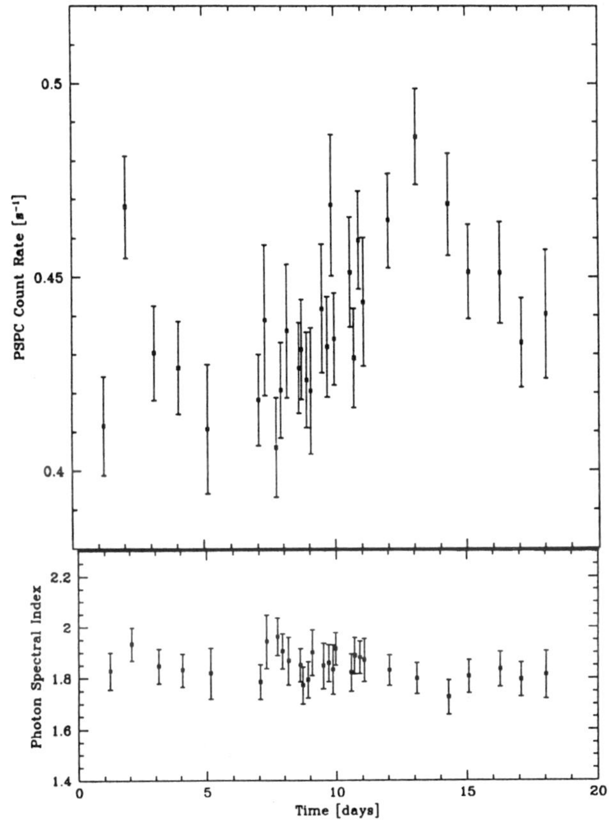

Figure 5. ROSAT PSPC observations of 3C279. The top panel gives the count rate in the 0.1–2.4 keV band and the bottom panel gives the photon number index assuming a simple powerlaw model with galactic absorption ($N_{\rm H} = 2.2 \times 10^{20}$ cm^{-2}), from Leach 1995. Day 0 = 26 December 1992.

was in a relatively low state and no useful information was obtained on timescales less than $\sim$ a day. Large amplitude (factor >2) variability on timescales of months/years has been reported by Makino et al. (1996) (see Section 4).

Other OVVs A small number of OVVs were monitored by EXOSAT, e.g., 1156+295 (Webb & M$^{\rm c}$Hardy 1986; see Section 4), 3C446 (Bregmann et al. 1988) and a summary of GINGA observations of OVVs is provided by Tashiro (1994). These observations confirm the pattern of large amplitude variability (factor >50%) on timescales as short as days but in general the sources are too faint for variability on timescales less than a few hours to be detected.

OVV quasars are by no means a complete set. For example the quasar PKS0558–504, which has not so far shown strong optical polarisation and so is not generally regarded as a blazar has been observed to undergo a factor of 2 increase in flux in less than 5 minutes (Remillard et al. 1991) in GINGA observations. Although apparent flaring activity observed in non-imaging detectors such as GINGA may not actually be associated with the supposed target, the authors could find no evidence of possible confusing sources, e.g., RS CVn systems. The rate of change of luminosity again implies an efficiency > 100% and so requires beaming to be invoked to restore physicality.

3.2. Spectral variability

3C273 In the winter 1992/93 ROSAT PSPC observations of 3C273, Leach et al. (1995) find that the spectra are best fitted by two independent power-law components. Leach et al. fix the spectral index of the harder component at 0.5, assuming it to be the extrapolation of the well-observed power-law of the same index which dominates the 2–10 keV band (e.g., Turner et al. 1990). The index of the soft component, which dominates below 0.5 keV, is then fitted as 1.7 and does not vary appreciably from observation to observation. Both components vary by the same amplitude (~20%) on the same timescales (~ a few days) but they vary *independently* (Figure 6). Thus we have very good evidence for 2 independent components to the 0.1–2.4 keV X-ray emission from at least one OVV. The origin of these two components is not yet clear but it is reasonable to suppose that the harder component may be associated with the jet. Based on a small number of contemporaneous UV observations which show quasi-simultaneous UV and very soft (0.1–0.3 keV) X-ray variations there is evidence that the softer component is an extension of the accretion disc emission which dominates the UV emission, although a hot corona (T ~10–15 keV) is also required to provide a good spectral fit (Leach et al. in preparation).

In the medium energy (2–10 keV) band, EXOSAT and GINGA spectra of 3C273 are presented by Turner et al. (1990). The spectral index range observed is real, but not large, $\Delta_\alpha \sim 0.13$. The plot of spectral index against time in Turner et al. looks rather similar to the plot of luminosity against time, giving the impression that the spectrum becomes softer as the source brightens. However the presence of a number of discrepant points makes a plot of spectral index against flux (again see Turner et al.) look confused with an indication of two separate tracks. This may be further indication of the presence of two separate components to the medium energy emission from 3C273, one of which hardens as the flux increases and one of which softens.

A weak ($\leq$ 50eV equivalent width) fluorescent iron K line is detected in 3C273. There is no evidence for variability in the strength of this line as the continuum flux varies but the errors on the line flux determination are large (Turner et al. 1990). If all the emission from OVVs is continuum emission from a relativistic jet and none escapes sideways we would not expect to find any fluorescent iron K lines, such as we commonly see in Seyfert galaxies. Thus the detection of such a line may provide some indication of a small Seyfert component. For a typical Seyfert 1 the iron K line equivalent width is ~ 250eV. Thus we can estimate that perhaps 20% of the luminosity of 3C273 arises in a Seyfert component of luminosity (2–10 keV) $\sim 2 \times 10^{45}$ ergs s^{-1}, which is itself a very high luminosity, more associated with (radio quiet) QSOs than with a Seyfert galaxy. Alternatively if we allow that some of the continuum jet emission does radiate sideways and illuminates surrounding cold material (along

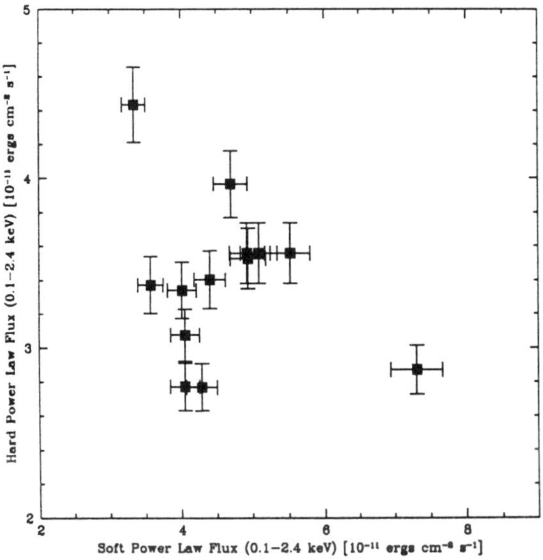

Figure 6. The model fitted fluxes of the soft and hard components of 3C273 from ROSAT PSPC observations (from Leach et al. 1995). There is no correlation between these two components.

the lines suggested by Ghisellini and Madau 1996) then we would expect a strong correlation between the continuum and the line flux. The present observations do not support this hypothesis but the errors are large and further observations are required to clarify this possibility.

3C279 The ROSAT 0.1–2.4 keV PSPC spectrum of 3C279 is a good fit to a single power law of energy index ~ 0.8 whereas the average 2–10 keV spectral index (e.g., Lawson 1995) is 0.6 and so an additional soft component may be present in the ROSAT band. It is noted, however, that, during the extensive winter 1992/93 PSPC observations, the spectral index remains constant (Leach 1995 and Figure 5) and so unless there is a strong correlation between the two possible components, one of them must dominate the PSPC band.

On timescales of years the medium energy spectral index measured from GINGA observations definitely varies, but only by a small amount $\Delta_\alpha \sim 0.2$. We are as yet unable to find a simple relationship between spectral index and luminosity (e.g., Lawson 1995) such as the spectral hardening, with hysterisis, which accompanies infrared and flaring in 3C279 (Litchfield et al. 1995). It is therefore likely that the 0.1–10 keV X-rays arise from scattering by relatively low energy electrons, well below the upper electron cut-off energy, where changes in the slope of the electron distribution are small.

Other OVVs Ohashi et al. (1989) present 2–10 keV spectral indices for 9 OVVs. With the exception of 3C454.3 which they detected to have a rather steep spectrum (α=1.7), the others were clustered around α=0.7. Lawson (1995) measures the spectral index for all quasars observed by GINGA (i.e., not just the OVVs). He finds a similar mean value and a real spread in intrinsic spectral indices between α=0.3 and 1.1 but, for any source, no correlation between spectral index and luminosity. The X-ray spectral indices are thus very close to the millimeter spectral indices, supporting an interpretation where the electrons responsible for the synchrotron millimeter emission are those which scatter photons into the X-ray band.

Ohashi et al. (1989) present GINGA observations of fluorescent lines in a number of OVVs but only uppper limits have been found. On one occasion an upper limit of only 20eV was placed on any line in 3C279. Thus the fraction of 'Seyfert' emission in 3C279 is noticeably less than in 3C273 but, using the same arguments as above, the upper limit on the luminosity of any such component is still quite large, $\sim 10^{45}$ ergs s^{-1}.

4. Relationship between X-ray and other waveband variability

The observational holy grail of blazar research is to find reliable lags between different wavebands as a diagnostic of the emission mechanism and the jet structure. The relationship between X-ray and other waveband variability for BL Lacs has already been discussed in the context of BL Lac spectral variability (Section 2.1.3) and so here we mention only a few relevant observations of OVVs where the X-ray spectra are not a straight extrapolation of the IR/optical/UV synchrotron component. We do not go into the subject of general multiwaveband variability of OVVs as this is discussed elsewhere in these proceedings (Webb, Urry).

In EXOSAT and optical monitoring, Webb and McHardy (1986 and Figure 7) found that the two bands definitely did not vary simultaneously, however it wasn't clear which band came first as only one X-ray flare was observed. Webb and McHardy show that the X-rays are therefore not the same component as the optical, which is almost certainly synchrotron radiation as its spectrum joins smoothly to the UV and IR, but can say little about whether the X-rays are SSC or ECS. However they note that the lags are of the order of weeks rather than the $\sim$ a day seen between optical and γ-ray observations of 3C279 (Hartman et al. 1996). Similarly in 3C446 (Bregman et al. 1988; McHardy 1985) a decrease in soft X-ray flux by factor ~ 3 over a week is not paralleled by any decrease in B-band flux.

In parallel with the ROSAT monitoring of 3C273, 1.1mm monitoring was carried out with the James Clerk Maxwell Telescope and with IRAM. If a substantial fraction of the X-ray flux arises from up-scattering of millimeter photons, as the SSC models predict, we expect to see some correlation. However the results are far from conclusive. In the winter 1992/3 observations, the same levels of variability ($\sim$ 20%) were detected in the millimeter observations as in the X-rays, and on similar timescales (days) and there was a suspicion that the X-rays might lead the millimeter by about 10 days (McHardy et al. 1993). However when the experiment was repeated in winter 1994/5, although similar patterns of variability were detected, the 10 day lag was not confirmed. Thus, on daily/weekly timescales, the correlation between the X-rays and the synchrotron component in OVVs is not yet clear.

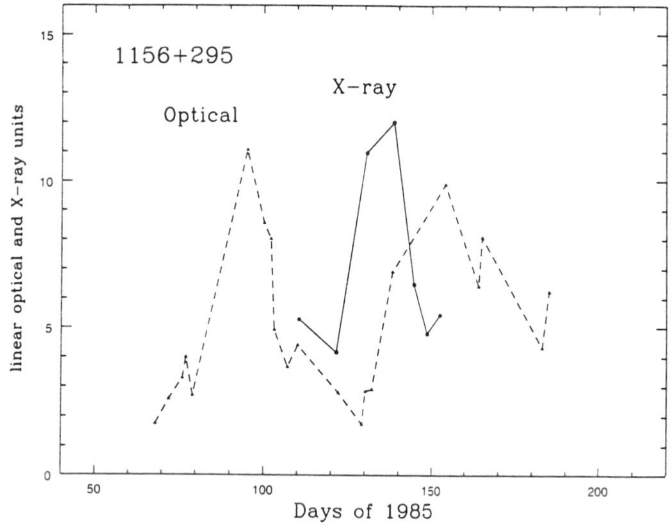

Figure 7. X-ray (solid line) and B-band (dashed line) observations of 1156+295 (from Webb & M[c]Hardy 1986). The ordinate gives the B-band flux in milliJanskies. The X-ray flux varies between 5 and 15 counts per 1000s through the thin lexan filter of the EXOSAT low energy (0.1–2.0 keV) imaging telescope. Both B-band and X-ray ordinates start at zero flux.

In case there might be an X-ray/millimeter correlation on longer timescales we have now examined the long-timescale X-ray variability of 3C273 as observed by BATSE (Paciesas, private communication) and have compared it to 1.1 mm observations (Robson, private communication). GINGA data are also included in the long timescale lightcurve, somewhat crudely scaled to the same mean level in counts as the BATSE data. Unfortunately X-ray coverage was rather sparse at the times of the millimeter outbursts in 1988 and late 1990 although the GINGA data from 1988–1990 do crudely resemble the millimeter lightcurve. However in general the correlation between X-rays and millimeter is, at best, poor. One should perhaps note that the BATSE energy range is somewhat higher (50–300 keV) than the ROSAT or GINGA bands and, in the SSC model, these higher energy X-rays might arise from scattering of far infrared rather than millimeter photons, but the 2–500 keV spectrum of 3C273 is well represented by a continuous powerlaw (e.g., Lichti et al. 1993) and so there is some evidence to suppose that the low and high energy X-rays might be related.

On longer timescales (months/years) Makino et al. (1996) show a correlation, albeit with rather few data points, between the radio and X-ray emission of 3C279. In the period 1987–91 they find a correlation between the radio and X-ray fluxes but the correlation disappears in observations in 1994–95. However there is a stronger correlation between the X-ray flux and the radio spectral index which holds up reasonably at all epochs. The correlation is in the sense that the X-ray flux rises as the 22–37 GHz radio spectral index hardens, i.e., as the

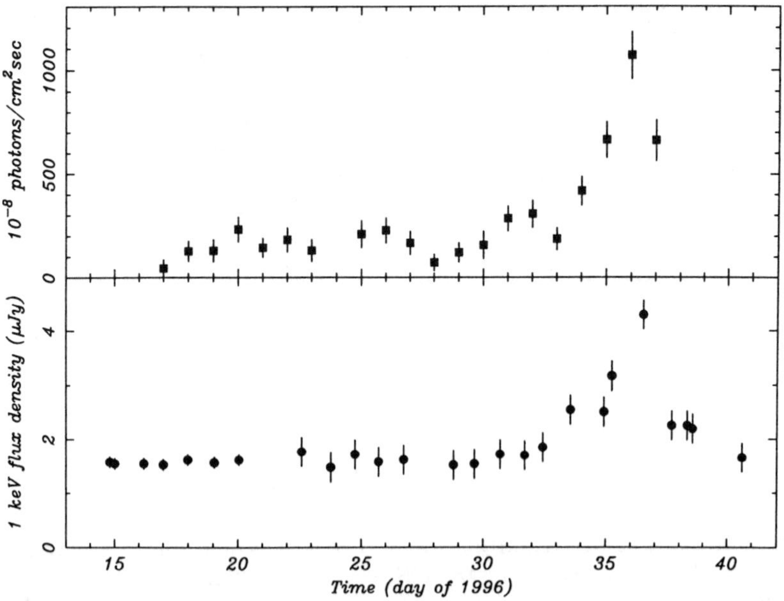

Figure 8. X-ray and γ-ray lightcurves of 3C279 covering the period 14 January to 11 February 1996. The γ-ray lightcurve (> 100) MeV s from EGRET. The X-ray observations from day 14–20 are from the ROSAT HRI and the X-ray observations from days 22–41 are RXTE observations. The ROSAT 0.1–2 keV count rate has been scaled to a flux density at 1 keV assuming α=0.8 as measured in that band and the XTE 2–20 keV count rate is scaled assuming α=0.6, as measured in the 2–20 keV band by GINGA.

synchrotron self-absorption turnover moves to lower frequencies, in agreement with SSC models.

A particularly interesting observation of X-ray variability in a blazar occurred simultaneously with this meeting. In Figure 8, I show the X-ray (from RXTE and ROSAT) and γ-ray (from EGRET on CGRO) lightcurves of 3C279 covering the period 14 January to 11 February 1996, revealing a dramatic flare in both bands. Each XTE observation lasted 10 minutes and the background was estimated as best we could from slew data, using our experience with the GINGA satellite as the RXTE background model is not yet ready. Thus the exact value of individual points may be revised slightly but the overall shape is unlikely to change significantly. We also caution that we do not yet know whether the soft X-rays as observed by ROSAT vary in exactly the same way as the medium energy X-rays as observed by RXTE and so including them both on the same plot may be marginally misleading.

This is the only OVV flare for which high quality X-ray and γ-ray observations are available and a number of points deserve attention. Regarding jet structure, the main point is that the X-ray and γ-ray flares are, to within the observational errors of about half a day, simultaneous, and the overall shapes and durations of the two flares are not too different. Thus complex spatially separated emission regions are not necessary and a simple model where the X- and γ-ray emission comes from the same place, as in synchrotron self-Compton emission from a shock, is favored. If we assume the previously measured average GINGA index of α=0.6, the X-ray spectrum extrapolates directly to the γ-rays, supporting a close γ-ray/X-ray link. Regarding the emission mechanism, in SSC models we would expect the γ-ray emission to slightly precede the X-ray emission. Our observations are consistent with this possibility, but do not confirm it. Daily observations undersample this flare. EGRET looked continuously and so a lightcurve with smaller time bin sizes, but larger flux errors, can be produced but nothing more can be done about the RXTE data. However observations of lags between the high energy bands and the synchrotron component are really needed to distinguish the emission possibilities but as the peak of the flare occured on the first day of this workshop many potential observers were not observing. Nonetheless some interesting observations have been made and the full dataset is discussed by Wehrle et al. (1996).

Other points worth noting are that the flare amplitude in the γ-rays is more than in the X-ray band indicating that emission from the flaring region is more important in the γ-rays than in the X-rays. Even before the flare, the γ-rays are more variable than the X-rays. It may also be worth noting that this γ-ray flare, to an all time historic high, occurred at a time when the quiescent level was also very high, supporting the general suspicion that flares occur when a source is already in a bright state.

5. Conclusions

We can summarise the main observational facts as follows. For the X-ray selected BL Lacs the smooth continuation of the X-ray spectrum from the optical/UV spectrum means that the 0.1–10 keV emission is almost certainly the high energy tail of the synchrotron component which dominates the lower energies. The X-ray spectrum hardens as the source brightens but there is considerable hysteresis in the spectral index/luminosity plane with the softer X-rays lagging behind the harder X-rays. The X-ray/UV spectrum pivots about ~0.1 keV and very little change occurs in the UV spectral index as the flux varies. Such behavior can be explained by a change in the upper cut-off of the electron energy distribution, an explanation which also accounts for the TeV flare seen in Mkn421.

In BL Lacs there is continual large amplitude variability on timescales of hours but no high frequency flickering such as is seen in Seyfert galaxies. The powerspectra are very steep. Thus the emission does not come from a large number of semi-independent discrete sub-regions as in Seyferts, but from a much smoother structure. If this structure is the quiescent jet, we have to explain why small-scale perturbations are damped out but larger amplitude perturbations are a permanent feature. On longer timescales occasional very large flares are seen and the lightcurves are non-linear, consistent with an interpretation of the flares as arising in shocks.

For OVVs the situation is more complicated and the observational foundation is less secure as OVVs are generally fainter X-ray sources than BL Lacs.

The X-ray spectra are all much flatter than an extrapolation of the UV spectrum and are generally interpreted in terms of Compton scattering of low energy seed photons. In the SSC model the seed photons would be millimeter photons and, indeed, the X-ray spectral indices are reasonably close to the millimeter spectral indices with little change of X-ray spectral index with flux as would be expected if the scattering electrons are far from the energies where spectral aging effects become important, i.e., the synchrotron millimeter regime. However although correlations have been found between the γ-ray and optical/IR variations, attempts to determine whether the seed photons for the X-ray emission arise externally (ECS) or internally (SSC) to the jet by means of finding lags between the X-rays and the lower energy seed photons have not, so far, been successful.

There are a number of examples where we see that the X-rays do not vary simultaneously with other bands (e.g., with the optical—1156+295 and 3C446; with the millimeter—3C273) but we are unable to determine any lags. In the cases of 1156+295 and 3C446 our lack of success may be due to poor sampling but for 3C273 we are reasonably well sampled and there is no simple correlation between the X-rays and millimeter fluxes. However we note the similar lack of a strong radio flux/X-ray flux correlation in 3C279 but a probable radio spectrum/X-ray flux correlation, as expected in SSC models. In 3C279 we see a strong correlation between the X-ray and γ-ray bands indicating they may well be the same component, and on a separate occasion simultaneous γ-ray/optical flaring has been seen. Thus we might expect to find simultaneous X-ray/optical–IR flaring in 3C279, unlike in 1156+295 or 3C446, if we can ever arrange the observations. This implies that not all OVVs are the same and so we caution against basing all our theoretical models on observations of 3C279.

It is probable that we are detecting X-ray emission from more than one region in OVVs. There is no particular surprise about detecting two components in the low energy ROSAT band from 3C273 where we might expect some emission from the high energy tail of the accretion disc spectrum as well as any jet emission. However it is probable that more than one emission component contributes to the medium energy band as well. The simplest explanation of the weak iron fluorescent line in 3C273 is that it comes from a quite luminous Seyfert-like component, although other explanations (jet emission scattering from surrounding material) are possible. The lack of variability of the iron line is consistent with a very luminous (i.e., large) Seyfert-like component.

We must also consider the relative contributions from the quiescent jet and from shocked regions in the jet. Shocks provide a reasonable explanation of the dramatic short timescale flares in the X- and γ-ray bands (e.g., Figure 8) and, from such shocks, we would expect also to see synchrotron flares at about the same time and there is at least some evidence that we do (Hartman et al. 1996). However a similar scenario does not easily explain the optical/X-ray activity in 1156+295. Alternative possibilities are that we have separated X-ray and optical emitting regions in the quiescent jet and that they respond at different times as rarefaction/compression disturbances roll down the jet. We also have the possibility of rapidly varying X-ray synchrotron radiation from the inner part of the quiescent jet and slower varying self-Compton X-ray emission from larger regions farther out. The latter may explain the observations of 3C279 by Makino et al. 1996 or those of 3C345 by Unwin and Wehrle (see elsewhere in these proceedings) who conclude that self-Compton emission might reasonably come from VLBI components which change on few month/year timescales.

6. Future observations

If we are considering the whole multiwaveband spectrum of blazars then the required observations are simple—at least twice daily in all bands for a few months, for a large sample of blazars. However if we are being a little more modest then, as far as the X-ray emission goes, the single most important measurement still missing is of any correlation with a lower-energy (synchrotron) band in OVVs. Such a measurement is vital if we are to distinguish the emission mechanism. In the favored SSC model the millimeter band is the most likely to provide the seed photons for the X-ray band but other bands should be checked just on the off chance that this model is wrong. We detect X-ray variations on a variety of timescales and so we have to observe on a variety of timescales. The final point to make is that we really do need these observations for more than one OVV as there is already evidence they do not all behave in the same way.

Acknowledgments. I thank Anthony Lawson for careful analysis of the RXTE observations of 3C279. I thank Ann Wehrle, Robert Hartman and the Egret Blazar consortium for permission to show the EGRET flare in 3C279 (Figure 8) and I thank Greg Madejski for permission to show the ROSAT data in the same figure. I thank Bill Paciesas for permission to show the BATSE lightcurve of 3C273 in Figure 4.

References

Baath, L. B. 1992, in Variability Of Blazars, ed. E. Valtaoja (Cambridge, Cambridge University Press), p. 229

Bregman, J., et al. 1988, ApJ, 331, 746

Brinkman, W., et al. 1994, A&A, 288, 433

Brodie, J., Bowyer, S., & Tennant, A. 1987, ApJ, 318, 175

Doxsey, R., et al. 1983, ApJ, 264, L43

Edelson, R., et al. 1995, ApJ, 438, 120

Fabian, A. C. 1979, Proc. R. Soc., 366, 449

George, I., Warwick, R., & Bromage, G. 1988a, MNRAS, 232, 793

George, I., Warwick, R., & McHardy , I. M. 1988b, MNRAS, 235, 787

Ghisellini, G., & Madau, P. 1996, MNRAS, 280, 67

Giommi, P., et al. 1991, ApJ, 356, 432

Grandi, P., et al. 1995, Advances in Space Research, 15, 23

Green, A. R., McHardy , I. M., & Lehto, H. 1993, MNRAS, 265, 664

Hall, R., et al. 1981, Space Sci.Rev., 30, 47

Hartman, R. C., et al. 1996, ApJ, in press

Hartman, R. C., these proceedings

Kawai, N., et al. 1991, ApJ, 382, 508

Lawson, A. J. 1995, Ph.D. thesis, University of Leicester
Leach, C. M. 1995, Ph.D. thesis, University of Southampton
Leach, C. M., McHardy , I. M., & Papadakis, I. E. 1995, MNRAS, 272, 221
Litchfield, S. J., et al. 1995, MNRAS, 274, 221
Macomb, D. J., et al. 1995, ApJ, 449, L99
Madejski, G., et al. 1996, ApJ, in press
Makino, F., et al. 1989, ApJ, 347, L9
Makino, F., et al. 1996, MPE Report 263, 413
Maraschi, L., et al. 1994 ApJ, 435, L91
Marscher, A. P., & Gear, W. K. 1985, ApJ, 298, 114
Marshall, N., Warwick, R. S., & Pounds, K. A. 1981, MNRAS 194, 987
McHardy , I. M. 1985, Space Sci.Rev., 40, 559
McHardy , I, M. 1989, ESA SP-296, p. 1111
McHardy , I. M., et al. 1993, Proc IAU Symposium 159, ed. T. Courvoisier (Dordrecht, Kluwer), p. 193
Morini, M., et al. 1986, ApJ, 306, L71
Mushotzky, R., et al. 1979, ApJ, 323, L17
Ohashi, T., et al. 1989, ESA SP-296, 837
Papadakis, I. E., & McHardy , I. M. 1995, MNRAS, 273, 923
Pollock, A. M. T., et al. 1985, Space Sci.Rev., 40, 607
Remillard, R, et al. 1991, Nature, 350, 589
Sambruna, R., Maraschi, L, and Urry, C. M. 1996, ApJ, in press
Scheuer, P. A. J., & Williams, P. J. S. 1968, ARA&A, 6, 321
Sembay, S., et al. 1993, ApJ, 404, 112
Shafer, R., et al. 1985, Space Sci.Rev., 40, 637
Snyder, W. A., et al. 1979, ApJ, 237, L11
Takahashi, N., these proceedings
Tagliaferri, G., et al. 1991, ApJ 380, 78
Tashiro, M. 1994, ISAS Research Note 549, Ph.D. thesis, University of Tokyo
Turner, M. J. L., et al. 1990, MNRAS, 224, 310
Unwin, S., & Wehrle, A., these proceedings
Vio, R., Crisiani, S., Lessi, O., & Provenzale, A. 1992, ApJ, 391, 518
Webb, J., & McHardy , I. M. 1986, Proc. Tucson AGN Meeting, p. 150
Wehrle, A., et al. 1966, in preparation

Blazar Continuum Variability
ASP Conference Series, Vol. 110, 1996
H. R. Miller, J. R. Webb, and J. C. Noble, eds.

ASCA Observation of X-ray/TeV Flare from Mkn421

T. Takahashi, J. Kataoka, T. Kii, H. Kubo, and F. Makino
ISAS, Sagamihara, Kanagawa, Japan

G. Madejski
NASA/GSFC, Greenbelt, Maryland, USA

M. Tashiro and K. Makishima
Univ. of Tokyo, Hongo, Tokyo, Japan

S. J. Wagner
Landessternwarte, D 69117 Heidelberg, Germany

Abstract. We observed BL Lac object Mkn 421 with the X-ray satellite *ASCA* in 1994 and 1995 as a part of a multi-frequency observation. In 1994, the observation was conducted one day after the onset of a TeV flare detected by the *Whipple* Observatory, and detected an X-ray flare, with no apparent variability in the optical, UV, and EGRET GeV flux. In the subsequent campaign performed in 1995, *ASCA* again detected very high level of activity, similar to that obtained in 1994. During the *ASCA* observation, *Whipple* telescope also detected a TeV flare. For both observations, the variability amplitudes are similar between X-rays and TeV. The correlation of the flux in the X-ray and the TeV bands indicates that a high energy tail of a single electron population is responsible for both X-rays and TeV γ-rays. In this presentation, we discuss about physical parameters derived from X-ray and TeV observations of flares. The results of the simultaneous observation in the optical band conducted in 1995 is also presented.

1. Introduction

The discovery that GeV γ-ray emission is a common property of blazars (see, e.g., Fichtel et al. 1994) provides us with an unprecedented insight into the physical processes operating in these sources. The high GeV flux implies that the energetics of many blazars is dominated by γ-rays, while the rapid, large amplitude GeV variability strongly suggests that their emission is likely to be Doppler-boosted towards us as a result of relativistic motion of the emitting matter. Among the GeV emitting blazars, Mkn 421 is unique as the first— and so far, the brightest blazar where the γ-ray emission extends up to the TeV energies at a level allowing detailed spectral and variability studies in the broadest possible range of observational wave bands. In fact, Mkn421 is one of the best known BL Lac objects and has been studied extensively for many years. It shows optical polarization, flat radio spectrum and significant time variability; characteristics of the blazar class of AGN (Makino et al. 1987; Tashiro 1992). It is bright in X-rays and has displayed the flare with dramatic flux and spectral changes (George et al. 1988). It is a GeV gamma-ray emitter, discovered in the full-sky survey by the *EGRET* gamma-ray detector (Lin et al. 1992). Whipple telescope has detected the emission above 500 GeV with the spectrum that connects smoothly with the *EGRET* spectrum (Punch et al. 1992).

The study of the distribution of the relativistic electrons is the key in establishing the origin of the GeV/TeV γ-rays. The simple continuum spectra of Mkn421 from the radio to the X-ray bands obtained by previous observations imply that the X-ray emission is due to the distribution of charged particle radiating via the synchrotron process (see, e.g., George et al. 1988), while the GeV and TeV emission is likely to be due to the Compton process from the same particles (see, e.g., Zdziarski & Krolik 1993). In any case, the X-ray emission and a comparison of it with that in the GeV/TeV bands is very important to study the emission mechanism of GeV/TeV photons. Furthermore, in order to examine the validity of X-ray synchrotron hypothesis and put constraints on the physical conditions in the synchrotron source, it is of basic importance to quantify the X-ray time variability. In particular, the correlation between the flux and spectral shape can be a valuable manifestation of non-steady synchrotron cooling process (Tashiro et al. 1995).

2. Observations

2.1. 1994 observation

The first multi-frequency observation from radio to TeV γ-rays, conducted in 1994, revealed a TeV flare detected by the *Whipple* Observatory (Kerrick et al. 1995), while the GeV γ-ray flux observed by *EGRET* maintained nearly constant level. The flux level at other wavelengths such as radio and UV was roughly the same as that observed in the quiescent level (Macomb et al. 1995; 1996). The 24-hr *ASCA* observation, started one day after the onset of the TeV flare, recorded a high level of 2–10 keV X-ray flux peaking at 3.7 $\times 10^{-10}$ ergs cm^{-2} s^{-1}, a 10-fold increase over the May 1993 value (Takahashi et al. 1995; 1996a), thus implying a flare-like behavior, coincident with the TeV flare. The time history of Mkn421 obtained from *ASCA* observation is shown in Fig. 1. The GeV flux measured by *EGRET* (Macomb et al. 1995; 1996) and the flux above 500 GeV measured by *Whipple* telescope (Kerrick et al. 1995) are also shown in the figure. The doubling time scale is about a half day. Observed flux is the brightest among the previous observations, about two times higher than that observed by EXOSAT in 1984 (George et al. 1988).

2.2. 1995 observation

In the 1995 observation, the eight observations of $\sim$ 10 ks were spaced between 1 and 3 days. *ASCA* again detected very high level of activity, similar to that obtained in 1994. The *ASCA* light curve is shown in Fig. 2, together with TeV

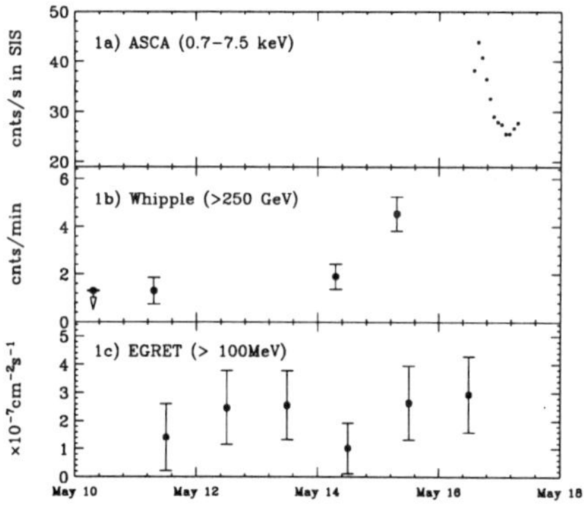

Figure 1. Time history obtained in 1a) X-ray (*ASCA*), 1b) TeV (*Whipple*), and 1c) GeV (*EGRET*) during the 1994 campaign

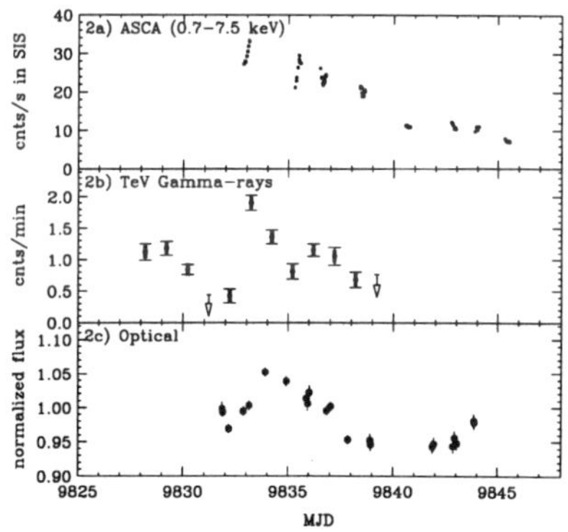

Figure 2. Time history during the 1995 campaign; 2a) X-ray (*ASCA*), 2b) TeV (*Whipple*), and 2c) the optical band.

data (Buckley et al. 1995) and optical data (Stefan et al. 1996). The variations shown in the figure are fairly similar in all three energy bands. The X-ray flux changed from 2.3 to 0.4 $\times 10^{-10}$ erg cm^{-2} s^{-1}. The variability amplitudes are similar between X-rays and TeV. On the other hand, the peak-to-peak amplitude is only 10% in the optical band. It is clear that the variability of the X-ray flux has two components, one on a $\sim$ one-day time scale and the other on a $\sim$ one-week time scale. Detail analysis of the cross correlation between different energy bands is presented in Stefan et al. (1996) and the spectral evolution is presented in Takahashi et al. (1996b).

3. X-ray spectral evolution during flare

A correlation between flux and the photon index as fitted to a simple power law with free absorption is shown in Fig 3. There is general tendency, both in one-

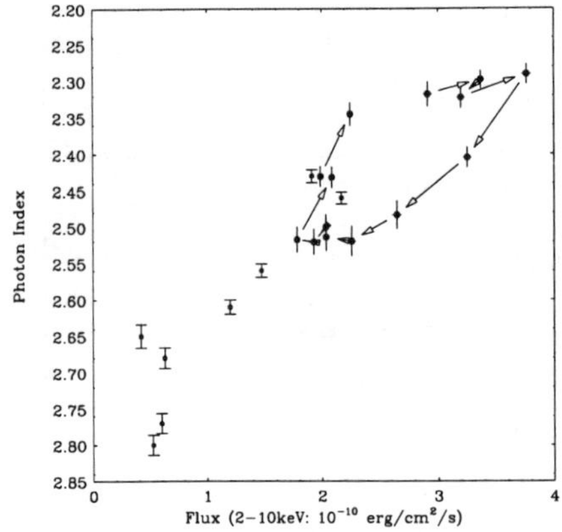

Figure 3. Correlation between photon index and the flux obtained in 1994 (filled circle) and 1995 (open circle).

day observation (1994) and two-weeks observation (1995), that the spectrum becomes hard when the source is bright.

In 1994, the "clockwise motion" (also observed in the X-ray flux of BL Lac object PKS 2155–304; Sembay et al. 1993) is clearly observed. The spectrum becomes steeper in the declining phase and becomes harder in the brightening phase. This indicates that the variations of the hard X-rays always lead these in the soft X-rays both during the increase and the decrease of the brightness of the source. In order to quantify the "soft lag," we measure it in several energy bands as compared to the 2–7.5 keV band using the cross correlation function discussed in Edelson and Krolik (1988). The results are shown in Fig. 4a, it appears that the hard X-ray variability leads that in softer X-rays by about 1 hour.

The discovery of the soft X-ray lag allows us to calculate the magnetic field from consideration of the synchrotron lifetime (cooling) of the relativistic electrons (see, e.g., Tashiro et al. 1995). The time $t_{\rm sync}$ when an electron loses a half of its energy would roughly be (in the observer's frame) 5×10^8 B^{-2} γ_{el}^{-1} δ^{-1} s (c.f., Rybicki & Lightman 1979). The peak observed frequency of the synchrotron emission ν_s by an electron with γ_{el} is given as $\nu_s \simeq 1.2 \times 10^6$ B γ_{el}^2 δ Hz (Rybicki & Lightman 1979). If E_{keV} is the observed energy in keV, $t_{\rm sync}$ is then expressed as $t_{\rm sync} \simeq 1.2 \times 10^3$ $B^{-3/2}$ $E_{keV}^{-1/2}$ $\delta^{-1/2}$ s. Since the lag is the difference of $\tau_{\rm sync}$ at the different X-ray energy. We obtain $\tau_{\rm sync} \sim 6000$ s at 1 keV (Fig. 4) and the magnetic field B of ~ 0.5 gauss. If we take a beaming factor into account, $\tau_{\rm sync}$ is given as $2 \times 10^3 B^{-3/2} E_{keV}^{-1/2} \delta^{-1/2}$ and B becomes $\sim$0.3 gauss for $\delta \sim 5$.

4. Discussion

The rapid variability of the TeV flux implies that the observed emission is cannot be emitted isotropically, and is likely to be Doppler-boosted due to a relativistic

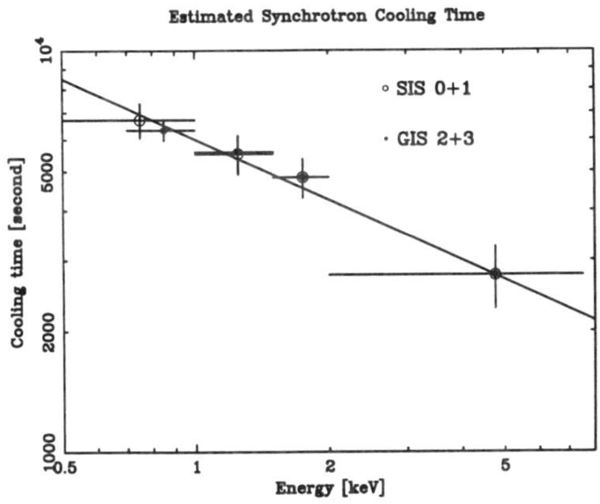

Figure 4. Synchrotron Cooling time estimated from the lag obtained from 1994 ASCA observation.

motion of the emitting material. This is because of the opacity seen by the TeV photons due to photon-photon pair creation, $\tau_{\gamma-\gamma}$ (c.f., Mattox et al. 1993; Takahara 1995), which for the TeV photons peaks in the optical band. The observed time variability $\Delta t = 0.5$ days implies that the size of the emission region is at most $R \sim \delta c \Delta t$, where δ is the relativistic beaming (Doppler) factor (defined via $\delta = \Gamma^{-1}(1-\beta\cos\theta)^{-1}$, where Γ is the Lorentz factor of the emitting material, and θ is the angle of motion to the line of sight, and $\beta = v/c$). Requiring that the $\tau_{\gamma-\gamma}$ for TeV photons < 1, and using the luminosity of $\sim 10^{44}$ erg/s at eV region, application of the formula from Takahara (1995) (where $r_{\rm pc}\delta^5 > 12E_\gamma L_{eV}/10^{47}$ and $r_{\rm pc} \equiv R/1\,{\rm pc}$) with $\Delta t = 0.5$ days and $E_\gamma = E/(1 \text{ GeV}) = 1000$ yields the beaming factor δ of at least ~ 5. Since the coincidence of the X-ray and TeV flares suggests that both are likely to arise from the same electron population, we have a strong evidence that *both* the TeV and keV emission must be beamed.

The shape of the broad-band continuum from the radio to X-rays suggests that a single electron population responsible for the emission in these bands (George et al. 1988). By the virtue of the correlated keV and TeV variability (while the optical, UV, and GeV emission was relatively steady; see Fig. 5), we conclude that the keV and TeV bands are produced by the most energetic end of a single relativistic electron distribution. In this scheme, X-ray emission is explained as due to the synchrotron process, while the TeV γ-rays are due to Comptonization. The Klein-Nishina limit requires that the seed photons have energies lower than X-ray; they probably arise from the IR band. This is further supported the relatively similar amplitudes of variability in the X-ray and TeV bands.

The similar flare-like behavior over an energy range of 13 orders of magnitude, the widest energy range over which simultaneous changes have been observed so far, was demonstrated in the multi-frequency campaign in 1995. The close correlation of the optical, X-ray and gamma-ray light curves suggest that the entire emission over the variable component over more than 13 decades in frequency is due to the same parent population of particles.

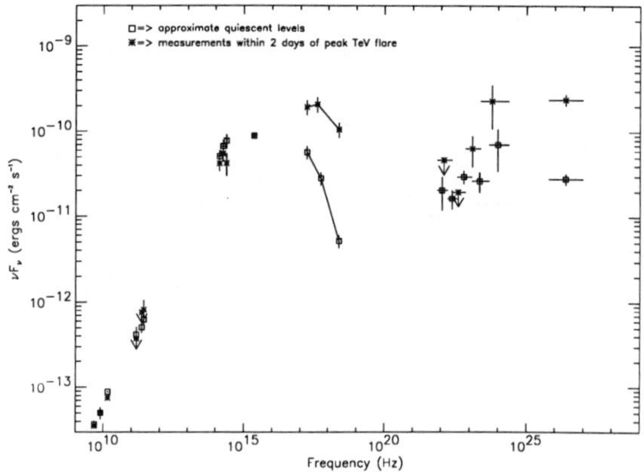

Figure 5. Multiband spectrum of Mkn421 within 2 days of peak TeV flare in 1994 campaign (Macomb et al. 1996).

References

Buckley J. et al. 1995, IAUC 6167

George, I. M., Warwick, R. S., & Bromage, G. E. 1988, MNRAS, 232, 793

Kerrick, A. D., et al. 1995, ApJ, 438, L59

Lin, Y. C., et al. 1992, ApJ, 401, L61

Makino, F., et al. 1987, ApJ, 313, 662

Macomb, D. J., et al. 1995, ApJ, 449, L99

Macomb, D. J., et al. 1996, ApJ, 459, L111

Mattox, J. R., et al. 1993, ApJ, 410, 609

Punch, M., et al. 1992, Nature, 358, 477

Rybivki, G. B., & Lightman, A. P. 1979, Radiative Processes in Astrophysics

Sembey, S., et al. 1993, ApJ, 404, 112

Stefan, W., Takahashi, T., Madejski, G., Mattox, J. R, et al. 1996, submitted

Takahara, F., et al. 1995, in Towards a Major Atmospheric Cerenkov Detector III, ed. T. Kifune (Tokyo, Universal Academy Press), p. 131

Takahashi, T., et al. 1995, in Multifrequency Behaviour of High Energy Cosmic Sources (proc. conf. in Vulcano, Italy, May 22–27, 1995), Mem. Soc. Astron. Ital., in press

Takahashi, T., et al. 1996, ApJL, in press

Takahashi, T., et al. 1996, in preparation

Tashiro, M. 1992, Ph.D. thesis, University of Tokyo

Tashiro, M., et al. 1995, PASJ, 47, 131

Zdziarski, A. A., & Krolik, J. H. 1993, ApJ, 409, L33

Blazar Continuum Variability
ASP Conference Series, Vol. 110, 1996
H. R. Miller, J. R. Webb, and J. C. Noble, eds.

Variability in the X-ray Flux of Quasar 3C345: Inverse-Compton Emission from the Parsec-scale Jet?

S. C. Unwin

Department of Physics, Mathematics and Astronomy, Caltech 220-47, Pasadena, CA 91125

A. E. Wehrle

Infrared Processing and Analysis Center, Jet Propulsion Laboratory, Caltech 100-22, Pasadena CA 91125

J. A. Zensus and A. P. Lobanov

National Radio Astronomy Observatory, Charlottesville, VA 22903

Abstract. We present the results of the first systematic study of *variability* in the X-ray emission from the 'superluminal' quasar 3C 345. Its power-law 1-keV X-ray emission varies by a factor of two on a timescale of years, but with no change in spectral index, closely following the high-frequency radio flux. Using VLBI images, we show that one of the superluminal 'knots' in the jet (at a distance of ≈ 15 pc from the nucleus), rather than the nucleus, produces most of the observed X-rays, via the synchrotron self-Compton process. We show that this knot accelerates as it moves away from the nucleus, along along a path at $\approx 10°$ from the line of sight.

1. Introduction

The quasar 3C 345 ($z = 0.595$) is a well-known 'superluminal' radio source (apparent proper motion $\beta_{\rm obs} > 1$). Its evolution has been monitored with VLBI imaging at cm-wavelengths since the late 1970s. All the VLBI images show a relativistic jet, which appears one-sided, due to Doppler-boosting. Figure 1 shows representative images; most of the identifiable parsec-scale components (or 'knots') persist several years, and yield measurable proper motions (Zensus, Cohen & Unwin 1995).

3C 345 is also a non-thermal X-ray source; the X-rays are believed to originate in the jet via inverse-Compton scattering of low-energy photons. Previous studies (e.g., Unwin et al. 1994) have shown that (1) superluminal motion, and (2) synchrotron self-Compton X-ray emission, both require the jet to be relativistic, from which a self-consistent picture of the jet kinematics can be built. By examining time-variability in the relation between X-ray and radio properties, we can eliminate many systematic errors resulting from over-simplified modeling—we are more interested in *changes* in derived parameters than their absolute values.

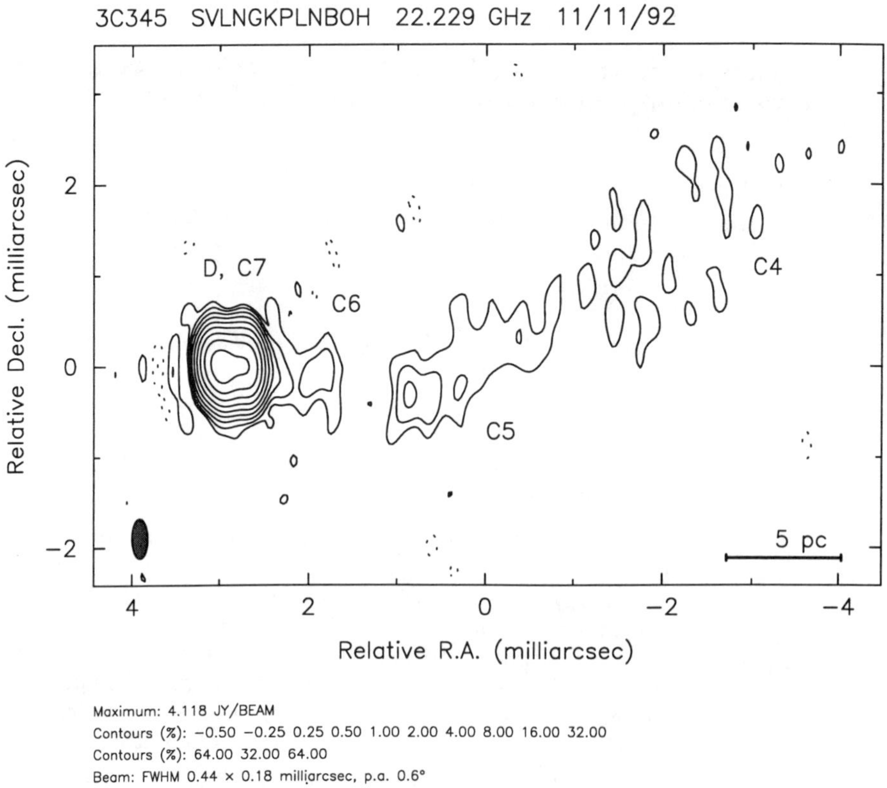

Figure 1. VLBI image of 3C 345 at 22 GHz, at a resolution of 0.7 pc, at epoch 1992.86. This image is representative of a long series of monitoring experiments at 5, 8, 11, and 22 GHz. Labeled features are the nucleus 'D', and superluminal jet components 'Cn', which are identified at other epochs and frequencies, allowing us to follow their spectral evolution.

Some of the questions we can answer with variability data are the following: How variable are the power-law X-rays from 3C 345, and is there any variation in spectral index? Are X-ray variations correlated with radio properties, such as total flux density at different frequencies? Are X-ray variations associated with individual components in the relativistic jet, and do they correlate with angular size or spectral shape? Can variability provide any new information on the geometry and dynamics of the parsec-scale jet?

2. X-ray monitoring results

3C 345 was observed by the ROSAT PSPC instrument (0.2–2.0 keV) instrument 5 times, during 1990–1993. All were reduced in a uniform way, which reduces systematic errors between epochs to a low level. Figure 2 shows the X-ray and

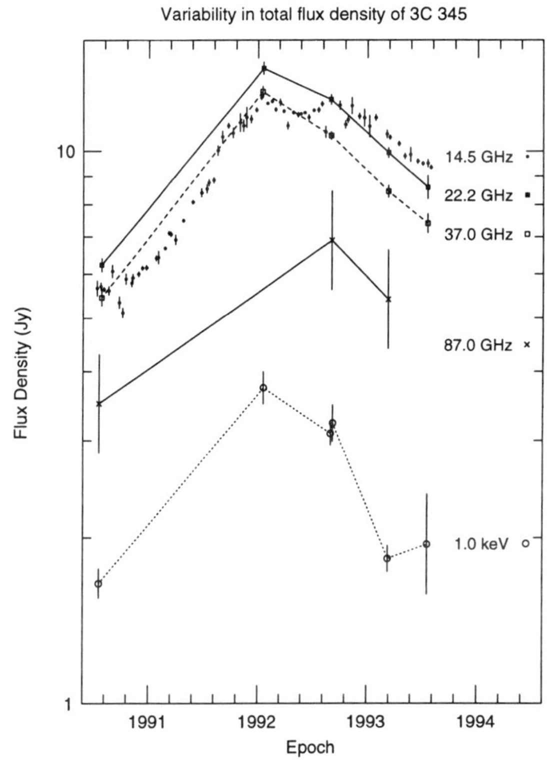

Figure 2. X-ray and radio flux monitoring of 3C 345, during 1990–1993. (X-ray flux densities scaled by 4×10^6 before plotting).

high-frequency radio flux density measurements vs. time. A constant Galactic column $N_{\rm H} = 7.4 \times 10^{19}\,{\rm cm}^{-2}$, and constant spectral index $\alpha = -0.75\ (S_\nu \propto \nu^\alpha)$ were assumed in fitting the X-ray spectra, since there was no evidence between epochs for variability in either parameter.

- At all epochs the photon count data in 0.2–2.0 keV are well fitted by a single power law, with constant absorbing column equal to the Galactic value, and weighted mean spectral index $\alpha = -0.75 \pm 0.17$.
- There is no evidence for spectral curvature, nor evidence for variability in the spectral index, during 1990–1993. This argues against the X-ray emission arising from two components, say, a hard and a soft component, each with different time histories.
- Two of the epochs each comprised data blocks separated by ≈ 1 week in time. We can place a limit of $< 5\%$ on any variability on that timescale.
- The X-ray flux history is directly correlated with the total flux density at 14.5 GHz and higher frequencies (Fig. 2), both in amplitude range and

variability timescale. Since the radio variability is known to arise in the parsec-scale components imaged directly with VLBI, this is strong evidence for a causal relation between VLBI and X-ray.

3. VLBI monitoring results

With VLBI imaging, we can track the spatial and spectral evolution of compact features in the parsec-scale jet (Fig. 1); more than a dozen VLBI images went into the spectral analysis we discuss here. Lobanov (1996), and Zensus et al. (in preparation) present most of the observational material, with an analysis of the component kinematics. We concentrated on the 'core' component ('D'), and the brightest most compact jet component ('C7'), since these are the most likely to produce significant inverse-Compton X-ray emission.

We fitted spheres to the components at each epoch to derive flux densities, and angular sizes, then interpolated the results to the epochs at which X-ray flux densities were measured. At each of these epochs, spectral turnovers were determined for each component by fitting homogeneous spheres to the 'knots', and a Königl (1981) model to the nucleus.

4. Inverse-compton X-ray calculation

From previous VLBI results, we know that the core ('D') is inhomogeneous, with a slowly-rising spectrum in the radio, whereas C7 and the other 'knots' can be represented by homogeneous spheres. We believe that shocks play a role in the knots, but the observations do not resolve them enough to justify a more-complicated model; therefore we computed the expected (self-) Compton X-ray emission from scattering of synchrotron photons in a homogeneous sphere. We assumed a spectral index $\alpha = -0.75$ for the optically-thin side of the sphere's spectrum; this is the same as the measured X-ray index, as required in this model; the predictions are a weak function of α.

For the core, we used the inhomogeneous conical-jet model of Königl (1981), in which the particle density and magnetic field fall as power-laws with radius. The rising synchrotron spectrum in the radio results from superposition of locally-homogeneous emission regions with varying 'turnover' parameters.

Core component ('D') For all epochs, the Königl model *under-predicts* the X-ray emission by more than 2 orders of magnitude for a wide variety of parameters. Only solutions with a very large angle to the line of sight ($\theta > 25°$) even approached the measured value, and we regard such a large angle as implausible for other reasons (Zensus, Cohen, & Unwin 1995).

Jet component ('C7') The homogeneous-sphere model *over-predicts* the X-ray emission from Component C7, unless there is bulk relativistic motion ($\delta \gg 1$). In terms of the jet energetics, the most conservative assumption is that C7 *is* the origin of the observed X-rays, and that the Doppler factor must be such as to lower the predicted flux to match the observed flux.

Since jet components are known to move 'superluminally' (with $\beta_{\rm obs} > 1$), we can derive the jet kinematics—the angle θ to the line of sight, and the bulk Lorentz factor γ. Figure 3 shows the geometric relation between these quantities.

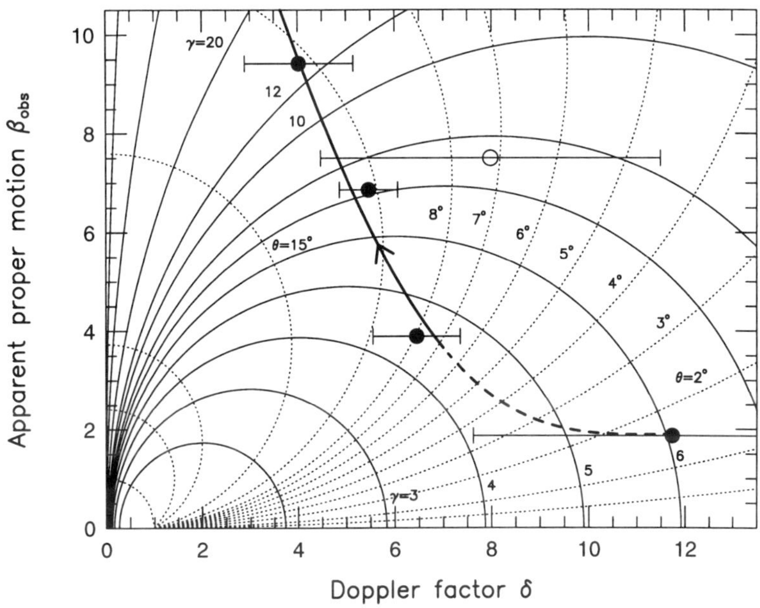

Figure 3. Geometry and kinematics of the jet in 3C 345. The Doppler factor δ (from X-rays) and β_{obs} (from superluminal motion) are *observable* quantities. The angle θ to the line of sight, and the bulk Lorentz factor γ are *derived* quantities. The track of C7 in this plane is indicated by the heavy line. The open circle represents C5, which dominated the jet in 1990 (Unwin et al. 1994).

We show measurements for C7 from our inverse-Compton X-ray calculations, using β_{obs} from Lobanov (1996). Clearly, fixed values of θ and γ for C7 are inconsistent with the X-ray and VLBI data: C7 must accelerate (γ increasing) as it moves away from the nucleus. The first point (at $\delta = 11.7$ from epoch 1992.05) suggests that C7 may have started out closer to the line of sight, then bent away. If other mechanisms, or other components, make a significant contribution to the X-ray flux, then the heavy line in Fig. 3 traces a lower limit to δ.

5. Conclusions

1. We have demonstrated a direct link between variations in parsec-scale jet structure and X-ray emission of a quasar for the first time—in 3C 345, a prototypical 'superluminal' radio source.

2. The core of 3C 345 (component 'D') can be modeled with a Königl jet, which explains the radio emission, but under-predicts the X-ray flux.

3. Jet component C7 (the brightest and most compact in 1992–1993) is the origin of the X-ray emission, unless the Doppler factor δ of the jet is even larger than the limits we derive.
4. The Doppler factor of C7 decreased from $\delta \approx 12$ to $\delta \approx 4$ during 1992–1993.
5. Combining X-ray and kinematic data for component C7, we conclude that C7 accelerates as it moves away from the nucleus, and its path probably bends away from the line of sight.
6. VLBI imaging with the VLBA, using several simultaneous frequencies, is now starting to provide better determinations of component spectra. With 43-GHz monitoring, we can probe the smallest jet scales, where we believe the X-ray emission is strongest.

Acknowledgments. We are grateful to Hugh and Margo Aller (U. Michigan), and Harri Teräsranta (U. Helsinki) for providing radio data in advance of publication. Greg Madejski (GSFC) helped with systematic analysis of the ROSAT data.

References

Königl, A. 1981, ApJ, 243, 700

Lobanov, A. P. 1996, Ph.D. thesis, New Mexico Institute of Mining and Technology

Unwin, S. C., Wehrle, A. E., Urry, C. M., Gilmore, D. M., Barton, E. J., Kjerulf, B. C., Zensus, J. A., & Rabaça, C. 1994, ApJ, 432, 103

Zensus, J. A., Cohen, M. H., & Unwin, S. C. 1995, ApJ, 443, 35

Blazar Continuum Variability
ASP Conference Series, Vol. 110, 1996
H. R. Miller, J. R. Webb, and J. C. Noble, eds.

The X-ray Spectrum of the BL Lacertae Object H1426+428

Rita M. Sambruna

NRC-GSFC, Code 662, Greenbelt, MD 20771

Ian M. George and G.M. Madejski

USRA, GSFC, Code 662, Greenbelt, MD 20771

C.Megan Urry and Joseph E. Pesce

STScI, 3800 San Martin Dr., Baltimore, MD 21218

L. Maraschi and A. Treves

Univ. di Milano, via Celoria 15, Milano, Italy

Abstract. We present ASCA, ROSAT, and BBXRT observations of the BL Lac object H1426+428, obtained at separate epochs. The three datasets confirm that the X-ray spectrum of this object is more complex than a single power law. The ASCA data exhibit spectral flattening below a few keV, consistent with either/both a low-energy feature or a convex continuum. An absorption feature around 0.6 keV is present in the BBXRT and possibly in the ROSAT data, indicating absorption by ionized Oxygen. The favored emission mechanism for the underlying continuum, which is modelled with a convex broken power law and exhibit correlated flux/slope variability, is the synchrotron process.

1. Introduction

The BL Lacertae (BL Lac) object H1426+428 is the optical counterpart of the bright ($F_{2-6\ \mathrm{keV}} \sim 3 \times 10^{-11}$ ergs cm^{-2} s^{-1}) X-ray source H1430+423, discovered in a hard (2–6 keV) X-ray range with the HEAO-1 Large Area Sky Survey experiment (Wood et al. 1984). It was classified on the basis of excess blue emission, featureless optical spectrum, and compact radio morphology (Remillard et al. 1989). The source optical light is polarized (polarization degree $\sim$ 2.5%; Jannuzi et al. 1993) and the underlying host galaxy is a giant elliptical at a redshift $z_{\mathrm{abs}} = 0.129$ (Abraham et al. 1991; Remillard et al. 1989). H1426+428 is a classical X-ray-selected BL Lac, with a spectral energy distribution from the radio through the X-ray band peaking in UV/soft-X-rays (Giommi et al. 1995). H1426+428 is indeed one of the 6 BL Lacs detected in the Extreme Ultraviolet band (Fruscione 1996).

The X-ray spectrum of H1426+428 appears more complex than a single power law. The EXOSAT spectrum in 0.1–10 keV is best-fitted with a downward-curved (convex) broken power law (Sambruna et al. 1993, 1994). The first moderate-resolution X-ray spectrum of the source obtained with BBXRT in 0.3–10 keV indicated an absorption feature around 0.6 keV (Madejski et al. 1992).

Motivated by the complexity of the X-ray emission of H1426+428, we observed the object with the ASCA and ROSAT satellites at two independent epochs. These data, together with the archival BBXRT observations, are presented here. A detailed report of this work is given in Sambruna et al. (1996; S96 in the following).

2. Results

2.1. ASCA

We observed H1426+428 with ASCA (0.4–10 keV) on February 6, 1994 for 30 ksec. The data from the two SIS and GIS were fitted together, with independent normalizations. Since residual calibration uncertainties affect the SIS data below 0.5 keV (ASCA Calibration Workshop, December 1995), the data below this energy were excluded from the analysis. During the ASCA pointing the source was in a low intensity state compared to previous observations (Table 1).

The ASCA data were first fitted with a single power law with absorption fixed to the Galactic value, $N_H = 1.41 \times 10^{20}$ cm^{-2} (Stark et al. 1992). The residuals of this model are shown in Figure 1. The fit is unacceptable; there is a deficit of counts at both low and high energies, which suggests downward curvature. Fitting the data with a power law and free absorption yields a very significant improvement (> 99% from the F-test), and yields a fitted N_H much larger than Galactic, $\sim 10^{22}$ cm^{-2}. Since H1426+428 is a copious EUV and soft X-rays emitter, the amount of cold absorption along the line of sight to the source in unlikely to significantly exceed the Galactic value. The spectral flattening in the ASCA data can only be due, thus, to either an intrinsically convex continuum, or to absorption by partially ionized matter, or both.

Indeed, the fit to the ASCA data with a convex broken power-law plus Galactic N_H model is acceptable and statistically equivalent to the fit with a single power law and free N_H. The fit yields a steepening of $\Delta\Gamma = 0.3$ above ~ 2 keV. See S96 for a discussion of a possible low-energy absorption feature.

2.2. ROSAT

The ROSAT PSPC (0.1–2.4 keV) observed H1426+428 on January 24, 1992 for a total of 9593 s. The residuals of the fit to the PSPC data with a single power law and Galactic N_H are shown in Figure 1. There is a count deficit centered around 0.6 keV, which suggests the presence of an absorption feature, at an energy similar to that detected in the BBXRT spectrum (see below). Adding an edge at 0.6 keV with optical depth ~ 0.5 improves the fit to the PSPC data with a significance > 99% (S96). However, such a feature could well be an artifact due to the secular gain drift of the PSPC (see Sambruna et al. 1995 for a simulation of this effect).

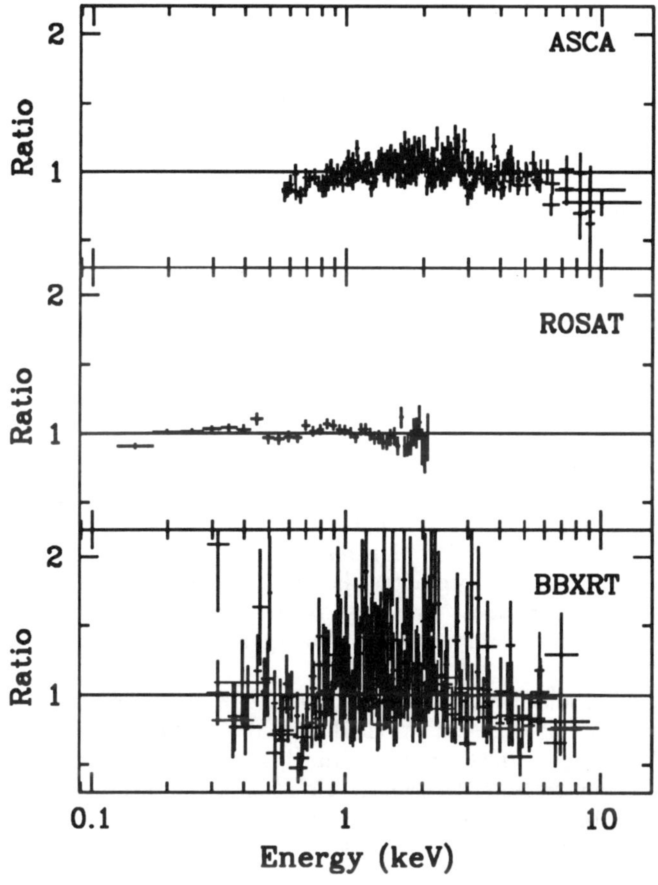

Figure 1. Residuals (ratio of data to model) of a fit with a power law plus Galactic N_H to the ASCA, ROSAT, and BBXRT spectra of the BL Lac H1426+428. The ASCA residuals show downward curvature, suggesting either a convex continuum, or absorption by ionized matter, or both. An absorption feature at $\sim$ 0.6 keV is clearly present in the BBXRT and possibly in the ROSAT data.

2.3. BBXRT

H1426+428 was observed with BBXRT in the range 0.3–10 keV on December 9, 1990 for a total exposure of 2192 s (Madejski et al. 1992). The residuals of the fit to the BBXRT data with a single power law plus Galactic N_H are shown in Figure 1. A narrow absorption feature is clearly present at $\sim$ 0.65 keV. In addition, the residuals suggest continuum curvature. Fitting the data with a broken power law plus Galactic N_H plus an edge at an energy $\sim$ 0.6 keV and with optical depth $\sim$ 1.1 yields a significant improvement (> 99% confidence) over a single power law fit. The fitted continuum is convex, with a steepening of $\Delta\Gamma \sim 0.8$ above 0.8 keV (S96).

3. Discussion

The ASCA data of H1426+428 show a significant spectral flattening of the X-ray continuum below a few keV, which is consistent with both a convex continuum and absorption by ionized matter along the line of sight. An absorption feature is definitely present in the BBXRT data around 0.6 keV, and is suggested in the ROSAT PSPC spectrum. Calibration uncertainties of the ASCA SIS at low energies and of the ROSAT PSPC at present hamper a detailed study of the feature in these datasets.

Our reanalysis of the BBXRT spectrum of H1426+428 confirm the earlier report by Madesjki et al. (1992). H1426+428 is thus the second BL Lac where a low-energy absorption feature has been detected so far. Indeed, an absorption line was first discovered in the high-resolution *Einstein* OGS spectrum of PKS 2155–304, which was interpreted as the redshifted Ly α resonant line of O VIII (Canizares & Kruper 1984). The large velocity ($v \sim 30{,}000$ km s^{-1}) associated to the line suggested substantial outflow in a relativistic wind (Krolik et al. 1985).

Different absorber conditions are indicated for H1426+428. The fitted energy of the feature is consistent with absorption by Oxygen in a lower ionization state, O V (E_{rest} = 0.645 keV) at the best-fit and O III–VI at 90% confidence (E_{rest} = 0.592–0.671 keV), in a stationary absorber. Oxygen densities of $N_O \sim 2 \times 10^{18}$ cm^{-2} are inferred, corresponding to $N_H \sim 3 \times 10^{21}$ cm^{-2} for cosmic abundances. A full discussion of the absorber conditions is given in S96. H1426+428 clearly qualifies as an excellent candidate for further higher-resolution spectroscopy with AXAF and XMM.

Table 1 reports the X-ray flux and spectral index history of H1426+428. The continuum photon index measured in 2–10 keV, where the effects of absorption are negligible, is Γ = 2.0–2.3, among the flattest measured for an X-ray-selected BL Lac at these energies (e.g., Sambruna et al. 1994). In addition, there is a trend for the continuum to harden when the source brightens, which is typical of X-ray-selected BL Lacs (Sambruna et al. 1994) and strongly suggests a synchrotron origin (George et al. 1988). A synchrotron emission is also supported by the continuum convexity, which may be related to radiative losses (Kardashev 1962).

Table 1. X-ray variability of H1426+428.

Instrument	Date	$Flux_{(2-10\ keV)}$ (10^{-11} ergs cm^{-2} s^{-1})	Luminosity (10^{44} ergs s^{-1})	Photon Index[a]
HEAO-1	1977–79	2.0	7.6	...
EXOSAT	Feb. 1985	5.0	19.0	2.10 ± 0.08
BBXRT	Dec. 1990	3.0	11.4	2.21 ± 0.10
ASCA	Feb. 1994	1.5	5.7	2.29 ± 0.03

[a]From the fit to the 2–10 keV continuum with a single power law + Galactic N_H model.

References

Abraham, R. G., McHardy, I. M., & Crawford, C. S. 1991, MNRAS, 252, 482

Canizares, C., & Kruper, J. 1984, ApJ, 278, L99

Fruscione, A. 1996, ApJ, 459, 509

George, I. M., Warwick, R. S., & Bromage, G. E. 1988, MNRAS, 232, 793

Giommi, P., Ansari, S. G., & Micol, A. 1995, A&AS, 109, 267

Jannuzi, B. T., Smith, P. S., & Elston, R. 1993, ApJS, 85, 265

Kardashev, N. S. 1962, Soviet Astronomy, 6, 317

Krolik, J., Kallman, T., Fabian, A., & Rees, M. 1985, ApJ, 295, 104

Madejski, G. M., et al. 1992, in Proc. 28th Yamada Conference on Frontiers of X-Ray Astronomy, ed. Y. Tanaka & K. Koyama (Tokyo, Universal Academy Press), p. 583

Remillard, R., Tuohy, I. R., Brissenden, R. J. V., Buckley, D. A., Schwartz, D. A., Feigelson, E. D., & Tapia, S. 1989, ApJ, 345, 140

Sambruna, R. M., et al. 1996, ApJ, submitted (S96)

Sambruna, R. M., Urry, C. M., Ghisellini, G., & Maraschi, L. 1995, ApJ, 449, 567

Sambruna, R. M., Barr, P., Giommi, P., Maraschi, L., Tagliaferri, G., & Treves, A. 1994, ApJ, 434, 468

Sambruna, R. M., Barr, P., Maraschi, L., Tagliaferri, G., & Treves, A. 1993, ApJ, 408, 452

Stark, A. A., Gammie, C. F., Wilson, R. W., Bally, J., Linke, R. A., Heiles, C., & Hurwitz, M. 1992, ApJS, 79, 77

Wood, K., et al. 1984, ApJS, 56, 507

Blazar Continuum Variability
ASP Conference Series, Vol. 110, 1996
H. R. Miller, J. R. Webb, and J. C. Noble, eds.

X-ray Novae: Galactic micro-Quasars?

Chris Shrader and Demos Kazanas
Laboratory for High Energy Astrophysics NASA Goddard Space Flight Center

Abstract. Recent observations have revealed X-Ray Novae (XRN) to be of considerable interest as laboratories for the study of astrophysical jets. In particular, two sources, GRS 1915+105 and GRO J1655–40, were detected as radio transients with peak fluxes exceeding 1 Jy and found, to exhibit superluminal motion. In the case of GRO J1655–40, the jet has an asymmetric morphology, and its temporal history consists of discrete plasmoid-ejection episodes which are correlated with hard-X-ray flaring. Other XRN have been radio emitters, but with much lower flux levels and without discernable spatial extent. In several cases such as GRO J1719–24, correlations between radio and hard-X-ray light curve, and the radio decay profiles are reminiscent of GRO J1655–40 suggestive of a common underlying physics. We present a compilation of some empirical properties of XRN and speculate on possible insights into Blazar AGN physics.

1. Introduction

X-ray Novae (XRN), sometimes called "soft X-ray transients," are a subclass of the galactic population known as low-mass X-ray binaries. They are distinct in that they exhibit extreme episodic behavior, spending the majority of their lifetimes in a quiescent state characterized by low accretion rates ($\dot{M} \sim 10^{12}$ gm/s), and occasionally, on a recurrence time scale of years to decades, undergo dramatic outbursts ($\dot{M} \sim 10^{17}$ gm/s). During outbursts their X-ray luminosity can increase by 6 orders of magnitude or more, with corresponding increases at other wavelengths. For example increases of ΔV=7 are common. The outbursts evolve rapidly, with typical light-curve e-folding times of $\sim$ 30 days; for a recent review see Chen, Shrader, and Livio (1996). Given their rapid and dramatic evolution, they provide a unique laboratory for studying accretion disk physics and related phenomena. Analogous phenomena in AGN would require an inordinate amount of time to observe, i.e., millennia! Since XRN can subsequently be examined in deep quiescence, the system binary parameters can be determined accurately, unlike in the steady low-mass X-ray binaries where emission from the accretion disk severely dilutes light from the secondary star. This allows the nature of the compact object to be determined, which has further revealed that a majority of these systems contain black holes.

XRN are of interest to studies of Blazars and other AGN subclasses for several reasons. First, as noted, a large fraction are known to contain black holes, which are of course the putative "central engine" for Blazar AGN; in fact 7 of the best 10 galactic black hole candidates are XRN. Like AGN they are broad-band emitters, detected from radio to gamma rays; GRO J0422+32 was detected at above 1 MeV for example. Second, since they are in low-mass binaries, that is they usually have late K or M main-sequence secondaries, they undergo disk-accretion rather than wind-accretion, or disk- plus wind-accretion as is likely in the high-mass X-ray binary systems such as Cyg X-1—thus, on a reduced scale they embody a key aspect of the standard AGN paradigm. Third, and most significant so far as this conference is concerned, several objects recently discovered by the Burst and Transient Source Experiment (BATSE) onboard the Compton Gamma Ray Observatory (CGRO) have exhibited superluminal, or near-superluminal radio jets, analogous to those known to be associated with Blazar AGN. The fact that XRN evolve rapidly, time dependent phenomena such as jet propagation and correlations between activity in the jet and in the central source, which require inordinately long periods of time to probe in AGN can be monitored.

In this contribution we present a current compilation of observational properties of XRN with emphasis on their radio emission and particularly on the super-luminal jet sources GRO J1655–40 and GRS 1915+105. We will then attempt to compare and contrast these objects with Blazars and attempt to identify what future XRN studies may provide insight into our understanding of the Blazar phenomena and AGN in general.

2. Summary of relevant observational properties

Historically, various XRN are known to have been radio emitters, although they are generally weak, with fluxes of order 1 mJy. Similarly, a small fraction of known persistent X-ray binaries are weak radio sources, among which are unusual cases such as Cyg X-3, which may be a Wolf Rayet Star—compact object binary, and Circinus X-1 for which it is unclear if it has high- or low-mass companion. LSI +61 303, which has been suggested to be the counterpart to the high-energy gamma-ray source 2CG135+1 is another radio emitter which is a poorly understood system. Radio emission has been detected from at least ten XRN (Table 1.)

Among the ~mJy radio emitters a "two-phase" radio behavior has often been seen. This consists of an initial fast (1-10 day) transient event, which has been interpreted as an expanding "synchrotron bubble" of charged particles. The sources then often exhibit a second phase, slowly varying radio source with an overall power-law like decay light curve. Often the initial synchrotron bubble event is not seen during its rising phase; also, it has been missed entirely if it is sufficiently rapid. More recently, several binaries, most notably GRO J1655–40 and GRS 1915+105, which have been among the brightest radio emitters have been spatially resolved into jet-like structures with VLA and VLBA interferometry (Hjellming & Rupen 1995; Mirabel 1995). The inferred jet-expansion velocities are $> 0.9c$, thus corresponding to Lorentz factors $\gamma \simeq 2.5$. GRO J1655–40 additionally exhibited a "cork-screw" like morphology, which has been kinematically modeled by Hjellming and Rupen (1995) as ejections from a precessing disk coupled to the 2.6-day binary orbital period.

Table 1. Radio Properties of X-Ray Novae

Object ID	5GHz Flx (mJy)	X-ray Peak (Crab)	Distance (kpc)	Log(R)
GRO J0422+32	10.0	3.0	2.2	-8.2
A0620-00	750.0	45.0	0.9	-7.5
Nova Mus 91	180.0	6.0	5.0	-7.2
Cen X-4	40.0	29.6	1.2	-8.6
GRO J1655-40	7000.0	1.1	3.2	-4.9
GRO J1719-24	13.0	1.4	2.4	-7.7
Aql X-1	0.7	1.1	2.5	-8.9
GRS 1915+105	700.0	0.3	10.0	-5.3
GS2000+25	35.0	12.0	3.0	-8.2
V404 Cyg	4200.0	20.0	3.5	-6.4

R == Log(νL_ν)/Log(L_x), where L_ν == the 5 GHz radio luminosity.

Do the weaker radio emitters also have jets, which, due to a combination of intensity and viewing geometry, we just fail to detect? As noted by Hjellming and Rupen (1995), the fast transients seen in historical XRN could have been collimated jet-ejection events rather than spherical plasma expansions. There is evidence which supports this, in the sense that it suggests a definitive link between disk activity and jet emission. An approximate 1-week lag time is seen between X-ray flaring and plasma ejection events in GRO J1655–40 (Hjellming & Rupert 1995). This pattern is now beginning to emerge, in other, lower intensity sources, such as GRO 1719–24 (Figure 1), although denser sampling is needed. GRO J0422+32 was too poorly sampled to determine a characteristic lag time, but clearly the radio emission responds to X-ray activity about 45 and 110 days post-outburst, and there is also an apparent spectral flattening associated with the event at 110 days, nominally consistent with the spectral versus flux evolution in GRO J1655–40 flaring (Shrader et al. 1994).

3. Comparison to Blazars

The question of similarity or lack thereof, between AGN and galactic XRN, tantalizing and significant as it may be, it is currently unknown. If indeed the mechanism responsible for these phenomena is the same, then the great dynamic range in mass between these two disparate systems ($1 M_\odot$ and $10^8 M_\odot$ respectively) provides a great opportunity for uncovering the scaling relations which are responsible.

In this respect, one should note that AGN show a clear demarcation in radio loud and radio quiet, according to the ratio of their radio to disk (or bolometric) luminosities, at a ratio $R = \nu L_{r,\nu}/L_b \simeq 10^{-4.5}$, with the radio loud being those with $10^{-2.5} > R > 10^{-4.5}$. The AGN exhibiting superluminal expansion are all in the category of radio loud AGN. On the other hand, the values of R associated with galactic sources are $10^{-5} > R > 10^{-8}$ (Figure 2).

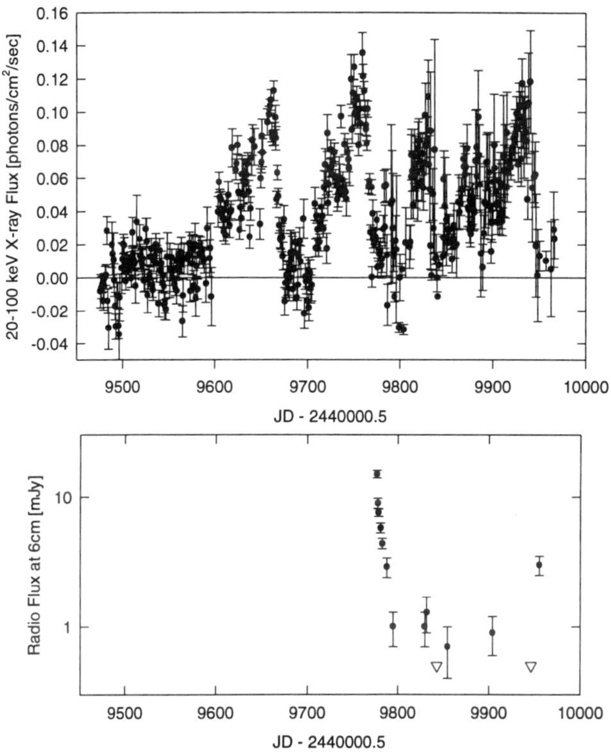

Figure 1. Radio and hard-X-ray light curves GRO J1719–24. Although much fainter than GRO J1655–40, and more poorly sampled, GRO J1719–24 appears to respond in a similar manner to X-ray flaring, i.e., hard X-ray flaring precedes radio activity by $\sim$ 10 days.

One then wonders whether the radio brightest among the XRN sources could be indeed the stellar equivalent of radio loud AGN. In such a case, one would be called to define the notion of "radio loudness" in a luminosity- (or mass-) dependent way which allows for GRO J1655–40 and GRS 1915+105 to be called radio loud while most of the rest would be classified as radio quiet. Unfortunately, XRNs are distributed almost uniformly in the interval 10^{-5} – 10^{-8}, with no additional particular characteristic which would classify them either way, unlike the AGN. There is an additional, significant difference between the galactic and extragalactic superluminals: the two galactic ones have bulk Lorentz factors $\Gamma \simeq 2.5$, while in the extragalactic ones Γ can be greater than 10. It thus appears that whatever the jet mechanism, it can convert accretion energy to directed motion more efficiently in extragalactic than in galactic sources.

4. Discussion

It has been suggested that the low values of Γ observed in the galactic sources are due to the physical size of the associated compact source (Contopoulos &

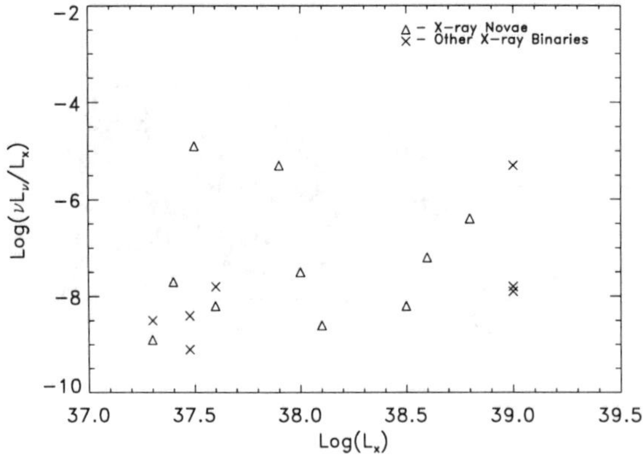

Figure 2. Radio-to-X-ray luminosity ratio plotted as a function of X-ray luminosity for 10 X-ray Novae and 7 other galactic X-ray binaries for which radio emission has been detected. Note that the most "radio loud" among the sample have radio-to-disk power ratios of $\sim 10^{-5}$.

Kazanas 1995). However, this fact may be only an artifact of the small number of such sources observed to date in the galaxy; if a sufficiently large sample is accumulated, values $\Gamma \simeq 10$ may be observed in galactic sources too. Related to the same issue of the statistics of small number of sources may also be the peculiar fact that the galactic jet sources appear to be at large angles to the observer's line of sight, with no cases at which we are "looking down" the jet known. Also, unlike Blazars, there have been no high-energy gamma-ray detections among the galactic jet sources, although three, 4U 1543–47, GRO 0422+32 and GRO J1655–40 have been partially covered by EGRET observations at or near their outburst peaks (Levinson & Mattox 1996). The most interesting case is GRO J1655–40, whose radio emission, however, was not at maximum at the time of the EGRET coverage. The lack of high energy gamma-ray emission may be consistent with the small value of the Lorentz factor of the outflow.

References

Chen, W. Shrader, C. R., & Livio, M. 1996, ApJ, submitted

Contopoulos, & Kazanas 1995, ApJ, 441, 521

Falcke, H., & Biermann, P. L. 1995, A&A, 308, 321

Falcke, H., Malkan, M. A., & Biermann, P. L. 1995, A&A, 298, 375

Harmon, B. A., et al. 1995, Nature, 374, 703

Hjellming, R. M., Rupen, M. P., & Shrader, C. R., 1996, ApJ, in preparation

Hjellming, R. M., & Rupen, M. P. 1995, Nature, 375, 464

Levinson, A., & Mattox, J. R. 1996, ApJ, 462, 67L

Mirabel, I. F., & Rodrigez 1996, Nature, 371, 46

Shrader, C. R., Wagner, R. M., Starrfield, S. G., & Hjellming, R. M. 1994, ApJ, 434, 698

Blazar Continuum Variability
ASP Conference Series, Vol. 110, 1996
H. R. Miller, J. R. Webb, and J. C. Noble, eds.

Gamma-ray Variability in Blazars

R. C. Hartman

Code 661, NASA/GSFC, Greenbelt, MD 20771

Abstract. This paper summarizes the currently available results on gamma-ray flux variations in blazars, primarily in the GeV (EGRET) energy range. Included are a description of the nature of the observations and analysis, followed by a sample of light curves illustrating the different timescales of variations that have been observed.

1. Introduction

Because the time variations of blazar gamma-ray fluxes are widely believed to probe the physical processes in the portion of the jet closest to the central engine, they play a key role in the developing understanding of these objects. In order to facilitate understanding of the observational results, a brief description is given of the EGRET properties relevant to studies of variability, and of the observation program and the EGRET data analysis.

2. EGRET capabilities and limitations

EGRET records individual photons in the energy range 30 MeV to about 30 GeV. However, the are several instrument characteristics that limit the energy range for which time variation investigations are viable. At the low end of the EGRET energy range, there are systematic uncertainties that make the spectral information marginally useful. In addition, deteriorating point spread function (psf) and energy resolution at low energies make the analysis there difficult and subject to still more uncertainties. For these reasons, blazar time variation investigations have utilized only photons with energies above 100 MeV.

At high energies, the systematic uncertainties are reduced, and the psf and energy resolution are more reasonable. However, because of the steeply falling spectra, few photons are detected above 5 GeV. Even for photons above 1 GeV the number is insufficient to permit studies of time variations in any but the brightest blazars. Thus almost all studies of blazar time variations utilize the integral flux above 100 MeV, but it must be kept in mind that the results are dominated by the more numerous 100–300 MeV photons, and that those above 1 GeV have little influence.

The angular resolution of EGRET is energy-dependent, varying from about 8° at 60 MeV to 0.4° above 3 GeV (68% containment). The accuracy of position determinations (68% confidence) varies from better than 0.1° for very bright sources to at least 0.5° just above the detection threshold. Since the number of bright (> 1 Jy) extragalactic flat-spectrum radio sources is roughly 20 times the 50 EGRET blazar detections, and the number of weaker radio sources which might be able to flare to an EGRET-detectable level is still larger, there is at times a significant problem with source confusion (although not as bad as for Galactic plane sources). A number of examples have occurred in which the identification of a source has been complicated by an apparent flare in an object within a few degrees. Sometimes this is apparent from the shape of the likelihood position contours, but in many cases there is no clear solution.

The dynamic range for most observations of blazar variations is fairly small. The threshold sensitivity (> 100 MeV) for a single observation ($1\text{–}3\times10^{-7}$ photon cm^{-2} s^{-1}) is only about a factor of 50–100 below the maximum blazar flux ever observed. Since most of the detections, even those described as "strong" by the EGRET Team, are within a factor of a few above threshold, many blazars are detected only in a flare state, and at other times disappear below the detection threshold.

One of EGRET's strong points is its large field of view. In terms of effective area, its sensitivity falls to half the value on-axis at about 18° from the axis, and it still retains useful sensitivity out to about 30°. (These numbers apply to the normal mode of operation. In the narrow-field lifetime-conserving mode, the corresponding angles are about 11° and 19°.) This wide field made it possible to do an all-sky survey during the initial 18 months (Phase 1) of the mission. Since that time, much of the sky has been observed, with varying levels of sensitivity, several times. Thus most blazars have been observed several times, and a few have been observed with good sensitivity as many as 7–8 times over the 5 years since the CGRO launch. A substantial fraction of the detected blazars were first detected more than 20° off-axis.

Another important feature of EGRET is its very low background. Because the electron-positron pair conversion of a gamma ray leaves a unique pattern in the instrument, there is essentially no internal background in the data. In principle, there could be a very small gamma-ray background due to interactions of cosmic-ray protons in the thin thermal blanket around the upper part of the instrument, but prelaunch calibration in a proton accelerator beam indicated that that background is well below the level of the extragalactic diffuse intensity.

3. The CGRO observing program

Because of the low count rates from gamma-ray sources, typical observation lengths are 1–3 weeks. During Phase 2, a number of 3–5 day observations were implemented, but these turned out to be almost useless for EGRET. A few observations have been as long as 4–5 weeks; these have been found to be very useful.

4. EGRET data analysis

The primary analysis tool for EGRET data is a package of maximum likelihood routines called LIKE (Mattox et al. 1996a). In most cases, these are applied to a set of standard maps for each observation period. The standard maps, all

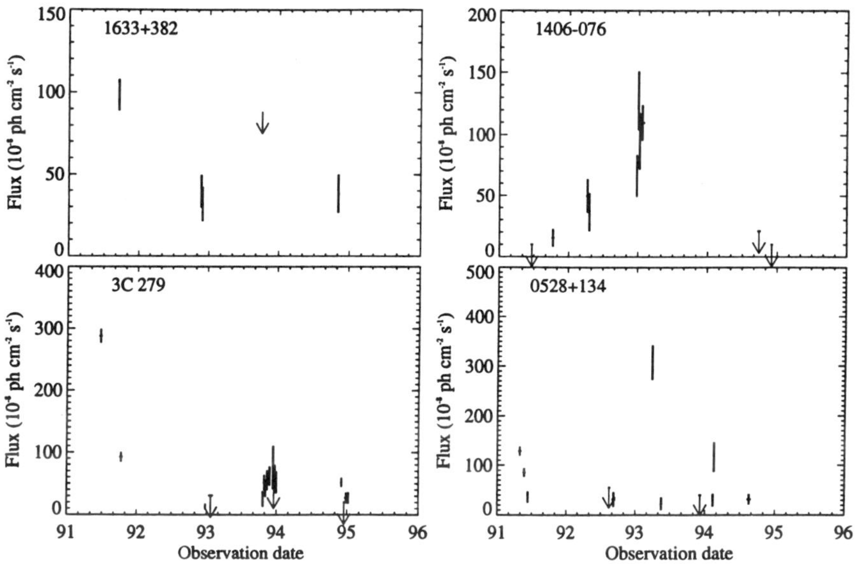

Figure 1. Examples of blazars with widely varying γ-ray flux

in half-degree bins, are generated routinely for a variety of energy intervals. All quicklook and initial analysis is done on > 100 MeV maps.

Once a strong source is detected in a particular observation, a search for variations on timescales shorter than the full observation often begins with a Kolmogorov-Smirnov (KS) test, which compares the rate of accumulation of exposure (proportional to the expected rate of accumulation of counts from a non-varying source) to the observed rate of accumulation of photons near the source position. This is only a preliminary tests, since it does no background subtraction, and therefore provides only a lower limit on variations. In a confused region, or one in which the diffuse gamma intensity is high, its effectiveness is diminished, since nearby non-varying sources tend to dilute any variation in a specific object of interest. If the KS test shows a possible interesting variation, full analysis requires making individual maps for short time intervals (1–7 days, depending upon how bright the detected source is), and doing the likelihood analysis on each of them. This process provides modeling of confusing sources as well as the diffuse background, and thus gives a more reliable estimate of possible variations.

5. Observed variations

5.1. Long-term variations

Many of the detected blazars have demonstrated large flux variations between different observations, up to a factor of about 25 (von Montigny et al. 1995). Several examples are shown in Figure 1. (The examples shown here may differ

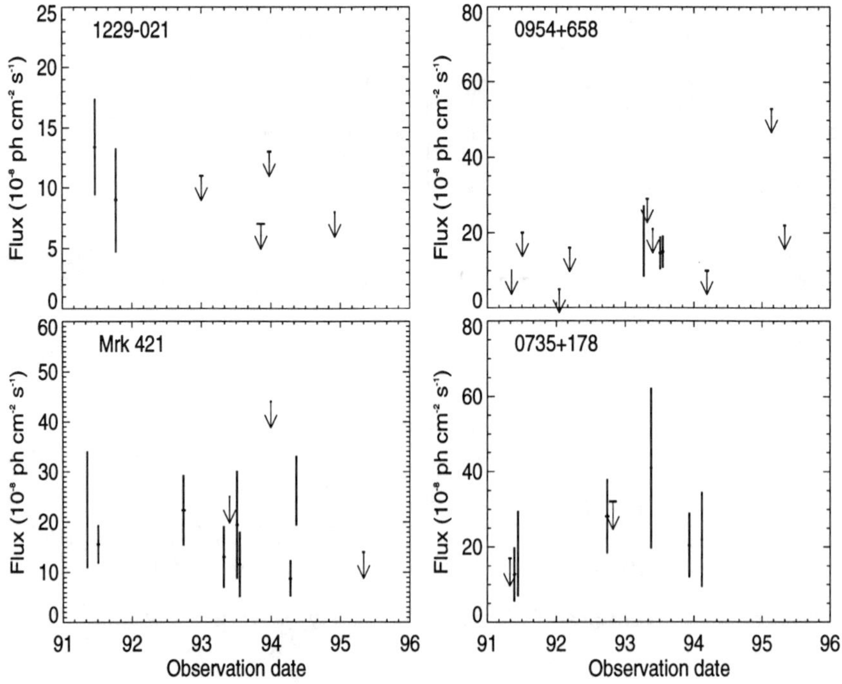

Figure 2. Examples of blazars with little observed γ-ray flux variation

slightly from the corresponding light curves in von Montigny et al. 1995 because of recent reprocessing and reanalysis of the EGRET database.)

Some blazars occasionally have been detected just above threshold, but at other times were undetected; in those cases the upper limits for non-detections do not conflict seriously with the detection fluxes, leaving no evidence for variations in those objects. A few objects have been detected several times with moderate fluxes, without showing any compelling evidence for large flux variations. Examples of these cases are shown in Figure 2. It is beginning to be apparent that a number of BL Lac objects, especially the nearer ones, tend to fall in this category; this suggests that they are not subject to the large variations seen in the quasars, or possibly that their gamma-ray outbursts are much less frequent than in the quasars.

The 2nd EGRET source catalog (Thompson et al. 1995) lists some blazar detections for objects that have never been statistically compelling in any single observation. These detections are based on LIKE analysis of maps which are sums of individual observation maps. Objects detected in this manner must be minimally variable, and must have fluxes not far below the EGRET single-observation detection threshold. Examples of such detections are those of 0954+556 and 1908-201.

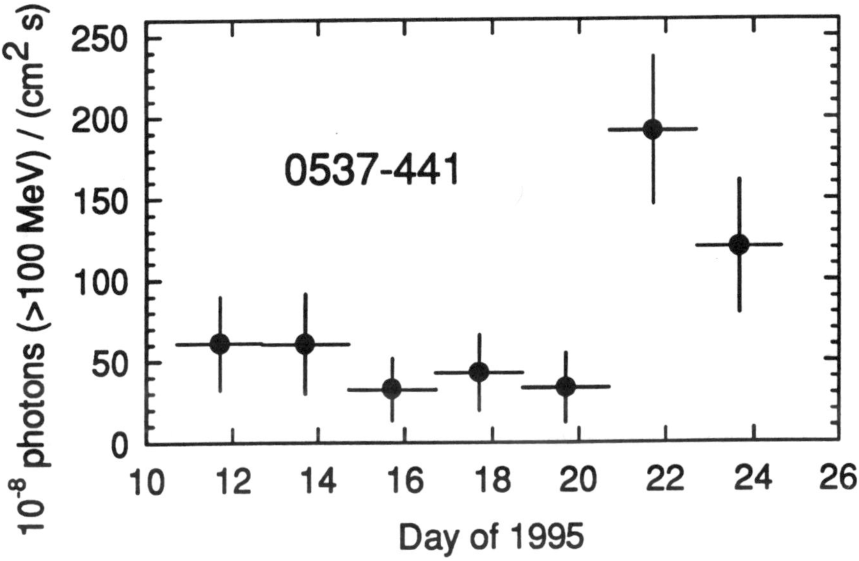

Figure 3. A 2–3 day flare in PKS 0537-441

5.2. Short-term variations

Only a handful of blazars have been sufficiently bright to permit a search for variations on timescales of a few days or less. In almost all such cases, few-day-timescale variations have been found to be statistically significant (e.g., Kniffen et al. 1993; Mattox et al. 1993; Hunter et al. 1993). The 1991 June flare of 3C 279 (Kniffen et al. 1993) does not seem to be typical. In that event, the gamma-ray flux appeared to increase gradually over about a week, then drop by a factor of 3.5 over about 2 days (Kniffen et al. 1993). There is growing evidence that typical flare timescales are 1–3 days (Figure 3 and Figure 4; see also Wagner et al. 1995), or maybe even less. It is worth noting that none of the closer BL Lac objects has ever been sufficiently bright during EGRET observations to permit a search for variations with timescales less than a week.

The two largest blazar gamma-ray flares observed by EGRET have been from PKS 1622-297, in June–July 1995 (Mattox et al. 1996b), and from 3C 279, in January–February 1996 (Wehrle et al. 1996). Those two events peaked at comparable gamma-ray fluxes, roughly 2.5–3 times the peak of the June 1991 flare in 3C 279, and each showed evidence for substantial gamma-ray flux changes on scales well under one day (Mattox et al. 1996b; Wehrle et al. 1996).

6. Results from higher and lower energy bands

In addition to the EGRET results on blazar variability, there are also results in both higher- and lower-energy bands. In the TeV range, two blazars have now been detected convincingly with the Whipple telescope: Mrk 421 and Mrk 501.

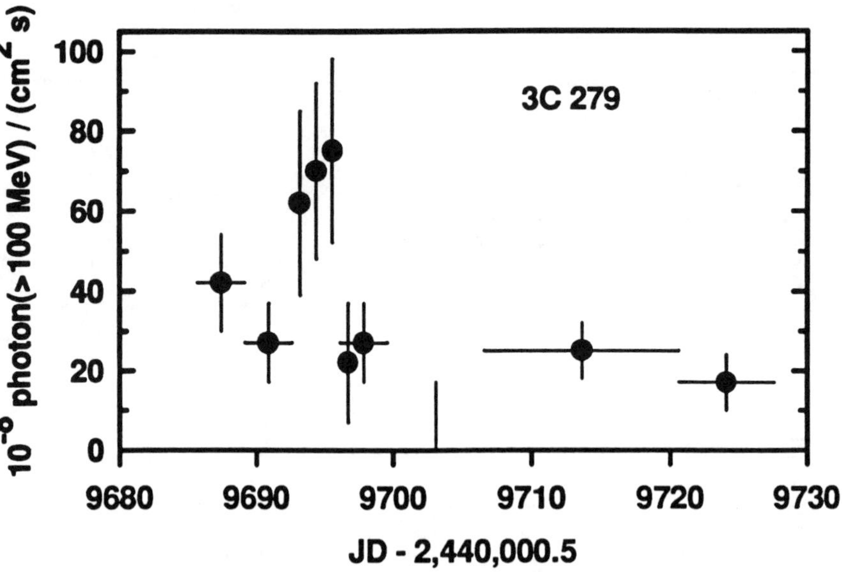

Figure 4. A small 3-day flare in 3C 279

Both of these objects have demonstrated major TeV flux variations, Mrk 421 by a factor of 2 over about 1 day (Kerrick et al. 1995; Buckley et al. 1996), and Mrk 501 by a factor of 2 over about 2 days (Quinn et al. 1996; Weekes, private communication). It is noteworthy that neither of the two Whipple-detected blazars is at all distinguished in the EGRET observations: Mrk 501 has never been detected at a significant level by EGRET, and Mrk 421, although reliably detected by EGRET, has never demonstrated more than minimal flux variations in the EGRET band.

Below the EGRET energy range, both COMPTEL and OSSE have detected a number of blazars. McNaron-Brown et al. (1995) have presented evidence for substantial flux variations between OSSE observations spaced weeks to months apart; the results are most significant in the 50–150 keV band, but there are indications of variations in the 150–500 keV and 500–1000 keV bands also. For 3C 273, the blazar that typically has been the brightest in the OSSE band, McNaron-Brown et al. (1995) and Johnson et al. (1995) have shown evidence for variations by at least a factor of 2 on timescales of one to several days.

7. Future EGRET plans and capabilities

By now (April 1996), about 80% is completed of a CGRO Cycle 5 multiwavelength campaign on 5 blazars previously seen to show bright gamma flares: 3C 454.3, PKS 0528+134, 3C 279, 1156+295, and 4C 38.41 . Although all 5 of these objects have been detected during Cycle 5, only 3C 279 has shown strong flaring behavior, as described above.

The remaining EGRET lifetime is quite limited: after the end of CGRO Cycle 5 (October 1996), the final gas fill will be 13 months old. To conserve lifetime, EGRET is operating with a narrow field-of-view throughout Cycle 5, and indeed is turned off during some pointings in which the EGRET field contains no objects of exceptional interest in the EGRET energy range (pointings motivated by COMPTEL/OSSE observations). During CGRO Cycle 6 (1 year starting October 1996), it is planned that EGRET will be turned off half the time, and that only observations of extreme importance will be implemented.

References

Buckley et al. 1995, IAU Circ. 6167
Buckley et al. 1996, in preparation
Hartman et al. 1993, ApJ, 407, L41
Hunter et al. 1993, ApJ, 409, 134
Johnson et al. 1995, ApJ, 445, 182
Kerrick et al. 1995, ApJ, 438, L59
Kniffen et al. 1993, ApJ, 411, 133
Mattox et al. 1993, ApJ, 410, 609
Mattox et al. 1996a, ApJ, 461, 396
Mattox et al. 1996b, ApJ, submitted
McNaron-Brown et al. 1995, ApJ, 451, 575
von Montigny et al. 1995, ApJ, 461, 698
Quinn et al. 1996, ApJ, 456, L83
Thompson et al. 1995, ApJS, 101, 259
Wagner et al. 1995, ApJ, 454, L97
Wehrle et al. 1996, May HEAD Meeting (San Diego)

Blazar Continuum Variability
ASP Conference Series, Vol. 110, 1996
H. R. Miller, J. R. Webb, and J. C. Noble, eds.

The High-energy Continuum of the γ-ray-loud Blazar PKS 0528+134

Rita M. Sambruna

NASA/GSFC, Code 662, Greenbelt, MD 20771

C. Megan Urry and Joseph E. Pesce

STScI, 3700 San Martin Dr., Baltimore, MD 21218

L. Maraschi

Oss. Astronomico di Brera, via Brera 28, Milano, Italy

R. Mukherjee, R. C. Hartman, and C. von Montigny

NASA/GSFC, Code 662, Greenbelt, MD 20771

S. J. Wagner

Landessternwarte Königstuhl, 69117 Heidelberg, Germany

Y. C. Lin

Standford University, HEPL, Stanford, CA 94305

A. E. Wehrle

IPAC, Caltech 100-22, Pasadena, CA 91125

Abstract. We present ASCA observations of the γ-ray-loud blazar PKS 0528+134 obtained at two separate epochs in 1994 August and 1995 March. The X-ray flux increased by a factor 4 in ~ 7 months, without appreciable change of the spectral shape. Both ASCA spectra are consistent with a single power law with column density $N_{\rm H} \sim 5 \times 10^{21}$ cm^{-2}, larger than Galactic. During the lower state of 1994 August, PKS 0528+134 was observed simultaneously in the optical and in γ-rays with EGRET. The source was in the lowest γ-ray state so far observed. In 1994 August, the extrapolation of the X-ray continuum to the γ-ray range requires a sharp spectral break at $\sim 10^{22}$ Hz. We present the spectral energy distribution of 1994 August and compare it with the literature multifrequency spectrum obtained during the flare of 1993 March.

1. Introduction

PKS 0528+134 is a compact radio-loud quasar at a redshift $z = 2.07$ (Hunter et al. 1993), with polarization $\sim$ 1–2% in the radio and $\leq 0.3\%$ in the optical (Perley 1982; Fugmann & Meisenheimer 1988). In the radio it is variable on timescales of years (Aller et al. 1985; Zhang et al. 1994), and exhibits superluminal motion, with $\beta_{\rm app} \sim 4$ (Pohl et al. 1995). The source is faint in the optical, with so far observed magnitudes $V = 19.5$ mag (Wall & Peacock 1985).

At MeV energies, PKS 0528+134 is one of brightest AGN, and the second most distant quasar, detected with CGRO EGRET (Hunter et al. 1993; Thompson et al. 1995). The emission in the 30 MeV–30 GeV energy band is variable, with flux changes of a factor 3 in $\sim$ 2 years and smaller-amplitude variations on a timescale of a few days. In addition, there is some evidence for accompanying spectral variability, with the EGRET spectrum becoming harder as the source brightens (Mukherjee et al. 1996). PKS 0528+134 was also detected with OSSE in 0.05–1.0 MeV (McNaron-Brown et al. 1995). The power emitted at MeV–GeV energies often exceeds the power at lower frequencies (von Montigny et al. 1995; Mukherjee et al. 1996), making the study of the high-energy continuum of PKS 0528+134 of great interest.

We observed PKS 0528+134 for the first time in a medium–hard X-ray range (0.4–10 keV) with ASCA on two separate occasions in 1994 August and 1995 March, during coordinated γ-ray pointings with EGRET. Here we present both ASCA observations of PKS 0528+134, together with the EGRET data and optical data simultaneous to the 1994 August observation. A full report of this work is presented in Sambruna et al. (1996; S96 in the following). Throughout this paper, $H_0 = 75$ km s^{-1} Mpc^{-1} and $q_0 = 0.5$ are assumed.

2. Observations

2.1. ASCA

ASCA observed PKS 0528+134 on 1994 August 28 and on 1995 March 19 for 20 ksec on each occasion. No flux variability was detected within each single ASCA exposure, and the probability that the light curves are constant is $\geq 11\%$ for the SIS and $\geq 40\%$ for GIS.

The ASCA data were fitted with a single power law with $N_{\rm H}$ both free and fixed to the total (atomic + molecular) HI column density along the line of sight, $N_{\rm H} = 3.9 \times 10^{21}$ cm^{-2}. At both epochs the fits with free $N_{\rm H}$ are preferred with high confidence ($\geq 98\%$ in 1994 August, $> 99\%$ for 1995 March from the F-test) over the fits with fixed column density. The fitted parameters are: column density $N_{\rm H} = 5.06^{+0.93}_{-0.70} \times 10^{21}$ cm^{-2}, photon index $\Gamma = 1.65^{+0.12}_{-0.08}$, $F_{1\ \rm keV} = 0.57 \pm 0.08$ μJy for the 1994 August observation, and $N_{\rm H} = 5.57^{+0.41}_{-0.39} \times 10^{21}$ cm^{-2}, $\Gamma = 1.79 \pm 0.05$, $F_{1\ \rm keV} = 2.26 \pm 0.16$ μJy for the 1995 March observation (errors at 90% confidence for one interesting parameter).

The fitted column density is in excess of the Galactic value at $\sim$ 95% confidence for the August 1994 and $> 99\%$ for the March 1995 observations (S96), confirming earlier reports of excess X-ray absorption based on ROSAT observations (Zhang et al. 1994; Mukherjee et al. 1996). Due the limited statistics of the ASCA data below 1 keV, it was not possible to constrain the location of the absorber. That is, there is no statistical preference among local ($z = 0$), rest-frame ($z = 2.07$), or intervening absorbers (see S96 for a discussion).

The fitted ASCA photon index is in good agreement with the average slopes of Highly Polarized/Optically Violently Variable Quasars at medium-hard energies measured with GINGA and EXOSAT (Ohashi et al. 1989; Sambruna et al. 1994). While the ASCA flux increased by a factor 4 from 1994 August to 1995 March, no significant change of the spectral slope is present.

2.2. Optical and γ-ray observations

The optical data were obtained at the 2.2-m telescope of the Calar Alto observatory, Spain, for nine nights overlapping the ASCA pointing of 1994 August 28. No variations down to the 6% level were found. Absolute calibrations were carried out during two nights of photometric quality. The large airmass during the observations of PKS 0528+134 introduced a large error. We derived an average brightness of $R = 20.8 \pm 0.3$ mag. Assuming $A_R = 2.4$, from the X-ray column density measured in 1994 August with ASCA, we derive a de-reddened optical flux of 0.13 ± 0.04 mJy.

PKS 0528+134 was observed with EGRET during viewing period VP337 (1994 August 9–29). The source was in the lowest state so far observed, with a flux in the range 100 MeV–30 GeV $(3.2 \pm 1.0) \times 10^{-7}$ ph cm^{-2} s^{-1}. The fitted photon index is $\Gamma = 2.68 \pm 0.44$.

3. The spectral energy distribution of PKS 0528+134

In Figure 1 we plot the spectral energy distribution from radio through γ-rays of PKS 0528+134. Apparent luminosities, obtained assuming isotropic emission, are plotted versus rest-frame frequencies, i.e., $(1+z)\nu_{\rm obs}$. The simultaneous ASCA, EGRET, and optical data of 1994 August are plotted as filled points and hatched areas. Contemporaneous (within 1 month) radio and millimeter flux densities were obtained from the on-line Michigan University database and from Mukherjee et al. (1996).

The radio through optical/UV emission has a peak somewhere in the region between the millimeter and the optical data, where unfortunately there is no data coverage at present. A parabolic interpolation yields a peak frequency $\nu_p = 1.1 \times 10^{13}$ Hz, peak luminosity $\nu_p L_p = 4.4 \times 10^{47}$ ergs s^{-1}, and bolometric luminosity $L_{\rm bol} = 2.4 \times 10^{48}$ ergs s^{-1} integrated in 10^{9-16} Hz. At higher energies, the ASCA and EGRET data form a second spectral component with a sharp peak at $\sim 10^{22}$ Hz. A simple integration of the area below the ASCA and EGRET data from 10^{17} to 10^{25} Hz yields a bolometric luminosity $\approx 10^{49}$ ergs s^{-1}. This shows that in PKS 0528+134 the power emitted at the higher energies is larger than at lower frequencies, even during a low state.

In Figure 1 we also plot the multifrequency spectrum obtained during the γ-ray flare of 1993 March (open symbols), from Mukherjee et al. (1996). The γ-ray flux decreased by an order of magnitude from 1993 March to 1994 August; variability with larger amplitude is present in the optical.

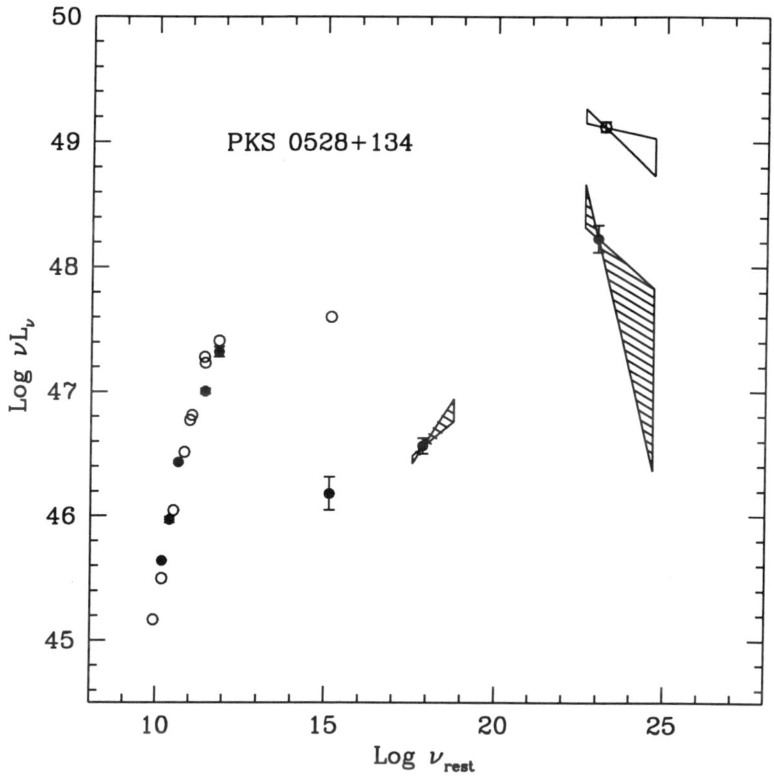

Figure 1. Spectral energy distribution of PKS 0528+134 from radio to γ-rays of 1994 August (filled circles and shaded areas). Optical, X-ray, and γ-ray data are simultaneous; the radio and millimeter data are within 1 month of the ASCA observation. Open symbols represent the multifrequency spectrum taken during the γ-ray flare of 1993 March (data from Mukherjee et al. 1996). The EGRET flux decreased by a factor 10 from 1993 March to 1994 August. Variability with larger amplitude occurred in the optical.

4. Discussion

There is now strong evidence that the smooth, highly variable, and polarized radio through optical/UV continuum of blazars is synchrotron emission from plasma in relativistic motion. The emission at higher energies is attributed to inverse Compton scattering of the same synchrotron photons (synchrotron-self Compton, SSC) or/and of seed photon external to the synchrotron-emitting region (external Compton, EC).

Irrespective of the source of the seed photons, however, in the observed spectra there must be a close correspondence between the synchrotron and Compton fluxes due to the fact that the same relativistic electrons radiate through both mechanisms (Maraschi 1996). In the context of the homogeneous model, if ν_S is the peak of the synchrotron component, there must be a second peak in the spectrum corresponding to the upscattered synchrotron photons at $\nu_C \propto \gamma_b^2 \nu_S$ in the SSC model and $\propto \gamma_b^2 \Gamma \nu_{ext}$ in the EC model, where γ_b is the electron energy and Γ and $\nu_{\rm ext}$ are the electron Lorentz factor and the frequency of the external photons, respectively. The variability of the synchrotron and Compton fluxes must be strongly correlated. Specifically, since the variability of the synchrotron spectrum is more pronounced above the peak, we expect large variability above the peak of the Compton emission.

In applying these arguments to the spectral energy distribution of PKS 0528+134, we identify $\nu_S = 1.2 \times 10^{13}$ Hz and $\nu_C = 3.7 \times 10^{21}$ Hz. The correspondence in the spectrum thus implies that the GeV photons are produced by the same electrons which emit synchrotron IR/optical photons, while the X-ray emission is linked to the submm/mm part of the synchrotron component. Since the EGRET data are above the peak of the emission, larger flux and spectral variability at MeV energies than in X-rays is expected, as is indeed observed. The large variability in the optical is also expected, since these data lie well above the synchrotron peak, which likely is at IR/far IR energies.

PKS 0528+134 has a spectral energy distribution qualitatively similar to 3C279. In both objects the synchrotron peak is located between the mm and IR regions, and the bulk of the bolometric power is emitted at MeV energies. In 3C 279 like in PKS 0528+134, large flux changes are observed at IR/optical energies together with simultaneous changes of the MeV flux (Maraschi et al. 1994; Hartman et al. 1996). An analogous behavior, but shifted forward by 2–3 decades, is exhibited by the X-ray-selected BL Lac Mrk 421. Simultaneous monitoring of Mrk 421 from radio through TeV frequencies showed large correlated variability in the X-ray and TeV bands, with little or absent variability from radio to UV and at MeV energies (Takahashi 1996). In Mrk 421, $\nu_S \sim 10^{16-17}$ Hz and ν_C is between the GeV and TeV region; again, as predicted by the synchrotron plus inverse Compton scenario, the variability above the synchrotron peak in the X-rays finds a correspondent variability in the region above the Compton peak.

Although there is a qualitatively similar behavior, different cooling mechanisms may be at work in different blazar classes (Ghisellini & Maraschi 1993). In BL Lacs like Mrk 421 the lower γ-ray-to-optical flux ratio and the absence of strong emission lines suggests that the SSC is the dominant cooling mechanism of the electrons up to TeV energies (Takahashi 1996). In quasar-like blazars, where strong optical/UV emission lines and excess blue emission ("blue-bump") are usually present, the external Compton scattering may play a more important role in the production of γ-rays. The latter may be the case in PKS 0528+134, as suggested by the large γ-ray-to-optical flux ratio in Figure 1 and by the presence of a prominent emission line in the optical spectrum (Hunter et al. 1993). See S96 for a detailed discussion.

References

Aller, H. D., et al. 1985, ApJS, 59, 513

Fugmann, W., & Meisenheimer, K. 1988, A&AS, 76, 145

Hartman, R., et al. 1996, ApJ, submitted

Hunter, S. D., et al. 1993, ApJ, 409, 134

Ghisellini, G., & Maraschi, L. 1993, in Proc. of the 2nd Compton Symp., AIP 304, ed. C. E. Fichtel, N. Gehrels, & J. P. Norris, p. 616

Maraschi, L. 1996, these proceedings

Maraschi, L., et al. 1994, ApJ, 435, L91

McNaron-Brown, K., et al. 1995, ApJ, 451, 575

Mukherjee, R., et al. 1996, ApJ, submitted

Ohashi, T., et al. 1989, in Proc. 23rd ESLAB Symp., Bologna, Italy (ESA SP-296, Nov. 1989)

Perley, R. A. 1982, AJ, 87, 859

Pohl, M., Reich, W., & Schlickeiser, R. 1995, A&A, 303, 383

Sambruna, R. M., et al. 1996, ApJ, submitted (S96)

Sambruna, R. M., Urry, C. M., Ghisellini, G., & Maraschi, L. 1995, ApJ, 449, 567

Sambruna, R. M., Barr, P., Giommi, P., Maraschi, L., Tagliaferri, G., & Treves, A. 1994, ApJ, 434, 468

Takahashi, T. 1996, these proceedings

Thompson, D. J., et al. 1995, ApJS, 101, 259

von Montigny, C., et al. 1995, ApJ, 440, 525

Wall, J. V., & Peacock, J. A. 1985, MNRAS, 216, 173

Zhang, Y. F., et al. 1994, ApJ, 432, 91

Blazar Continuum Variability
ASP Conference Series, Vol. 110, 1996
H. R. Miller, J. R. Webb, and J. C. Noble, eds.

High Energy Gamma Radiation from PKS 0528+134 Observed by EGRET

R. Mukherjee[1]

NASA/GSFC (USRA), Code 661, Greenbelt, MD 20771

W. K. Gear

Royal Observatory, Blackford Hill, Edinburgh EH9 3HJ, UK

A. P. Marscher, E. M. Moore,[2] J. P. Travis, and Y. F. Zhang[3]

Department of Astronomy, Boston University, Boston, MA 02215

E. I. Robson and J. A. Stevens

Joint Astronomy Center, University Park, Hilo, HI 96720

H. Teräsranta and M. Tornikoski

Metsahovi Radio Research Station, SF-02540 Kylmala, Finland

S. J. Wagner

Landessternwarte Königstuhl, 69117 Heidelberg, Germany

Abstract. A summary of EGRET observations of PKS 0528+134 is given, with special emphasis on the 1993 March observations when the source flared in gamma-rays. The flux history of PKS 0528+134 as well as spectra of the source are presented. Multiwavelength observations of PKS 0528+134 during the flare are also given. A detailed relativistic SSC jet model agrees well with the 1993 March multiwavelength spectrum. However, the data are insufficient to discriminate between this and other emission models.

1. Introduction

High energy gamma-rays from PKS 0528+134 were first detected by EGRET during the early pointings of CGRO from April to June, 1991 (Hunter et al. 1993). PKS 0528+134 is one of the brightest active galactic nuclei (AGN) de-

[1]For the EGRET Team

[2]Current address: Dept. of Phys. & Astr., Rutgers Univ., Piscataway, NJ 08854

[3]Current address: Center for Astrophysics, 60 Garden Street, Cambridge MA 02128

tected by EGRET. It is located near the Galactic anticenter at $l = 191.37°$, $b = -11.01°$ ($\alpha = 5^h28^m6.759^s$, $\delta = +13°29'42.19''$ J2000). It has a redshift of $z = 2.07$ (Hunter et al. 1993), and a mean optical brightness of $m_v = 20$. Since the exposures in mid 1991, EGRET has made several subsequent observations of the anticenter region. In 1993, March 23–29, PKS 0528+134 was seen to flare and gamma-rays were detected at a level approximately three times greater than the observed intensity during the early part of the GRO mission (Sreekumar et al. 1993). In this paper we report on these observations of PKS 0528+134.

2. Gamma-ray observations

2.1. Flux history

Table 1. EGRET observations of PKS 0528+134

Viewing Period	Observation Dates	Flux $\times 10^{-7}$ ph cm^{-2} s^{-1}	Significance σ	Inclination Angle of Source (deg)
0.2-0.5	1991 Apr 22–May 07	12.9 ± 0.9	20.3	8.0
1.0	1991 May 16–30	8.5 ± 0.8	13.5	6.3
2.1	1991 Jun 08–15	3.6 ± 1.1	3.8	5.1
36.0+36.5	1992 Aug 11–20	2.3 ± 1.5	1.8	22.0
39.0	1992 Sep 01–17	3.2 ± 1.4	2.6	23.9
213.0	1993 Mar 23–29	30.8 ± 3.5	13.6	9.1
221.0	1993 May 13–24	2.3 ± 1.2	2.2	6.4
310.0	1993 Dec 01–13	$< 4.0^a$	1.2	15.7
321.1+321.5	1994 Feb 08–17	4.9 ± 1.2	5.0	12.9
337.0	1994 Aug 09–29	3.2 ± 1.0	3.6	13.5
412.0	1995 Feb 28–Mar 07	GI data		13.1
413.0	1995 Mar 07–21	GI data		7.7
419.1	1995 Apr 04–11	GI data		17.4
419.5	1995 May 09–23	GI data		20.9
420.0	1995 May 23–Jun 06	GI data		9.8
426.0	1995 Aug 08–22	GI data		8.5
502.0	1995 Oct 17–31	5.7 ± 8.4^b	8.5	0.86

[a] 2σ upper limit
[b] Marscher et al. 1996

Table 1 lists the dates of observation and viewing periods during which PKS 0528+134 was within 30° of the EGRET instrument axis. The integrated flux above 100 MeV as well as the significance of detection of PKS 0528+134 in each viewing period (VP) is also given in Table 1. Consecutive VPs with similar pointing directions (0.2, 0.3, 0.4 and 0.5; 36.0 and 36.5; 321.1 and 321.5) were added together in order to improve the significance of detection. Figure 1 shows the history of the gamma-ray flux of PKS 0528+134 above 100 MeV until 1995 October. The individual points correspond to the average flux for the VPs listed in Table 1. The highest flux from PKS 0528+134 was observed during 23–29 March, 1993, in VP 213.0. During the other observations of the source made in 1992 and 1993, the source was detected in the 'low' state (Nolan et al. 1993). Over the period covered by the EGRET observations, the high energy flux from PKS 0528+134 varied over the range (2.3 ± 1.2 to 30.8 ± 3.5) $\times 10^{-7}$

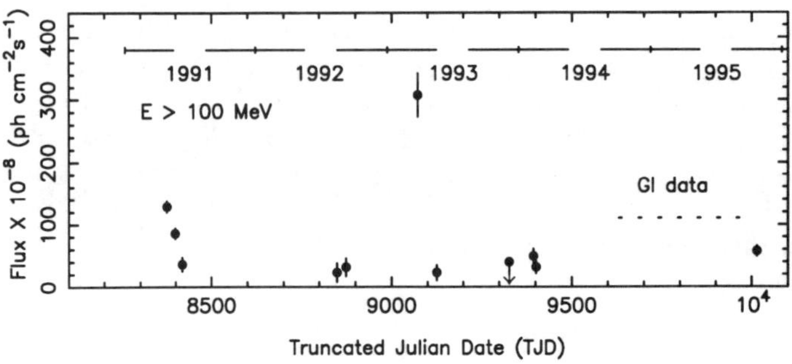

Figure 1. Flux of gamma-rays above 100 MeV from PKS 0528+134 over the period 1991 April to 1995 October (Phases 1 through 5). The Phase 5 data point is from Marscher et al. (1996). 2σ upper limits are shown as downward arrows.

photons cm^2 s^{-1}. A χ^2 test of the data yields a probability of $< 0.1\%$ that the fluctuations are consistent with a constant flux.

2.2. Spectrum

Figure 2 shows the photon spectra of PKS 0528+134 as observed during VP 213.0, when the source was in the 'high' state, and during VP 1.0. The best fit using a single power law model of the form $F(E) = k\left(E/E_0\right)^{-\alpha}$ photons cm^{-2} s^{-1} MeV^{-1} is shown as superimposed solid lines in the figure. Here α is the photon spectral index, and k is the coefficient. The energy normalization factor, E_0, is chosen so that the statistical errors in the power law index and the overall normalization were minimally correlated.

Table 2. EGRET spectral analysis

Viewing Period	Spectral Index	k $\times 10^{-9}$ ph cm^{-2} s^{-1} MeV^{-1}	E_0 MeV	χ^2/n_f of Source
0.2-0.5	2.27 ± 0.07	3.26 ± 0.19	199	1.53
1.0	2.48 ± 0.09	3.33 ± 0.26	172	0.83
213.0	2.21 ± 0.10	8.39 ± 0.77	198	0.78
321.1+321.5	2.45 ± 0.23	2.85 ± 0.46	160	0.70
337[a]	2.68 ± 0.44	2.15 ± 0.57	135	0.52

[a]Sambruna et al. 1996

The best fit parameters for the spectra of PKS 0528+134 in VPs 0.2–0.5, 1.0, 213.0, 321.1+321.5, and 337 are indicated in Table 2. The spectrum of PKS 0528+134 in VP 1.0 is found to be steeper than that in VP 213.0. The results on the spectral indices suggest that there is marginal correlation between the spectral index and the measured flux, with a tendency towards spectral hard-

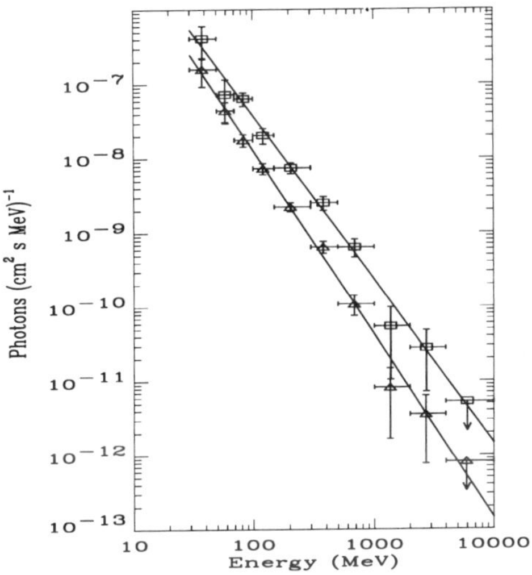

Figure 2. Photon spectra of PKS 0528+134 as observed during VP 1.0 (triangle) and VP 213.0 (square), when the source was in an outburst. The solid lines are the best fits to a single power law model.

ening when the flux from the source is higher. The linear correlation coefficient between the flux and the spectral index, measured for those VPs when the detection significance of PKS 0528+134 was $> 6\sigma$, is -0.87, and the probability that the two variables are not correlated is 13%. A χ^2 test indicates that a slope of zero can be ruled out with a probability of $\sim 93\%$ (Mukherjee et al. 1996).

3. Radio light curves

Figure 3 shows the light curves of PKS 0528+134 measured at several different radio frequencies. The flux density measurement at 10.55 GHz was made with the Effelsberg 100-m telescope over a period of 3 years (Pohl et al. 1995). Data at 22 GHz and 37 GHz were obtained at the Metsahovi Radio Research Station, Finland. The radio light curves at 90 GHz and 230 GHz were measured with the Swedish-ESO Submillimeter Telescope (SEST). The epoch of the EGRET 'high' state (VP 213.0) is shown as a broken line. From the figure it is seen that at the time of the gamma-ray flare in VP 213.0 (1993 March), the flux at 32 GHz was at a high level of about 7 Jy. In mid 1993, following the gamma-ray flare, the flux densitiy at 32 GHz reached a maximum of 9 Jy. The flux density at 10.55 GHz remained approximately constant at a high level until the end of 1993. The gamma-ray flare occurred during a time when all except the lowest radio frequencies were rising, which is claimed to be a common occurrence by Valtaoja and Teräsranta (1995). Based on the data presented here, however, we cannot make any conclusive comments on the correlation of the gamma-ray and radio data.

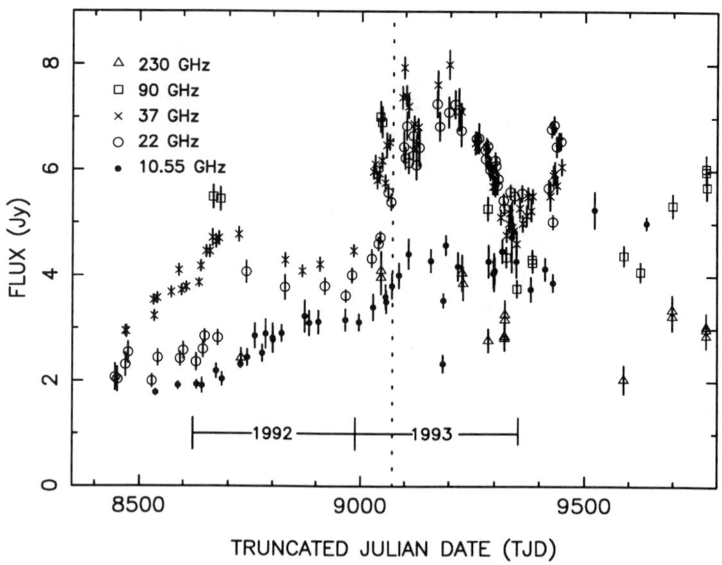

Figure 3. Flux from PKS 0528+134 at radio frequencies (10.55 GHz, 22 GHz, 37 GHz, 90 GHz, and 230 GHz). The broken line corresponds to the period of the EGRET observation (VP 213.0) when the source was in a flaring state.

4. Multiwaveband spectrum

The power per log frequency interval emitted by PKS 0528+134 is shown in Figure 4. Two sets of observations are shown in the figure: those that were simultaneous with the EGRET flare (VP 213.0) are shown as solid circles, while data taken during the gamma-ray 'low' state of PKS 0528+134 (VP 39.0) are depicted as open squares. Details of the individual observations shown in the figure may be found in Mukherjee et al. (1996). The multiwavelength spectrum of the source during the March 1993 flare was fit with the relativistic jet synchrotron self-Compton spectrum model of Marscher and Travis (1996). The superimposed solid lines in Figure 4 correspond to the synchrotron and first-order self-Compton emission from an adiabatic relativistic jet with a truncated cone geometry. The model is found to reproduce the data well. Details of the fit parameters, as well as a discussion of models suggested to explain the production of high energy gamma-rays from quasars is given in Mukherjee et al. (1996). On the basis of the spectrum alone of PKS 0528+134, it is not possible to distinguish between the different models of the production of high energy gamma-rays from quasars.

Acknowledgments. The EGRET team gratefully acknowledges support from the following: Bundesministerium für Forschung und Technologie, Grant 50 QV 9095 (MPE); NASA Grant NAG5-1742 (HSC); NASA Grant NAG5-1605

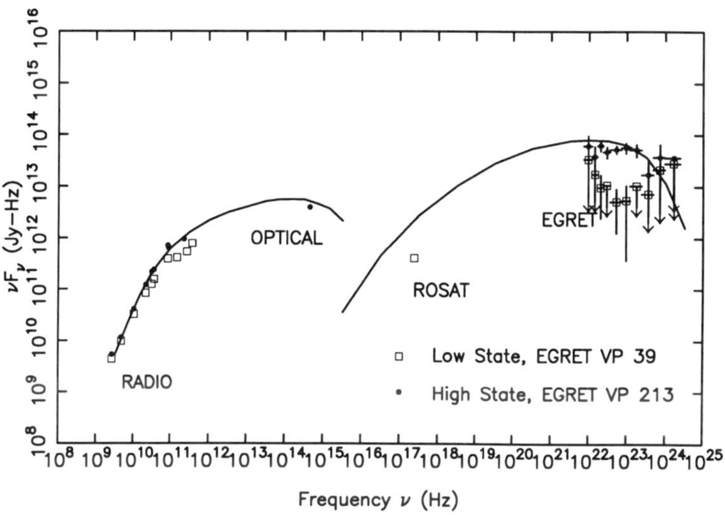

Figure 4. Multiwavelength spectrum of PKS 0528+134 during the 1993 March flare along with a model fit synchrotron self-Compton spectrum from a relativistic jet. EGRET upper limits are shown as downward arrows.

(SU) and NASA Contract NAS5-31210 (GAC). The JCMT is operated jointly by the UK Particle Physics and Astronomy Research Council, the Netherlands Organization for the Advancement of Pure Research (ZWO), the Canadian National Research Council (NRC), and the University of Hawaii.

References

Hunter, S. D., et al. 1993, ApJ, 409, 134
Marscher, A. P. & Travis, J.P. 1996, A&A, in press
Marscher, A. P., et al. 1996, in preparation
Mukherjee, R., et al. 1996, ApJ, submitted
Nolan, P. L., et al. 1993, IAU circular 5802
Pohl, M., et al. 1995, A&A, in press
Sambruna, R., et al. 1996, these proceedings
Sreekumar, P., et al. 1993, IAU circular 5753
Valtaoja, E., & Teräsranta, H. 1995, A&A, 297, L13

Blazar Continuum Variability
ASP Conference Series, Vol. 110, 1996
H. R. Miller, J. R. Webb, and J. C. Noble, eds.

An Intense Gamma-ray Flare of PKS 1622−297

J. R. Mattox

Astronomy Dept., University Of Maryland, College Park, MD 20742 and Space Astronomy Program, Universities Space Research Association

S. J. Wagner

Landessternwarte Heidelberg–Königstuhl, Königstuhl, D-69117 Heidelberg, Germany

Abstract. We have observed a spectacular flare of PKS 1622−297 with the *Compton Gamma Ray Observatory.* A peak flux of $(21 \pm 7) \times 10^{-6}$ cm^{-2} s^{-1} (E > 100 MeV) was observed. The corresponding isotropic luminosity is 3.6×10^{49} erg/s. We find that PKS 1622−297 exhibits γ-ray intra-day variability. A flux increase by a factor of 7 was observed to occur in less than 7.1 hours (with 99% confidence). Assuming an exponential rise, the corresponding doubling time is less than 2.5 hours. A flux decrease by a factor of 2 in less than 6.1 hours is similarly obtained.

1. Introduction

The Energetic Gamma Ray Experiment Telescope (EGRET) aboard the *Compton Gamma Ray Observatory* is sensitive in the energy range 30 MeV to 30 GeV (Thompson et al. 1993). It has detected ~50 AGN (Montigny et al. 1995, Thompson et al. 1995) in the blazar class (by which we mean the ensemble of BL Lac objects, high polarization quasars (HPQ), and optical violently variable (OVV) quasars). Mattox et al. (1996a) have developed a method for an objective assessment of the reliability of the identification of EGRET high-energy γ-ray sources with radio counterparts using deep 5 GHz radio catalogs.

The absence of pair-production absorption in the EGRET spectra and the fact that only sources which show parsec scale radio jet structure appear to be identified as EGRET sources indicate that the hard γ-rays are emitted in a relativistic jet directed toward us. A plethora of models for the γ-ray emission have appeared. See Urry et al. (1996) for a review. Most models feature inverse Compton scattering as the γ-ray emission mechanism, but there is not a consensus as to the origin of the low energy photons which are scattered. It has been suggested that the low energy photons might originate within the jet as synchrotron emission (Maraschi, Ghisellini, & Celotti 1992; Bloom & Marscher 1993). This is designated as the synchrotron self-Compton (SSC) process. Another possibility is that the low energy photons come from outside of the jet. This is designated as the external Compton scattering (ECS) process. Dermer, Schlickeiser, and Mastichiadis (1992) suggested that they come directly from an accretion disk around a blackhole at the base of the jet. It was subsequently proposed that the dominant source of the low energy photons for scattering could be re-processing of disk emission by broad emission line clouds (Sikora, Begelman, & Rees 1994; Blandford & Levinson 1995, Levinson & Blandford 1995; Levinson 1996). Ghisellini and Madau (1996) suggest that the dominant source of low energy photons for scattering is broad-line-region re-processing of jet synchrotron emission. Hartman et al. (1996) find that the multiwavelength spectra of 3C 279 can be adequately fit with either a SSC model or an ECS model in both the high and low states.

1.1. The pedigree of PKS 1622−297

PKS 1622−297 has not received much attention previously (probably because it is located in the Galactic center region, $l = 348.82°$, $b = 13.32°$). It is not cataloged by Hewitt and Burbidge (1987, 1989). No optical polarization measurement, nor search for rapid optical variability has been previously reported. However, the radio properties indicate that it belongs to the blazar class. A 5 GHz flux density of 1.92 Jy and a spectral index of $\alpha = +0.07$ ($S(\nu) \propto \nu^{\alpha}$) were reported by Kühr et al. (1981). It was optically identified by Torres and Wroblewski (1984) at 21 mag. A red shift of $z = 0.815$ is reported in the PKS catalog (Wright & Otrupcek 1990). Steppe et al. (1993) report 90 GHz flux densities at three epochs of 1.5, 1.8, and 2.0 Jy; and one 230 GHz observation at a flux density of 1.0 Jy. Preston et al. (1985) report a VLBI correlated flux density at 2.29 GHz of 0.29 Jy, 13% of the total. PKS 1622−297 was detected by ROSAT during the sky survey at a flux of $3.2 \pm 0.8 \times 10^{-13}$ erg cm^{-2}s^{-1} in the 0.1–2.4 keV energy band (Voges 1996).

2. The observation of PKS 1622−297 with the Compton Gamma Ray Observatory

PKS 1622−297 was deeply exposed by EGRET during the sky survey. A likelihood analysis (Mattox et al. 1996b) of the sum of EGRET exposure for the first half of the mission (4/22/91–10/04/94, a total exposure of 1.4×10^{9} cm^2 s) yields a 95% confidence upper limit of 0.10×10^{-6}cm^{-2}s^{-1} (E > 100 MeV). It was much brighter during our 1995 observation.

2.1. The EGRET light curve

The cycle 4 observations have been analyzed to obtain an EGRET light curve for the event energy selection E > 100 MeV. The exposure was binned according to the quality of the EGRET sensitivity and the strength of the emission to obtain as much time resolution as possible given the statistical limitations imposed by the very sparse EGRET data. The EGRET sensitivity to PKS 1622−297 was limited because it was observed at a substantial off-axis angle (except during VP 423.5) and the response of EGRET at the time was about half of what it originally was due primarily to degradation of the spark-chamber gas. Maps of counts and exposure were constructed for adjoining time intervals. A likelihood analysis (Mattox et al. 1996b) of these maps gave flux estimates.

The EGRET light curve of PKS 1622−297 is shown in Figure 1. The flux is observed to vary dramatically during this observation. A major flare occurred at MJD 49894 (MJD ≡ JD − 2400000.5). The maximum EGRET flux observed

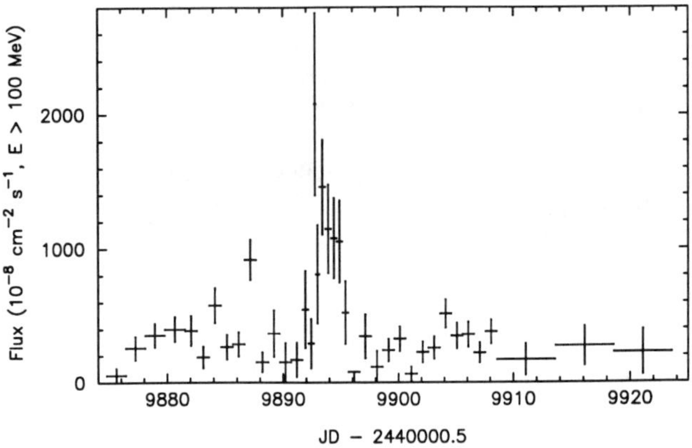

Figure 1. The EGRET light curve for PKS 1622−297. The horizontal bars show the extent of the time bins. The vertical bars indicate the ranges of the 68% confidence flux estimates.

from PKS 1622−297 ($21 \pm 7 \times 10^{-6}$cm^{-2}s^{-1}, E $>$ 100 MeV) is the largest ever seen from any EGRET source (other than γ-ray bursts and solar flares). It is a factor of 5 larger than the flux of 3C 279 at the peak of the 1991 flare (Kniffen et al. 1993) and a factor of 2 larger than the flux of 3C 279 at the peak of the 1996 flare (Wehrle et al. 1996).

The determination of the minimum time scale of variability of the high-energy γ-ray flux of blazars is not straight forward because of the sparse EGRET statistics. The available information is summarized in the time history of cumulative counts shown in Fig. 2. In addition to the cumulative counts, this figure shows the cumulative counts history expected for an invariant flux at the average value. Mattox et al. (1996c) establish that the onset of the MJD 49894 flare was very rapid. A Kolmogorov-Smirnov test, applied to the 7.1-hour interval (MJD 49892.517–49892.812) shown in Fig. 2, indicates a flux increase with a significance of 9×10^{-3}. Assuming an exponential rise by a factor of 7 in less than 7.1 hours, the doubling time is less than 2.5 hours with 99% confidence. Mattox et al. (1996c) show that potential trials do not significantly dilute the significance of this result. Fig. 2 also shows the interval (MJD 49887.593–49887.996) for which Mattox et al. (1996c) establish with 99.9% confidence that the flux decreased in less than 9.7 hours by a factor of 3. Assuming an exponential decrease, the corresponding time for a flux decrease by a factor of two is less than 6.1 hours.

3. The spectrum of PKS 1622−297

We have obtained simultaneous detections by BATSE, OSSE, COMPTEL, and EGRET during a target of opportunity pointing during VP 423.5. A multiwavelength spectrum is shown in Fig. 3. When the OSSE and EGRET data are fit simultaneously with a single power law, a photon spectral index of -1.87 ± 0.02 is obtained. The fit is not excellent, but acceptable with a reduced $\chi^2 = 1.14$. For comparison, the reduced χ^2 for the total OSSE and EGRET data for the

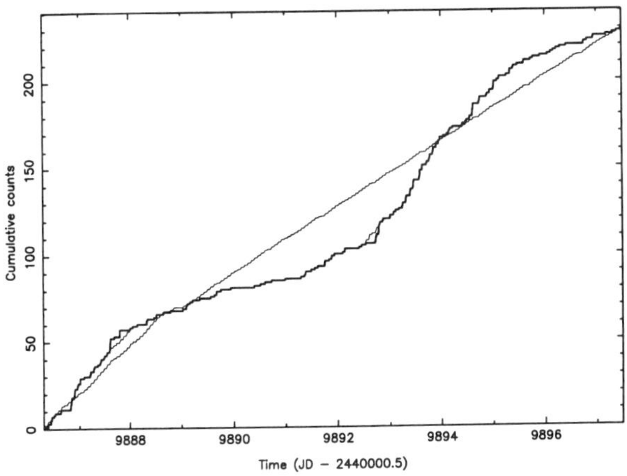

Figure 2. The time history of cumulative EGRET counts for PKS 1622−297 is shown with the bold line. Events with E > 100 MeV were selected from an energy dependent cone which includes 68% of the PSF. The cumulative count time history expected for an invariant flux for the entire interval is shown with the continuous thin line. This is an integral of the EGRET effective area which was determined through an examination of the effective area of each trigger mode for each orbit. The steps are due to Earth occultation during ~1/2 of each 90 minute orbit. Perturbations due to SAA passages and the occasional unavailability of TDRSS satellites for telemetry downlink are also apparent. The two short thin lines show the time histories of invariant flux expected for shorter intervals corresponding to KS tests described in the text.

two separate power-law fits described above is 0.86. The F-test indicates that the separate fits do not offer a significant (0.14) improvement. The overall GRO spectrum suggests that there is a break at ~10 MeV to a harder X-ray spectrum, as reported for 4 of the 5 blazars previously detected by both EGRET and OSSE (McNaron-Brown et al. 1996). However, the lowest energy OSSE points show a hard spectral index which contradicts this hypothesis. It is unfortunate that the anti-solar position of PKS 1622−297 at the time of the γ-ray flare prevented simultaneous X-ray observation by either ROSAT or ASCA, and the *Rossi* XTE satellite had not yet been launched. A measurement of the intensity and spectrum of the X-ray emission would have been very helpful for understanding this source.

4. Multiwavelength observations of PKS 1622−297

The large flux of PKS 1622−297 was apparent to the EGRET team during a routine "quicklook" analysis of the VP 421 data. The news of the detection was conveyed to us on 6/14/95. This triggered our ToO program. The detection was announced in an IAU Circular (Mattox et al. 1995). We also notified directly a group of multiwavelength observers who had been previously enlisted.

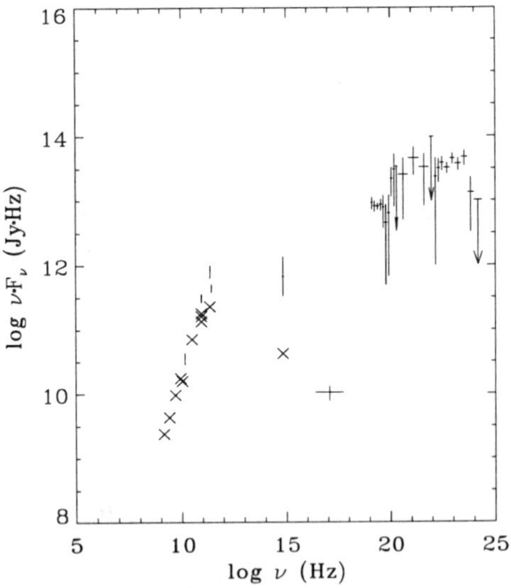

Figure 3. The spectral energy distribution of PKS 1622−297. The gamma-ray data is from the OSSE, COMPTEL, and EGRET instruments aboard GRO during VP 423.5 is shown with crosses. The vertical bars indicate the 1σ statistical uncertainties of the flux estimates. The X's are quiescent flux densities. The crosses in the optical and radio ranges are contemporaneous measurements. The vertical bars in this case indicate the range of variability. The X-ray cross is from the ROSAT sky survey and the vertical bar indicates statistical uncertainty.

Fortuitously, this followed the first week of a 5-week observation of the Galactic center region which included PKS 1622−297, and this source was opposite the sun. Detailed multiwavelength results, including multiwavelength spectra, will be published latter.

Extensively optical monitoring of PKS 1622−297 commenced when we learned of the EGRET detection. We found that PKS 1622−297 was 3 magnitudes brighter than its quiescent state, and that the brightness was observed to vary by as much as 150% in less than 24 hours. Thus, PKS 1622−297 displays optical intra-day variability (IDV) as do all other blazars detected by EGRET which have been observed frequently enough for IDV to be detected (Wagner 1996). Several radio observations in the mm band occurred which showed a flux density significantly higher than previously observed. It was also detected by IUE (Bonnell et al. 1995) during observations on July 1, 5, 6, 8, and found to be variable. We will also report subsequently on a possible correlation between the EGRET and OSSE flux in VP 423.5. Unfortunately, due to bad weather in Europe, and poor communication with collaborating observers in South America, optical observations did not occur during the major γ-ray flare at MJD 49894. In collaboration with Alan Marscher, we made a VLBA observation of PKS 1622−297 on July 25th. Subsequent observations are in progress which will provide for a search for a new VLBI component ejected at the time of the gamma-ray flare, and a measurement of its velocity if it becomes apparent.

5. Discussion

The maximum EGRET flux of $21 \pm 7 \times 10^{-6}\text{cm}^{-2}\text{s}^{-1}$, E > 100 MeV corresponds to a peak energy flux of 2×10^{-8} ergs $\text{cm}^{-2}\text{s}^{-1}$ for the observed energy range 30 MeV to 10 GeV (assuming a power law spectrum with photon index of −1.9). The corresponding isotropic luminosity is 3×10^{49} erg/s assuming a Friedmann universe with $q_0 = 1/2$ and $H_0 = 75$ km $\text{s}^{-1}\text{Mpc}^{-1}$. For this luminosity to be less than the standard Eddington limit, the central black hole mass must exceed $2\times10^{11}\ M_\odot$. However, Dermer and Gehrels (1995) point out that the standard Eddington limit does not pertain to the γ-ray luminosity because the Compton scattering cross-section in the Klein-Nishina regime is smaller than the Thomson cross-section. Using their expression for the cross-section in the Klein-Nishina regime for an EGRET flux of $70 \times 10^{-6}\text{cm}^{-2}\text{s}^{-1}$ (E > 30 MeV), the lower limit for the mass of the PKS 1622−297 black hole is $8\times10^{8}\ M_\odot$. The corresponding limit on the Schwarzschild radius is $R_g > 2.5\times10^{14}$ cm.

We apply the "Elliot-Shapiro argument" (Elliot & Shapiro 1974; Dermer & Gehrels 1995) to establish that the γ-ray emission must occur in a relativistic jet by showing that a contradiction follows otherwise. Under the assumption that the γ-ray emission is unbeamed, the observed upper limit on the doubling time of $\tau = 2.5$ hours implies that the extent of the emission region is less than $c\tau/(1+z) = 1.5\times10^{14}$ cm. Because this upper limit is less than lower limit for R_g, the γ-ray emission must occur in a medium which is moving relativistically with an extent less than $c\tau\delta/(1+z) = 1.5\times10^{15}(\delta/10)$ cm, where $\delta \equiv \gamma^{-1}(1 - \beta\cos\theta)^{-1}$ is the relativistic Doppler factor, θ is the angle between the jet axis and the line of sight, and $\gamma \equiv (1-\beta^2)^{-1/2}$, where β is the bulk velocity in units of c. This argument assumes that the accretion of optically thin material is the immediate source of power for the γ-rays. It is possible that this is not the case. Accretion energy could be stored in the rotation of a black hole, or in magnetic fields to power the γ-ray flare; or the γ-ray flare could result from the accretion of a star. Also, for a Kerr blackhole, emission could occur as close as $R_g/2$.

Also, if the emission does not take place in a relativistic jet, the γ-ray emission would be absorbed by γ-ray/X-ray pair production. Mattox et al. (1996c) find that the optical depth would be

$$\tau = 500\left(\frac{E_\gamma}{GeV}\right)^{0.7}$$

assuming that the ROSAT sky survey flux for PKS 1622−297 was present at the time of the flare, and was produced in the same location as the γ-rays. No indication of absorption at high energy is apparent in the spectrum of Figure 3. A lower limit for the Doppler factor (for $h_{75} = 1$, and $\tau < 1$ for $E_\gamma = 3$ GeV) is $\delta \geq 8.7$. We note that the X-ray observation was not simultaneous. If it were a factor of 3 lower during the γ-ray flare, the lower limit on δ would decrease to 7.1. However, examination of Figure 3 in the context of the proposed models suggests that the X-ray flux could have been a factor of ~100 larger during the γ-ray flare. This would inply $\delta \geq 20$.

Acknowledgments. We are grateful to the EGRET team for the timely processing of telemetry and data to obtain the "quicklook" detection of PKS 1622−297. J. Mattox acknowledges support from NASA Grants NAG5-2833 and NAGW-4761.

References

Blandford, R. D., & Levinson, A. 1995, ApJ, 441, 79

Bloom, S. D., & Marscher, A. P. 1993, in Proceedings of the Compton GRO Symposium, AIP Conf. Proc. #280, p. 578

Bonnell, J. T., et al. 1995, IAU circular 6186

Dermer, C. D., Schlickeiser, R., & Mastichiadis, A. 1992, A&A 256, L27

Dermer, C. D., & Gehrels, N. 1995, ApJ, 447, 103; and erratum, ApJ, 456, 412

Elliot J. L., & Shapiro, S. L. 1974, ApJ, 192, L3

Ghisellini, G., & Madau, P., 1996, MNRAS, in press

Hartman, R. C., et al. 1996, Ap J, in press

Hewitt, A., & Burbidge, G. 1987, ApJS, 63, 1

Hewitt, A., & Burbidge, G. 1989, ApJS, 69, 1

Kniffen, D. A., et al. 1993, ApJ, 411, 133

Kühr, H., Witzel, A., Pauliny-Toth, I. I. K., & Nauber, A. 1981, A&AS, 45, 367

Levinson, A., & Blandford, R. D. 1995, ApJ, 449, 86

Levinson, A. 1996, ApJ, in press

McNaron-Brown, K., et al. 1995, ApJ, 451, 575

Maraschi, L., Ghisellini, G., & Celotti, A. 1992, ApJ 397, L5

Mattox, J. R., Wagner, S., McGlynn, T. A., Malkan, M., & Schachter, J. F. 1995, IAU circular 6179

Mattox, J. R., et al. 1996a, ApJ, in preparation

Mattox, J. R., et al. 1996b, ApJ, 461, 396

Mattox, J. R., et al. 1996c, ApJ, submitted

von Montigny, C., et al. 1995, ApJ, 440, 525

Nolan, P. L., et al. 1993, ApJ, 409,697

Preston R. A., et al. 1985, AJ, 90.1599

Sikora, M., Begelman, M. C., & Rees, M. J. 1994, ApJ, 421, 153

Steppe, H., et al. 1993, A&AS, 102, 611

Torres, C., & Wroblewski, H. 1984, A&A, 141, 271; and erratum, A&A, 160, 406

Thompson, D. J., et al. 1993, ApJS, 86, 629

Thompson, D. J., et al. 1995, ApJS, 101, 259

Urry, M. 1996, in preparation

Voges, W. 1996, private communication

Wagner, S. 1996, A&AS, in press

Wehrle, A., et al. 1996, IAU circular 6294

Wright, A., & Otrupcek, R. 1990, PKSCAT90 catalog

Blazar Continuum Variability
ASP Conference Series, Vol. 110, 1996
H. R. Miller, J. R. Webb, and J. C. Noble, eds.

On the Soft γ-ray Variability of 3C273

G. E. Romero[1]

Instituto Argentino de Radioastronomía, CC5, (1894) Villa Elisa, Argentina

Laboratorio de Física Teórica, UNLP, CC67, (1900) La Plata, Argentina

Abstract. Observations with the COMPTEL instrument on the Compton Gamma Ray Observatory (CGRO) have shown evidence of γ-ray variability in the superluminal quasar 3C273 in the $0.75 - 8.0$ MeV range during 1991 (Williams et al. 1995a). Simultaneous or quasi-simultaneous cm–mm–IR outbursts have been reported by several authors. In addition, VLBI observations have revealed the formation of a new superluminal component in the source by the same epoch. In the present paper I outline a shocked beam model in which the interaction of a relativistic shock identified with the new VLBI component strikes a particle density inhomogeneity in the pc-scale pair beam of the object, producing an enhancement of the blue-shifted annihilation radiation. The lower frequency light curves can be understood in terms of synchrotron emission of the shocked particles and emission from the shock precursor. Several physical parameters of the inner source are estimated in this context.

1. Introduction

The COMPTEL instrument on the CGRO has revealed the existence of several blazar-type objects with very strong emission in the $1 - 10$ MeV energy range (Bloemen et al. 1995; Blom et al. 1995a, b; Williams et al. 1995a, b). Among these blazars, the well-known superluminal quasar 3C273 presents particular interest. This source has been observed during two different viewing periods by the COMPTEL detector in 1991. The first observations were performed in June 1991 (Hermsen et al. 1993), whilst the second observing session was carried out in October of the same year (Williams et al. 1995a). Strong emission was detected in June over the $0.75 - 8$ MeV range. The photon flux in this energy range reached $259 \pm 62 \times 10^{-6}$ ph cm^{-2} s^{-1}, with a peak near ~ 2 MeV. The energy spectrum can be fitted by a power law E^{-p}, with $p = 2.4$. The absolute isotropic luminosity in the mentioned energy band was $\sim 5 \times 10^{46}$ erg s^{-1}. This luminosity decreased at least $\sim 26\%$ in the October observing period. This means a variation $\Delta L_{\gamma}(0.75 - 8.0) \sim 1.3 \times 10^{46}$ erg s^{-1} over a timescale of $t_{\rm v} \sim 9.7 \times 10^{6}$ s.

[1] Fellow of CONICET

By the same epoch, 3C273 showed a remarkable increase in its flux density at all radio frequencies above 10.5 GHz (Steppe 1992; Reich et al. 1993). Robson et al. (1993) have presented detailed light curves in the IR–mm–cm range for the period 1987–1991. Strong variability is observed in all light curves during 1991. The curves at 3.3, 8.1 and 13.5 mm are peaked around August 1991. These observations are particularly important because during 1989 the source was in a relatively long period of inactivity which allows a determination of its quiescent spectrum. For this spectrum, the mm spectral index is found to be 0.86 ± 0.02, whilst the near IR has a slope of 1.71 ± 0.02. A spectral break can be appreciated at a frequency $\nu_b \sim 10^{14}$ Hz. At lower frequencies, the quiescent synchrotron turnover frequency ν_m and the corresponding flux density S_m are $\nu_m \approx 70$ GHz and $S_m \approx 25$ Jy, respectively (Unwin et al. 1985).

VLBI observations of 3C273 at different frequencies have shown several compact superluminal knots moving away from the core (e.g., Unwin et al. 1985; Zensus et al. 1988, 1990; Krichbaum et al. 1990; Abraham et al. 1994). At least ten superluminal components, labelled from C1 to C10 according to the identification epoch, seem to have been successively ejected from the nucleus. The younger superluminal emission knot (C10) was detected in 1990.17 at a distance of 0.3 ± 1 mas from the core (Abraham et al. 1994). The same knot appeared at 0.75 ± 0.25 mas in 1991.15, implying an apparent velocity $\beta_{app} \sim 4$ (in units of c).

The superluminal component C10 can be identified with the post-shock region of a relativistic shock front which is propagating along the initial pc-scale beam of the object. In the general framework of the two-fluid model of jets, this shock can also be responsible for both the MeV and radio variability of the source when interacts with density inhomogeneities produced by Kelvin-Helmholtz instabilities developed in the interface between the relativistic pair beam and the nonrelativistic $e^- - p$ jet of the object. I shall briefly discuss this point in the next section.

2. Model outlines

In the two-fluid model of extragalactic radio jets originally proposed by Sol et al. (1989) a relativistic $e^+ - e^-$ beam moves throughout a channel in a nonrelativistic jet of $e^- - p$ plasma wind originated in the surface of the central accretion disk. The jet is responsible for the formation of the hot spots in the extended radio sources because it carries most of the mass and kinetic energy ejected by the AGN. Conversely, the pair beam produces observable effects at just pc-scale distances from the core. Kelvin-Helmholtz instabilities are expected to develop in the contact surface between both fluids. Romero (1995) has presented a linearized analysis of the fluid dynamical equations for a cylindrical pair beam in pressure equilibrium with its confining jet, taken into account the effects of the axial magnetic fields in the beam. The derived dispersion relation can be used to study the stability of conical beams because density and temperature gradients do not affect the local relation (Hardee 1982, 1986). Changes in the behavior of the flow are introduced by the modification of the physical parameters along the beam. The unstable modes can be characterized by two integer numbers $(n,\ l)$, usually called the fundamental and reflection mode numbers. Modes with structure ($n = 0$, l) are axisymmetric and produce a "pinching" effect on the particle density of the beam. Modes with ($n = 1$, l) are nonaxisymmetric and lead to a helical twisted beam. Just these two sets of modes can originate observable alterations in the beam structure.

Axial magnetic fields tend to stabilize the pair beam. At short wavelengths and density ratios in the $10^{-1} - 10^{-3}$ range, all modes are stable provided that the flow be sub-Alfvénic. The critical magnetic field is:

$$B_{\rm c}^{\rm KH} = [4\pi n_{\rm b} m_{\rm e} c^2(\gamma_{\rm b}^2 - 1)]^{1/2}\gamma_{\rm b}^{-1}, \tag{1}$$

where $n_{\rm b}$ is the particle density in the beam, $m_{\rm e}$ is the electron rest mass, and $\gamma_{\rm b}$ is the bulk Lorentz factor of the beam flow. Instabilities will develop when $B_{\rm b} < B_{\rm c}^{\rm KH}$. In the hypersonic regime (Mach number $M_{\rm b} > 5$) the instability is dominated by the ($n = 0$, $l = 1$) pinching mode. A change in the Mach number with increasing distance to the origin of the beam may cause a dominance of the helical modes when the beam enters in the transonic regime ($M_{\rm b} \sim 2 - 5$). This could explain why superluminal components in 3C273 tend to move ballistically when they emerge from the core and then display helical patterns in their trajectories (Abraham et al. 1994, 1996).

If a relativistic shock propagates through the beam, it will produce an enhancement in the blue-shifted $e^- - e^+$ annihilation line when strickes a density inhomogeneity. Two components can be distinguished in the total γ-ray luminosity originated in pair annihilations: one due to processes occurring in the quiescent beam, and other produced in the shocked region. Since the particle density is increased in the shocked gas (see Blandford and McKee 1976 for the relativistic jump conditions at the shock front), the second component can contribute with a significant part of the total luminosity. Time variability is introduced by the variations in the particle density arisen in the Kelvin-Helmholtz instabilities. The observed luminosity is:

$$L_{\gamma,{\rm obs}}^{\rm (ann)}(t) = \delta_{\rm b}^{3+\alpha_{\rm b}} L_{\gamma,{\rm b}}^{\rm (ann)} + \delta_{\rm ps}^{3+\alpha_{\rm s}} L_{\gamma,{\rm s}}^{\rm (ann)}(t), \tag{2}$$

where $\delta_{\rm b}$ and $\delta_{\rm ps}$ are the Doppler factors of the beam and the post-shock region, respectively, and $\alpha = (p-1)/2$, if the particles (e^- and e^+) have a power law energy distribution $f_\pm(E_\pm) \propto E_\pm^{-p}$. The second term on the right side of Eq. (2) introduces the variability. The intrinsic luminosity is:

$$L_\gamma^{\rm (ann)} = \int (E_+ + E_-)\, R_\pm\, f(E_-) f(E_+)\, dE_+ dE_- dV. \tag{3}$$

Here $R_\pm$ is the annihilation rate given by Coppi and Blandford (1990). The magnetic field in the beam decreases as (Marscher 1980):

$$B_{\rm b} = B_0 \left(\frac{R_{\rm b}}{R_{\rm b_0}}\right)^{-s}, \qquad s = 2. \tag{4}$$

The beam radius $R_{\rm b}$ can be estimated if the magnetic field and the beam radius at the particle injection point are known (Marscher & Gear 1985). Thus, using values obtained from VLBI observations for the kinematic parameters of the shocked region and simultaneous or quasi-simultaneous COMPTEL temporal and spectral data, it is possible to obtain an estimate for the lower limit of the pair density in the stricken feature of the beam (i.e., where $B_b \sim B_c^{KH}$). A summary of the input and derived parameters of the model in the case of 3C273 is presented in Table 1.

Table 1. Summary of the main assigned and derived values in the model for 3C273

Parameter	Symbol	Value
Beam Lorentz factor	γ_b	2
Beam Mach number	M_b	5.5
Beam magnetic field	B_b	0.5 G
Beam number density	n_b	1.4×10^4 cm^{-3}
Beam radius	R_b	0.2 pc
Compression ratio	q	8.2
Distance from the beam apex to the shock	z	14.3 pc
Doppler post-shock factor	δ_{ps}	4.5
Energy power law index	p	2.4
Linear size of inhomogeneities	l	1.7 pc
Superluminal velocity	β_{app}	4
Opening angle	ϕ	0.8°
Post-shock Lorentz factor	γ_{ps}	4.1
Quiescent turnover frequency	ν_m	70 MHz
Quiescent flux density at ν_m	S_m	25 Jy
γ-ray enhancement (0.75–8.0 MeV)	$\Delta L_{\gamma,obs}$	1.3×10^{46} erg s^{-1}
γ-ray luminosity from the shock (0.75–8.0 MeV)	$L_{\gamma,s}^{ann}$	5×10^{43} erg s^{-1}
Redshift	z_s	0.158
Shock Lorentz factor	γ_s	5.8
Thickness of the shocked region	Δz	0.5 pc
Timescale	t_v	$\sim 9.7 \times 10^6$ s
Viewing angle	θ	12.8°

3. Discussion

The same population of $e^- - e^+$ pairs that produces a flare in the emission at MeV energies of a shocked beam can be responsible for simultaneous or quasi-simultaneous outbursts at lower frequencies. The particles will emit synchrotron radiation in the magnified magnetic field of the post-shock region and a flare will appear at all optically thin frequencies. The basic evolution has been described by Marscher and Gear (1985) and Marscher (1990). Firstly, Compton losses are severe. X-ray and hard γ-ray emission are produced at this stage (Königl 1981). Synchrotron losses soon become dominant and the spectrum shifts towards lower frequencies (i.e., optical–mm). Finally, at the so-called adiabatic-loss stage, the flare extends to centimeter wavelengths and the spectrum flattens ($S_\nu \propto \nu^{(p-1)/2}$).

This is, of course, an oversimplified scenario. Gallant et al. (1992) have shown that irrespective of the energy fraction of incoming flow that is radiated as incoherent synchrotron emission in the post-shock region, a substantial fraction of the upstream energy can be converted into radio emission at the shock in the form of a radiative coherent precursor (for coherent processes in blazars see Benford 1992). This precursor has been found in numerical simulations of relativistic shock propagation through pair plasmas (Gallant et al. 1992, Hoshino et al. 1992).

The precursor originates in collective emission of gyrating charge bunches created by MHD instabilities at the shock front. If a parameter σ is introduced by:

$$\sigma \equiv \frac{B_{\rm b}^2}{4\pi\gamma_{\rm b}^2 n_{\rm b} m_{\rm e} c^2} \tag{5}$$

in order to quantify the MHD effects in the beam, the numerical simulations show that, for a $0.003 \leq \sigma \leq 0.3$ range, the precursor flux can reach $\sim 4\%$ of the total incoming energy (in the shock frame). In this way, it is possible that a radio flare appears quasi-simultaneously with the MeV-flare. However, the radio flare will evolve with larger timescales owing to its later states will roughly follow the adiabatic states described by the shocked jet model.

4. Conclusions

Annihilations in a pair beam can produce a strong emission in the form of soft γ-rays as anticipated by Henri et al. (1993). The very recently reported variability in MeV blazars can arise when strong relativistic shocks strike features in the beam. These features are naturally expected as a consequence of the developing of Kelvin-Helmholtz instabilities. The reports of radio–mm–IR bursts associated to MeV variability tend to support the idea of intervening shocks. VLBI observations of some sources, like 3C273, provide additional support because they exhibit superluminal knots moving away from the core. These knots can be understood as the expanded post-shock region of relativistic shock fronts.

Acknowledgments. It is a pleasure to thank P. Benaglia and A. P. Marscher for valuable comments. Travel support granted by UNLP is gratefully acknowledged. This work has been partially supported by CONICET and the University of La Plata.

References

Abraham, Z., Carrara, E. A., Zensus, J. A., & Unwin, S. C. 1994, in Compact Extragalactic Radio Sources, ed. J. A. Zensus & K. I. Kellerman (Green Bank, NRAO), p. 87

Abraham, Z., Carrara, E. A., Zensus, J. A., & Unwin, S. C. 1996, A&A, in press

Benford, G. 1992, ApJ, 391, L59

Blandford, R. D., & McKee, C. F. 1976, Phys. Fluids, 19, 1130

Bloemen, H., et al. 1995, A&A, 293, L1

Blom, J. J., et al. 1995a, A&A, 295, 330

Blom, J. J., et al. 1995b, A&A, 298, L33

Coppi, P. S., & Blandford, R. D. 1990, MNRAS, 245, 453

Gallant, Y. A., et al. 1992, ApJ, 391, 73

Hardee, P. E., 1982, ApJ, 257, 509

Hardee, P. E., 1986, ApJ, 303, 111

Henri, G., Pelletier, G., & Roland, J. 1993, ApJ, 404, L41
Hermsen, W., et al. 1993, A&ASS, 97, 97
Hoshino, M., Arons, J., Gallant, Y. A., & Langdon, A. B. 1992, ApJ, 390, 354
Königl, A. 1981, ApJ, 243, 700
Krichbaum, T. P., et al. 1990, A&A, 237, 3
Marscher, A. P., 1980, ApJ, 235, 386
Marscher, A. P., & Gear W.K. 1985, ApJ, 298, 114
Marscher, A. P., 1990 in Parsec-Scale Radio Jets, ed. J. A. Zensus & T. J. Pearson (Cambridge, Cambridge University Press), p. 236
Reich, W., et al. 1993, A&A, 273, 65
Robson, E. I., et al. 1993, MNRAS, 262, 249
Romero, G. E. 1995, Ap&SS, 234, 49
Sol, H., Pelletier, G., & Asséo, E. 1989, MNRAS, 237, 441
Steppe, H. 1992, A&A, 259, 61
Unwin, S. C., et al. 1985, ApJ, 289, 109
Williams, et al. 1995a, A&A, 298, 33
Williams, et al. 1995b, A&A, 297, L21
Zensus, J. A., Bååth, L. B., & Cohen, M. H. 1988, Nature, 334, 410
Zensus, J. A., Unwin, S. C., Cohen, M. H., & Biretta, J. A. 1990, AJ, 100, 1777

Blazar Continuum Variability
ASP Conference Series, Vol. 110, 1996
H. R. Miller, J. R. Webb, and J. C. Noble, eds.

Differences between X-ray-selected and Radio-selected BL Lacertae Objects in EGRET Observations

Y. C. Lin, P. F. Michelson, P. L. Nolan, and T. D. Willis

W. W. Hansen Experimental Physics Laboratory and Department of Physics, Stanford University, Stanford, CA 94305

D. L. Bertsch, B. L. Dingus, J. A. Esposito, C. E. Fichtel, R. C. Hartman, S. D. Hunter, C. von Montigny[1], R. Mukherjee, and D. J. Thompson

NASA/Goddard Space Flight Center, Code 661, Greenbelt, MD 20771

G. Kanbach and H. A. Mayer-Hasselwander

Max-Plank-Institut für Extraterrestrische Physik, D-85748 Garching b. München, Germany

D. A. Kniffen

Hampden-Sydney College, P. O. Box 862, Hampden-Sydney, VA 23943

E. J. Schneid

Northrop Grumman Corporation, Mail Stop A01-26, Bethpage, NY 11714

Abstract. In the course of the Compton Observatory mission, the EGRET telescope has detected 16 BL Lac objects in high-energy gamma rays through Phase 4. According to the BL Lac sample in Ciliegi, Bassani, and Caroli (1995), 3 of these 16 sources are X-ray-selected (XBL), 11 are radio-selected (RBL), and 2 are not listed in this sample. An examination of the X-ray fluxes and radio fluxes for those BL Lac objects that are detected by EGRET reveals the fact that, while the radio fluxes of the EGRET-detected BL Lac's, either XBL or RBL, are at least moderately strong, there does not seem to be a similar correlation between the high-energy gamma-ray fluxes and the X-ray fluxes for these sources. Furthermore, for the BL Lac objects listed in Ciliegi et al. (1995) that are not detected by EGRET, co-adding of the EGRET data for the undetected RBL, in a way that the source positions coincide with each other, yields a flux excess at a 3-σ significance level. But a similar co-adding of the EGRET data for the undetected XBL yields no flux excess.

[1]Present address: Landessternwarte Heidelberg-Königstuhl, 69117 Heidelberg, Germany

1. Introduction

BL Lacertae Objects form a class of rare objects among Active Galactic Nuclei. Hewitt and Burbidge (1993) designates 90 sources in their catalog as BL Lac objects. Véron-Cetty and Véron (1993) lists 171 BL Lac objects. The most recent catalog of BL Lac objects of Padovani and Giommi (1995) lists 233 objects. These numbers are rather small compared to the number of known quasars (e.g., Hewitt & Burbidge 1993) or flat-spectrum radio sources (Fichtel et al. 1994 and references there in). Yet in the EGRET-detected AGN, all blazars so far as we know, BL Lac objects constitute about 25% of the detected sources (Fichtel et al. 1994; Thompson et al. 1995). This high success rate for the detection of BL Lac objects in high-energy gamma rays, especially those sources that are nearby, led to a detailed study of the probability that a close BL Lac object can be seen by EGRET (Lin et al. 1996). In the study of Lin et al. (1996), it was found that the BL Lac objects that are listed in Hewitt and Burbidge (1993) with $z \leq 0.2$ have a probability of 25% to be detected by EGRET, and for those with $z \leq 0.2$ that are not detected by EGRET, the 2-σ upper limit in the photon flux for E > 100 MeV was found to be 0.17×10^{-7} cm^{-2} s^{-1}, calculated from the sum EGRET data of Phase 1 and Phase 2.

In this paper we carry out the analysis one step further: We want to see if there is any systematic difference between the X-ray-selected BL Lac objects (XBL) and the radio-selected BL Lac objects (RBL) in the EGRET data.

2. The sample

Again we use Hewitt and Burbidge (1993) to select BL Lac objects. But for the classification as XBL or RBL, we further use the BL Lac sample in Ciliegi et al. (1995).

For $z < 0.3$, the BL Lac samples in Hewitt and Burbidge (1993) and in Ciliegi et al. (1995) overlap with each other to a large extent. Out of the 31 BL Lac's in Hewitt and Burbidge (1993) with $z < 0.3$, 26 can be found in Ciliegi et al. (1995) for information of XBL or RBL. We list these 31 sources in Table 1 according to the redshifts. The redshifts and the V-magnitudes in Table 1 are taken from Hewitt and Burbidge (1993). The EGRET fluxes can be found in Thompson et al. (1995) except 2155−304 which is published in Vestrand et al. (1995). The fluxes quoted here are the ones that were first found at the time of their detection. The X-ray fluxes are taken from Ciliegi et al. (1995), and the radio fluxes are further found in Véron-Cetty and Véron (1993). The letter "X" or "R" in parentheses indicate whether the source is regarded as XBL or RBL in Ciliegi et al. (1995).

For $z > 0.3$ or no measured redshift value, we can only use the sources IN Ciliegi et al. (1995) because most of the BL Lac's with $z > 0.3$ or with no known redshifts in Hewitt and Burbidge (1993) have no information on whether they are XBL or RBL. In Table 2, we list the BL Lac objects with $z > 0.3$ or with no known redshifts that are included in the sample of Celiegi et al. (1995) according to the source designations. (Two of the redshift values in Table 1 of Ciliegi et al. (1995) are incorrect: the redshift for 0829+046 should be 0.18 and the redshift for 1E 1402+0416 should be 0.200. See NED database for further information.

Table 1. BL Lacertae objects in Hewitt and Burbidge (1993) with redshifts below 0.3

Object	Other name	z (gal)	V	EGRET flux[a]	X-ray flux[b]	Radio flux[c]
1722 + 119	4U	0.018?	16.6		1.15(X)	0.095
1101 + 384	MRK 421	0.031	13.82	1.5 ± 0.4	3.67(X)	0.696
1652 + 398	MRK 501	0.034	14.44		4.10(X)	1.313
1133 + 704	MRK 180	0.046	14.49		1.04(X)	0.295
1514 − 241	AP LIB	0.049	15		0.13	1.94(R)
1807 + 698	3C 371	0.050	14.22		0.23	1.74(R)
0521 − 365	PKS	0.055	14.5	2.1 ± 0.6	0.46	8.89(R)
1727 + 502	OT 546	0.055	16.70		0.89(X)	0.17
0548 − 322	PKS	0.069	15.5		2.77(X)	0.23
2200 + 420	BL LAC	0.069	15.14	4.0 ± 1.2	0.27	2.96(R)
2005 − 489	PKS	0.071	15.3	1.8 ± 0.5	2.59(X)	1.19
1219 + 285	W COM	0.102	16.5	0.8 ± 0.2	0.05	0.72(R)
2155 − 304	PKS	0.117	13.58	2.7 ± 0.7[d]	3.36(X)	0.31
1426 + 428	HEAO	0.129	16.45		4.41(X)	0.038
1218 + 304	RS 4	0.130	16.45		2.41(X)	
0323 + 022	H	0.147	17.48		0.89(X)	0.075
1418 + 546	OQ 530	0.152	15.91		0.12	1.09(R)
1229 + 645	1E	0.170	16.89			0.042
0829 + 046	OJ 049	0.180	16.66	1.9 ± 0.6	0.09	0.70(R)
1101 − 232	HEAO	0.186	16.55			
0317 + 185	1E	0.190	18.12		0.80(X)	0.017
2254 + 074	OY 091	0.190	17.03		0.07	0.48(R)
1011 + 496		0.200	16.15			0.286
1402 + 042	1E	0.200	17.5		(X)	0.018
1215 + 303	ON 325	0.237	15.73		0.51	0.42(R)
1415 + 259	1E	0.237	15.73		(X)	0.054
1400 + 162	4C 16.39	0.244	16.37		0.07	0.528(R)
1413 + 135	PKS	0.260	20.0		0.01	1.2(R)
0836 + 182		0.280	16.8			0.325
1704 + 607	1E	0.280	19.4			0.002
0414 + 009	1H	0.287	17.59		1.99(X)	0.083

[a] for E > 100 MeV, in 10^{-7} cm^{-2} s^{-1}

[b] at 3 keV, in 10^{-29} erg cm^{-2} s^{-1} Hz^{-1}, quoted from Ciliegi et al. (1995)

[c] at 5 GHz, in Jy, quoted from Véron-Cetty & Véron (1993)

[d] Vestrand et al. (1995)

Table 2. BL Lacertae objects in Ciliegi et al. (1995) with redshifts greater than 0.3 or no value

Object	Other name	z	V[a]	EGRET flux[b]	X-ray flux[c]	Radio flux[d]
$0219+428$	3C 66A	0.444	15.58	1.9 ± 0.5	0.07	1.04(R)
$0235+164$	AO	0.940	15.5	8.3 ± 0.9	0.15	2.79(R)
$0537-441$	PKS	0.894	16.48	3.6 ± 0.9	0.09	3.96(R)
$0556-3838$	EXO		17.1[e]		2.28(X)	0.064
$0716+714$	S5		15.5	2.2 ± 0.5	0.02	1.12(R)
$0735+178$	PKS	0.424	15.10	1.3 ± 0.4	1.06	1.99(R)
$0808+019$	PKS		17.5		0.14	0.67(R)
$0851+202$	OJ 287	0.306	14	Detected[f]	0.46	2.61(R)
$0912+297$	OK 222		16		0.23	0.20(R)
$0954+658$[a]	S4	0.368	16.7	1.4 ± 0.4		1.46
$1034-293$	PKS	0.311	16.46		0.19	(R)
$1538+149$	4C 14.60	0.605	18.31		0.08	1.95(R)
$1604+159$[a]	MC 3	0.357	18.7	4.0 ± 1.2		0.50
$1717+178$	OT 129		18.5		0.08	0.94(R)
$1749+096$	OT 081	0.322	17.88		0.34	(R)
$1803+784$	S5	0.684	17		0.12	(R)
$2007+777$	S5	0.342	16.5		0.04	1.28(R)
$2032+107$	MC 3	0.601	18.6	1.4 ± 0.4		0.77(R)

[a]Hewitt & Burbidge (1993)

[b]for $E > 100$ MeV, in 10^{-7} cm^{-2} s^{-1}

[c]at 3 keV, in 10^{-29} erg cm^{-2} s^{-1} Hz^{-1}, quoted from Ciliegi et al. (1995)

[d]at 5 GHz, in Jy, quoted from Véron-Cetty & Véron (1993)

[e]Padovani & Giommi (1995)

[f]Shrader et al. (1996)

One source in this table, IZW 187, also known as OT 546, is mistakenly listed as IZW 186. See NED for clarification.) The V-magnitudes are taken from Hewitt and Burbidge (1993). The EGRET fluxes, the X-ray fluxes, and the radio fluxes are listed in the same way as Table 1 above. Tables 1 and 2 form the basis for the analysis of this paper.

3. Analysis and results

We examine the sources that are detected by EGRET and those that are not detected by EGRET as listed in Tables 1 and 2 separately.

3.1. EGRET-detected BL Lac objects

Out of the 16 BL Lac objects that are detected by EGRET, as indicated in Tables 1 and 2, 14 have information of XBL or RBL in Ciliegi et al. (1995). Three of these 14 sources are regarded as XBL, and 11 as RBL. The average radio flux of the 3 XBL at 5 GHz is 0.73 Jy, the lowest value being 0.31 Jy, compared with an average value of 2.50 Jy for the 11 RBL, according to the radio fluxes quoted in Vèron-Cetty and Vèron (1993). But the average X-ray flux of the 11 RBL at 3 keV, in the units of 10^{-29} erg cm^{-2} s^{-1} Hz^{-1}, is 0.27, the lowest value being 0.02 and with 5 values below 0.1, compared with an average value of 3.21 for the 3 XBL, as quoted in Ciliegi et al. (1995). Thus while the radio fluxes of the EGRET detected BL Lac's, either XBL or RBL, are at least moderately strong, there does not seem to be a similar correlation between the high-energy gamma-ray fluxes and the X-ray fluxes for these sources.

3.2. BL Lac objects not detected by EGRET

For the BL Lac objects listed in Tables 1 and 2 that are not detected by EGRET, we do a co-adding of the EGRET data in a way that the source positions coincide with each other. We carry out this analysis separately for XBL and RBL. Not all the undetected sources in Tables 1 and 2 are used for this analysis. A BL Lac is not included if there is a known gamma-ray source either too close or too bright such that the subtraction of the known source leaves a large uncertainty in the residual flux in the region of the BL Lac position. If a BL Lac is situated too close the Galactic plane, that BL Lac is not included in the analysis either due to an excessively high contanmination of the Galactic diffuse emission in the sum data. In this way, we are left with 10 XBL and 12 RBL to do the co-adding of the EGRET data.

As in Lin et al. (1996), we only use Phase 1 and Phase 2 data for this analysis. What we find is this: When the data of the 12 RBL are co-added in a way that the source positions coincide with each other, the sum data show a clear flux excess above the background level at the source position. This flux excess becomes more pronounced when events of higher energies are selected. A flux excess can be obtained at 3-σ significance level, with the value of $0.3(1.0 \pm 0.4) \times 10^{-7}$ cm^{-2} s^{-1} for E > 100 MeV. But for the 10 RBL, co-adding of the EGRET data yields no flux excess at the source position. All we obtain is a flat background level extending all the way to the source region. We can obtain an average 2-σ flux upper limit as 0.3×10^{-7} cm^{-2} s^{-1}.

4. Conclusions

What we see in this paper is a clear distinction between the X-ray selected and the radio-selected BL Lac objects as far as high-energy gamma-ray emission is concerned. It seems that a high X-ray flux is not essential for the emission of high-energy gamma rays from a BL Lac object, at least to the EGRET detection limit. It is only the radio flux that seems to be always present in strength whenever a BL Lac is detected by EGRET.

We need to point out here that what we see as a correlation between the EGRET detection and the radio flux is only limited to the sense that strong radio fluxes are invariably associated with EGRET detections. Whether the values of the fluxes are also correlated is still an open question (Mücke et al. 1996).

The flux excess in the RBL sum data points to a potentially very important possibility. A 3-σ significance level is still not high enough to qualify for a true EGRET detection (Thompson et al. 1995). However, if this flux excexx can, be substnatiated under further studies, it indicates that substantially more BL Lac objects can be detected in single viewing periods if the detector were just a factor of 3 or 4 more sensitive. Any future high-energy gamma-ray missions will certainly provide detectors much more sensitive than EGRET. This paper provides direct evidence for great success in the field of AGN study for such missions.

Acknowledgments. The EGRET team gratefully acknowledges support from the following: Bundesministerium fur Forschung und Technologie, grant 50 QV 9095 (MPE); NASA grant NAG5-1742 (HSC); NASA grant NAG5-1605 (SU); and NASA contract NAAS 5-31210 (NGC). This work used the NED data retrieval system which is supported by the IPAC at California Institute of Technology. This work also used information from the SIMBAD data retrieval system at the University of Strasbourg, Strasbourg, France.

References

Ciliegi, P., Bassani, L., & Caroli, E. 1995, ApJ, 439, 80

Fichtel, C. E., et al. 1994, ApJS, 94, 551

Hewitt, A., & Burbidge, G. 1993, ApJS 87, 451

Lin, Y. C., et al. 1996, in Proc. The Third Compton Symp., June 12–14, 1995, München, submitted

Mücke, A., et al. 1996, A&A, submitted

Padovani, P., & Giommi, P. 1995, MNRAS, 277, 1477

Shrader, C. R., et al. 1996, in Proc. The Third Compton Symp., June 12–14, 1995, München, in press

Thompson, D. J., et al. 1995, ApJS, 101, 259

Vestrand, W. T., et al. (1995), ApJ, 454, L93

Véron-Cetty, M. -P., & Véron, P. 1993, ESO Sci. Report No. 13

Blazar Continuum Variability
ASP Conference Series, Vol. 110, 1996
H. R. Miller, J. R. Webb, and J. C. Noble, eds.

Low-state Gamma-ray Emission from Blazars and the Gamma-ray Background

Eric Perlman

Mail Code 660.2, Goddard Space Flight Center, Greenbelt, MD 20771

Demosthenes Kazanas

Mail Code 661, Goddard Space Flight Center, Greenbelt, MD 20771

Abstract. We examine the contribution of blazars to the Diffuse Gamma Ray Background (DGRB) by using their well-studied radio $\log N - \log S$. The basic assumption of our study is that all blazars are γ-ray emitters, with γ-ray luminosity directly proportional to radio luminosity, i.e., $(\nu F_\nu)_{100\mathrm{MeV}} = f \times (\nu F_\nu)_{5\mathrm{GHz}}$. For blazars to produce the DGRB, a mean ratio $\langle f \rangle \simeq 70$ is required, compared to a mean observed value of $\langle f \rangle_{\mathrm{obs}} \simeq 750$ seen among *EGRET*-detected blazars. This result implies that quiescent (i.e., low-state) γ-ray emissions at a level typically $\sim 0.1\times$ that found in a typical *EGRET* detection, must dominate the integrated γ-ray emissions from the ensemble of blazars at any time.

1. Introduction

The existence of diffuse, presumably extragalactic, γ-ray emission at energies > 30 MeV was demonstrated by SAS-2 (Fichtel et al. 1978; Thompson & Fichtel 1982) and has been confirmed by *EGRET* (Osborne et al. 1994, Sreekumar et al. 1996). Prior to the launch of *CGRO* several models were proposed to explain the DGRB, including matter-antimatter annihilation (Stecker et al. 1971), Active Galactic Nuclei (hereafter AGN) (Bignami et al. 1979; Kazanas & Protheroe 1983 and references therein) and annihilation of supersymmetric particles comprising the putative dark matter haloes of galaxies (Silk & Srednicki 1984). Of these, only the AGN model remains viable today (e.g., Stecker 1989).

To date, *EGRET* has discovered GeV γ-ray emissions from over 50 AGN (Thompson et al. 1995), all of which are blazars. Blazars are highly variable, radio loud AGN, quite often exhibiting superluminally moving components, with "flat" (i.e., $\alpha < 0.5$; $F_\nu \propto \nu^{-\alpha}$) spectra in the 1–100 GHz range. These observations have prompted a number of recent studies addressing the contribution of AGN to the DGRB (Padovani et al. 1993; Stecker, Salamon, & Malkan 1993; Salamon & Stecker 1994; Stecker & Salamon 1996).

Padovani et al. (1993) first showed a strong link between the radio and γ-ray emission in blazars. By assuming a constant ratio of γ-ray to radio emission, equal to the mean of the sixteen then detected sources, and integrating the Dunlop and Peacock (1990) luminosity function, they found that blazars should contribute $\simeq 70\%$ of the flux of the DGRB, leaving little room for contribution by another class of source.

This finding was disputed by Stecker et al. (1993), whose analysis included *EGRET* upper limits, leading them to estimate a ratio of radio to γ-ray luminosities a factor ~ 0.1 that assumed by Padovani et al. (1993), yielding a much smaller contribution by blazars to the DGRB. Both these studies, however, used as their basis the older extrapolation of the DGRB from SAS-2 data (Thompson & Fichtel 1982), and incomplete data from *EGRET* .

Salamon and Stecker (1994) used more recent *EGRET* observations and the luminosity function of Dunlop and Peacock (1990) to derive a γ-ray luminosity function. To account for the fact that most of the ~ 300 blazars with radio fluxes > 1 Jy have not been detected by *EGRET*, they assumed that the opening angle of the radio jet is larger than that of the γ-ray jet. Finally, Stecker and Salamon (1996) explored the contribution of a "quiescent," i.e., non-flaring, state to the DGRB using their previously derived γ-ray luminosity function. They find that if the average blazar γ-ray flux increases by a factor of $\simeq 5$ during a flare and if the average flare duty cycle is $\simeq 0.03$, the entire DGRB flux could be accounted for by the ensemble of blazars.

Here we investigate the contribution of blazars to the DGRB by using the radio $\log \mathrm{N} - \log \mathrm{S}$ of flat-spectrum radio sources (FSRS). This approach is simpler and subject to fewer assumptions than the use of the luminosity function. It also allows us to directly estimate the mean ratio of γ-ray to radio emission required in order for the class of blazars to produce the DGRB.

2. Approach and justification

The spectra of most *EGRET* -detected AGN are power-laws of slope $\sim E^{-2}$, similar to that of the DGRB (Sreekumar et al. 1996), over at least 2 decades in frequency. This observation guarantees that, if blazar emission is responsible for the DGRB, its observed spectrum is unlikely to be an envelope of breaks in the spectra of individual sources, redshifting across the *EGRET* bandpass. Thus if all blazars are γ-ray emitters, with γ-ray luminosity proportional to their radio luminosity, one can compute the unresolved γ-ray emission from flat spectrum radio sources to the DGRB by integrating their radio $\log \mathrm{N} - \log \mathrm{S}$ distribution, without any reference to their redshifts and hence their luminosity function. A particular advantage of this method is its simplicity; it does not involve any complicated inversions or threshold cuts and the associated pitfalls (e.g., Stecker et al. 1996).

We believe that this approach introduces less uncertainty than the use of the luminosity function. While luminosity functions are constructed from complete, flux-limited samples, they are subject to the assumption that all populations are adequately represented therein. For example, the Dunlop and Peacock (1990) luminosity function was constructed using samples complete to $\simeq 100$ mJy. However, as shown in §3, if the ensemble of blazars produces the DGRB, the largest contribution must be made by blazars of $F \lesssim 100$ mJy. It is also likely that the Dunlop and Peacock (1990) sample is not representative of the entire population of blazars at all redshifts. In particular, the Dunlop and Peacock (1990) sample contains few objects ($< 20\%$) at or below the knee of the BL Lac radio luminosity function (10^{26} W Hz^{-1}, Stickel et al. 1991), contains numerous objects which are neither flat-spectrum quasars or BL Lacs (e.g., 0500+019, which is a gigaHertz peaked-spectrum source), and has redshifts for only $\sim 50\%$. Further, one would expect a higher proportion of BL Lacs among blazars at lower radio fluxes since BL Lacs are less luminous in the radio than are quasar-type blazars. (the fraction at 1 Jy is $\sim 10\%$, Kühr et al. 1981, Stickel et al. 1991). In addition, measuring redshifts for BL Lacs (which have (if any) very faint emission or absorption lines) are difficult, further contributing to the uncertainty of the luminosity function method. The use of the γ-ray (Salamon & Stecker 1994), rather than the FSRS, luminosity function, should introduce even more uncertainty, considering the additional assumptions involved in its determination from the observed population of FSRS.

3. Results and discussion

In Figure 1, we show the FSRS $\log \mathrm{N} - \log \mathrm{S}$ at 4.86 GHz (generously provided to us by J. Condon). We have integrated this function numerically to obtain the total flux per steradian observed from this class of sources:

$$\begin{aligned} \int_0^\infty N(>S)\,dS &= 742.7 \text{ Jy sr}^{-1} \\ &= 3.61 \times 10^{-11} \text{ erg cm}^{-2} \text{ s}^{-1} \text{ sr}^{-1}. \end{aligned} \tag{1}$$

In Figure 2, we show the contribution of sources with fluxes up to S to the diffuse emission, i.e., the integral $\int_0^S N(>S)\,dS$ *vs.* the flux S.

The flux $F_{\rm DGRB}$ of the DGRB at 100 MeV is (Sreekumar et al. 1996)

$$\begin{aligned} F_{\rm DGRB}(100\rm MeV) &= 1.6 \text{ keV cm}^{-2} \text{ s}^{-1} \text{ sr}^{-1} \\ &= 2.56 \times 10^{-9} \text{ erg cm}^{-2} \text{ s}^{-1} \text{ sr}^{-1}. \end{aligned} \tag{2}$$

If our basic assumption—that all blazars are 30–3000 MeV γ-ray emitters—is correct, then the ratio of the above two fluxes should provide the mean value of the γ-ray to radio flux ratio for the typical blazar. This ratio is

$$f_B \equiv \langle \frac{(\nu F_\nu)_{\rm 100MeV}}{(\nu F_\nu)_{\rm 5GHz}} \rangle = \frac{F_{\rm DGRB}(100\rm MeV)}{\int_0^\infty N(>S)\,dS} = 71. \tag{3}$$

By comparison, the analyses of both Padovani et al. (1993) and Salamon and Stecker (1994) show that the mean observed value for *EGRET* detected AGN is $\langle P_{\rm 100MeV}/P_{\rm 5GHz}\rangle \simeq 10^{-9.8}$, where power is expressed in units of W Hz^{-1} sr^{-1}. In our units, this corresponds to a value of $\langle f\rangle_{\rm obs} = 760$.

The discrepancy between $f_{\rm obs}$ (the mean value observed in *EGRET* detections of blazars) and f_B (the mean value required for blazars to produce the DGRB) implies the existence of quiescent blazar γ-ray emission at $\sim 10\%$ of that observed in a typical *EGRET* detection. In order for blazars to produce the DGRB, this quiescent component must dominate the total flux emitted by the ensemble of blazars at any given time. A future instrument with 5–10 times

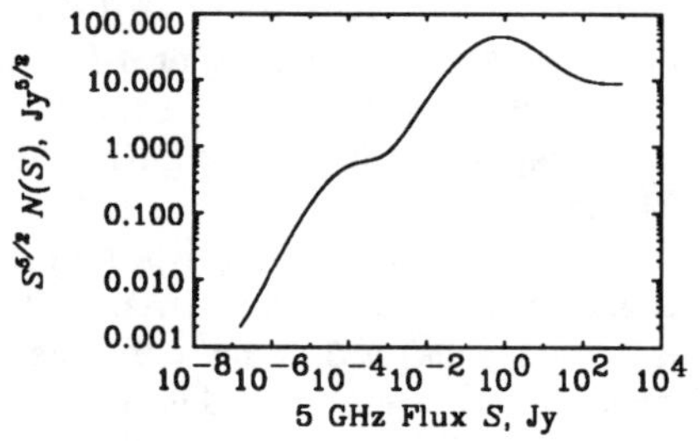

Figure 1. The $\log N(S) - \log S$ distribution for flat-spectrum radio sources at 4.86 GHz (from Condon 1984).

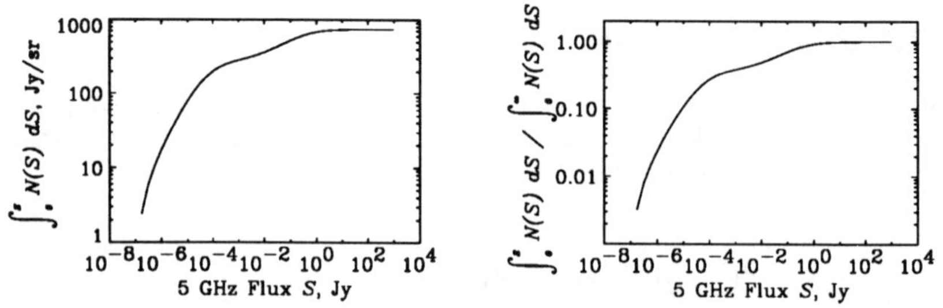

Figure 2. The integrated flux from the ensemble of flat-spectrum sources at 4.86 GHz. At left (Fig. 2a), the total flux; at right (Fig. 2b), as a fraction of the total flux.

the sensitivity of *EGRET* should detect the majority of the 1 Jy blazars in such a quiescent state. Failure to do so will imply that another population of objects must be invoked to produce the DGRB.

A direct implication of this result is that *EGRET* is detecting only those blazars which are brighter than average in γ-rays and which are (during the *EGRET* observation) in a "flaring" state. This provides a natural explanation for the fact that only $\sim$ 15% of the 300 FSRS with $F_r > 1$ Jy have been detected at significance $> 4\sigma$. Indeed, it can easily be shown that a 1 Jy blazar, emitting at $P_{100\text{MeV}} = f_B \times P_{5\text{GHz}}$ would not be detected by *EGRET* in a typical observation. The 4σ detection threshold at 100 MeV in a long ($\sim$ 4 weeks) *EGRET* integration is $F \simeq 10^{-7}$ ph cm^{-2} s^{-1} $= 1.44 \times 10^{-11}$ erg cm^{-2} s^{-1}. By comparison, the flux of a 1 Jy radio source at 5 GHz is 5×10^{-14} erg cm^{-2} sr^{-1}, indicating that a mimimum value of $f_D \simeq 300$ is required to detect a 1 Jy blazar. It is therefore likely that the mean flare duty cycle is small, a conclusion which has also been reached by Salamon and Stecker (1996) using a very different approach. Unfortunately, the current γ-ray data do not sufficiently constrain either the duty cycle or mean flare amplitude of γ-ray emitting blazars. Both of these are critical pieces of the DGRB puzzle.

One would also expect an increase in the fraction of *EGRET* detected FSRS with increasing radio flux and a corresponding decrease at lower fluxes. The *EGRET* data appear to support this suggestion, although the statistics are poor and the sensitivity threshold is not constant. For example, at $F_\nu > 5$ Jy, 5 out of 13 sources in the Kühr flat-spectrum source catalogue have been detected, a fraction which increases to 4 out of 6 for sources > 10 Jy (including Cen A, which is not a blazar, but does have a flat radio spectrum, and has a possible *EGRET* detection; see Thompson et al. 1995). For objects with fluxes $F_\nu < 1$ Jy, there are only three firm *EGRET* detections, while there are ~ 40 unidentified high latitude sources. Even if all of the unidentified sources are indeed blazars, the *EGRET* detected fraction has certainly significantly decreased.

If indeed unresolved, quiescent blazar emission is responsible for the DGRB, the largest contribution to this component must come from blazars of $F_r < 1$ Jy (Figure 2). It is interesting to note that there have been only three *EGRET* detections among objects of this subset of blazars. One of these, Mkn 421, is unique among GeV γ-ray sources in showing no evidence of γ-ray variability. Therefore, the next generation of γ-ray missions should reveal a significant variety in blazar γ-ray characteristics only hinted at by the *EGRET* data.

It is natural to inquire as to the appropriateness of our original assumption that all FSRS are γ-ray emitters. We can explore this question in two ways. First, what if a significant fraction (R) of the FSRS population are not in fact blazars, but other types of objects (e.g., starburst galaxies, Windhorst et al. 1985), which are unlikely to be γ-ray emitters? In this case if all blazars are still γ-ray emitters, then $f_B \propto (1-R)^{-1}$. The observational data constrain R to small values, certainly < 0.5, and probably considerably smaller in the 1–1000 mJy range, since otherwise the class of blazars cannot make the DGRB (note that the mean value of f for all blazars cannot rise more than a factor of ~ 2 over f_B and still be consistent with the observation that only $\sim 1/6$ of > 1 Jy blazars are detected by *EGRET*). A second possibility is that some blazars are not γ-ray emitters. However, a similar argument as above applies, *i.e.*, these cannot be $> 1/2$ of the total number of blazars and still be consistent with both the *EGRET* data and the hypothesis that the class of blazars produce the DGRB. To sum up, the observational data require that at the very least 50% of all flat-spectrum sources must be γ-ray emitters in order for blazars to make the DGRB. Similarly, if there is no quiescent γ-ray emission, i.e. AGN emit γ-rays only in flares, with a duty cycle ~ 0.03 (e.g., Stecker & Salamon 1996) the number density of γ-ray emitting objects at any given time is insufficient to account for the flux of the DGRB.

4. Conclusions

Our findings require the existence of low-state "quiescent" γ-ray emission from blazars, at a level $\sim 1/10$ of the typical *EGRET* detected blazar. We therefore believe that those blazars detected by *EGRET* are in "flaring" or high states and/or have very high ratios of gamma-ray to radio flux. This also naturally explains *EGRET* 's failure to detect the majority of > 1 Jy blazars. As a result of the procedure we have used, we can constrain the blazar fraction among the sub-Jansky population (which has not been fully explored) to be considerably greater than 50%.

Since γ-ray emitting blazars typically emit most of their power in the *EGRET* band, the issue of γ-ray production is critical to understanding the blazar class. Our results place constraints upon models of γ-ray production in AGN, making clear that these models must eventually encompass both flaring and the quiescent blazar behavior. In addition, these arguments provide constraints on the ratio between the "quiescent" and the flaring emission, which must also be accounted for by these models. Further exploration of the γ-ray sky by future instruments and the results of current efforts to identify the flat-spectrum populations at lower fluxes will allow us to not only explore the nature of the DGRB and FSRS populations but also more tightly constrain the duty cycle and amplitude of AGN variability in the γ-rays.

References

Bignami, G. F., Fichtel, C. E., Hartman, R. C., & Thompson, D. J. 1979, ApJ, 232, 649

Dunlop, J. S., & Peacock, J. A. 1990, MNRAS, 247, 19

Fichtel, C. E., Simpson, G. A., & Thompson, D. J. 1978, ApJ, 222, 833

Kazanas, D., & Protheroe, J. P. 1983, Nature, 302, 228

Kühr, H., Witzel, A., Pauliny-Toth, I. I. K., & Nauber, U. 1981, A&AS 45, 367

Osborne, J. L., Wolfendale, A. W., & Zhang, L. 1994, J. Phys. G 20, 1089

Padovani, P., Ghisellini, G., Fabian, A. C., & Celotti, A. 1993, MNRAS, 260, L21

Salamon, M. H., & Stecker, F. W. 1994, ApJ, 430, L21

Silk, J., & Srednicki, M. 1984, Phys.Rev.Lett, 53, 624

Sreekumar, P., et al. 1996, in preparation

Stecker, F. W., Morgan, D. L., & Bredekamp, J. 1971, Phys.Rev.Lett, 27, 1469

Stecker, F. W. 1989, Nucl. Phys. B, 10B, 93

Stecker, F. W., Salamon, M. H., & Malkan, M. A. 1993, ApJ, 410, L71

Stecker, F. W., & Salamon, M. H. 1996, ApJ, in press

Stickel, M., Fried, J. W., Kuhr, H., Padovani, P., & Urry, C. M. 1991, ApJ 374, 431

Thompson, D. J., & Fichtel, C. E. 1982, A&A, 109, 354

Thompson, D. J., Bertsch, D. L., Dingus, B. L., Esposito, J. A., Etienne, A., Fichtel, C. E., Friedlander, D. P., Hartman, R. C., Hunter, S. D., Kendig, D. J., Mattox, J. R., McDonald, L. M., von Montigny, C., Mukherjee, R., Ramanamurthy, P. V., Sreekumar, P., Fierro, J. M., Lin, Y. C., Michelson, P. F., Nolan, P. L., Shriver, S. K., Willis, T. D., Kanbach, G., Mayer-Hasselwander, H. A., Merck, M., Radecke, H. D., Kniffen, D. A., & Schneid, E. J. 1995, ApJS, 101, 259

Windhorst, R. A., Miley, G. K., Owen, F. N., Kron, R. G., & Koo, D. C. 1985, ApJ 289, 494

Blazar Continuum Variability
ASP Conference Series, Vol. 110, 1996
H. R. Miller, J. R. Webb, and J. C. Noble, eds.

A Lighthouse Model for Blazar Jets—Variability and Spectrum

Oliver Dreissigacker and Max Camenzind

Landessternwarte, Königstuhl, 69117 Heidelberg, Germany

Abstract. We explain the overall continuous Gamma Ray Blazar spectrum from the synchrotron turnover to the *EGRET* GeV detections by means of Comptonization in the parsec scale jet's substructures. While making use of the constraints on the synchrotron spectrum and other measurable quantities, no exotic particle acceleration is needed to achieve the high energy output. We show, that the "Lighthouse Model" of blobs of relativistic electrons, travelling with the jet plasma at relativistic speeds, produce both, correct timescales and shapes for the lightcurve, and correct ratios and slopes of the synchrotron, X-ray and γ-ray branches.

1. Introduction

The parsec-scale jets in elliptical galaxies can be considered as normal plasma flows driven away magnetically from the immediate vicinity of the horizon of rapidly rotating black holes (Appl & Camenzind 1993a). Magnetic fields amplified in the inner disk launch the outflows and collimate them on the scale of a few hundred Schwarzschild radii into more or less perfect cylinders. Since the plasma is injected from a rapidly rotating accretion disk, it also carries away angular momentum (Camenzind 1990), the amount of which is determined by the position of the Alfvén point (near the light cylinder $R_{\rm LC}$ for relativistic flows). In the collimated region, the plasma therefore follows helical trajectories, built up by the poloidal velocity component $v_p \simeq c$ and the much smaller toroidal component $v_\phi \simeq c\, R_{\rm LC}/R$ (Camenzind & Krockenberger 1992).

We assume, that magnetic reconnection is responsible for creating the areas of relativistic electrons, in the following called "blobs". Their varying Doppler factor with respect to an observer creates the rapid variability of the object. In a later stage, the blobs become visible in the VLBI as radio knots, c.f., Krichbaum (1990) for 3C 273.

The overall form of the optical lightcurve indicates that the jet is almost cylindrical on the parsec scale (opening angle $\phi_0 \approx 0.05°$), but later opens to $\phi = 1.6°$ as seen in VLBI observations (e.g., Zensus 1992, 1994). The distance, at which the transition occurs, is identical with the expected core radius of the host galaxy (Kormendy 1994).

2. The model: ingredients and derivations

This model is based on the standard model for active galactic nuclei: an accretion disk which is feeding a supermassive black hole and a prominent bipolar outflow. Along with environmental matter, the magnetic field of the host galaxy is accreted towards the center of the galaxy, serving as a seed field for a dynamo (Khanna & Camenzind 1992, 1994). These fields are amplified in the inner disk around a rapidly rotating black hole. The magnetic pressure in the disk launches outflows consisting of normal disk plasma (Khanna & Camenzind 1994). The jet plasma is confined between magnetic flux surfaces which bend towards the symmetry axis after crossing the light cylinder at $R_{\rm LC}$. The jet reaches a self-collimated state at a radius of $R_0 \approx \Gamma R_{\rm LC}$, where Γ is the bulk motion Lorentz factor (Appl & Camenzind 1993b).

In general, this configuration is not stationary. The dynamo continously creates magnetic flux up to a critical value where the magnetic pressure exceeds the turbulent pressure. The axisymmetric structure is being perturbed and islands of reconnection are dragged along the jet. The electrons in the plasma are being accelerated and a terrestrial observer detects an increase in synchrotron flux in the radio and optical regime. Ongoing reconnection drives this increased emission for a few years, reaccelerating the particles which cool on a much shorter time scale by radiative processes.

Short time variability in Blazars can now be explained in terms of a geometrical effect. The variation of the Doppler factor $\mathcal{D}(t) = \{\Gamma(1-\beta\cos\theta(t))\}^{-1}$ of the blob moving with relativistic speed $\beta = v_{\rm blob}/c \approx 1$, $\Gamma > 1$ at an angle $\theta(t)$ between velocity vector $\vec{v}_{\rm blob}(t)$ and line of sight can give rise to a (quasi) periodic and symmetric lightcurve.

Figure 1 shows a blob's model path for 3C 345, see Schramm et al. (1993). Near the point where the jet opens, the blobs become distinguishable radio knots in VLBI measurements. The curved blob trajecory, predicted by our model is in accordance to measurements of previous knots of this quasar. The same considerations hold for recent measurements of knots in PKS 0420–014 (Krichbaum, private communication). A blob's motion on such a trajectory leads to a lightcurve, as it is displayed in Fig. 1 for PKS 0420–014, c.f., Wagner et al. (1995).

3. Theoretical lightcurves and parameters for bright quasars

The two objects PKS 0420–014 and 3C 345 have been monitored over the last years (Schramm et al. 1993, Camenzind 1994, and Wagner et al. 1995). They are almost twins according to the features and timescales of their lightcurves. Hence, the derived parameters might be typical for this type of bright Quasars.

The black hole mass is derived to be $\simeq 4\times10^{10} M_\odot$. The jet's axis inclination angle is $1° \leq \iota \leq 5°$. The jet's opening angle in the innermost part $x < 0.3$ mas is $\phi_0 < 0.1°$ and on the VLBI scale $x \geq 0.5$ mas it is $\phi = 1-2°$. The Lorentz factor of the bulk motion was derived as $\Gamma = 16$, leading together with the mass to a jet diameter at the base of $R_0 \approx 1$ lyr and to varying Doppler factors $10 \leq \mathcal{D}(t) \leq 20$.

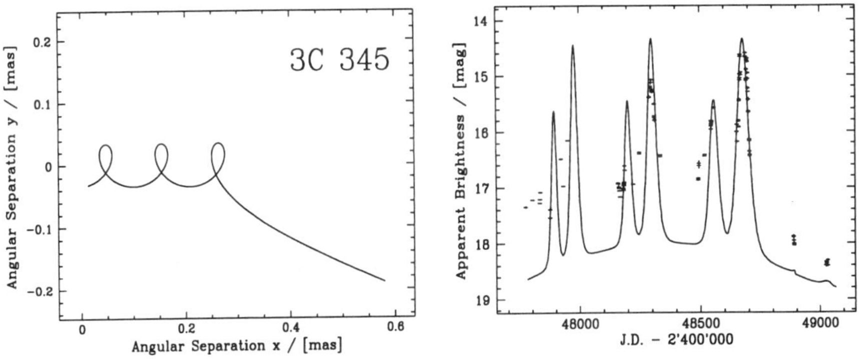

Figure 1. **Left:** Projection of a blob's trajectory in the jet of the Quasar 3C 345 at a redshift $z = 0.595$ onto the sky plane ($H_0 = 50$ km/s/Mpc, $q_0 = 0.5$). The angle of the jet axis to the line of sight is $\iota = 1.75°$. C.f., Schramm et al. 1993. **Right:** Calculated lightcurve of a bone-shaped blob with two peaks in the density of relativistic electrons. The free parameters were chosen to match the data of PKS 0420–014. The solid line describes two blobs of different emissivity traveling along the cylindrical (inner) part of the jet. Unmatched data points may indicate the presence of additional density peaks with less emissivity. C.f., Wagner et al. 1995.

Based on these kinematic parameters, the following spectral properties are derived for PKS 0420–014. The electron spectrum has an initial spectral index $p = 2.0$ with $\gamma_\vee = 10$ to $\gamma_{\rm break} = 70$. The spectrum extends to a maximum electron energy corresponding to a Lorentz factor of $\gamma_\wedge = 3000$. The total number density in the blob is $5 \times 10^9 m^{-3}$ and the volume of the knot is $V = 6 \times 10^{43} m^3$. The magnetic field in the knot is $150\ \mu T = 1.5\ G$. The thermal spectrum being inverse Compton scattered by the blob has an effective temperature of 4700 K in the Blazar's rest frame.

4. High energy spectra

Since the reconnection is not homogeneous in the jet, there will be one or more regions with electrons accelerated to Lorentz factors of $\gamma = E/mc^2 = 10^3 - 10^4$. Their spectrum will be a power law with a break due to the interplay between cooling and acceleration (Kardashev 1962):

$$N(\gamma) = \begin{cases} N_0\, \gamma_\vee^p\, \gamma^{-p} & , \quad \gamma_\vee \le \gamma \le \gamma_{\rm b} \\ N_0\, \gamma_\vee^p\, \gamma_{\rm b}\, \gamma^{-p-1}, & \quad \gamma_{\rm b} \le \gamma \le \gamma_\wedge \\ 0 & , \quad \text{else.} \end{cases} \tag{1}$$

The Lorentz factor $\gamma_{\rm b}$ at which the break in the spectrum occurs, is a function of the cooling time τ, the acceleration time which is expressed as the light travel time through the blob and the magnetic field strength: $\gamma_b \propto \frac{c}{R} \tau B^{-2}$.

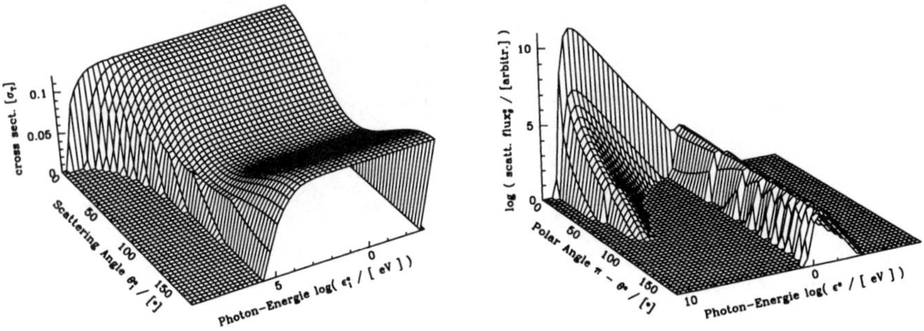

Figure 2. **Left:** Differential cross section in units of the Thomson cross section as a function of energy and scattering angle as used in the calculations. **Right:** Emission characteristics in arbitrary units for a blob with a Lorentz factor $\Gamma = 10$ and a maximum electron Lorentz factor $\gamma = 10^3$. The emission is completely confined into the direction of motion, except for the remainder of the initial isotropic seed field.

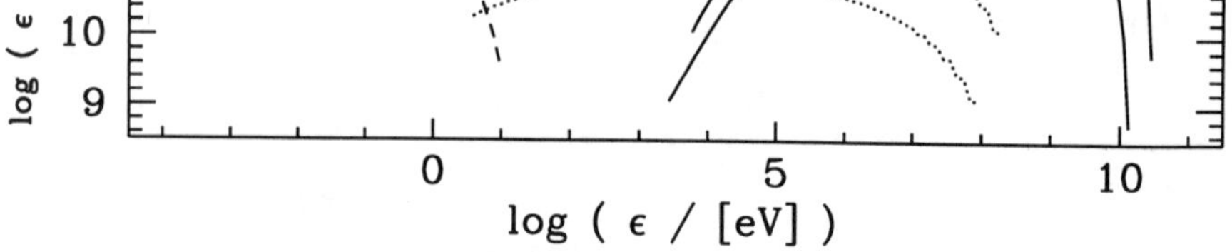

Figure 3. Plot of the dominant radiation fields in PKS 0420–014: optically thin synchrotron radiation (dashed line), synchrotron self Compton (dotted line), and Comptonized dust radiation (solid line). The upper and lower lines refer to different Doppler beaming towards the observer with $\mathcal{D} = 10$ and $\mathcal{D} = 19$. The upper and lower displayed data points (boxes) are the historically documented extremal values and are not at all simultaneous between the different frequencies! (Wagner et al. 1995)

The calculation of the spectrum follows Blumenthal and Gould (1970). Since their analytical results hold for a single index power law electron distribution only, all integrations are done numerically.

If one calculates the flux [in Watts], the complete expression for inverse-Compton-scattering in the co-moving frame is:

$$\epsilon_1 L_{\epsilon_1} = \frac{3}{2}\frac{r_o^2}{R^2}\epsilon_1^2 \int_{\gamma_\vee}^{\gamma_\wedge}\int_{\epsilon_\vee}^{\epsilon_\wedge} N(\gamma)\frac{\epsilon L_\epsilon}{\gamma^2\epsilon^3} I(\epsilon,\epsilon_1,\gamma)\mathrm{d}\epsilon\mathrm{d}\gamma \tag{2}$$

denoting the quantities after the scattering with the subscript "1," and using the abbreviations: ϵ = Photon energy [in eV], r_o = classical electron radius, R = linear source scale, $\gamma = E/mc^2$, dimensionless electron energy, and $I(\epsilon,\epsilon_1,\gamma) = I(\Gamma,q) = 2q\ln q + (1+2q)(1-q) + \frac{1}{2}\frac{(\Gamma q)^2}{1+\Gamma q}(1-q)$, energy dependent part of the cross section, with $\Gamma = \frac{4\gamma^2\epsilon}{\gamma mc^2}$, $\eta = \frac{\epsilon_1}{\gamma mc^2} \leq 1$, $\varepsilon = \frac{\epsilon_1}{4\gamma^2\epsilon} \leq 1$, $q = \frac{\varepsilon}{1-\eta}$, and $\Gamma q = \frac{\epsilon_1}{\gamma mc^2 - \epsilon_1}$.

Our code can handle simultaneously synchrotron and IC emission as scattering of relativistic electrons on (virtual) photons of the magnetic field, or on photons of a 'real' radiation field—synchrotron radiation, inverse Compton radiation, diffuse ambient radiation from stars, dust, etc. The relevant cross section $I(\epsilon,\epsilon_1,\gamma)$ for the scattering is included without approximations, in order to account for the energy dependence (Thomson limit–intermediate case–Klein-Nishina limit).

The electron distribution is assumed to be isotropic, since the most energetic electrons (with $\gamma \geq \sqrt{m_p/m_e}$) are being re-isotropized by scattering off Alfvén waves at timescales shorter than the radiation cooling times (Melrose 1980). Hence, the electrons see—on average—a randomly orientated magnetic field.

For the inverse-Compton-scattering, the origin of the initial photons has to be considered. In the case of SSC (synchrotron-self-Compton) the photon field is isotropic. In the case of scattering ambient photons, the angle between the blob's relativistic motion and the photon source has to be taken into account. The Lorentz transformation into the blob's rest frame redshifts and attenuates photons coming from behind, e.g., from the accretion disk. Photons from a diffuse field seen face on by the blob are blueshifted and intensified. Therefore, ambient diffuse radiation will be the prime candidate for being upscattered by the blobs.

5. Optical–gamma time lags?

Recent observations report time lags between the optical and gamma lightcurves, e.g., in S5 0716+714 (Wagner, Heines, private communication), but they are of the order of one day, and more information is needed to decide, whether these lags are intrinsic in some way, or environmental effects.

However, the working emission mechanisms have different angular dependencies, as soon as relativistic boosting is involved. Depending on the environment of the emitting region, the scattering depth may be angular dependent, and thus

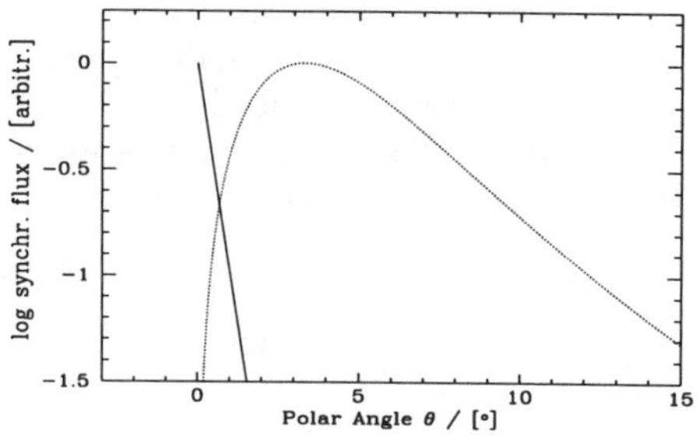

Figure 4. Comparison of the angular dependence of the beaming pattern for inverse-Compton-upscattered ambient radiation at 1 GeV (solid line) and synchrotron radiation at optical wavelengths (dotted line), for a knot with bulk Lorentz factor $\Gamma_{\text{bulk}} = 10$.

leading to an emission asymmetry around the direction of motion, mimicking a non-simultaneity of optical- and gamma-brightness-peaks.

6. Conclusions

This model is capable of explaining the lightcurves observed in the optical and gamma regimes, requiring that synchrotron (radio–optical), synchrotron-self-Compton (X-ray) and inverse Compton scattered ambient thermal radiation (gamma-ray) are produced in the blob. Thus, simultaneous flux variations in these optically thin regimes are predicted.

References

Appl, S., & Camenzind, M. 1993a, A&A, 270, 71

Appl, S., & Camenzind, M. 1993b, A&A, 274, 699

Blumenthal, G. R., & Gould, R. J. 1970, Rev.Mod.Phys., 42, 237

Camenzind, M. 1990, Rev.Mod.Astron., 3, 234

Camenzind, M. 1994, IAU Symp. 159, 257

Camenzind, M., & Krockenberger, M. 1992, A&A, 255, 59

Dreissigacker, O., & Camenzind, M. 1997, in prep.

Kardashev, N. S. 1962, Soviet Ast., 6, 317

Khanna, R., & Camenzind, M. 1992, A&A, 263, 401

Khanna, R., & Camenzind, M. 1994, ApJ, 435, L129

Kormendy, J. 1994, in Nuclei of Normal Galaxies: Lessons from the Galactic Centre, ed. A. Harris & R. Genzel (Dordrecht, Reidel)

Krichbaum, T. 1990, in Parsec-Scale Radio Jets, ed. J. A. Zensus & T. J. Pearson (Cambridge, Cambridge University Press), p. 83

Melrose, D. B. 1980, in Plasma Astrophysics: Non-thermal Processes in Magnetized Plasmas, Vol. II, (New York, Gordon & Breach)

Schramm, K.-J., et al. 1993, A&A, 278, 391

Wagner, S. J., et al. 1995, A&A, 298, 688

Zensus, J. A. 1992, in Parsec-Scale Radio Jets, ed. J. A. Zensus & T. J. Pearson (Cambridge, Cambridge University Press), p. 28

Zensus, J. A. 1994, in NRAO Workshop Compact Extragalactic Radio Sources, 153

This document is available at the Landessternwarte's WWW account via URL `http://www.lsw.uni-heidelberg.de/~odreissi/`

Blazar Continuum Variability
ASP Conference Series, Vol. 110, 1996
H. R. Miller, J. R. Webb, and J. C. Noble, eds.

Unified Models of AGN Accretion Disks and Blazars

M. Kafatos, E. Ramos, P. Becker, P. Subramanian, and R. Yang

Center for Earth Observing and Space Research, Institute for Computational Sciences and Informatics, George Mason University, Fairfax, VA 22030

Abstract. Accretion disks around supermassive black holes are expected to be the power sources in all AGNs, including blazars. To date, though, little direct evidence for such disks exists in AGN observations. In blazars, the intense relativistic beaming would mask any underlying disk component. Here we present relevant work that our group at GMU is carrying out on unified aspects of disk accretion onto supermassive black holes and the possible coupling of thick disks to beams in the inner regions.

1. Introduction

The physical nature of the central power source in AGNs has remained one of the most challenging problems in modern astrophysics. At present the most favored model in all AGNs, including the special class of blazars, is the supermassive black hole hypothesis (Rees 1984). In this scenario, gas falling onto the central black hole radiates with a luminosity that can approach the Eddington limit. Despite the attractiveness of this picture, the discovery of the underlying disks still eludes us. Moreover, in blazars one has the additional problem that the brightness of the relativistic jet would obscure any disk contribution even if it were otherwise present. The discovery of powerful γ-ray emitting blazars by EGRET at the beginning of CGRO observations (Hartman et al. 1992) and in the years following, has presented challenges for any non-jet source mechanisms. The observed luminosity from 3C 279 implies a central mass close to $10^{10}\,M_\odot$ if the emission is unbeamed and the source is radiating near the Eddington limit (Becker, Kafatos, & Maisack 1994). The viability of disk models for blazar emission does not, however, imply that disks are not present. In fact, it is difficult to imagine any scenaria utilizing supermassive object where disks would not be present. As such, the issue of developing unified models which would involve both disks and relativistic beams is still a forefront problem in AGN astrophysics. At GMU, we are working on unified models.

2. Disk models

We are utilizing a particular disk model first developed by Shapiro, Lightman and Eardley (1976) and extended by Eilek and Kafatos (1983) to include full GR effects and γ-ray production. The advantage of these hot, ion-dominated disks is that they naturally produce copious amounts of gamma-rays, extending to $\sim$ GeV energies, as well as relativistic electrons and positrons (with Lorentz factors $\sim$ 300) and are, therefore of interest to γ-ray observations. The inner region of an ion-dominated disk is quasispherical, h/r $\sim$ 1 and $T_i \gg T_e \sim 10^9$ K. If Coulomb collisions is the only means of energy exchange between protons and electrons, protons do not have enough time to establish a common temperature with electrons, the latter radiate much more efficiently and consequently protons become very hot. Proton-proton collisions in the ion-dominated region produce pions which subsequently result in electron/positron production with positrons dominating the production outcome. Although the disk formalism is applied here as well (with the usual assumptions of slow infall velocities, small or zero vertical flow velocities and spatially-thin disk structure) it is obvious that the disk formalism is strained in these regions. For example, if h/r $\sim$ 1, one would expect appreciable flows in the z- and θ-directions which may take on the form of circulation patterns as revealed in numerical simulations. Finally, the material quickly falls onto the central object at radii close to the marginally stable radius, which occurs at 1.2 r_g for canonical Kerr black holes and 6 r_g for Schwarzschild black holes. A schematic diagram is shown in Figure 1.

The hot inner region eventually connects to a standard, radiation- dominated Shakura and Sunyaev region (Shapiro, Lightman, & Eardley 1976), expected to occur at $\sim$ 30-50 r_g (Eilek & Kafatos 1983). There, electrons and ions achieve a common temperature, $T \sim 10^7$ K or less.

A crucial element in many disk scenaria, including the ion-dominated disk, is the operation of Comptonization (Sunyaev & Titarchuk 1980). In the ion-dominated regions, Comptonization of softer, cool disk radiation by the hot electrons produces a power-law photon spectrum extending to hard X-rays/soft γ-rays, $\sim$ several 100 keV. Recently, Chakrabarti and Titarchuk (1995) have examined a variety of Comptonization scenaria in the inner accretion disk regions. They find that shocks may form in the transition region between the two-temperature, ion-dominated and cooler, standard disk. Recently, Becker (in progress) has been examining the physics of strongly radiating shocks. It is not immediately obvious where shocks could be located and the possible locations have been examined in idealized fully-relativistic inviscid flows (Kafatos & Yang 1994; Yang & Kafatos 1995). The existence of shocks may be an element to account for the mysterious QPO behavior seen in galactic black hole candidates (c.f., Yang & Kafatos 1995). In Chakrabarti and Titarchuk's picture, the shocks provide the requisite seed photons for Comptonization (Figure 1).

3. Viscosity mechanisms

Ever since the seminal work of Shakura and Sunyaev (1973), the precise physical mechanism for viscous angular momentum transfer and energy loss has remained a mystery. Fortunately, the emitted spectrum is not strongly dependent on the details or actual values of the viscosity parameter, α (c.f., Pringle 1981). It has been generally assumed that turbulent viscosity may be operating in the standard Shakura and Sunyaev scenaria with precise values being unknown, although $\alpha < 1$. Other viscosity mechanisms have been proposed, including ra-

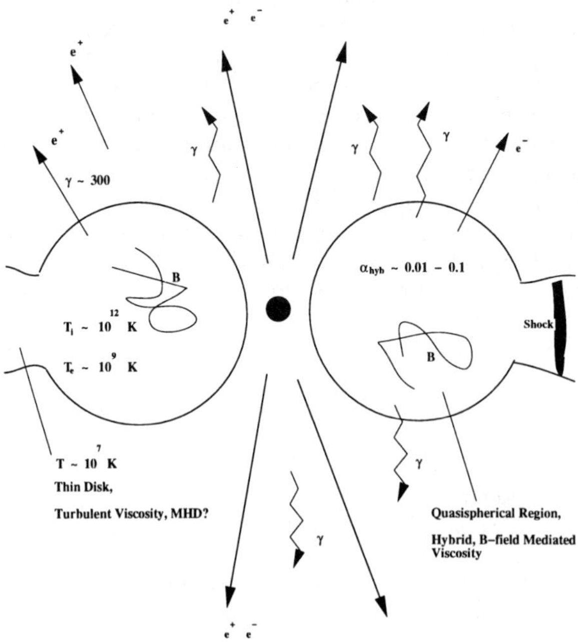

Figure 1. Unified disk/beam scenario in a blazar

diation viscosity (Loeb & Laor 1992) and MHD-driven viscosity (Eardley & Lightman 1975).

Recently, we have achieved a breakthrough in viscosity mechanism applicable to hot ion-dominated disks (Subramanian, Becker, & Kafatos 1996). We have examined the viscosity associated with the shear stress exerted by ions in the presence of a tangled magnetic field (Figure 1). We have taken a self-similar field with a coherence length that is a fixed fraction of the disk scaleheight h. We have arrived at a "hybrid" formulation for the viscosity in which the tangled magnetic field acts as an intermediary in the transfer of momentum between different layers in the shear flow. Consequently, the ions play a much larger role in the transfer of momentum in the presence of B-fields than was previously thought. The hybrid viscosity can dominate over conventional magnetic viscosity for fields that are tangled on sufficiently small scales.

4. Towards unified models of AGNs and Blazars

The observational situation with respect to blazars generally indicates the existence of synchrotron/self-Compton (SSC) mechanism (see articles in present volume). BL Lac spectra like that of Mk 421 (Macomb et al. 1995) indicate the presence of two broad components, centered in the microwave–optical and X-ray–γ-ray regions (synchrotron and self-Compton, respectively). One blazar, however, may be prototypical of the presence of both a jet and a disk, 3C 273 (Ramos, Kafatos, & Fruscione 1996). The UV/EUV peak seen in 3C 273 may be indicative of the presence of an underlying accretion disk.

If indeed this is the case, it may point to a path towards a unification of AGN models. Whereas relativistic beaming is probably operating in blazars, it is hard to see how just a beaming can explain the whole picture without invoking another component, the disk, which is the ultimate power source of the central engine and probably contains the majority of the energy. Becker and Kafatos (1995) have shown that the $\gamma - \gamma$ transparency constraints in extreme blazar sources like 3C 279 allow the EGRET γ-rays to be emitted close to the central black hole, within 45 GM/c^2, perhaps in the active plasma located above the central funnel of the accretion disk (Becker, Kafatos, & Maisack 1994). The steps towards unification in AGNs and specifically blazars involve the following:

- Comptonization/two-temperature disk $\longrightarrow$
- stochastic MHD acceleration in evacuated funnel $\longrightarrow$
- Lorentz factors $\sim$ 2,000 $\longrightarrow$
- γ-ray focusing along axis of rotation $\longrightarrow$
- pair escape in the high-radiation environment of the accretion disk $\longrightarrow$
- focusing of pairs along axis of rotation $\longrightarrow$
- boundary conditions at base of beam near thick accretion disk $\longrightarrow$
- observed properties?

In this prescription, one has an orderly progression to the phenomenology of jets based on principles and mechanisms that are reasonable.

References

Becker, P., Kafatos, M., & Maisack, M. 1994, ApJS, 90, 949
Becker, P., & Kafatos, M. 1995, ApJ, 453, 83
Chakrabarti, S., & Titarchuk, L. G. 1995, ApJ, 455, 620
Eardley, D. M., & Lightman, A. P. 1975, ApJ, 200, 187
Eilek, J., & Kafatos, M. 1983, ApJ, 271, 804
Hartman, R. C. 1992, ApJ, 385, L1
Kafatos, M., & Yang, R. 1994, MNRAS, 268, 925
Loeb, A., & Laor, A. 1992, ApJ, 384, 115
Macomb, D. J., et al. 1995, ApJ, 449, L99
Pringle, J. E. 1981, ARA&A, 19, 137
Ramos, E., Kafatos, M., & Fruscione, A. 1996, these proceedings
Rees, M. J. 1984, ARA&A, 22, 471
Shakura, N. I., & Sunyaev, R. A. 1973, A&A, 24, 337

Shapiro, S. L., Lightman, A. P., & Eardley, D. M. 1976, ApJ, 204, 187
Subramanian, P., Becker, P., & Kafatos, M. 1996, ApJ, in press
Sunyaev, R. A., & Titarchuk, L. G. 1980, A&A, 86, 121
Yang, R., & Kafatos, M. 1995, A&A, 295, 238

Part IV
Multi-frequency Variability

Blazar Continuum Variability
ASP Conference Series, Vol. 110, 1996
H. R. Miller, J. R. Webb, and J. C. Noble, eds.

An Overview of Blazar Variability

C. Megan Urry

Space Telescope Science Institute, 3700 San Martin Drive, Baltimore, MD 21218

Abstract. Blazars are characterized by rapid variability at virtually all wavelengths from radio through TeV gamma-rays. The challenge since their discovery has been to understand the origin of their luminous, apparently nonthermal, nuclear emission. Considerable progress has been made in recent years thanks to a handful of multiwavelength monitoring campaigns with high enough temporal sampling to resolve the most rapid variations. The best data for a few objects have shown a variety of behaviors, for the most part commensurate with synchrotron and Compton-scattered emission from a relativistic jet, though better data for more blazars are still clearly needed. In particular, the origin of the seed photons that are upscattered to gamma-ray energies remains unclear. The latest multiwavelength light curves for the BL Lac object PKS 2155–304 appear to rule out synchrotron emission from a homogeneous source.

1. Introduction

Soon after blazars were discovered and identified, they were selected as interesting targets for long-term monitoring programs, by pioneers like Alex Smith among others, because of their extreme characteristics: they were the most variable, the most luminous, the most polarized, and in some sense the most exciting type of Active Galactic Nuclei (AGN). Understanding them at first seemed the key to understanding the AGN phenomenon. Ironically, blazars were eventually perceived as less interesting for the very same reason—because they were so unusual, so the reasoning went, they must not be relevant to the greater body of AGN (i.e., Seyfert galaxies and quasars). Blazar research was seen as a special field and blazars as arcane oddities.

Now in the mid-90s, we have come full circle. Because blazars are rare geometric manifestations of a general phenomenon (assuming they are relativistic jets pointing directly at us; Urry & Padovani 1995), they must be a quite common kind of AGN, and when they are not pointing at us, we simply call them radio galaxies. This makes blazars highly relevant to understanding AGN as a whole. Specifically, the enormous energy of a relativistic jet, its emanation from the vicinity of the putative central black hole, and its high degree of collimation over many orders of magnitude in scale, offer direct clues to the extraction of energy from the black hole. Blazars thus reveal the energetic processes occurring in the very centers of active galaxies, while in the more common radio galaxies, jet radiation (and hence information) is beamed away from us. With blazars, then, the goal is to understand black hole physics through understanding the physics of the jet. This in turn can be deduced from multiwavelength spectral characteristics, most notably correlated variability across the spectrum.

2. Blazars as relativistic jets

Here I summarize the arguments for believing blazars are relativistic jets. First, they commonly exhibit superluminal motion (Vermeulen & Cohen 1994; see also Wehrle et al., these proceedings) which, while it could arise from pattern rather than bulk relativistic velocity, is at least suggestive. Second, blazars can exhibit extremely high brightness temperatures; in at least some cases, intraday variations are observed at optical wavelengths, ruling out an extrinsic (scintillation) explanation for the variability, although the implied bulk Lorentz factors are uncomfortably high (Wagner & Witzel 1995). Third, the characteristically high and variable polarization of blazars is explained naturally by an aligned jet (Smith et al., these proceedings). Fourth, multiwavelength radio variability is well explained by shocks in a jet (Aller, these proceedings).

Fifth and most compelling, the strong and variable gamma-ray emission observed in many blazars (Hartman, these proceedings) implies such a high compactness that the gamma-ray source would inevitably be dominated by pair production unless the emission is relativistically beamed (Dondi & Ghisellini 1995). While there is some uncertainty about the degree of beaming required (it depends on the ambient X-ray photon density, as X-gamma interactions are the most likely pair-production mechanism), the argument for some beaming is fairly tight.

Blazars can be defined in various ways, via their rapid variability, their compact flat-spectrum radio emission, their superluminal motion, their polarization, and now their gamma-ray brightnesses, and in fact these characteristics occur in the same sources. That is, those sources that are superluminal have flat radio spectra and are highly polarized, and so on.[1] An even more direct link has been seen in at least two cases. Ten years of VLBI maps of 3C 279 (Wehrle et al., these proceedings) show a new VLBI component being "born" (extrapolating the observed position backward with the observed velocity) at the time of the bright gamma-ray flare in June 1991 (Wehrle et al. 1994). The same phenomenon has been observed in PKS 0528+134 (Pohl et al. 1995). In these two blazars, the production of (beamed) gamma-rays is directly related to superluminal motion of the radio source. That these various blazar characteristics are all closely linked, statistically and in some cases directly, is a strong argument that the underlying cause is relativistic beaming.

---

[1] An exception is that some highly polarized quasars (HPQ), largely radio-quiet, have continuum emission polarized by scattering rather than intrinsic processes like synchrotron radiation. These obviously do not have blazar characteristics like superluminal motion or rapid variability.

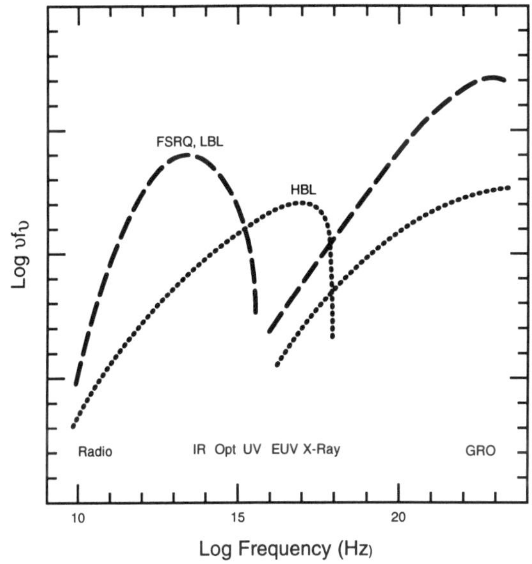

Figure 1. Schematic broad-band spectra of blazars from radio through TeV gamma rays. The low-energy component is probably due to synchrotron radiation and the high-energy component to Compton scattering of lower-energy seed photons, possibly the synchrotron photons or ambient UV/X-ray disk or line photons. Two different curves represent the average spectral shapes (Sambruna et al. 1996) of HBL (High-frequency peaked BL Lac objects; dotted line) and LBL (Low-frequency peaked BL Lac objects; dashed line) as defined by their ratios of X-ray to radio flux (see footnote 2). Strong emission-line blazars (i.e., flat-spectrum radio quasars, or FSRQ) have continua like LBL (Sambruna et al. 1996).

3. Multiwavelength spectra and monitoring

Observations in individual wavebands have established the viability of the relativistic beaming hypothesis for blazars but have led to at best cursory understanding of the physical state of the jet. For this, multiwavelength variability holds the key. Blazar spectra span an extremely broad range of energies, from radio through GeV gamma-rays and perhaps through TeV gamma-rays. The emission consists of two distinct spectral components, a low-energy synchrotron bump and a high-energy Compton-scattered bump (Figure 1).

The temporal evolution of each spectral component and the correlation between them are critical to understanding the underlying emission mechanisms, which is why a number of large multiwavelength monitoring campaigns have been carried out in recent years.

It should be noted that the difficulty of arranging these multiwavelength monitoring observations, typically coordinated among several satellites and many more ground-based telescopes, has kept us from obtaining good data on more than a few of the brightest blazars. More importantly, sensitivity limits have introduced significant target selection effects. Specifically, among BL Lac objects, "High-frequency peaked BL Lacs" (HBL) and "Low-frequency peaked BL Lacs" (LBL) have distinctly different continuum shapes.[2] These may indeed be opposite extremes of a continuous distribution because current BL Lac samples come from radio or X-ray surveys with fairly high flux limits and so are naturally dominated by LBL or HBL, respectively. The strong emission-line blazars (i.e., flat-radio-spectrum quasars, or FSRQ), also generally radio-selected, have continua like LBL (Sambruna et al. 1996).

In any case, the spectral energy distributions of HBL and LBL/FSRQ differ in several ways (Fig. 1). The peak wavelength of the synchrotron component is in the infrared-optical band for LBL and FSRQ, whereas it peaks in the extreme ultraviolet to soft X-ray range for HBL. Also, LBL and FSRQ have a higher ratio of gamma-ray to synchrotron flux than the HBL (Sambruna et al. 1996), and most of the EGRET blazars are in fact FSRQ and LBL. Note that the synchrotron emission is most variable above the peak in νF_ν, where the shortest wavelength component becomes optically thin (Ulrich, Maraschi, & Urry 1996).

Only two blazars, Mrk 421 and Mrk 501, both HBL, have been detected at TeV energies (Punch et al. 1992, Quinn et al. 1996). It may be that these particular objects were detected because they are relatively nearby ($z \sim 0.03$ in both cases) so that the ultra-high-energy gamma-rays have little path length along which to produce pairs via scattering of intergalactic microwave photons, but it is probably also significant that the peak Compton emission in HBL is likely at much higher energies than in LBL/FSRQ. The luminosity represented by the extension of the blazar spectrum to GeV/TeV energies is phenomenal; clearly, understanding the production of this emission is central to understanding the blazar.

Because of these systematic spectral differences, the observational details of the multiwavelength study dictate the type of blazar studied. To study blazars above the synchrotron peak, where they are most variable, means selecting UV- and X-ray-bright targets, which are inevitably HBL. To study correlated intraday variability at radio and optical wavelengths means selecting LBL/FSRQ. To correlate with GeV gamma-rays, one looks primarily at LBL/FSRQ; to correlate with TeV gamma-rays, one looks instead at HBL. With higher sensitivities, this artificial distinction will disappear, but it is important to remember that as presently observed, the radio-optical intraday variables are systematically different objects than the highly variable UV/X-ray-bright sources.

With multiwavelength monitoring of blazars, there are two critical questions we are in the process of addressing. First, where do the gamma-rays come from, and second, what is the structure of the jet itself? By figuring out what the particle density is, what the magnetic fields are, how each varies along the jet, and what causes flaring behavior, we can ultimately understand what created

---

[2]The two sub-classes of BL Lac object are defined by their ratio of X-ray to radio flux, which anticorrelates with the wavelength of their peak synchrotron emission. A High-frequency peaked BL Lac (HBL) has α_{rx} (between 5 GHz and 1 keV) less than 0.75, while a Low-frequency peaked BL Lac (LBL) has $\alpha_{rx} > 0.75$ (Padovani & Giommi 1995).

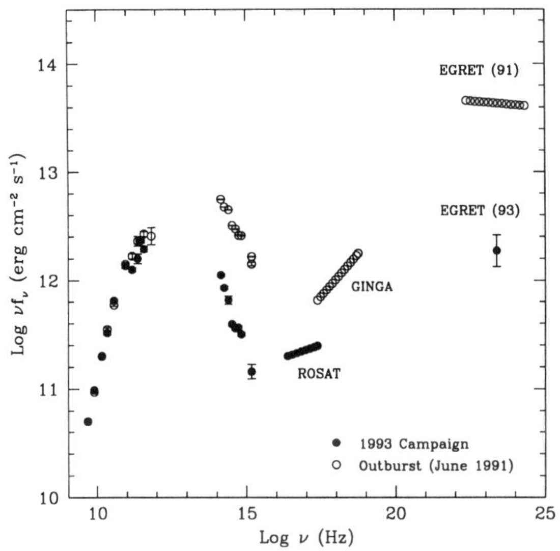

Figure 2. Multi-epoch broad-band spectra of 3C 279, in a high state in June 1991 and in a low state in January 1994 (Maraschi et al. 1994). While the UV decreased by a factor of 3–4, the gamma rays decreased by a factor of $\sim$ 10 or more. As is typical for blazars, there is little or no variability below the peak wavelength of the synchrotron component, the peak of which is in the unobserved far-infrared (now accessible with ISO).

the jet, how it was formed, and what is happening down at the center where we cannot observe directly. For the rest of this paper, I discuss only two objects, 3C 279 and PKS 2155–304, which illustrate some of the best available data (see also Takahashi et al., these proceedings) and thus the limits of what we can learn about blazars at this point.

4. Multi-epoch flaring in the superluminal quasar 3C 279

3C 279 is the brightest gamma-ray blazar in the sky. [During the week of the Miami blazar meeting, it was undergoing a major outburst, rising an order of magnitude above its previous highest gamma-ray state, with substantial variations at other wavelengths.] Figure 2 shows the broad-band spectra at two epochs, the high state in June 1991 when it was first discovered and a low state in January 1994; the greatest change in intensity occurs at gamma-ray wavelengths, with lesser but still substantial variations at UV and X-ray wavelengths, and little change at radio to sub-millimeter wavelengths (Maraschi et al. 1994).

Note that the gamma-ray flux can dominate the bolometric luminosity, particularly in the flare state. Maraschi et al. (1994) suggested that the larger increase in gamma-rays relative to synchrotron emission was consistent with the synchrotron self-Compton (SSC) model, although a change in Doppler factor of the jet or concurrent increases in seed photons and scattering electrons could also cause the observed variability. Clearly, sampling these flares more finely is critical to identifying the origin of the seed photons and thus the production of the dominant spectral component. The multiwavelength observations of the 1996 flare should help resolve the ambiguity about the identity of the seed photons (Wehrle et al., in preparation).

The basic blazar models currently under consideration have one common element, a synchrotron-emitting jet filled with energetic electrons and perhaps positrons. In addition, there may be ambient UV and X-ray emission from the vicinity of an accretion disk, and broad-line emission from clouds farther out, photoionized either by the central UV/X-ray source or by the jet itself. The debate turns on which photons are Compton-scattered to gamma-ray energies. Photons impinging on the jet from the side are boosted in the frame of the jet electrons, and so constitute a very intense flux of seed photons even when the directly observed non-jet UV/X-ray flux is low.

Ghisellini and Madau (1996) present a nice comparison of the principal ideas (see references therein as well) and conclude two interesting points. One is that the inner part of the jet has to be dissipationless. If the energy density of gamma-rays in an optically thick inner region of the jet were high, there would probably be enough local UV photons to generate a pair cascade, transferring much of the gamma-ray energy into X-rays, contrary to what is observed. So the principal mode by which energy is transferred from the black hole to the jet must not be via energetic photons. The second point is that, for plausible numbers, the illumination of the broad-line clouds by the beamed continuum can be an important contribution to the UV flux impinging on the jet. The bulk of the gamma-rays could therefore come from scattered broad-line photons which were photoionized by the jet itself. This might be the dominant mechanism in FSRQ, say, while SSC emission dominates in the weaker-lined HBL.

It would be extremely interesting to monitor simultaneously the variability of the broad lines and the synchrotron continuum. As far as I am aware this has not yet been done, at least with sufficient sampling. We have looked at the archival UV data for 3C 279 (Koratkar et al. 1996) but even there, with a very well-observed source, the data are not sufficient to determine the photoionization source unambiguously. The 1996 campaign on 3C 279 will help and there are a number of other blazar campaigns planned for the upcoming year which may contribute to solving this problem. Unfortunately, the loss of IUE, with its long, sustained monitoring campaigns, is a major blow for this kind of study.

5. Multiwavelength variability of the HBL PKS 2155–304

PKS 2155–304 is the brightest BL Lac object at ultraviolet wavelengths and one of the brightest in the X-ray as well. So it is the obvious choice for UV/X-ray monitoring. It is an HBL, with peak synchrotron emission near 10^{17} Hz, which is four orders of magnitude higher in frequency than the peak synchrotron emission in 3C 279. We have every reason to expect, therefore, that the physics of the emission from PKS 2155–304 and 3C 279 differ in significant ways.

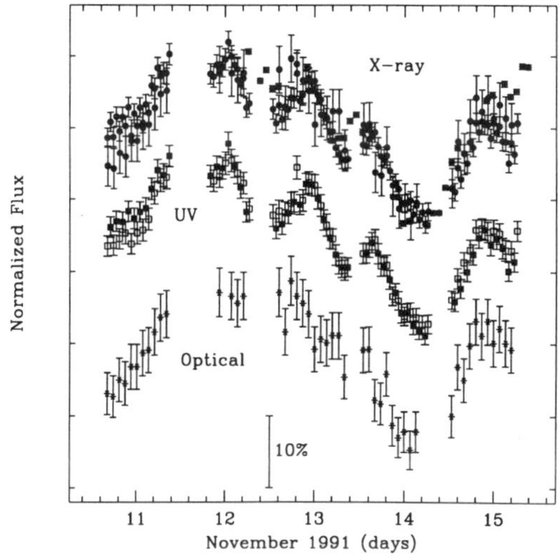

Figure 3. Normalized X-ray, UV, and optical light curves of PKS 2155–304 from the first intensive multiwavelength monitoring campaign, in mid-November 1991 (Edelson et al. 1995). The emission is well-correlated, with comparable amplitude independent of wavelength, and the X-rays appear to lead the UV by $\sim$ 2–3 hours. Even with extensive coverage from the ground, the best optical data were obtained with the FES monitor on IUE, a simple star-tracking device, which highlights the value of adding modest optical devices to high-energy satellites.

There have been two intensive multiwavelength campaigns to observe PKS 2155–304. [There was a third three months after the Miami blazar meeting.] The first was in November 1991 and lasted for one month. At that time, no one even knew whether the UV and X-ray emission were related, nor what the fastest time scale for variability was. We observed PKS 2155–304 with IUE once per day throughout November 1991 (thinking that was probably overkill) and then in the middle of the month, at the insistence of Rick Edelson (who rightly realized it was not overkill), we observed it for nearly 5 days continuously. The UV variations we detected were indeed fast enough that the daily sampling was insufficient and only the continuous observations were useful for multiwavelength cross-correlations (Urry et al. 1993). We also had $\sim$ 3.5 days of continuous X-ray observations (Brinkmann et al. 1995) overlapping with most of the intensive UV coverage, as well as considerable ground-based radio, infrared, and optical observations (Smith et al. 1992, Courvoisier et al. 1995).

Results from the November 1991 campaign are shown in Figure 3.

The X-ray, UV, and optical light curves are well-correlated, arguing for a common origin of the optical through X-ray emission, and the X-rays lead the UV by $\sim$ 2–3 hours (Edelson et al. 1995). The optical/UV emission can not be produced by thermal emission from a viscous accretion disk because they

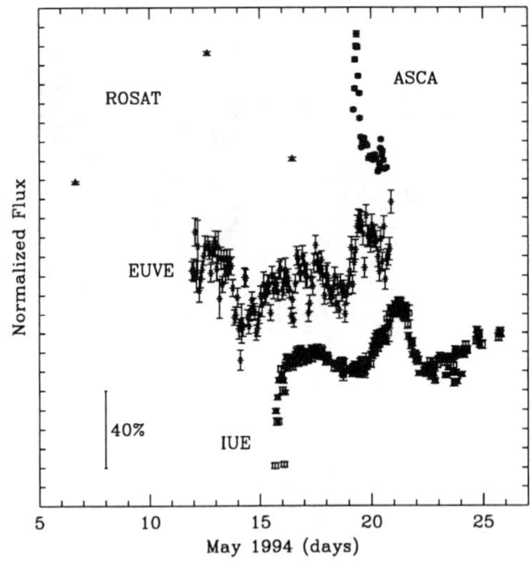

Figure 4. Normalized X-ray, EUV, and UV light curves of PKS 2155–304 from the second intensive multiwavelength monitoring campaign, in May 1994 (Urry et al. 1996). The ASCA data show a strong flare, echoed one day later by EUVE and two days later by IUE. The amplitude of the flares decreases and the duration increases with increasing wavelength.

should arise at very different radii, implying a large lag between the two. The amplitude of variation is independent of wavelength, a result contrary to what is expected from a synchrotron flare caused by an increase in energetic electrons (Celotti, Maraschi, & Treves 1991).

Several aspects of the 1991 results were intriguing enough to inspire us to repeat the experiment for a longer period and with more extensive wavelength coverage. First, the light curves consisted of a series of peaks modulated by an overall decline in flux; while not strictly periodic (Edelson et al. 1995), they suggested possible repetition through five cycles. Second, the UV and X-ray light curves, which tracked very well for most of the observation, appeared to diverge at the very end of the overlapping period.

We therefore arranged a second intensive monitoring campaign in May 1994, with IUE for 10 days continuously (Pian et al. 1996), EUVE for 9 days (Marshall et al., in prep.), and ASCA for 2 days (Kii et al., in prep.), and with additional Rosat (Urry et al. 1996) and ground-based data (Pesce et al. 1996). Figure 4 shows the results from this second set of multiwavelength observations of PKS 2155–304 (Urry et al. 1996).

A very sharp flare is seen in X-rays, followed by an EUV flare one day later and a broader, lower amplitude UV flare two days later. The X-ray flux doubles in ~ 0.2 days and is approximately symmetric. The Rosat data, while sparse, show that large X-ray flares were not unusual but were occurring throughout the week prior to the ASCA observations. The EUV flux increased by $\sim 50\%$ in

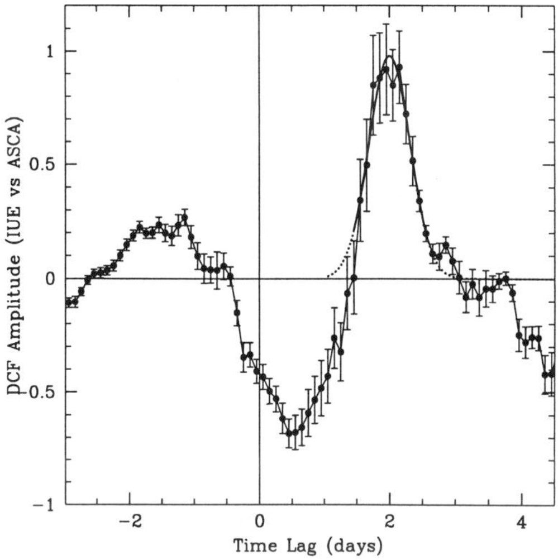

Figure 5. Cross correlation of May 1994 ASCA and IUE light curves of PKS 2155–304, showing that the X-ray flux leads the UV by $\sim$ 2 days (Urry et al. 1996).

less than a day (the time scale is difficult to estimate given the errors and the untimely end of the data train), and the UV flux rose by $\sim$ 35% with a doubling time scale of $\sim$ 3 days. The duration of the flare increases from less than a day in the X-rays to nearly 4 days in the UV.

The delays between X-ray, EUV, and UV light curves are easily measured with cross-correlation functions. Formally, the EUV flux leads the UV by 1.1 days and the X-ray flux leads the UV by 2.0 days. The cross-correlation between IUE and ASCA light curves is shown in Figure 5; the cross-correlation of X-ray versus EUV light curves is not well defined due to the small temporal overlap. (Fig. 5).

The IUE data also reveal a complex and extremely rapid flare at the beginning of the observation, with doubling times as fast as 1 hour, the fastest ever observed at ultraviolet wavelengths and comparable to the fastest doubling times seen in the X-ray (Pian et al. 1996). The event is seen in both LWP and SWP cameras, and is undersampled by both; it has larger amplitude in the LWP in part because the LWP integration times are less than half the SWP integration times. It is also possible to see similar structure in the EUV light curves, although the EUVE data are relatively noisy, $\sim$ 1–2 days in advance of the IUE event.

Figure 6 shows an expanded view of the EUVE and IUE light curves, with the EUVE curve shifted forward by 1.25 days.

There are strong differences between the May 1994 and November 1991 light curves of PKS 2155–304. In the second epoch, the amplitude and duration depend strongly on wavelength and the lags are considerably longer than in the first. In addition, there is little of the low-amplitude repetitive variability, at

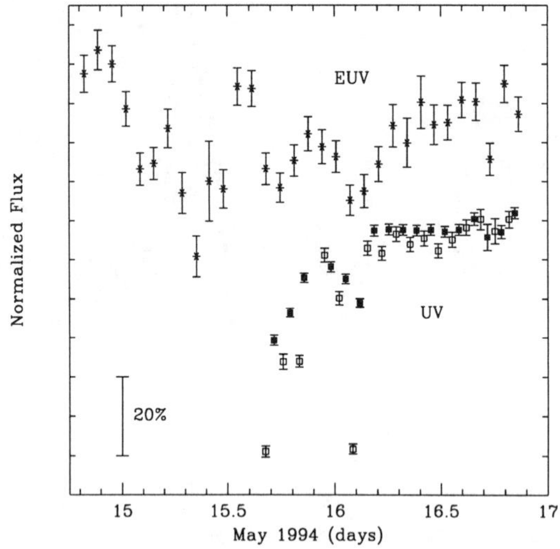

Figure 6. Expanded view of the beginning of the IUE light curve of PKS 2155–304 from May 1994, with the EUVE light curve shifted forward by 1.25 days. The extremely rapid UV variability on May 16, undersampled in both LWP and SWP cameras, is similar to earlier, lower amplitude variations in the EUVE light curve.

least in the first half of the IUE observation. These differences mean either that there are two different mechanisms operating or that the relevant physical parameters have changed considerably. The variability in 1991 was probably not a synchrotron flare, since the observed time scales were energy independent. Instead, it is possible that it was caused by microlensing: for a dense star cluster at the redshift of a known Lyman-alpha absorption system (about half way to the BL Lac), and assuming relativistic motion of the BL Lac jet, the amplitudes and time scales are approximately correct, and the achromatic nature is automatic as long as the source size is independent of wavelength. This is a plausible but not proven explanation.

The 1994 data are much richer in extent and wavelength coverage. These data are consistent with a synchrotron flare in a jet but the clear delay between X-ray and UV flares rules out the homogeneous case (Urry et al. 1996). The reasoning is as follows. The simplest causes of a flare in the homogeneous case would be a sudden uniform increase in the injection of energetic electrons or the instantaneous and uniform enhancement of the magnetic field, perhaps via compression of a charged plasma. In both cases, the flux at all (optically thin) wavelengths would rise simultaneously, with amplitude increasing with decreasing wavelength, while the duration of the flare (if due to energy losses only) would go as $\lambda^{1/2}$. That is, the long-wavelength emission would last longer than the short-wavelength emission but the (instantaneous) peaks in the light curves would be simultaneous. The delay in the flare onset and the longer rise time for the UV flare compared to the X-ray flare clearly rule out the simplest homogeneous case, unless *ad hoc* dependences on time and energy of the injection rate are postulated.

Thus, some energy stratification of the synchrotron-emitting plasma is required, either behind a shock or in the jet structure itself. The X-ray decay is faster than the UV but perhaps by less than a factor $(\lambda_{UV}/\lambda_X)^{1/2} \sim 20$, suggesting the decay is dominated by geometry rather than synchrotron losses. The flare duration increases with wavelength, possibly indicating that the size of the emitting region is increasing with wavelength. The observed lags are also comparable to the flare durations. Both these results are as expected for a shock propogating outward in an inhomogeneous jet, successively passing from X-ray- to EUV- to UV-emitting regions.

6. Future multiwavelength monitoring

As this review has shown, the best available data are still insufficient for determining jet structure but enough to indicate the kind of data needed. We should repeat the kind of multiwavelength monitoring done for PKS 2155–304 for many more blazars, including LBL, which will require a new ultraviolet capability. We also need to study further the correlation of gamma-ray and optical/IR variability. In all cases, long and intensive time sampling is critical: two light curves that are well correlated could, if sampled at few points or for less time than the characteristic lag, appear uncorrelated. Thus the single epoch approach is no longer valuable for adding information. The requirement for intensive monitoring and for larger samples of objects points to the need for a multiwavelength platform with modest, very simple, very inexpensive optical/UV telescopes paired with X-ray and gamma-ray detectors.

Acknowledgments. I am grateful to my many collaborators for their contributions to our multiwavelength monitoring of blazars, particularly those with whom I have worked most closely in recent years, Laura Maraschi, Joe Pesce, Elena Pian, Rita Sambruna, Aldo Treves, and Ann Wehrle, and those who were PIs for the second campaign on PKS 2155–304, Tsuneo Kii, Herman Marshall, and Greg Madejski.

Discussion

Al Marscher: In the 1994 PKS 2155–304 flare, you say that the fast rise is from the time scale for particle acceleration but that the slower decay is geometric. Light travel effects should limit the shorter time scale, not the longer. Also I disagree that the longer time scales and lags at lower frequencies require gradients in the underlying jet to be the root cause: shocks should be frequency-stratified and give the behavior you observe (Marscher, Gear & Travis 1992, in *Variability of Blazars*.

Meg Urry: You are right about the geometric limit for the fastest time scale. We agree that the jet needs to be inhomogeneous. Whether the inhomogeneity arises from the natural stratification due to energy-dependent losses and diffusion from a shock, or from some other means, is not known at this point.

Hugh Aller: I am worried about involving different physical processes to explain two different flare events (1991 versus 1994) in PKS 2155–304. Isn't a single process which permits both events to be explained ultimately simpler?

Meg Urry: I can see how it would seem that way but frankly, the character of the variations in 1991 and in 1994 is so different that the required change in parameters of the single process you might prefer is more or less equivalent to a distinct process. If, for example, we decide that the 1994 flares are the signature of a shock propagating through an inhomogeneous jet, with the expected dependence of amplitude on wavelength and delays of $\sim$ 1 day from X-rays to EUV to UV-emitting regions, then how can this model explain the 1991 variability, with its achromaticity and negligible lags? In a sense these are opposite behaviors, which naturally suggest unrelated causes.

References

Brinkmann, W., et al. 1994, A&A, 288, 433

Celotti, A., Maraschi, L., & Treves, A. 1991, ApJ, 377, 403

Courvoisier, T. J.-L., et al. 1995, ApJ, 438, 108

Dondi, L., & Ghisellini, G. 1995, MNRAS, 273, 583

Edelson, R. A., et al. 1995, ApJ, 438, 120

Ghisellini, G., & Madau, P. 1996, MNRAS, 280, 67

Koratkar, A., Pesce, J. E., Urry, C. M., & Pian, E. 1996, in preparation

Maraschi, L., et al. 1994, ApJ, 435, L91

Padovani, P., & Giommi, P. 1995, ApJ, 444, 567

Pesce, J. E., et al. 1996, ApJ, submitted

Pian, E., et al. 1996, ApJ, submitted

Pohl, M. et al. 1995, A&A, 303, 383

Punch M., Akerlof C. W., Cawley M. F., Chantell M. ., Fegan D. J., Fennell S., Gaidos J. A., Hagan J., Hillas A. M., & Jiang Y. 1992, Nature, 358, 477

Quinn J., et al. 1996, ApJ, 456, L83

Sambruna R., Maraschi L., & Urry C. M. 1996, ApJ, 463, 444

Smith, P. S., Hall, P. B., Allen, R. G., & Sitko, M. L. 1992, ApJ, 400, 115

Ulrich, M.-H., Maraschi, L., & Urry, C. M. 1996, ARAA, in preparation

Urry, C. M., et al. 1993, ApJ, 411, 614

Urry, C. M., et al. 1996, ApJ, submitted

Urry, C. M., & Padovani, P. 1995, PASP, 107, 803

Vermeulen R., & Cohen M. 1994, ApJ, 430, 467

Wagner S., & Witzel A. 1995, ARA&A, in press

Wehrle, A. E., Zook, A. C., Unwin, S. C., Urry, C. M., & Madejski, G. 1994, in The Multi-Mission Perspective (High Energy Astrophysics Division Meeting of the American Astronomical Society, Napa Valley, November 1994), p. 27

Blazar Continuum Variability
ASP Conference Series, Vol. 110, 1996
H. R. Miller, J. R. Webb, and J. C. Noble, eds.

Multifrequency Continuum Variability of Blazars: Observations

James R. Webb

Dept. of Physics, Florida International University and SARA Observatory, Kitt Peak

Abstract. The multifrequency (MF) approach to studying the continuum emissions of blazars has proved to be a necessary step in understanding the physics of these sources. The very nature of the observations requires large numbers of observers and observatories to obtain the necessary frequency and temporal coverage. We first summarize the results of previous multifrequency observations of blazars, discussing observational approaches, the objects studied, and the important results. We then present recent multifrequency variability observations of blazars during outbursts.

1. Introduction

Comprehensive MF programs designed specifically to observe blazars have been organized since 1981. Most of the original "truly multifrequency" campaigns were coordinated around observing time with spacecraft such as IUE. These programs required many people, many instruments, and time-critical scheduling. Observational windows were extended in frequency and in time as it became clear that any hopes of solving the problem of blazar continuum emission would require truly simultaneous multifrequency observing in the strictest sense. The earliest attempts at compiling MF spectra of blazars resulted in observations with time delays between frequency bands on the order of months or weeks. The most recent MF campaigns attempt to get "truly simultaneous" observations, or at least "quasi-simultaneous" observations with observations separated by days or hours.

The sections below summarize the basic observational approaches that have been used to obtain MF spectra of blazars. The blazars that are frequently targets of MF campaigns are discussed and an attempt is made to summarize the most important results of these programs. The remainder of the paper will present some of the MF observations currently analyzed by our group.

2. Observational approaches

There are three basic observational approaches to obtaining MF spectra of blazars: *Archival Spectral Composition* (ASC), a *Pre-Coordinated Campaign* (PCC), and *Target of Opportunity* (TOO).

The first type of program, the *ASC*, involves gathering previously published data (and sometimes unpublished data) which is not necessarily simultaneous, or which happens to be roughly contemporaneous with other observations from the literature (Impey & Neugebauer 1988, Sadun 1992, Hartman et al. 1996). From this data, one can construct a quasi-contemporaneous MF spectrum. The problem with this type of study is that it requires a great deal of good fortune since the probability that many observers will be looking simultaneously at the same object independently drops substantially for lesser known blazars. Also, pivotal high energy data are usually not present for these types of programs since satellite observing time is very limited.

The second approach, the *PCC*, has been used for a number of successful campaigns. These programs are popular with Seyfert galaxy observers as well, but examples of blazar programs are: Bregman et al. 1981, 1984, 1986, 1990, Wills et al. 1983, Glassgold et al. 1983, Worrall et al. 1984, Landau et al. 1986, Maraschi et al. 1991, Kawai et al. 1991, Maraschi et al. 1994. These campaigns usually cover the important high frequency bands using pre-arranged observing time on satellites and large ground-based facilities, supplemented by monitoring data and optical photometric observations from smaller telescopes. One can get excellent spectral and temporal coverage with *PCC* programs, the primary draw back being possiblity that the source is faint when the observations are scheduled to occur. Since most blazars are fainter at high frequencies, they are usually close to the limit of detectability of the high energy detectors. It is difficult to study extreme variations and outbursts with *PCC* programs, except in a few sources that frequently show extreme flaring activity.

The final type of program, the *TOO*, ensures that the source is bright and active. The *TOO* programs require a large number of observers willing to change their observing targets relatively quickly so that reasonable spectral and temporal coverage can be obtained. Usually a trigger criteria is agreed upon (optical, X-ray or γ-ray brightness level) and the source is monitored until the trigger criteria is fulfilled and the program is initiated. The biggest advantage of the *TOO* programs are that they guarantee the source is in an active, bright phase. Other advantages of *TOO* programs are that during periods of rapid, high amplitude variations, it is much easier to discern whatever correlations may be present between different frequency bands. It is much easier to plan and make high energy observations when the source is bright. Finally, interesting and extreme physics is occuring during these energetic outbursts, which help constrain theoretical models. The disadvantages of *TOO* programs are that most satellite telescopes are difficult to use with short notice, although a few satellites such as IUE are tailor-made for such observations. Observations of the major outburst of 3C 345 in 1991 is an example of such a *TOO* program (Webb et al. 1994).

3. Multifrequency programs and target objects

There have been a large number of MF campaigns in the past 10 years. Table 1 lists some of the blazars that have been the targets of multifrequency programs. Column 1 gives the source common name, columns 2 and 3 give the Right Ascension and Declination of the source. Column 4 gives a brief indication of the frequency bands covered in the references while column 5 gives the date the observations were made. Column 6 gives a recent MF reference for each source. This list is not intended to be exhaustive, but a guide to work that has been done. An excellent review, based on blazars observed with IUE, and with additional references can be found in Bregman, Maraschi and Urry (1987). The source 3C 120, although not generally considered a blazar, is sufficiently interesting to merit inclusion in this table.

Table 1. Popular Multifrequency Blazar Targets

NAME	R.A. (1950)	DEC. (1950)	FREQ	DATE	REF
3C 120	04^h30^m	$+05^d14^m$	γ-ray–radio	1994	Maraschi et al. 1991
PKS 0537−441	05^h37^m	-44^d06^m	X-ray–IR	1985	Tanzi et al. 1986
PKS 0735+174	07^h35^m	$+17^d49^m$	X-ray–radio	1981	Bregman et al. 1984
OJ 287	08^h51^m	$+20^d17^m$	UV–radio	1993	Pian et al. 1995.
PKS 1156+295	11^h56^m	$+29^d31^m$	UV–radio	1981	Glassgold et al. 1983
3C 279	12^h53^m	-05^d30^m	γ-ray–radio	1991	Hartman et al. 1996
1413+135	14^h13^m	$+13^d30^m$	X-ray–radio	1980	Bregman et al. 1981
3C 345	16^h41^m	$+39^d54^m$	X-ray–optical	1981	Webb et al. 1994
3C 371	18^h07^m	$+69^d48^m$	UV–optical	1983	Worrall et al. 1984
BL Lac	22^h00^m	$+42^d02^m$	X-ray–radio	1988	Kawai et al. 1991
3C 446	22^h23^m	-05^d12^m	X-ray–radio	1984	Brown et al. 1986

4. General MF results

What have we learned from all of these multifrequency campaigns? The answer is: "not enough." Three basic types of information can be gleaned from the multifrequency studies:

1. The "snapshot" spectral energy distribution
2. The temporal correlations between various frequency bands
3. The general change in character of the MF spectrum as a function of time.

As an example, Figure 1 shows the multifrequency spectrum of 3C 279 at several different epochs compiled from the literature and re-plotted on the same scale. This figure illustrates the amount of variability seen in each of the observed spectral bands. These points are not absolute maxima and minima, but rather an indication of the range of variability seen in this source during some of the multifrequency campaigns. The data shown are from Hartman et al. 1988, Maraschi et al. 1994, Brown et al. 1989, and Makino et al. 1991. Close examination of this figure shows that the γ-rays and X-rays have relatively large amplitudes of variability along with the UV, optical, and IR. On this scale, the radio variability is only evident shortward of 10^{11} Hz. Figure 2 shows a similar plot for 3C 345. This plot is composed of MF spectrum from Bregman et al. 1984 and Webb et al. 1988. Unfortunately this plot does not contain much high energy data, but it does illustrate the variability seen in the radio, mm, IR and optical.

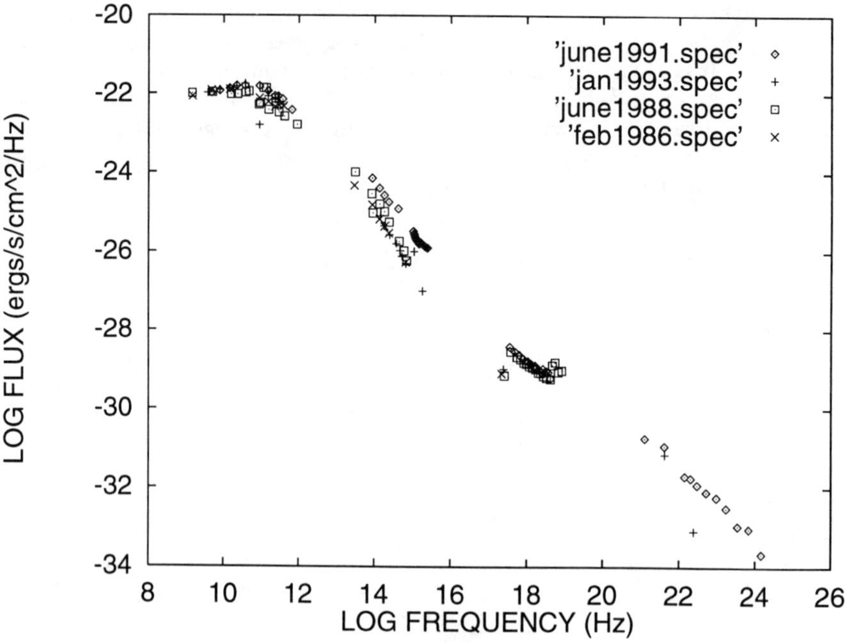

Figure 1. MF Spectral Variability of 3C 279

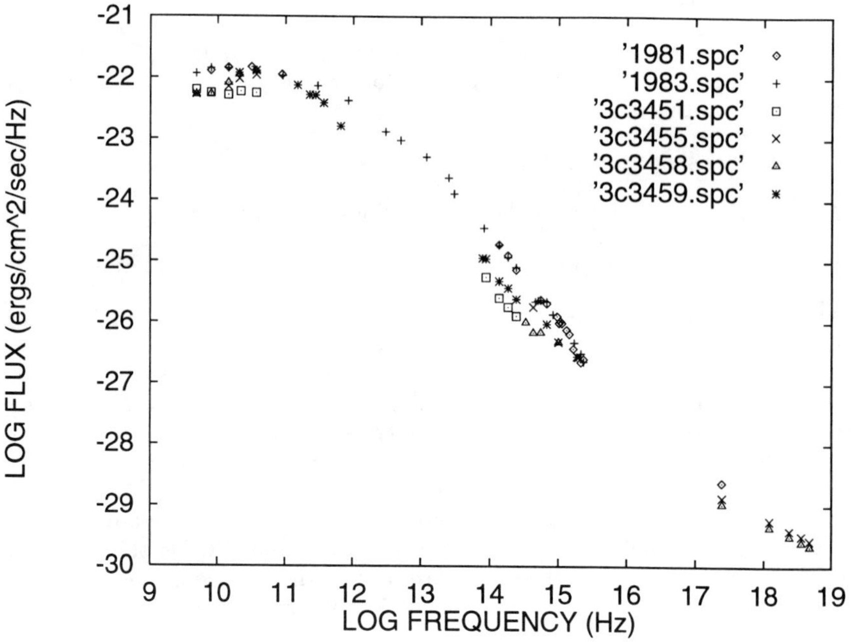

Figure 2. MF Variability of 3C 345

4.1. General spectral shape

The general shapes of blazar MF spectra are presented by Impey and Neugebauer 1988, and Landau et al. 1986. These spectra are not strictly simultaneous (δt = 2–3 mo.) and the spectral shapes can be somewhat deceiving for blazars, but these spectra show discontinuities in some cases between the mm-cm nonthermal continuum (Robson 1992). The situation is somewhat less clear through the far-IR since the IRAS observations are limited, but they appear consistent with the near-IR spectral component. The near IR–UV may be a composite of several components since the spectral slopes change significantly with time (Brown et al. 1989, Webb et al. 1990), and the polarization changes rapidly (Smith 1986). Whether these components are all from the jet, or whether some are also from the disk is a question still under discussion (Maraschi these proceedings). The high energy observations are still inconclusive about the origin and number of components that emit in the X-ray and γ-ray spectral bands. It is generally accepted that the X-ray band is made up of Compton-scattered IR-optical photons, and quite possibly the γ-ray bands can be interpreted as higher-order scattering. The high energy, quasi-simultaneous observations also indicate that these sources are, at least in cases such as 3C 279 and OJ 287, much more luminous in γ-rays than in any other spectral region. Figure 3 shows the MF spectrum of 3C 279 plotted as νF_ν observed during June 1991 and October 1991 and of OJ 287 during the recent 1994 outburst. The 3C 279 plot clearly indicates the range of variability in the GeV region of the spectrum during the outburst. Converting the flux to luminosity using standard formula and the values of the cosmological parameters H_0 = 55 km/sec/Mpc and q_0 = 0) and then integrating under the light curves gives an idea of the energetics involved in the outbursts. The total luminosity of 3C 279 between radio and γ-ray spectral bands (extrapolated between unsampled spectral regions) was 5.6×10^{48} ergs/sec in June, and 2.6×10^{48} in October. This represents a factor of 2.2 change between the γ-ray outburst and the subsequent low state. The luminosity in the γ-ray region alone is 3×10^{48} ergs/sec in June and 8.3×10^{47} ergs/sec in October, a decrease by a factor of 3.6. It is evident that the γ-ray region dominated the energetics of this source during the active period.

The OJ 287 MF spectrum is shown in Figure 3 during the 1994 optical outburst. The γ-ray flux is not nearly as dominate as for 3C 279. OJ 287 was previously undetected by CGRO before this active period in despite a number of other pointings, indicating that the source was bursting in γ-rays as well as the optical–UV spectral bands during that time period even though it is not as important energetically in this spectrum.

4.2. Correlations

Correlations between radio and optical variations have been studied using many techniques. As the length of light curves has increased in these spectral bands, the reliability of the results has also increased (Pomphrey et al. 1976, Balonek 1982, Hufnagel & Bregman 1992, Clements 1995), but few real cases of substantial correlations have been found. Although there has been some evidence for radio–optical correlations in a few sources (OJ 287 being one), the relationship is not clear, and if any correlations do exist there are significant time delays involved, especially between major optical flares and the associated radio flares

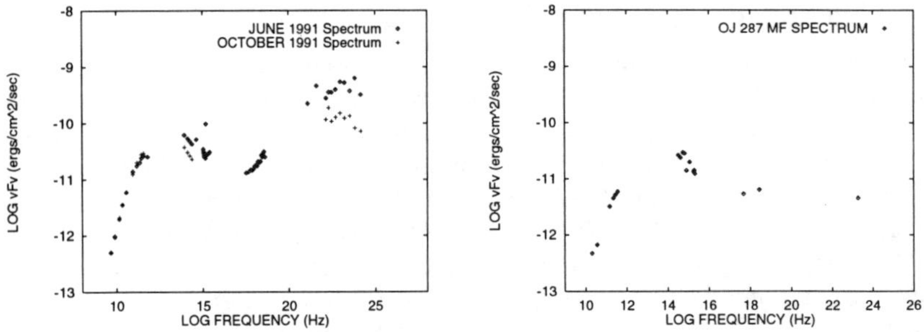

Figure 3. MF Variability of 3C 279 and OJ 287

(Webb et al. 1994). For 3C 345, the 1991 outburst was first seen in the optical and then later at lower frequencies. Delays between the optical and GHz bands were on the order of months. The character of the rapid radio and optical variations seem to be similar in some sources (Quirrenbach et al. 1992), suggesting a common origin for the optical and the radio photons.

The optical/Near IR spectrum is generally steeper than the optical and is highly correlated with optical flux variations (Brown et al. 1994). The UV seems so strongly correlated with the optical that optical measurements are successfully used to determine IUE exposure times.

The UV spectrum is nearly always steeper than the IR–optical spectrum (possibly due to Synchrotron break?) and there seems to be modest UV slope variation. There is some evidence that slope variations might be correlated with flux level in 3C 279 (Bonnell, Vestrand, & Stacy 1994) in the sense that the brighter the source, the flatter the spectrum, similar to the correlation seen in the optical.

The optical/X-ray correlations are extremely uncertain, although there are some suggestions of correlated outbursts (Webb & McHardy 1991, Makino et al. 1993). The γ-ray/X-ray and γ-ray/optical correlations are still unknown for blazars as a class, due primarily to the small number of blazar outbursts detected to date with EGRET (see Hartman, this volume).

In general, the average timescale of variability decreases and amplitudes increase with higher frequencies. The spectral slope in the optical bands are highly variable, but do not seem to be correlated with burst duration. There is evidence of "flux level" dependence in the sense that the spectra are flatter when the source is brighter and the spectrum is steeper when the source is faint (Landau et al. 1986 and Brown et al. 1989). This cannot be considered a general rule, since counter examples exist (3C 279 Webb et al. 1990). The lack of consistent correlations between UV-optical-IR spectral shapes and outburst durations and/or flux levels is probably an indication that the emission in this region is a blend of several synchrotron components and/or thermal emission (from a disk?). A further complication might be the location of the break frequency which might fall in this region for some sources. If the frequency of this "break" is variable as well, then this adds complications to the flux level dependency of the spectral slope.

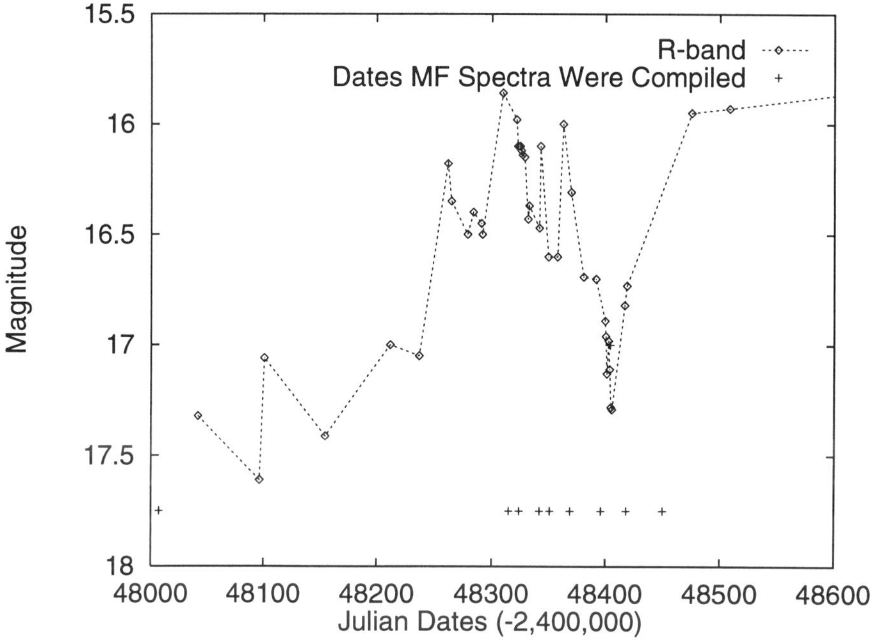

Figure 4. The 1991 Outburst of 3C 345

5. Spectral evolution

The *TOO* approach to analyzing MF spectral evolution was used to compile data on 3C 345 during the 1991 flare (Webb et al. 1994). The optical flare was detected by ground-based observers, then the IUE *TOO* program was activated. Many observers at many different observatories participated in data acquisition for this project. The program resulted in a series of multifrequency spectra throughout the duration of the outburst from X-ray to radio frequencies. Figure 4 shows the optical light curve of 3C 345 during the outburst. The plus signs along the bottom of the lightcurve indicate the times at which quasi-simultaneous spectra were compiled. Figures 5 through 7 shows examples of the multifrequency evolution of 3C 345 during the outburst. The solid lines in the figures represent a reference spectrum for comparison with the 1991 outburst data. The diamonds are data points observed during the course of the flare. The dates in the upper right hand corner of each plot indicate when the data were gathered for each spectrum (see Webb et al. 1988). The spectral sequence shows the flux level in the radio and optical well below the reference spectrum on 90-APR-23–26 (Fig. 5), then the rise in the high frequency GHz region and optical on 91-FEB-26–28 (Fig. 6), followed by the subsequent decline of the outburst (Fig. 7). Although the frequency coverage is lacking the multifrequency nature of the outburst is evident from the series of spectra. The importance of these observations is their ability to constrain theoretical models when they are taken together instead of analyzed individually.

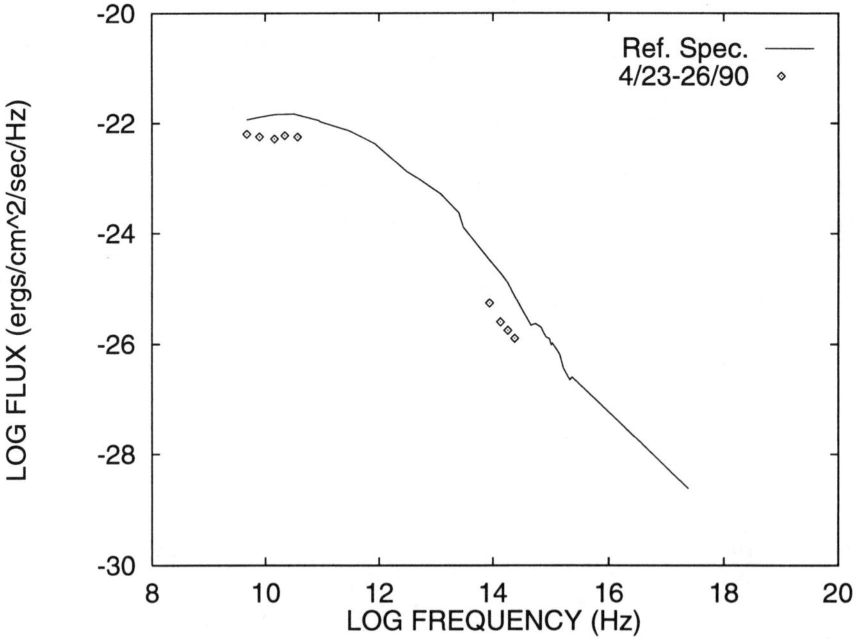

Figure 5. The pre-Outburst MF Spectrum in 1990

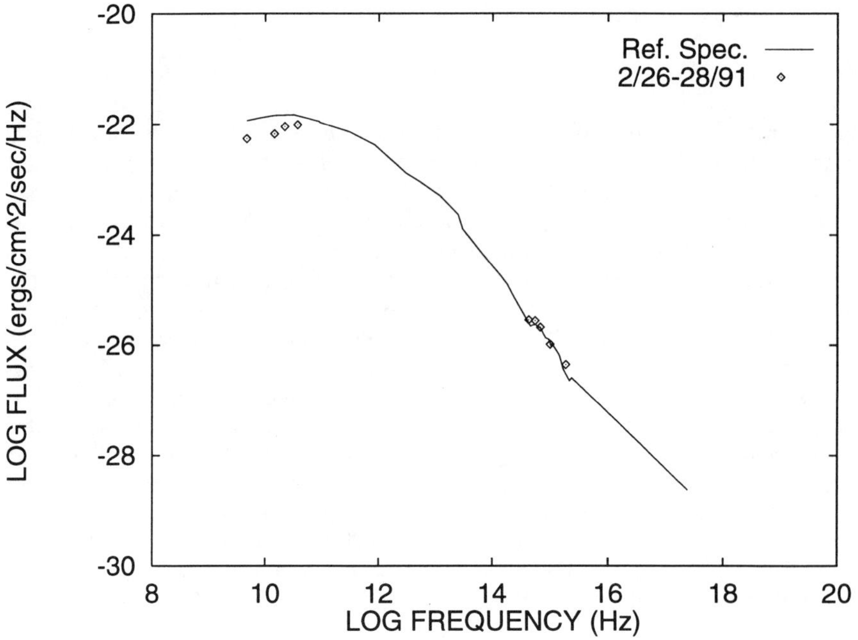

Figure 6. The Outburst MF Spectrum in February 1991

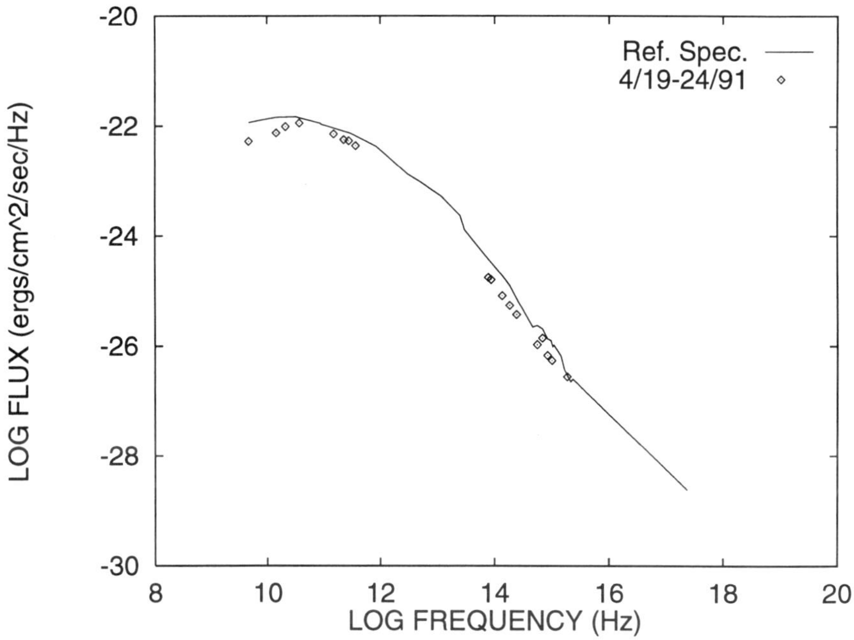

Figure 7. The post MF spectrum in June 1991

5.1. Multifrequency modelling

The usual models of blazars focus mainly on the jets and possible scenarios for illuminating the synchrotron electrons with seed photons for scattering. Whether the seed photons come from the jet, the disk, or a disk corona, these models have a multitude of parameters that allow them to fit any individual MF spectrum. Even for a simple jet model there are some parameters that are virtually unconstrained, and many that are only weakly constrained by the observations. Table 2 lists a set of model parameters for typical jet and disk models. The first column contains the physical variable and the second column its definition. The third and fourth columns give observational parameters that constrain the model parameters and their explanations.

The larger number of model parameters and the small number of observational constraints from a *single* MF spectrum makes the comparison of competing models ineffective. On the basis of physical arguments, some of the parameters should not vary on short timescales (e.g., mass of BH). The procedure followed in the case of 3C 345 is as follows:

1. fit the MF spectrum by varying the model parameters,
2. fit each subsequent MF spectrum using the initial model parameters and varying the fewest number of parameters possible,
3. vary only parameters that might physically be changing.

Table 2. Model and Observational Parameters

θ	angle WRT line of sight	F_{SC}/F_{OTS}	Ratio of scattered
ϕ	Opening angle of the jet		to opt. thin flux
K_1	Electron normalization	F_{ot}	Opt. thin flux
m	Electron radial parameter	α_{ot}	Opt. thin slope
Γ_u	upper elec ener cutoff	ν_{break}	Break freq.
Γ_l	lower elec ener cutoff	$\nu_{turnover}$	Turnover freq.
B_1	Mag. Field norm.	$F_{turnover}$	Turnover flux
n	Mag. Field radial par.		
R_m	inner radius		
δ	Doppler Boost factor	F_{SC}/F_{OTS}	Flux Ratio
M	accretion disk mass	$\alpha_{bluebump}$	Blue bump slope
$\dot{M}$	Accretion rate	$F_{bluebump}$	Blue bump flux

This method is an attempt to reduce the degrees of freedom using the existing observational parameters and several MF spectra. It was found that reasonable fits were obtained after the initial model fit by varying only the accretion rate $\dot{M}$ and the electron normalization N_1. (It was assumed that the "blue bump" was from thermal emission from the disk.)

This approach can be utilized to differentiate between the many models that can all reproduce the single-epoch MF spectrum equally well. This approach also demands that several "quasi-simultaneous" MF spectra be acquired during an outburst. Light curves in each of the widely separated frequency bands, radio through γ-rays, also must be utilized in order to differentiate between models such as the SSC, external SC, and disk-SSC models (Marscher 1996).

5.2. MF work in progress

An outburst in the blazar PKS 1156+295 was observed in 1994 with our IUE *TOO* program. Optical and UV data were gathered and the analysis of these data are now proceeding. Unfortunately, the outburst was not particularly well-covered in frequency space.

Multifrequency observations of BL Lac were made between August and November of 1994 during a one magnitude outburst. Observations include IUE LWP spectra, BVRI optical data, mm and GHz data during this period. BL Lac is notorious for undergoing rapid outbursts that last days or weeks, giving very little time to organize *TOO* observations, but the outburst was fairly well observed in this instance. BL Lac is also known to remain faint a large portion of the time and thus makes it difficult to observe with *PCC* programs as well. The 1994 flare was well observed in the optical and UV bands, but very little data exists in the mm where it is critical to estimate the optically thin synchrotron power-law slope.

Due to the enormous amount of work done by the OJ-94 consortium in coordinating observations of OJ 287, a wealth of multifrequency data has been compiled on this source. MF observations of OJ 287 which cover the entire accessible spectrum were made during the 1994 flare of OJ 287. The outburst was observed with excellent spectral and temporal coverage and includes radio, optical, IUE, ASCA, ROSAT and CGRO data.

A preliminary report of the status of these observations was given at the OJ-94 Conference held at Oxford University (Webb et al. 1995) and Pian et al. 1995).

6. Energetics and polarization

Blazar optical outbursts typically involve maximum luminosities on the order of 2×10^{48} ergs/sec in the optical region alone, and total observed energies on the order of 2×10^{55} ergs (Pollock 1982 and Webb 1992), the most luminous being AO 0235+164 with luminosities exceeding 1.2×10^{49} ergs/sec (Pollock 1982). The tremendous energies released during typical outbursts require that the disk component play an important role in energy storage in these sources, even if the disk component is not sufficiently luminous to contribute to the multifrequency variability. The disk would undergo catastrophic changes during a major outburst if the energy originated in the disk and these changes might contribute to the flux variations in the MF spectrum. A good way of isolating the disk emission from the jet emission is through simultaneous MF polarization observations (Smith et al. 1987, Hagen-Thorn et al. 1994). Polarization observations might be pivotal in determining the relative importance of the many physical models being investigated.

7. Conclusions

In summary, the three types of observational programs for observing the multifrequency spectra of blazars are all important and all can yield necessary data. The *Archival Spectral Composition* method, although useful, is not the ideal way of constructing MF spectra. The *Pre-Coordinated Campaign* is the best way to ensure broad spectral coverage and dense spectral coverage, but is less effective if the is source faint. The *Target of Opportunity* program is difficult to schedule, requires much effort to get adequate frequency and temporal coverage, but offers an important way of obtaining MF spectra during extreme activity in the source.

Although the number of blazars that have been properly observed using the MF approach is small, the trend is increasingly going toward cooperation among observers to participate in these types of programs. As these proceedings are being edited, several programs observing number of blazars are in progress (Pesce 1996).

The first lesson we learned from simultaneous MF programs of blazars was that simultaneity means within hours, not days, and certainly not months. Although the relativistic jet models seem capable of fitting the data, the variations place needed restrictions on models. The lack of MF variability data does not yet allow us to distinguish between sources of the scattered radiation, or determine whether a disk component is important to the emission.

Correlations between widely separated spectral regions are difficult to determine, although some evidence exists for these various correlations (see Hartman, this volume). Consistent monitoring is the key to establishing or ruling out these various correlations. Correlations between spectral indices and flux levels have been noted in the optical and UV, in the sense that the brighter the source, the flatter the spectral slope. These correlations are noted for only a small number of sources.

Modelling the MF spectral evolution is perhaps the only way to eventually differentiate between competing jet/disk models. This requires many MF spectra during a period of extreme activity to place additional observational constraints on the many-parameter models and to impose some physical restrictions on which parameters may be variable. Polarization is also a key to deconvolving the possibly complicated emission sources in the UV–optical portion of the spectrum.

Acknowledgments. JRW would like to acknowledge the NASA JOVE program for supporting various phases of the *TOO* multifrequency work, a succession of NASA IUE *TOO* grants and the RAs and TOs at GSFC who helped to obtain the *TOO* observations of 3C 345 and BL Lac. I would also like to acknowledge the ESA grant SQ 080, Heinz Andernach and the staff at Vilspa for supporting the observations of OJ 287.

Most importantly I would like to acknowledge the work of all of my collaborators in these MF programs, including Chris Shrader, Tom Balonek, Margo and Hugh Aller, Alberto Sadun, Ian Robson, and all of the OJ-94 consortium, especially Leo Takalo and Mark Kidger.

References

Balonek, T. J. 1982, Ph.D. thesis, University of Massachusetts

Bonnell, J. T., Vestrand, W. T., & Stacy, J. G. 1994, ApJ, 420, 545

Bregman, J. N., Lebofsky, M. J., Aller, M. F., Reike, G. H., Aller, H. D., Hodge, P. E., Glassgold, A. E., & Huggins, P. J. 1981, Nature 293, 29

Bregman, J. N., Glassgold, A. E., Huggins, P. J., Aller, M. F., Aller, H. D., Hodge, P. E., Reike, G. H., Lebofsky, M. J., Pollock, J. T., Pica, A. J., Leacock, R. J., Smith, A. G., Webb, J. R., Balonek, T. J., Dent, W. A., O'Dea, C. P., Ku, W. H., Schwartz, D. A., Miller, J. S., Rudy, R. J., & LeVan, P. D. 1984, ApJ, 276, 454

Bregman, J. N., Glassgold, A. E., Huggins, P. J., and Kinney, A. L. 1986, ApJ, 301, 708

Bregman, J. N., Maraschi, L., & Urry, C. M. 1987, in The Scientific Accomplishments of IUE, ed. Y. Kondo, p. 685

Bregman, J. N., Glassgold, A. E., Huggins, P. J., Neugebauer, G., Soifer, B. T., Matthews, K., Elias, J. H., Webb, J. R., Pollock, J. T., Leacock, R. J., Smith, A. G., Aller, H. D., Aller, M. F., Hughes, P. A., Maccagni, D., Garilli, B., Giommi, P., Miller, J. S., Stephens, S., Balonek, T. J., Dent, W. A., Kinsel, W., Wisniewski, W. A., Williams, P. M., Brand, P. W. J. L., & Ku, W. H.-M. 1990, ApJ, 352, 574

Brown, L. M. J., Robson, E. I., Gear, W. K., Crosthwaite, R. P., McHardy, I. M., Hanson, C. G., Geldzahler, B. J., Webb, J. R. 1986, MNRAS, 219, 671

Brown, L. M., Robson, E. I., Gear, W. K., Hughes, D. H., Griffen, M. J., Geldzahler, B. J., Schwartz, P. R., Smith, M. G., Smith, A. G., Sheppard, D. W., Webb, J. R., Valtaoja, E., Teräsrantra, H., & Salonen, E. 1989, ApJ, 340, 129

Falmo, R., Treves, A., Chiappetti, L., Maraschi, L., Pian, E., & Ranzi, E. G. 1993, ApJ, 402, 532

Clements, S. D. 1995, Ph.D. thesis, University of Florida

Glassgold, A. E., Bregman, J. N., Huggins, P. J., Kinney, A. L., Pica, A. J., Pollock, J. T., Leacock, R. J., Smith, A. G., Webb, J. R., Wisniewski, W. Z., Jeske, N. , Spinrad, H., Henry, R. B. C., Miller, J. S., Impey, C., Neugebauer, G., Aller, M. F., Aller, H. D., Hodge, P. E., Balonek, T. J., Dent, W. A., & O'Dea, C. P. 1983, ApJ, 274, 101

Grandi, et al. 1995, ApJ

Hagen-Thorn, V. A., Yakoleva, V. A., Takalo, L. O., and Sillampää, A. 1994, A&A, 290, 1

Hartman, R. C., Webb, J. R., Marscher, A. P., Travis, J. P., Dermer, C. D., Aller, H. D., Aller, M. F., Balonek, T. J., Bennett, K., Bloom, S. D., Fujimoto, R., Hermsoen, W., Hughes, P., Jemkins, P., Kii, T., Kurfess, J. D., Makino, F., Mattox, J. R., von Montigny, C., Ohashi, T., Robson, I., Ryan, J., Sadun, A., Schönfelder, V., Smith, A. G., Teräsrantra, H., Tornikoski, M., & Turner, M. J. L. 1996, ApJ, 461, 698

Hufnagel, B. R., & Bregman, J. N. 1992, ApJ, 386, 473

Impey, C. R., & Neugebauer, G. 1988, AJ, 95, 307

Kawai, N., Matsuoka, M., Bregman, J. N., Soifer, B. T., Balonek, T. J., Webb, J. R., Moriarty-Schieven, G. H., Chambers, K. C., Hughes, V. A., Clegg, R. E. S., Roche, P. F., Miley, G. K., Valtaoja, E., Goodrich, R., Miller, J. S., Hereld, M., Clements, S. D., Leacock, R. J., Smith, A. G., Aller, H. D., Aller, M. F., Hughes, P. A., Hoare, M. G., Wills, B. J., Wills, D., Ohashi, T., Balbus, S. A., and Thronson, H. A. 1991, ApJ, 382, 508

Landau, R., Golisch, B., Jones, T. J., Jones, T. W., Pedelty, J., Rudnick, L., Sitko, M. L., Kennet, J., Roellig, T., Salonen, E., Urpo, S., Schmidt, G., Neugebauer, G., Matthews, K., Elias, J. H., Impey, C., Clegg, P., & Harris, S. 1986, ApJ, 308, 78

Makino, F., Kii, T., Hayashida, K., Ohashi, T., Turner, M. J. L., Sadun, A. C., Urry, C. M., Neugebauer, G., Matthews, K., Tersärantra, H., & Aller, M. F. 1990, in Varaibility of Active Galactic Nuclei, ed. Miller & Wiita (Cambridge)

Maraschi, L., Chiappetti, L., Falamo, R., Garilli, B., Malkan, M., Tagliaferri, G., Tanzi, E. G., & Treves, A. 1991, ApJ, 368, 138

Maraschi, et al. 1994, ApJ, 435, L91

Marscher, A. 1996, this volume

Pesce, J. 1996 (private communication)

Pian, et al. 1995, Adv. Space Res., 16, No.3, 57

Pollock, J. T. 1982, University of Florida Dissertation

Pomphrey, R. B., Smith, A. G., Leacock, R. J., Olsson, C. N., Scott, R. L., Pollock, J. T., Edwards, P., & Dent, W. A. 1976, AJ, 81, 489

Quirrenbach, A., Witzel, A., Wagner, S., Sanchez,-Pons, F., Kirchbaum, T. P., Wegner, R., Anton, K., Erkens, U., Haehnelt, M., Zensus, J. A., & Johnston, K. J. 1991, ApJ, 372, L1

Robson 1992, Variability of Blazars, ed. Valtaoja and Valtonen (Cambridge)

Sadun, A. C., 1992, JRASC, 86, 15

Saunders, et al. 1989, ApJ, 347, 29

Smith, P. S., Blaonek, T. J., Elston, R. Heckert, P. A. 1987, ApJS, 64, 459

Smith, P. S. 1986, Ph.D. thesis, University of New Mexico

Stevens, J. A., Litchfield, S. J., Robson, E. I., Gear, W. K., Teräsranta, H., Valtaoja, E., & Tornikoski, M. 1994, ApJ, 437, 91

Tanzi, E. G., Barr, P., Bouchet, P., Chiappetti, L., Cristiani, S., Falomo, R., Giommi, P., Maraschi, L., & Treves, A. 1986, ApJ, 311, L13

Webb, J. R., & McHardy, I. 1986, Proc. Workshop on the Continuum Emission of Active Galactic Nuclei, ed. M. Sitko (Tucson)

Webb, J. R., Carini, M. T., Clements, S., Fajardo, S., Gombola, P. P., Leacock, R. J., Sadun, A. C., & Smith, A. G. 1990, AJ, 100, 1454

Webb, J. R. 1992, in Variability of Blazars, ed. Valtaoja & Valtonen (Cambridge)

Webb, J. R., Shrader, C., Balonek, T. J., Crenshaw, D., Kazanas, D., Clements, S., Smith, A. G., Nair, D., Leacock, R. J., Gombola, P. P., Sadun, A., Miller, H. R., Robson, I., Fujimoto, R., Makino, F., Kii, T., Aller, H., Aller, M., Hughes, P., Valtaoja, E., Tersärantra, H., Salonen, E., Tornikoski, M., and Chism, W. 1994, ApJ, 422, 570

Webb, J. R., Shrader, C., Nair, D., Balonek, T., Aller, M., Aller, H., Bonnell, J., Heinamaki, P., Hewit, N., Hughes, P., Jenkins, P., Kidger, M., Lehto, H., McCollum, B., McHardy, I., Okyuda, Persimo, T., Pian, E., Poyner, G., Del Rio, L., Sadun, A., Sillanpaa, A., Smith, A., Robson, I., Takalo, L., Teräsrantra, H., Treves, A., Valtaoja, E., Worraker, W. 1996, Workshop on OJ 287 (Oxford)

Wills, B. J., Pollock, J. T., Aller, H. D, Aller, M. F., Balonek, T. J., Barvainis, R. E., Binzel, R. P., Chaffee, F. H., Dent, W. A., Douglas, J. N., Fanti, C., Garrett, D. B., Gregorini, L., Henry, R. B., Hill, R. E., Howard, R., Jeske, N., Kepler, S. O., Leacock, R. J., Mantovani, F., O'Dea, C. P., Padrielli, L., Perley, P., Pica, A. J., Puschell, J. J., Sanduleak, N., Shields, G. A., Smith, A. G., Thuan, T. X., Wade, C. M., Wasilewski, A. J., Webb, J. R., Wills, D., & Wisniewski, W. Z. 1983, ApJ, 274, 62

Worrall, D. M., Puschell, J. J., Bruhweiler, F. C., Miller, H. R., Rudy, R. J., Ku, W. H.-M., Aller, M. F., Aller, H. D., Hodge, P. E., Matthews, K., Neugebauer, G., Soifer, B. T., Webb, J. R., Pica, A. J., Pollock, J. T., Smith, A. G., & Leacock, R. J. 1984, ApJ, 278, 521

Blazar Continuum Variability
ASP Conference Series, Vol. 110, 1996
H. R. Miller, J. R. Webb, and J. C. Noble, eds.

Multiwavelength Monitoring of the BL Lacertae Object PKS 2155–304: the May 1994 IUE Campaign

Elena Pian, C. Megan Urry, and Joseph E. Pesce

Space Telescope Science Institute, 3700 San Martin Drive, Baltimore, MD 21218

Laura Maraschi and Aldo Treves

Dpt. of Physics, University of Milan, via Celoria 16, I-20133 Milan, Italy

Steve Penton and J. Michael Shull

University of Colorado, JILA, Campus Box 440, Boulder, CO 80309-0440

Abstract. PKS 2155–304, the brightest BL Lac object in the ultraviolet sky, was monitored with both cameras of the IUE satellite at ~1 hour time-resolution for ten nearly uninterrupted days in May 1994. The campaign, which was coordinated with EUVE, ROSAT, and ASCA monitoring, along with optical and radio observations from the ground, yielded the largest set of spectra and the richest rapid variability information ever gathered for a blazar at UV wavelengths. The source flared dramatically during the first day, with an increase by a factor ~2.2 in an hour and a half. In subsequent days, the flux maintained a nearly constant level for ~5 days, then flared with ~40% amplitude for two days. These results are compared with previous IUE intensive monitoring of PKS 2155–304 in November 1991.

1. Introduction

The luminous emission from blazars spans the range from radio to γ-ray energies, and exhibits more rapid and higher amplitude variability than other AGN (Bregman 1990; Wagner & Witzel 1995). Therefore, simultaneous multiwavelength monitoring of blazars is particularly useful to estimate the sizes of the emitting regions (as a function of wavelength) and to understand, through correlated variability at different frequencies, the physical processes responsible for the radiation.

In PKS 2155–304, the brightest BL Lac object in the UV, the synchrotron power per decade (νF_ν) extends up to the medium X-ray range (Kii et al. 1996) and has a maximum between the UV and soft X-ray range (Edelson et al. 1995). The spectral steepening is attributed to radiative energy losses in the single shock model and to the decreasing volume of the region emitting at higher frequencies in the inhomogeneous jet model (Marscher 1980; Königl 1981; Ghisellini,

Maraschi, & Treves 1985; Hutter & Mufson 1986). In both cases, at or above the frequency of the peak power output, the highest amplitude synchrotron variability is expected to be observed, since small changes in the electron acceleration substantially alter the spectral emission.

Monitoring with IUE of PKS 2155–304 probed its variability in the far-UV domain (1200–3000 Å) on a range of timescales from years down to hours (Maraschi et al. 1986; Urry et al. 1988; Treves et al. 1989; Edelson et al. 1991; Urry et al. 1993, henceforth U93). The IUE campaign in November 1991 (U93), which was coordinated with ROSAT observations, had adequate time coverage (30 days) and sampling to probe significant interday variability on an extended time interval, and even intraday variability during the continuous observing period (~5 days out of 30). The presence of recurrent flares on a ~0.7-day timescale superimposed on a 10-day flux doubling prompted the need for further IUE intensive monitoring in May 1994, coordinated with EUVE (Marshall et al. 1996) and ASCA (Kii et al. 1996), plus additional ROSAT observations (Urry et al. 1996a). Radio, near-IR, and optical coverage was secured as well from ground telescopes (Pesce et al. 1996a,b). The aim of the IUE campaign was to obtain the longest and best ever sampled UV light curve, in order to test the shortest possible variation timescales, within the flux integration limit of the IUE instruments, and to explore the correlation with emission at other wavelengths (Urry et al. 1996a,b).

2. Observations and data analysis

IUE was scheduled for continuous observations (three 8-hr shifts per day) from 1994 May 15 to 25 inclusive, with some gaps due to Earth occultation and to other programs intersecting with ours, which however did not affect critically the information on variability. The SWP and LWP cameras were exposed alternately for 55 and 25 minutes respectively, to achieve comparable S/N in both cameras. In the absence of any operational problem, we obtained one pair of spectra during each 96 minutes. This time interval was chosen to phase with the ASCA satellite orbital period to allow cleaner cross-correlation analysis between the UV and X-ray light curves. Spectra were extracted from each of the 236 good IUE images using the TOMSIPS routine (Ayres 1993; Ayres et al. 1995), a modified version of the Slit-Weighted Extraction Technique (SWET) of Kinney, Bohlin, and Neill (1991). For consistency with U93 we corrected the data for interstellar extinction using $A_V = 0.1$ and the extinction curve of Seaton (1979).

For the spectral analysis we followed a similar procedure as used in U93. Through an iterative, chi-squared minimization fitting routine, the dereddened SWP spectral flux distributions (1230–1950 Å) were fitted in wavelength space to a simple power-law model of the form $F_\lambda \propto \lambda^{-\beta}$, ignoring the regions 1277–1281 Å, 1286–1290 Å, 1660–1666 Å, and 1780–1800 Å, which are affected by camera artifacts (Crenshaw, Bruegman, & Norman 1990), as well as regions in individual spectra affected by cosmic ray hits. Similarly, a power-law spectral fit was attempted to the LWP spectra in the 2100–2800 Å region, which is supposedly not heavily affected by the solar scattered light. We also fitted a power-law model to the pairs of SWP and LWP spectra taken close together in time. The fit was done over the wavelength range 1230–2800 Å, after excluding

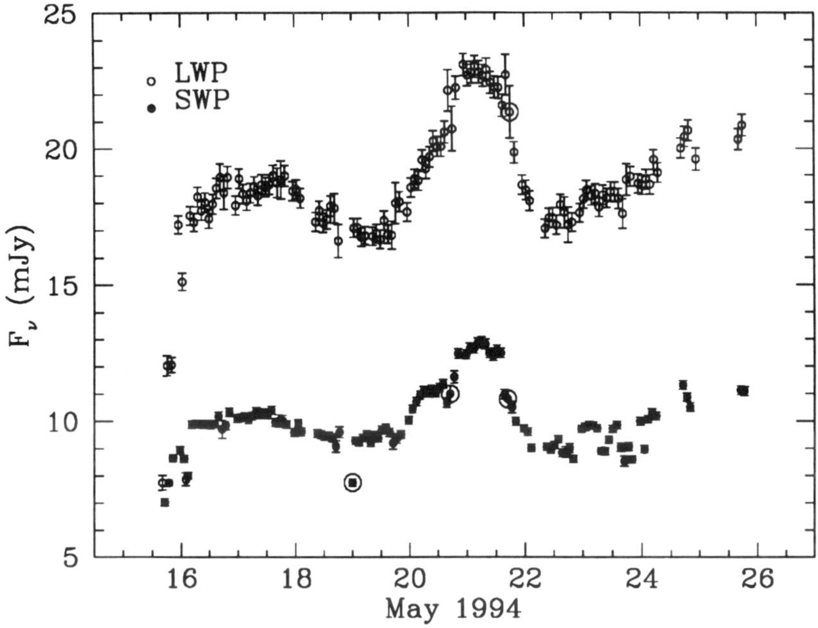

Figure 1. Dereddened light curves at 1400 Å (filled dots) and 2800 Å (open dots). The circled points correspond to underexposed spectra. Error bars represent the sum in quadrature of 1-σ uncertainties and a 1.25% photometric error.

the same regions which were discarded in the SWP spectra, and the 1900–2150 Å interval, which is affected by large errors.

3. Results and discussion

The result of the IUE campaign consisted in the best sampled IUE light curves ever obtained for an AGN, with 2.2 times the temporal extension of the intensive part of the November 1991 campaign (4.6 days, see U93).

Both IUE light curves clearly show strong fast variability (Fig. 1). The most striking result of the campaign is the high amplitude, extremely rapid flux variability detected during the first day of monitoring. At 2800 Å, flux variations as large as a factor 2.2 were found in 1.5 hours (i.e., the sum of the LWP and SWP integration times), implying that significant variability can occur on time scales shorter than the typical IUE integration time. The flux at 1400 Å also varied remarkably in the first part of the monitoring, though with less amplitude (~25% in 1.5 hours). This behavior is entirely new and unexpected, since the many previous IUE observations showed only minor differences in SWP and LWP variability, usually in the sense of a larger amplitude in SWP (Edelson 1992). There can be no doubt that the rapid fluctuation detected with the LWP is real and the smaller amplitude variations seen at shorter UV wavelengths

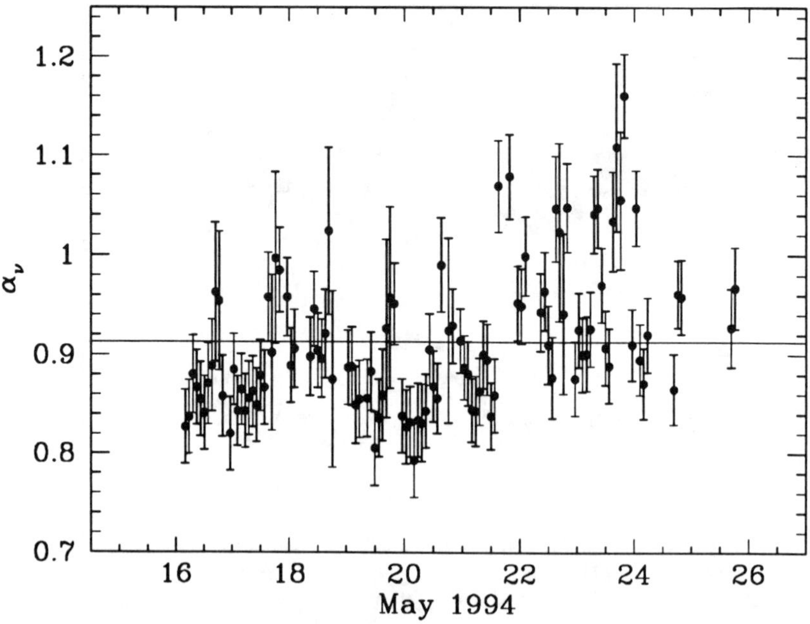

Figure 2. Spectral indices referring to the dereddened flux distributions in the 1230–2800 Å band ($\alpha_\nu = 2 - \beta$). The energy index was neither computed for underexposed spectra nor for the spectra during the first part of the monitoring, where the extremely rapid flux variability does not allow any meaningful match between SWP and LWP exposures. The horizontal solid line represents the average energy index.

could be ascribed to the longer integration time and the limited time resolution which does not allow us to adequately sample, in both cameras, the extremely rapid intrinsic variability.

In both light curves, after the very active period recorded at the beginning of the campaign, the flux stays relatively constant (after May 16), with a gentle increase and decline of ~10% over 3.5 days, and then there is a prominent 2-day flare, with a total increase of ~40%, starting on May 19. Further variation in the last four days of the monitoring is $\lesssim$ 20% in both bands. Throughout the monitoring, small (5–10%) and rapid (~ hours) flux variations are superimposed on the more dramatic flares. Spectral variations are detected (Fig. 2) but they are not clearly correlated with flux.

The completely resolved central flare, together with the ~50% and 170% correlated flares detected at the extreme UV and X-ray wavelengths respectively (Marshall et al. 1996; Kii et al. 1996; Urry et al. 1996a), is consistent with a variability amplitude monotonically increasing with energy, as expected for a synchrotron flare in an inhomogeneous jet (Celotti, Maraschi, & Treves 1991; Georganopoulos & Marscher 1996).

On average, the state presently detected was slightly lower than in 1991 both in SWP (∼22%) and in LWP (∼7%), with a steeper average spectral index (Fig. 2) and the observed flux variability behavior was remarkably different. In particular, we do not see the ∼0.7-day timescale quasi-periodicity seen in 1991 (see U93), which has therefore to be regarded as random or transitory.

Compared to the historical IUE light curve (Edelson et al. 1992; Pian & Treves 1993) the May 1994 UV flux was about average, whereas the spectral index is higher (steeper spectrum) than the mean value for PKS 2155–304, as derived in Pian and Treves (1993) and in Edelson et al. (1992). Note that the spectral indices given in Pian and Treves (1993) apply to the spectra dereddened with $A_V = 0.1$, while those in Edelson et al. (1992) are uncorrected for extinction.

Conspicuous spectral variations corresponding to the big central flare are not seen, a behavior reminiscent of other BL Lacs in optical and UV (e.g., OJ 287, Pian et al. 1996; Sillanpää et al. 1996), but different with respect to OVV blazars, like 3C 279 (Bonnell, Vestrand, & Stacy 1994; Shrader et al. 1994).

Defining for the present data a flux doubling timescale as $\tau = \frac{F_1}{F_2 - F_1}\Delta t$ one obtains $\tau_{\rm SWP} = 6.87$ hr and $\tau_{\rm LWP} = 1.36$ hr as minimum values at 1400 Å and 2800 Å respectively during the monitoring. This is the shortest observed doubling timescale for PKS 2155–304 either in the UV or optical range, where typical values are of the order of days (see U93; Carini & Miller 1992; Miller 1996). In X-rays, Morini et al. (1986) reported a doubling timescale of ∼1 hr based on EXOSAT observations. Similar events were observed in the X-ray band in X-ray bright blazars (see e.g., H 0323+022, Feigelson et al. 1986; PKS 2155–304, Treves et al. 1989, Sembay et al. 1993; PKS 0716+714, Cappi et al. 1994; Mkn 421, Macomb et al. 1995). Thus, these monitoring data have shown that the variability timescales in UV are comparable to those seen at the X-ray energies. A complete and more extensive analysis and discussion of the data presented here will appear in a paper currently in progress (Pian et al. 1996b).

References

Ayres, T. R. 1993, PASP, 105, 538

Ayres, T. R., et al. 1995, ApJS, 96, 223

Bonnell, J. T., Vestrand, W. T., & Stacy, J. G. 1994, ApJ, 420, 545

Bregman, J. N. 1990, A&A Rev., 2, 125

Cappi, M., Comastri, A., Molendi, S., Palumbo, G. G. C., Della Ceca, R., & Maccacaro, T. 1994, MNRAS, 271, 438

Carini, M., & Miller, H. R. 1992, ApJ, 385, 482

Celotti, A., Maraschi, L., & Treves, A. 1991, ApJ, 377, 403

Crenshaw, M. D., Bruegman, O. W., & Norman, D. J. 1990, PASP, 102, 463

Edelson, R. A., et al. 1991, ApJ, 372, L9

Edelson, R. A. 1992, ApJ, 401, 516

Edelson, R. A., Pike, G. F., Saken, J. M., Kinney, A., & Shull, J. M. 1992, ApJS, 83, 1

Edelson, R. A., et al. 1995, ApJ, 438, 120
Feigelson, E. D., et al. 1986, ApJ, 302, 337
Georganopoulos, M., & Marscher, A. P. 1996, these proceedings
Ghisellini, G., Maraschi, L., & Treves, A. 1985, A&A, 146, 204
Hutter, D. J., & Mufson, S. L. 1986, ApJ, 301, 50
Kii, T., et al. 1996, in preparation
Kinney, A. L., Bohlin, R. C., & Neill, J. D. 1991, PASP, 103, 694
Königl, A. 1981, ApJ, 243, 700
Macomb, D. J., et al. 1995, ApJ, 449, L99
Maraschi, L., Tagliaferri, G., Tanzi, E. G., & Treves, A. 1986, ApJ, 304, 637
Marscher, A. P. 1980, ApJ, 235, 386
Marshall, H., et al. 1996, in preparation
Miller, H. R. 1996, these proceedings
Morini, M., et al. 1986, ApJ, 306, L71
Pesce, J. E., et al. 1996a, in preparation
Pesce, J. E., et al. 1996b, these proceedings
Pian, E., & Treves, A. 1993, ApJ, 416, 130
Pian, E., Treves, A., Webb, J., Kazanas, D., Maraschi, L., McCollum, B., Shrader, C., & Wamsteker, W. 1996a, in Proceedings of the OJ-94 Project Meeting, Tuorla Observatory Reports, No. 176, ed. L. O. Takalo, p. 26
Pian, E., et al. 1996b, in preparation
Seaton, M. J. 1979, MNRAS, 187, 73p
Sembay, S., Warwick, R. S., Urry, C. M., Sokoloski, J., George, I. M., Makino, F., & Ohashi, T. 1993, ApJ, 404, 112
Shrader, C. R., et al. 1994, AJ, 107, 904
Sillanpää, A. 1996, these proceedings
Treves, A., et al. 1989, ApJ, 341, 733
Urry, C. M., Kondo, Y., Hackney, K. R. H., & Hackney, R. L. 1988, ApJ, 330, 791
Urry, C. M., et al. 1993, ApJ, 411, 614 (U93)
Urry, C. M., et al. 1996a, in preparation
Urry, C. M., et al. 1996b, these proceedings
Wagner, S. J., & Witzel, A. 1995, ARA&A, 33, 163

Blazar Continuum Variability
ASP Conference Series, Vol. 110, 1996
H. R. Miller, J. R. Webb, and J. C. Noble, eds.

Multiwavelength Observations of PKS 2155–304 in May 1994: the Ground-based Campaign

J. E. Pesce, C. M. Urry, and E. Pian

Space Telescope Science Institute, 3700 San Martin Drive, Baltimore, MD 21218

L. Maraschi

Osservatorio Astronomico di Brera, via Brera 28, I-20121 Milan, Italy

A. Treves

SISSA/ISAS, strada Costiera 11, I-34014 Trieste,Italy

P. Grandi

IAS/CNR, via Enrico Fermi 23, CP67, I-00044 Frascati, Italy

R. I. Kollgaard

Dept. of Astronomy and Astrophysics, Penn State University, University Park, PA 16802

H. Marshall

MIT, Room 37-667a, Cambridge, MA 02139

P. S. Smith and E. J. Hooper

Steward Observatory, University of Arizona, Tucson, AZ 85721

H. D. Aller and M. F. Aller

Dept. of Astronomy, Univ. of Michigan, Ann Arbor, MI 48109-1090

A. J. Barth and A. V. Filippenko

Department of Astronomy, University of California, Berkeley, CA 94720

D. A. H. Buckley, D. Kilkenny, F. Marang, F. van Wyk, and P. A. Whitelock

SAAO, P.O. Box 9, 7935 Observatory, Western Cape, South Africa

E. Covino

Osservatorio Astronomico di Capodimonte, via Moiariello 16, I-80131 Naples, Italy

M. D. Joner

Dept. of Physics and Astronomy, FB, Brigham Young University, Provo, UT 84602

L. B. G. Knee

Swedish-ESO Sub-mm Tel., ESO, Casilla 19001, Santiago 19, Chile

M. Kunkel

Max-Planck-Institut für Astronomie, Königstuhl 17, 69117 Heidelberg, Germany

A. C. Layden

Dept. of Physics and Astronomy, McMaster University, Hamilton, ON L8S 4M1 Canada

A. M. Magalhães, V. E. Margoniner, and A. Pereyra

Instituto Astron. e Geofísico, Univ. de São Paulo, São Paulo SP 01065-970, Brazil

C. Palma

Dept. of Astronomy, University of Virginia, PO Box 3818, Charlottesville, VA 22903-0818

C. V. Rodrigues

INPE-DAS, Caixa Postal 515, São José dos Campos, SP 12201-970, Brazil

A. Schutte

Siemens Telecom., 270 Maggs Street, Waltloo, Pretoria, South Africa

M. L. Sitko

Dept. of Physics, University of Cincinnati, Cincinnati, OH 45221-0011

M. Tornikoski

Metsahovi Radio Research Station, Metsahovintie 114, FIN-02540 Finland

J. van der Walt

Space Research Unit, Potchefstroom University, Potchefstroom 2520, South Africa

S. J. Wolk

SUNY, Stony Brook, New York 11794-2100

Abstract. The rapid variability of blazars is an important clue to the origin of their unusual continuum emission. Optical, near-infrared, and radio observations of PKS 2155–304 were obtained, simultaneous to a continuous UV/EUV/X-ray monitoring campaign in May 1994. Further optical observations were obtained regularly throughout 1994. The results from these observations are discussed. The source is seen to 'flicker' in the optical by 0.2–0.3 mag over several days throughout the observation period. The optical polarization and V-band flux correspond closely to the ultraviolet flux.

1. Introduction

This conference discusses blazars and their structure, in particular the scenario wherein they are dominated by relativistic jets aligned to the line of sight. The structure of the relativistic jet, however, remains largely unknown, as models are under-constrained by single epoch spectra (e.g., Königl 1981). This is why multiwavelength monitoring is needed, and to this end, the bright BL Lac object, PKS 2155–304 was the subject of a large multiwavelength campaign in November 1991 (Urry et al. 1993; Brinkmann et al. 1994; Courvoisier et al. 1995; Edelson et al. 1995). The results were intriguing but inconclusive. Therefore, in May 1994, PKS 2155–304 was again observed during a longer, more extensive multiwavelength campaign. Here we present the ground-based data. The UV (Pian et al. 1996 and these proceedings), EUV (Marshall et al. in prep.), X-rays (Kii et al. in prep.), and overall campaign (Urry et al. 1996 and these proceedings) are presented elsewhere.

2. Satellite observations

The high energy observations during the multiwavelength campaign commenced 11 May 1994 (MJD 9484) with the Extreme Ultraviolet Explorer satellite (9 days duration), followed four days later by IUE (10 days duration) and eight days later by ASCA (2 days duration). Rosat provided three brief observations on 6, 12, and 16 May. The continuous IUE, EUVE, and ASCA observations all show a prominent flare between 19–22 May.

3. Summary of ground-based observations

The May 1994 ground-based campaign to observe PKS 2155–304 shows variability at all wavelengths. The radio data (VLA, University of Michigan 26m single dish telescope) exhibit variability of a few percent at all frequencies (Figure 1). The high frequency radio data appear to increase by about 10%, with the 8.5 and 15 GHz data peaking together ~4 days after a similar peak in the 22.5 GHz data. No variability was apparent at millimeter wavelengths, as observed by SEST, although variations of ~30% could be hidden by the large errors.

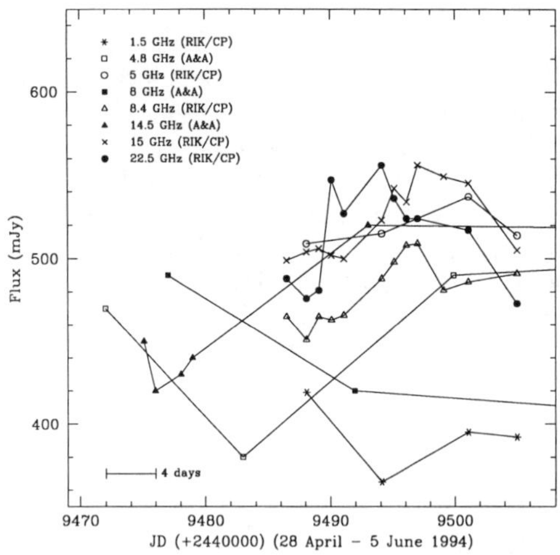

Figure 1. The radio light curves. Errors on the VLA data (RIK/CP) are ~10% and representative errors for the UMRAO data (A&A) are 40, 80, and 20 mJy at 4.8, 8.0, and 14.5 GHz, respectively. A scale bar is shown for comparison with the other figures.

The near-infrared (*JHKL*) bands were observed during 20–27 May, and an increase of ~ 0.3 mag is seen in all bands. Better coverage, though not ideal, was obtained in the optical *UBVRI* bands from late April to November. Figure 2 shows the optical light curves during the May campaign. The small increases of 0.2–0.4 mag (V- and R-bands) after MJD 9492 correspond to increases in the UV, with no apparent lag.

The entire April–November light curves are shown in Figure 3. All optical wavebands vary together, and the source exhibits flickering of 0.2–0.4 mag over timescales on the order 2 days. In early June (MJD $\sim$ 9500), a large "flare" is seen in all bands. The amplitude of this feature is 0.3–0.4 mag with a rise time of about 10 days and a duration of about 20 days. The maximum variation in any one waveband is about 0.3 mag in one day. This is all superposed on a general brightening by ~ 0.7 mag of the source from April to September (MJD 9610) and a drop of ~ 0.7 mag from September to our final observation in November. While none are apparent, we lack sufficient sampling to detect lags to better than a few days.

As can be seen in Figure 4, the V-band polarization dips by a factor of two (MJD 9486) and there is a 3-fold increase in all bands during 19–21 May (MJD 9492–9494). The increasing portions of the polarization curves occur simultaneously with the two increases in the UV flux (Figure 5).

Figure 6 shows the 3 June 1994 (MJD 9506.9875) Lick spectrum of PKS 2155–304. A power law of index $\alpha = -0.71 \pm 0.02$ (where $f_\nu \propto \nu^\alpha$) provides a good fit to the spectrum, which is featureless to ~ 0.2Å. Such a slope is typical for this object.

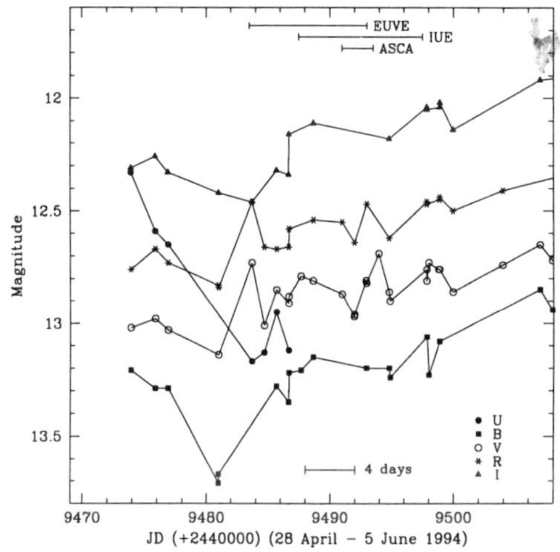

Figure 2. The optical light curves for May 1994. The durations of the EUVE, IUE, and ASCA experiments are shown at the top of the figure. Errors are the size of the points. A scale bar is shown for comparison with the other figures. The feature visible in the V- and R-bands at MJD ~9492 corresponds to the main UV flare (see Figure 5).

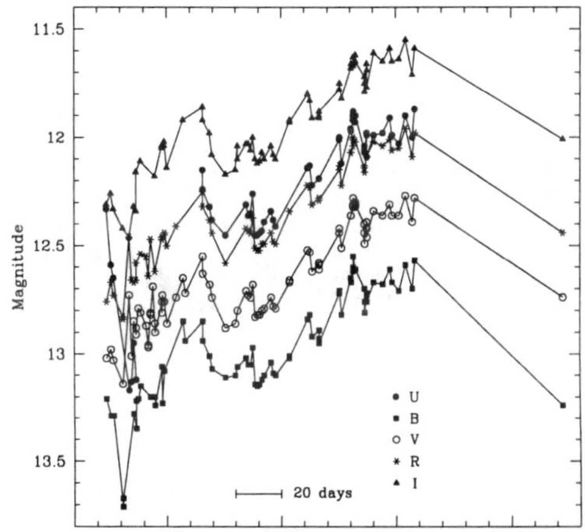

Figure 3. The complete optical light curves from April to November 1994. Errors are typically the size of the points. A scale bar is shown for comparison with the other figures. Variations, from short-scale "flickering" to the longer-term trends, are of similar amplitude and show no delay at all wavebands.

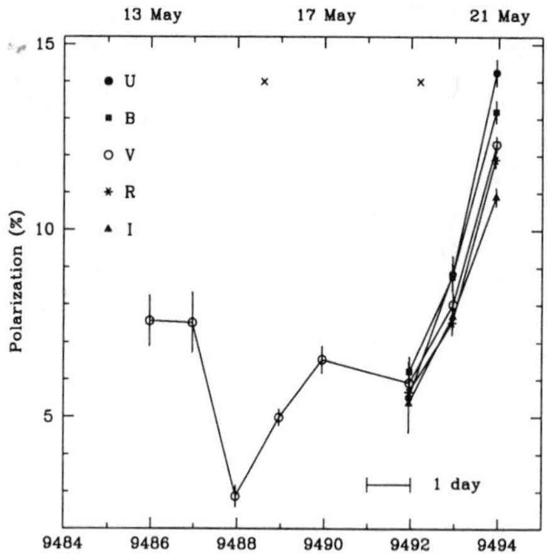

Figure 4. The polarization light curves from May 1994. A scale bar is shown for comparison with the other figures. The increase in polarization after MJD 9492 is similar in all wavebands, though stronger in the blue. The UV flaring events are marked with an 'X'. See also Figure 5.

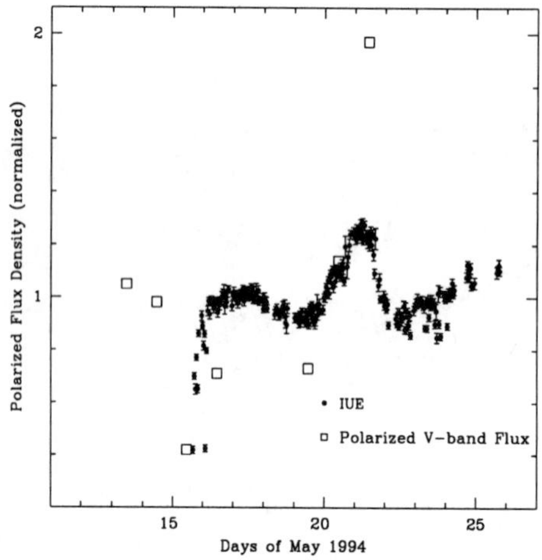

Figure 5. The normalized V-band polarized flux density from May 1994 compared to the normalized IUE SWP and LWP light curves. Note that the rise in polarization after 15 May (MJD 9488) coincides with an increase of UV flux at the beginning of the IUE monitoring period and the large increase in polarization after 19 May (MJD 9492) occurs at the start of the UV flare.

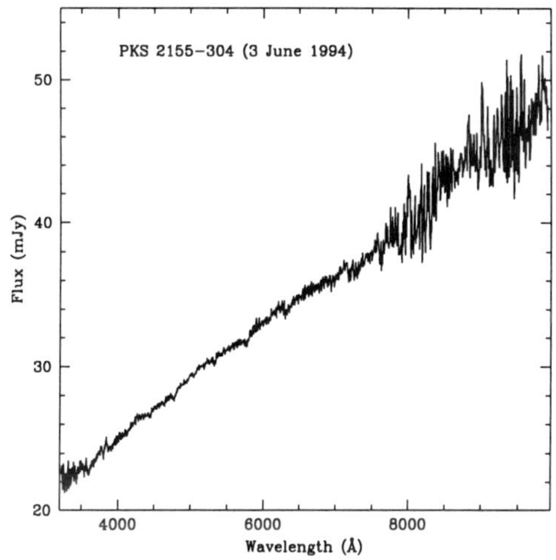

Figure 6. The optical spectrum of PKS 2155–304, from Lick (MJD 9506.9875). The high frequency oscillations most noticeable redward of 7500Å are produced by incompletely flattened CCD interference fringes. A power law ($\alpha = -0.71$) is a good representation of the featureless spectrum.

References

Brinkmann, W., et al. 1994, A&A, 288, 433

Courvoisier, T. J.-L., et al. 1995, ApJ 438, 108

Edelson, R. A., et al. 1995, ApJ 438, 120

Königl, A. 1981, ApJ, 243, 700

Pian, E., et al. 1996, ApJ, submitted

Urry, C. M., et al. 1993, ApJ 411, 614

Urry, C. M., et al. 1996, ApJ, submitted

Blazar Continuum Variability
ASP Conference Series, Vol. 110, 1996
H. R. Miller, J. R. Webb, and J. C. Noble, eds.

VLBI Imaging and Multiwavelength Variability of 3C279

A. E. Wehrle
Infrared Processing and Analysis Center, Jet Propulsion Laboratory and California Institute of Technology

S. C. Unwin
California Institute of Technology

A. C. Zook
Pomona College

C. M. Urry
Space Telescope Science Institute

A. P. Marscher
Boston University

H. Teräsranta
Metsähovi Radio Research Station, Helsinki University of Technology, 02540 Kylmälä, Finland

Abstract. We present 10 years of VLBI imaging of 3C279 at 22 GHz. We relate the appearance of a new blob to flaring in millimeter wavebands in 1991, and speculate on its possible connection to the June 1991 γ-ray flare. If a similar blob is associated with the dramatic γ-ray flare of February 1996, we should be able to track its emergence within a year.

1. Introduction

3C279 was the first quasar in which superluminal motion was discovered (Whitney et al. 1971). Along with 3C345 and 3C273, it was the object of VLBI monitoring programs using a global array of radio telescopes during the 1970's and 1980's (Cotton et al. 1979; Unwin et al. 1989; Carrara et al. 1993). The observations took place roughly once a year when the appropriate receivers were installed in the participating antennas of US and European VLBI Networks. By 1995, the routine operation of the National Radio Astronomy Observatory Very Long Baseline Array enabled more frequent observing sessions. We present 10 years of VLBI imaging at 22 GHz, and discuss the prospects for identifying changes in the parsec-scale structure with flares observed at millimeter and shorter wavelengths.

2. Observations

The first eight images (1984.75–1993.13) were made using networks of five to fifteen European and American telescopes. The 1993.85 and 1994.45 images were made using seven stations of the VLBA. The 1995.22 image was made with all 10 VLBA antennas. The resolution in the images is approximately 0.2 mas (0.8 pc) east-west, and varies from 0.6 mas to 3 mas north-south (FWHM of the synthesized beam). All the data were recorded using 2 MHz bandwidth in left circular polarization, except the last observation which used 16 MHz bandwidth.

The data were initially calibrated using system temperatures and gain curves of the antennas. The overall flux scale was established using the 22 GHz flux monitoring data from the Metsähovi antenna. Comparison of the Metsähovi total flux monitoring and NRAO Very Large Array imaging (de Pater & Perley 1983) showed that the flux density in structure resolved out by 22 GHz VLBI is about 2 Jy; the lack of short spacings in the uv-plane prevents adequate reconstruction of the radio emission on angular scales larger than a few milliarcseconds. Maps made from the 1984–1988 data were initially presented by Unwin et al. (1989) and Carrara et al. (1993). We have reanalyzed the data using the "difference mapping" technique implemented in Caltech VLBI software "Difmap" to improve the fidelity of the images. All images shown have been convolved with a Gaussian beam of 0.2 x 0.2 mas; they are "super-resolved" in the north-south direction.

3. Discussion

The images show that the parsec-scale radio structure is composed of a barely resolved core and a one-sided jet (Figures 1 and 2). The core dominates the 22 GHz emission at all epochs. There is a stationary feature with variable intensity, possibly a standing shock, about 1 mas from the core, which has persisted for at least 10 years. Leppanen, Zensus, and Diamond (1995) have reached the same conclusion by comparing a single very high quality VLBA image to those of Wehrle et al. (1994). A fairly strong blob is located about 2.5 mas from the core in the 1994.45 image; its motion can be traced back over at least five, possibly ten years. A new blob emerged from the core in mid-1991, the same year in which a strong γ-ray flare was detected by the Compton Gamma Ray Observatory EGRET instrument (Hartman et al. 1992). The new blob moves much more slowly (estimated at 0.15 mas/year) than the strong blob ejected in about 1968 and tracked in 1970–1971 (0.5 mas/year; Whitney et al. 1971). The old fast blob is not visible in the 22 GHz images. In Figure 3, we show the 1991.48 image and a newer, 10-station VLBA image from 1995.22; the new blob can be seen as a complex feature about 0.6 mas from the peak (core). The coincidence of the γ-ray flare and the emergence of the new blob in 1991 suggest that the two phenomena may have their origin in a common physical event, perhaps the passage of an unusual ensemble of relativistic charged particles through a shock located near the base of the relativistic jet. However plausible such an association might be, the interpretation rests on sparsely sampled observations. A similar association has been claimed for the γ-ray blazar 0528+134 where complex, rapid changes in the parsec-scale jet make it more difficult to track a

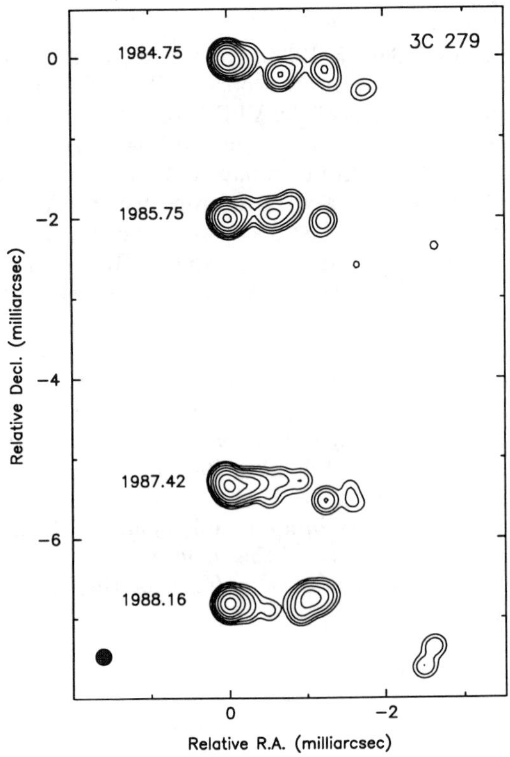

Figure 1. 22 GHz images of 3C279. Contour levels are 1, 2, 4, 8, 16, 32, and 64% of peak. Images were rotated counter-clockwise and "super-resolved" with 0.2 x 0.2 mas FWHM beam.

single component back to its origin (Krichbaum et al. 1995; see also Pohl et al. 1995).

Two major millimeter wavelength flares (1985, 1991) visible in the Metsähovi data (see light curves in Stevens et al. 1994 and Litchfield et al. 1995) may have been accompanied by the births of new blobs from the core. The identification of a particular flare/blob pair is difficult because it requires assumption of linear motion at a constant speed between epochs of VLBI observations. However, not every millimeter flare has resulted in the birth of a new blob which could be tracked, given the infrequent scheduling of 22 GHz global VLBI prior to the advent of the VLBA. We suggest that some millimeter flares give birth to very short-lived blobs which fade before they move more than about 0.2 mas from the core (0.8 pc in projected distance); the short lifespan would be only a few months. It is not known if this behavior is common in blazars because very little VLBI monitoring data at 22 GHz (or higher frequencies) on other blazars exists.

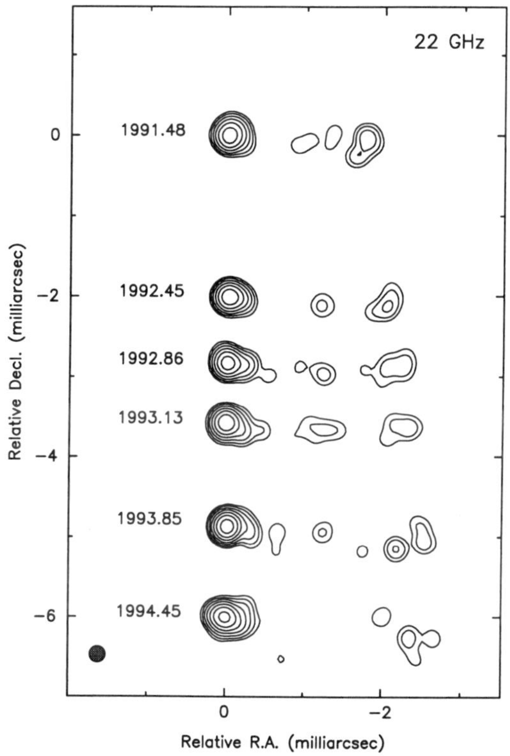

Figure 2. 22 GHz images of 3C279. Contour levels are 1, 2, 4, 8, 16, 32, and 64% of peak. Images were rotated counter-clockwise and "super-resolved" with 0.2 x 0.2 mas FWHM beam.

Litchfield et al. (1995) have used extensive millimeter-infrared monitoring data on 3C279 to model flares as phases in the evolution of a shocked relativistic jet, after Marscher and Gear (1985). The three phases may be distinguished by energy loss mechanism: inverse-Compton (growth), synchrotron, and adiabatic (decay). We don't know in which phase new VLBI blobs appear because our observations do not have enough spatial or temporal resolution.

Traditionally, superluminal blobs have been interpreted as shocked regions in the underlying flow of a relativistic jet. Leppanen, Zensus, and Diamond (1995) have shown that two blobs within the inner 0.5 mas in their 22 GHz image (epoch 1994.33) have high polarization with electric field vectors parallel to the jet direction. These two blobs are interpreted as regions where initially tangled magnetic fields have been compressed such that the resulting magnetic fields are perpendicular to the overall flow. Farther down the jet, the magnetic field changes direction by nearly 90 degrees just beyond the stationary blob (about 1 mas from the core); here a longitudinal magnetic field becomes dominant for

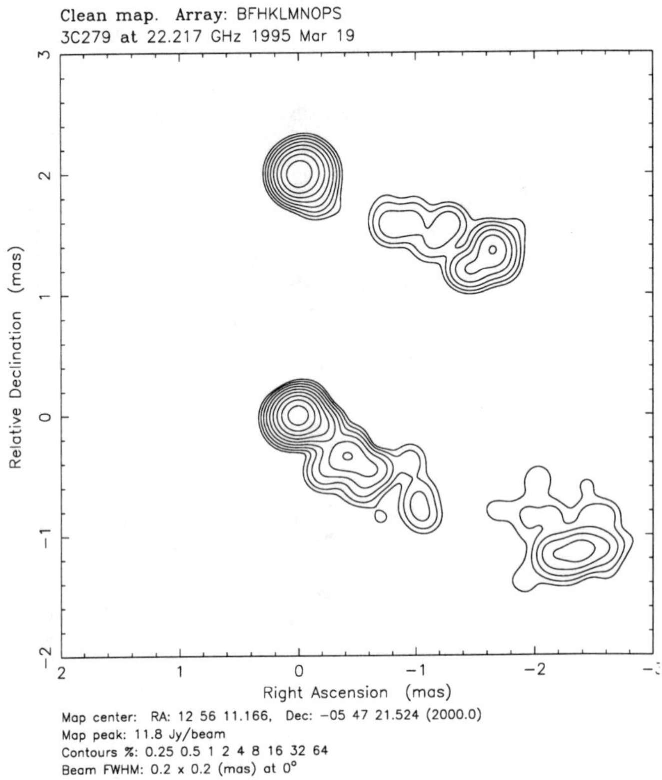

Figure 3. 22 GHz VLBI images of 3C279, epochs 1991.48 and 1995.22, restored with a " super-resolved" 0.2 x 0.2 mas FWHM beam.

unknown reasons. The blob about 2.0–2.5 mas from the core is also substantially polarized with magnetic field perpendicular to the jet; it may represent a shocked "working surface" with compressed magnetic field. Tracking the evolution of the polarization intensity and direction as the blobs age will show if the blobs change character abruptly at the location of the stationary feature or if they evolve gradually as they move along the jet path. Our VLBA monitoring observations now routinely include polarization measurements.

4. Prospects

In January 1996, 3C279 became significantly more active at optical and shorter wavelengths, including the EGRET γ-ray band (Wehrle & Hartman 1996; Hartman et al. 1996; Makino et al. 1996; Cooray & Fletcher 1996). On the first day of this conference, 3C279 produced a γ-ray flare three times brighter than the 1991 flare. Observations by XTE showed that X-ray and γ-ray flares occurred simultaneously within one day (see McHardy et al., these proceedings). This means that the X-rays and γ-rays must be produced co-spatially. Models which require a significant time delay between the two bands appear to be ruled out by the simultaneity of the flares. The short timescale of the flare (about 1–2 days) also provides a difficult constraint for models. Analysis of the flare data (including coordinated observations using ISO, IUE, ROSAT, ASCA, XTE, CGRO, serendipitously-scheduled HST and several dedicated ground-based telescopes, led by various observers) is in progress. VLBI observations were obtained in January 1996 with more observations scheduled within the next few months. Any new blob moving at the same speed as the 1991 blob should be trackable within the next year.

Acknowledgments. We wish to acknowledge the participating observatories of the US and European VLBI Networks who made this work possible: the National Radio Astronomy Observatory (US) , Metsähovi (Finland), Onsala (Sweden), Effelsberg (Germany), Owens Valley Radio Observatory (California Institute of Technology, US), Haystack Radio Observatory (NEROC, US), Medicina and Noto (CNER, Italy), and the staff of the Caltech Block II VLBI Correlator. We are grateful to Esko Valtaoja for stimulating discussions.

References

Carrara, E., Abraham, Z., Unwin, S. C., & Zensus, J. A. 1993, A&A, 279, 83

Cooray, A. R., & Fletcher, A. B. 1996, IAU Circular 6313

Cotton, W. D., et al. 1979, ApJ, 229, 115

Hartman, R. C., et al. 1992 ApJ, 385, 1

Hartman, R. C., Wehrle, A. E., & the Blazar Consortium 1996, BAAS, in press

Krichbaum, T. P., Britzen, S., Standke, K. J., Witzel, A., Schalinski, C. J., & Zensus, J. A. 1995, in Publ. of National Academy of Sciences 92, 11377

Leppanen, K., Zensus, J. A., & Diamond, P. J. 1995, AJ, 110, 2479

Litchfield, S. J., Stevens, J. A., Robson, E. I., & Gear, W. K. 1995, MNRAS, 274, 221

Makino, F., Inoue, H., Kii, T., Nagase, F., & Yamashita, A. 1996, IAU Circular 6302

Marscher, A. P., & Gear, W. K. 1985, ApJ, 298, 114.

McHardy, I., et al. these proceedings

de Pater, I., & Perley, R. 1983, ApJ, 273, 64

Pohl, M., et al. 1995 A&A, 303, 383

Stevens, J. A., Litchfield, S. J., Robson, E. I., Hughes, D. H., Gear, W. K., Teräsranta, H., Valtaoja, E., & Tornikoski, M. 1994, ApJ, 437, 91

Unwin, S. C., Cohen, M. H., Biretta, J. A., Hodges, M. W., & Zensus, J. A. 1989, ApJ, 340, 117

Wehrle, A. E., & Hartman, R. C. 1996, IAU Circular 6294

Whitney, A. R., et al. 1971, Science, 173, 225

Blazar Continuum Variability
ASP Conference Series, Vol. 110, 1996
H. R. Miller, J. R. Webb, and J. C. Noble, eds.

High Energy Variability and Blazar Emission Models

G. Ghisellini and L. Maraschi

Osservatorio Astronomico di Brera, via Brera 28, 20121 Milano Italy

Abstract. Recent observations of γ-ray emission and variability from blazars bear fundamental implications on the physical conditions of the emitting plasma. The available data seem to outline a simple picture, which can be basically understood using simple one-zone models involving synchrotron and inverse Compton radiation from the same population of relativistic electrons. We show how simple approximations allow to estimate physical parameters and variability correlations for the synchrotron and inverse Compton emission. We discuss specific models for the two best studied sources, 3C 279 and Mkn 421, for which simultaneous multifrequency data at different epochs are available. For 3C 279 we compare models involving different sources of seed photons for the Compton scattering process: synchrotron photons (SSC) or, alternatively, photons outside the jet, emitted and/or scattered in a Broad Line Region (EC), as seeds for inverse Compton scattering. A third possibility is a mirror model, where seed photons are generated in response to illumination of surrounding gas by the beamed radiation (MC). We conclude that all three alternatives are viable at present. For Mkn 421 we show that both the spectrum and variability can be understood within the SSC model. For both sources inhomogeneous jet models yield results in good agreement with one-zone models taking into account that the dominant emission region in the jet is at the transition between the parabolic and conical approximation to the jet geometry. The fact that the observed spectral variability of two sources as diverse as 3C 279 and Mkn 421 can be understood within a "general" synchrotron + inverse Compton scheme, irrespective of the dominant source of soft photons, strongly supports this picture. Well sampled multifrequency light curves are essential to further discriminate between different models.

1. Introduction

The discovery that many blazars emit large powers in rapidly variable high energy γ-rays (e.g., Hartman, this volume) has profound implications for theoretical models. The γ-ray emitting region must be transparent since γ-rays are indeed observed, yet the size limit imposed by the rapid γ-ray variability implies very high optical depths to photon-photon interaction, unless the γ-rays are relativistically beamed (Maraschi, Ghisellini, & Celotti 1992, Dondi & Ghisellini 1995, Becker & Kafatos 1995). This conclusion reinforces the arguments that the blazar continuum from radio to γ-ray frequencies is produced in relativistic jets (e.g., Blandford & Rees 1978).

The spectral energy distributions (SEDs) of γ-ray emitting blazars (νF_ν vs. ν) show two broad peaks, separating the continuum into two components, which are generally interpreted as arising from different emission mechanisms or from different particle populations (e.g., von Montigny et al. 1995). The first component peaks at different frequencies in different objects, for instance in the far infrared for 3C 279 and in the soft X-ray band in Mkn 421. In the X-ray band either one or the other component can prevail (the first in the case of Mkn 421, the second in the case of 3C 279), or a mix of the two can be present (see Sambruna, Maraschi, & Urry 1996, for a systematic study). The second component generally appears to peak in the γ-ray range. However there are large differences between different objects: the peak falls in the GeV range for 3C 279 and is close to the TeV range for Mkn 421 (see Fig. 1 and 2 below). It is worth noting that in these two best studied objects the second peak seems to "remember" the first, that is the data are consistent with a fixed separation between the two (on a $\log \nu$ scale).

Present observations indicate that γ-ray variations are correlated with variations at lower frequencies. This supports models involving a single population of relativistic electrons which emit synchrotron radiation up to the UV or X-ray band and Compton scatter soft photons up to the GeV or TeV range respectively. Some general remarks on spectra and variability in this scenario are presented in Section 2. We then discuss in detail spectral models for the two prototype objects mentioned above, using first a simple one-zone approximation and considering different sources of seed photons (Section 3). These are maximally constrained models, which imply strong correlations between the variations of the two components.

We discuss three alternatives. If synchrotron radiation dominates the energy density, the high energy flux is produced by self-Compton radiation (SSC) (Maraschi, Ghisellini, & Celotti 1992). The energy density of radiation produced externally to the blob may also be important for Compton scattering (EC) since it is enhanced by relativistic effects in the blob frame. Various contributions have been proposed: the radiation produced by an accretion disk (see Dermer, Schlickeiser, & Mastichiadis 1992), or reprocessed by the Broad Line Region (BLR) and/or scattered by material surrounding the jet (Sikora, Begelman, & Rees 1993, 1994, Blandford 1993 and Blandford & Levinson 1995). Finally, a portion of the BLR could be illuminated by the beamed jet radiation, provided that some broad line clouds (or scattering material) exist in the vicinity of the jet (Ghisellini & Madau 1996). We will call this model the mirror Compton model (MC). Possible criteria to discriminate between different types of seed photons through variability properties are summarized in Section 2. Finally we discuss how multi-zone (inhomogeneous) jet models compare with the simpler ones (Section 4).

2. One-zone synchrotron + Inverse Compton models: general considerations

Adopting a one-zone approximation, the spectral shape of the synchrotron emission reflects the energy distribution of the radiating electrons, which must therefore be curved to explain the observed spectra. The simplest case is a broken power law, where the break energy, $E_{\rm b} = \gamma_{\rm b} m_e c^2$, can result from a balance between radiative losses and expansion or escape. The inverse Compton spectrum mimics the shape of the synchrotron spectrum provided the seed photon spectrum is not too broad. Then the two peaks in the SED at $\nu_{\rm S}$ and $\nu_{\rm C}$ respectively must both be due to electrons with the break energy.

We also assume that the emitting blob moves with bulk Lorentz factor Γ, and that the variability timescale indicates its radius. Taking into account the usual relativistic correction, the size of the emitting region is assumed to be $R = ct_{\rm var}\delta/(1+z)$, where δ is the Doppler factor.

2.1. Spectral constraints

With the above assumptions one can derive approximate relations which allow to solve for the physical parameters of the blob in terms of five "observables:" the peak frequency and peak luminosity of the synchrotron emission ($\nu_{\rm S}, L_{\rm S}$), the peak frequency and peak luminosity of the inverse Compton emission ($\nu_{\rm C}, L_{\rm C}$) and the variability timescale $t_{\rm var}$. If the seed photons are the synchrotron photons themselves (SSC) the parameters of the emitting region (B, R, $N_{\rm el,b}, \gamma_{\rm b}, \delta$) can be unequivocally determined.

If the seed photons are produced outside the jet, (we consider here only the case of an external isotropic radiation, appropriate for reprocessed or scattered radiation, see Section 1) the external radiation energy density and its typical frequency are additional parameters which have to be specified to solve the system.

A compendium of these formulae can be found in Ghisellini, Maraschi and Dondi (1996) and a more extended discussion is in preparation. Here we only recall two important relations based on the fact that *the maximum powers per decade emitted via the synchrotron and inverse Compton mechanisms are both due to the same electrons with energy $\gamma_{\rm b}$*.

For SSC models we derive:

$$\gamma_{\rm b} \simeq \left(\frac{3\nu_{\rm C}}{4\nu_{\rm S}}\right)^{1/2}, \tag{1}$$

$$B\delta \simeq (1+z)\frac{\nu_{\rm S}^2}{3.7\times 10^6 \nu_{\rm C}} . \tag{2}$$

Isotropic ambient photons with typical frequency $\nu_{\rm o}$, will appear boosted in the blob frame to

$$\nu_0' = \Gamma\nu_0 = 10^{15}\Gamma\nu_{0,15} \quad \text{Hz}. \tag{3}$$

Scattering of this radiation by the relativistic electrons of energy $\gamma_{\rm b}$ will produce γ-rays at the observed frequency

$$\nu_{\rm C} = \frac{4}{3}\gamma_{\rm b}^2\nu_0\frac{\Gamma\delta}{1+z} . \tag{4}$$

The ratio ν_C/ν_S in this case yields

$$B = 3.6 \times 10^8 \, \Gamma \frac{\nu_S \nu_{0,15}}{\nu_C} \text{ Gauss.} \tag{5}$$

With the above formulae it is possible to compare the electron energies and magnetic fields obtained applying the same model to different objects or the SSC and the EC models to the same object. For instance using SSC for both 3C 279 and Mkn 421 and assuming $\nu_S = 10^{14}$ Hz, $\nu_C = 10^{22}$ Hz for 3C 279 and $\nu_S = 10^{17}$ Hz, $\nu_S = 10^{25}$ Hz for Mkn 421, one can estimate, for $\delta \simeq 10$, $B = 0.025$ G in 3C 279 and $B = 25$ G in Mkn 421, and the same $\gamma_b \simeq 10^4$ for both sources. However the elementary formulae above fail in the Klein-Nishina regime and for Mkn 421 higher electron energies are needed to explain a Compton peak at 10^{25} Hz. Detailed spectral models are necessary.

Comparing an SSC and an EC model for 3C 279, we can see that the magnetic field derived in the latter case is about 10^3 times larger than in the SSC model, implying that electrons emitting at the peak frequency have much lower energies. Self-absorption will substantially modify the synchrotron emission from these low energy electrons, again requiring detailed spectral calculations. Note that the mirror Compton model does not differ from an EC model as for the spectral properties.

We wish to stress however, that irrespective of the seed photons, in 3C 279 the IR photons are close to the peak of the synchrotron component and the EGRET γ-ray photons are at the peak of the Compton component, so these two should be strongly correlated. The X-ray emission in 3C 279 clearly belongs to the Compton component and is on the low frequency side of its peak. Therefore it should be due to Compton scattering by low energy electrons whose synchrotron emission in the cm–mm range may be self-absorbed. Instead, in Mkn 421, the X-ray photons are close to the first peak (on the high energy side), clearly due to the synchrotron mechanism and the TeV photons are close to the second peak on the high energy side, so the two should be strongly correlated.

2.2. Variability

The simplest approach to variability is to compare stationary models with different parameters. This is certainly adequate when comparing different "states" of the source, i.e., SEDs measured at intervals of months and years. A much more difficult problem, briefly mentioned in the last section, is to model multifrequency light curves taking into account some physical description of energy input and evolution of the particles in the jet.

The simple hypotheses considered above lead to definite predictions on the variations of the Compton scattered flux associated with a given variation of the synchrotron flux, which can be easily computed varying the model parameters. Again some general relations can be derived if we restrict ourselves to synchrotron and Compton photons originating from electrons with the same energy, i.e specifically to the two peaks in the power per decade. Given the dependence of the synchrotron flux on electron density N_{el}, magnetic field, and bulk Lorentz factor:

$$F_S \propto \Gamma^4 N_{el} B^2, \tag{6}$$

and the dependence of the Compton flux on the relevant parameters for the three possible sources of seed photons

$$F_{\rm SSC} \propto \Gamma^4 N_{\rm el}^2 B^2, \tag{7}$$

$$F_{\rm EC} \propto \Gamma^6 N_{\rm el} U_{\rm ext}\,, \tag{8}$$

$$F_{\rm MC} \propto \Gamma^8 N_{\rm el}^2 B^2, \tag{9}$$

it is easy to derive that for a variation of the electron density the first and third mechanism predict a variation $\Delta F_{\rm C} \propto \Delta^2 F_{\rm S}$ while for the second the relation is linear.

The full range of possibilities is summarized in Table 1. Note that for both SSC and MC a variation of the Compton flux always implies a variation of the (corresponding) synchrotron flux and vice-versa, while this is not true in the EC model, where a simple magnetic field compression could enhance the synchrotron flux without affecting the Compton one or a varying density of external photons could cause the Compton flux to vary without a variation of the synchrotron flux. It is important to stress that if more than one parameter varies, the behavior will be intermediate (e.g., in between quadratic and linear for SSC and SEC if the magnetic field varies together with the electron density and so on).

Table 1. Correlations between synchrotron and Compton fluxes

	$N_{\rm el}$	B	Γ	$U_{\rm ext}$
SSC	$F_{\rm C} \propto F_{\rm S}^2$	$F_{\rm C} \propto F_{\rm S}$	$F_{\rm C} \propto F_{\rm S}$	$F_{\rm S} = $ const. $F_{\rm C} = $ const.
EC	$F_{\rm C} \propto F_{\rm S}$	$F_{\rm C} = $ const. $F_{\rm S} \propto B^2$	$F_{\rm C} \propto F_{\rm S}^{3/2}$	$F_{\rm C} \propto U_{\rm ext}$ $F_{\rm S} = $ const.
MC	$F_{\rm C} \propto F_{\rm S}^2$	$F_{\rm C} \propto F_{\rm S}$	$F_{\rm C} \propto F_{\rm S}^2$	$F_{\rm C} = $ const. $F_{\rm S} = $ const.

In a homogeneous model the variability at $\nu_{\rm S}$ and $\nu_{\rm C}$ should be simultaneous for both the SSC and EC models. These two could be however distinguished from the different degree of correlation between $F_{\rm C}$ and $F_{\rm S}$. On the other hand, in the 'mirror' model, which mimics SSC producing larger variations in the γ-ray band than in the optical, the γ-rays should lag the optical photons. Thus multifrequency light curves especially on short timescales will be crucial to test different models.

The relations in Table 1 need to be corrected if the Klein-Nishina regime sets in. We stress that a comparison of *variability at frequencies not matched by a common electron energy requires more detailed computations.* Moreover to identify the peaks of the power distribution from the available data, which always have wide gaps in the frequency coverage, is to some extent arbitrary. In the following we therefore discuss specific models for the SEDs and variability of the two best studied γ-ray blazars, 3C 279 and Mkn 421.

3. Specific models for the high and low states of 3C 279 and Mkn 421

3.1. 3C 279

We tried to reproduce the observed SEDs of 3C279 in high (June 1991) and low (January 1993) state using first the SSC and secondly the EC scenarios.

In the SSC case we assumed a broken power law energy spectrum for the electrons allowing the slopes and normalization to vary between the two states. The spectral indices of the two power laws are chosen to match the slopes of the synchrotron SED below and above the peak frequency. Their values have a minor influence on the model parameters. The results are shown in Fig. 1. The particle distribution is a broken power law, with the break at $\gamma_b = 1.2 \times 10^4$. $R = 2 \times 10^{17}$ cm, $B = 0.03$ Gauss, and the Doppler factor $\delta = 16$. The two states are obtained by varying the normalization of the particle distribution at γ_b by a factor 3.5, and by varying the slopes n_1 and n_2 below and above γ_b, respectively $[N(\gamma) \propto \gamma^{-n_{1,2}}]$. For the high state, $n_1 = 2$, $n_2 = 4.2$, while for the low state $n_1 = 2.4$, $n_2 = 5$.

We note that at low frequencies additional contributions are required, possibly from different (larger) regions of the jet. The model reproduces rather well the two states of the source varying only the normalization at γ_b and the slopes of the electron energy spectrum.

In the EC case we followed the model of Sikora, Begelman and Rees (1993 & 1994) considering a particle distribution resulting from continuous injection and incomplete cooling. In the emitting region, of size of $R = 10^{17}$ cm, relativistic electrons are continuously injected with a spectrum $Q(\gamma) \propto \gamma^{-1.7}$ [$\mathrm{cm}^{-3}\ \mathrm{s}^{-1}$] up to $\gamma_{\mathrm{max}} = 3500$ and power $L_{\mathrm{inj}} \sim 7.4 \times 10^{43}$ erg/s (in the comoving frame). The equilibrium particle and photon distributions are calculated taking into account radiation losses, Coulomb collisions, Klein-Nishina effects, particle escape and pair production. The result approximates a broken power law with a break energy at which the escape time equals the cooling time. The model is shown in Fig. 1.

As anticipated the magnetic field is much larger than in the SSC case, and the break energy is rather low, $\gamma_b \sim 200$. As a result the flat injected spectrum does not show up in the synchrotron component due to self-absorption. The steeper, cooled electron spectrum extends to $\gamma_{\mathrm{max}} = 3500$. The very steep IR to UV spectrum is due both to the steepening of the electron distribution and to the synchrotron tail emitted by the most energetic electrons.

Note that the Compton scattered spectrum rises rather abruptly at 10^{18} Hz. This represents the minimum frequency shift of the UV photons scattered by the lowest energy electrons. This should not be a serious problem as the "hole" is partially filled in by the SSC component. It could also conceivably be filled in if a broader spectrum was assumed for the seed photons. Note also that Compton

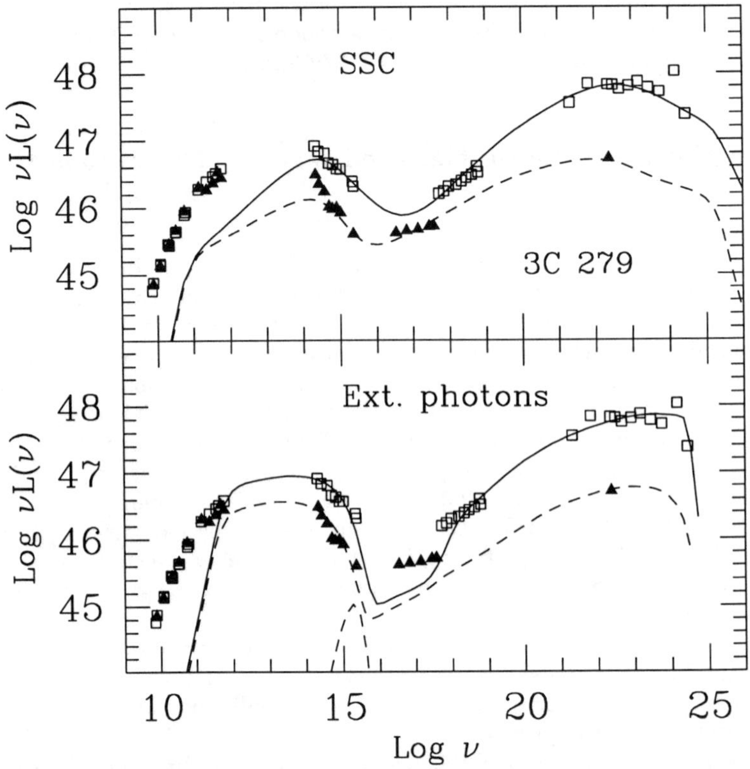

Figure 1. Overall spectral distribution of 3C 279 in June 1991 (squares) and during the 1993 campaign (triangles). *Top panel:* solid dashed lines refer to a homogeneous SSC model. $R = 2 \times 10^{17}$ cm, $B = 0.03$ Gauss, and the Doppler factor $\delta = 16$. The particle distribution is a broken power law, with the break at $\gamma_{\rm b} = 1.2 \times 10^4$. *Bottom panel:* fit with the model of Sikora et al. (1994). In the emitting region, of size of $R = 10^{17}$ cm, relativistic electrons are continuously injected with a spectrum $Q(\gamma) \propto \gamma^{-1.7}$ [cm^{-3} s^{-1}] up to $\gamma_{\rm max} = 3500$ and power $L_{\rm inj} \sim 7.4 \times 10^{43}$ erg/s (in the comoving frame). The ratio of external to internal radiation energy densities is higher in the high state, but still larger than one in the low state. This is achieved by changing the bulk motion Lorentz factor between the two models, from 18 to 11, and the magnetic field value seen in the comoving frame, from 1.5 to 1.9 Gauss. In this case the curved SED of the model is due to the incomplete radiative cooling of the particles. The synchrotron spectrum corresponding to the rising part of the inverse Compton spectrum is not visible because it is self-absorbed.

scattered photons from X-rays up to low energy γ-rays derive from electrons whose synchrotron emission is not visible due to self-absorption. Finally, an intrinsic feature of the model is the high-energy cut-off which appears at $\simeq 10^{25}$ Hz, due to absorption of the most energetic γ-rays by the external radiation field. $L_{\rm BLR} = 10^{44}$ erg/s, $R_{\rm BLR} = 10^{18}$ cm.

To reproduce the larger variation in γ-rays between the two states we varied the bulk Lorentz factor, from 18 to 11, so as to change the photon energy density seen by the blob. We also varied the magnetic field in the comoving frame from 1.5 to 1.9 Gauss. All other parameters are unchanged. The ratio of external to internal radiation energy densities is already larger than one in the low state and increases in the high state.

With respect to the SSC scenario this model can account for a larger fraction of the IR to optical emission as due to the same region and predicts a sharper high energy cut off. The cut off should shift to lower energies with increasing intensity, eventually producing a steeper γ-ray spectrum in the EGRET range. In both models the size of the emission region is of the order of 10^{17} cm, consistent with a variability timescale of few days, once Doppler contraction is taken into account.

4. Mkn 421

We modeled the overall emission of the BL Lac Mkn 421 with a homogeneous SSC model, in which the equilibrium particle distribution is found balancing continuous injection, cooling and particle escape. Here the synchrotron peak is observationally constrained to lie between the UV and soft X-ray bands while the IC peak is between the GeV and TeV regions, therefore requiring an extremely high $\gamma_{\rm b} > 10^5$. In this case Klein-Nishina effects are important, and only synchrotron photons up to the IR–optical bands undergo efficient up-scattering with the most energetic electrons. We consider two states of the source as reported by Macomb et al. (1995). The high state refers to simultaneous data obtained during the flare in May 1994, the low state is a composite of previous observations. In Fig. 2 we show the spectra resulting from two homogeneous SSC models with different rates of particle injection and different maximum energy of the injected particles.

The reduced scattering efficiency in the Klein-Nishina regime yields a Compton luminosity comparable to the synchrotron one despite the fact that the internal radiation energy density is larger than $U_{\rm B}$. In practice only photons below the Klein-Nishina threshold $(m_e c^2)^2/E_{\rm el}$ are effective in the scattering process. This makes the effective energy density substantially lower than the total one, which peaks in the UV–soft X-ray range. For the same reason external UV photons would be extremely inefficient for up-scattering in Mkn 421 due to their even higher frequency in the jet frame. A high external radiation density in the IR would be needed which we consider unlikely. See however Dermer and Schlickeiser 1993 for an alternative view.

The Klein-Nishina regime affects also the variability properties. The non linear relation between the variations of the inverse Compton and synchrotron intensities remains valid for energy ranges below the Klein-Nishina limit, that is up to the optical for the synchrotron component and up to the GeV range for the

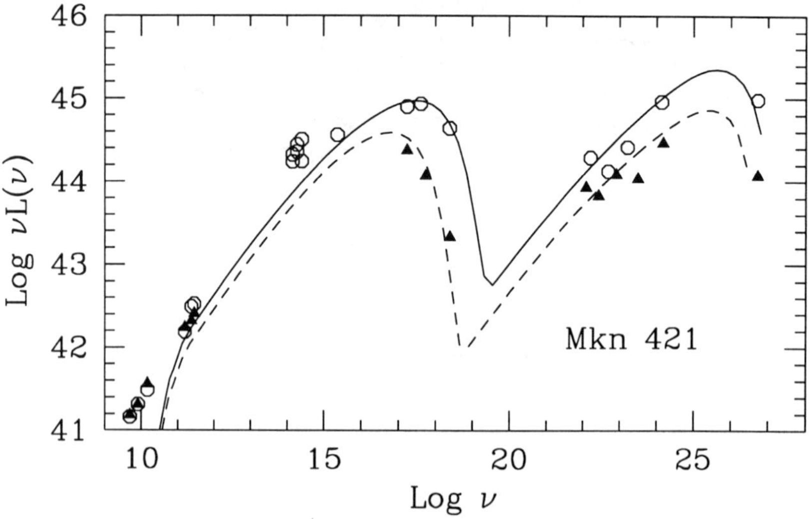

Figure 2. Overall spectral distribution of Mkn 421 during the May 1995 flare (circles) and during the preceding quiescent period (filled triangles) (from Macomb et al. 1995). The solid and dashed lines refer to a homogeneous SSC model calculated assuming that relativistic electrons are continuously injected throughout the source. The size of the region is $R = 10^{16}$ cm, and the injection rate $Q(\gamma) \propto \gamma^{-1.7}$ up to $\gamma_{\max}$, corresponding to an intrinsic power of 2×10^{42} erg/s for the high state (with $\gamma_{\max} = 8 \times 10^5$), and half this value for the low state (with $\gamma_{\max} = 3.3 \times 10^5$). For both models $\delta = 11$ and $B = 0.16$ Gauss are assumed. As for the model in Fig. 1, the equilibrium particle distribution is derived self consistently, and the curved SED is achieved because of incomplete radiative cooling of the electrons.

Compton component. In the higher energy branches (synchrotron X-rays and TeV Inverse Compton) the situation is different. In fact, the reduced scattering cross-section of high energy electrons ($\gamma > \gamma_b$) with photons above the IR range limits the increase of the (TeV) Inverse Compton flux to be approximately linear with the (X-ray) synchrotron flux as can be seen clearly in the figure. We conclude that the SSC scenario explains naturally the spectral and variability properties of Mrk 421.

5. Inhomogeneous jet models

Although the IR to γ-ray spectra of blazars are basically consistent with a homogeneous emission region, the recent discovery of lags in the multifrequency light curves of Mkn 421 (of the order of an hour between soft and medium energy X-ray photons; Makino et al. 1996, see also Takahashi this volume) and PKS 2155 (a day between EUV and X-rays, two days between UV and X-rays, see Urry this volume) is not easy to reconcile with a one-zone model even for the high energy emission from blazars. Energy stratification in the source is required. This may result from radiation losses limiting the path of electrons of different energy injected at a specific site (shock) or from a more diffuse acceleration process yielding a characteristic energy dependent on radius (inhomogeneous jet). The fact that strong spectral variability is observed on the steep branch of the synchrotron spectrum (at frequencies higher than the peak) where lags have been observed suggests that we are close to studying in detail the processes of energy injection and the dynamics of particle spectra in the jet. Understanding these processes will be a major step forward in our knowledge about relativistic jets.

Here we discuss inhomogeneous jet models [Ghisellini, Maraschi & Treves 1985 (GMT) and Ghisellini & Maraschi 1989 (GM)] for the SEDs and variability of the same sources considered above with the scope of evidentiating similarities and differences between the one-zone and multi-zone approach. Spectral models for the high and low states of 3C 279 and Mkn 421, based on the accelerated jet model fully described in GM are shown in Fig. 3. The geometry of the jet is modeled as a paraboloid in the inner regions where bulk acceleration occurs, going over to a cone where the flat radio spectra are formed from the convolution of self-absorption cut offs at different frequencies.

The parameters at the intermediate zone (between the paraboloid and the cone) are given in Table 2. The first thing to note is that the physical parameters at the wide end of the paraboloid (i.e., $B, N_{\rm el}, \gamma_{\rm p}$) are very close to those of the corresponding homogeneous models. This happens because the peak in the power per decade in the inhomogeneous model is due to this region, which therefore must match the characteristics of the one-zone approximation.

However, in the multi-zone approach, the emission at frequencies higher than the peak comes from progressively smaller regions, therefore it can vary on shorter timescales. Moreover if the cause of variability travels from the inner to the outer regions of the paraboloid, variations in the higher frequency emission can lead those at lower frequencies (Celotti, Maraschi, & Treves 1991). The dot-dashed lines in the figures show separately the contribution of a region one-tenth the size of the full paraboloid.

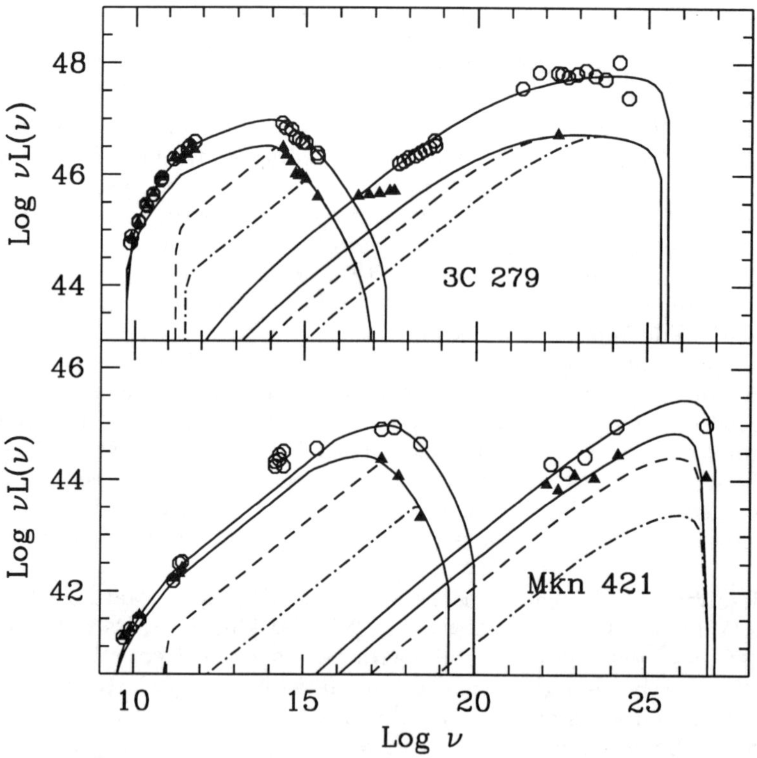

Figure 3. The overall spectra of 3C 279 (*top panel*) and Mkn 421 (*bottom panel*) are fitted with an inhomogeneous jet model. Values of the various parameters at the wide end of the paraboloid are listed in Table 2. Solid lines refer to the emission of the entire jet (paraboloid+cone). Dashed lines refer to the emission of the paraboloid only, while the dash-dotted lines are produced in a region 10 times smaller than the entire paraboloid.

Table 2. Parameters of jet models

	Mkn 421 high	Mkn 421 low	3C 279 high	3C 279 low
R_p	1e17	1e17	6e18	6e18
r_p	7.1e15	7.1e15	9.5e16	9.5e16
B_p	0.14	0.14	0.11	0.11
$N_{el,p}$	7.1e3	4.5e3	2.9e3	1.2e3
m	1.0	1.0	0.8	0.8
n	1.5	1.5	1.8	1.8
γ_p	6.8e5	3e5	9.4e3	8.2e3
δ_p	10.9	10.9	15.6	15.6
Γ_p	8.3	8.3	12.0	12.0
θ	5.0	5.0	3.5	3.5

The synchrotron and Compton emission from the same spatial region should vary together as in the one-zone models. A non negligible difference is that in the inhomogeneous models each jet slice Compton scatters only photons produced in that slice and does not "see" the others. In addition the ratio of radiation to magnetic energy densities varies along the jet. For these reasons the spectrum of the Compton scattered emission from the jet is less "similar" to the synchrotron spectrum than is the case in a homogeneous model.

6. Conclusions

Let us summarize the observational and theoretical points most relevant for this discussion:

- For 3C 279 the X-ray to GeV energy distribution is similar to that from the sub-mm to IR–optical–UV ranges. Variations in γ-rays seem to correlate with variations in the IR to UV ranges but have higher amplitude.
- For Mkn 421 the GeV to TeV energy distribution is similar to that from the IR–UV to the X-ray range. Variations in the TeV emission correlate with X-ray variations and have similar amplitude.

  It is remarkable that these apparently different properties can be accounted for by simple one-zone synchrotron + inverse Compton models. We consider this as a strong observational support for the synchrotron + inverse Compton scenario in general, irrespective of the source of seed photons.

More specifically :

- a one-zone SSC interpretation of 3C 279 is strongly constrained and marginally consistent with the rapid variability observed in γ-rays.

- External photons may be important in this source. An EC model allows a higher magnetic field and a smaller emission region. A high frequency cut off is predicted at 10^{24-25} Hz shifting to lower frequencies at higher intensity.

- The larger amplitude of γ-ray variations is naturally explained in the SSC model varying only the electron spectrum. If external photons are important the variability properties require either a variation of the bulk Lorentz factor Γ or self-illumination as in the mirror model of Ghisellini and Madau (1996).

  Well sampled multifrequency light curves especially on short timescales could discriminate between the different models. The data obtained during the 1996 January February campaign are very promising.

- For Mkn 421, we consider it unlikely that radiation outside the jet provides the seeds for Compton up-scattering. The detected TeV radiation cannot be produced by Compton scattering of UV photons (this would occur in the inefficient Klein-Nishina regime), but must originate in scatterings involving IR seed photons. The TeV radiation is therefore linked with the IR emission (seed photons) and with the highest energy electrons, emitting via synchrotron in the X-ray band. The SSC model accounts "naturally" for these features.

- In both sources and irrespective of the chosen models the particle energies are required to be higher in higher intensity states (flatter energy distribution or higher γ_b, or γ_{max} or Γ). Understanding the physical processes underlying this spectral dynamics requires more theoretical work, taking into account energy stratification in the emission region.

Finally, the differences in the SEDs of the two prototype sources, which may be relevant for a comparison of BL Lacs and HPQs at large, remain to be explained. The models considered here point to intrinsic differences, in the sense of higher particle energies and or higher magnetic fields in Mrk 421, yielding a *shift of the synchrotron peak and of the Compton peak to higher frequencies.* There is no indication that the Doppler factors should be different in the two sources as would be the case if the differences were due to a different orientation of the jet with respect to the line of sight. Allowing for the frequency shifts, the phenomenology of the two sources appear closely similar, suggesting that basically similar processes are at work. It is tempting to speculate that the same type of engine channels energy into a relativistic jet, while the intrinsic power of the jet and the ambient density of photons and/or gas determine the spectral energy distribution.

References

Becker, P. A., & Kafatos M. 1995, ApJ, 453, 83

Blandford, R. D., & Rees, M. J. 1978, in Pittsburgh Conference on BL Lac objects, ed. A. M. Wolfe, p. 328

Blandford, R. D. 1993, in Compton Gamma-ray Observatory, AIP Proceedings 280, ed. M. Friedlander, N. Gehrels, & D. J. Macomb (New York, AIP), p. 533

Blandford, R. D. & Levinson, A. 1995, ApJ, 441, 79

Dermer, C. D. 1995, ApJ, 446, L63

Dermer, C., Schlickeiser, R., & Mastichiadis, A. 1992, A&A, 256, L27

Dondi, L., & Ghisellini, G. 1995, MNRAS, 273, 583

Ghisellini, G., Maraschi, L., & Treves, A. 1985, AA, 146, 204 (GMT)

Ghisellini, G., & Maraschi, L. 1989, ApJ, 340, 181 (GM)

Ghisellini G., & Maraschi L. 1994, in The Second Compton Symp., ed. C. E. Fichtel, N. Geherels, & J. P. Norris (New York, AIP), 304, p. 616

Ghisellini, G., & Madau P. 1996, MNRAS, 280, 67

Ghisellini, G., Maraschi, L., & Dondi, L. 1996, AA, in press

Macomb D. J., et al. 1995, ApJ 449, L90

Makino, F., et al. 1996, in Rontgenstrahlung from the Universe, MPE report 263, ed. H. U. Zimmerman, J. E. Trumper, & H. Yorke, p. 413.

Maraschi, L., Ghisellini, G., & Celotti, A. 1992, ApJ 397, L5

Maraschi, L., Ghisellini, G., & Celotti, A. 1994, in Multifrequency Continuum Emission of AGN, IAU Symp. No. 159, ed. T. J.-L. Courvoisier, & A. Blecha, p. 221

Maraschi L., et al. 1994, ApJ 435, L91

Sambruna, R., Maraschi, L., & Urry, C. M. 1996, ApJ, 463, 444

Sikora, M., Begelman, M. C., & Rees, M. J. 1993, in The First Compton Gamma-ray Observatory Symp., AIP Proceedings 280, ed. M. Friedlander, N. Geherels, & D.J. Macomb (New York, AIP), p. 598

Sikora, M., Begelman, M. C., & Rees, M. J. 1994, ApJ 421, 153

von Montigny, C., et al. 1995, ApJ, 440, 525

Blazar Continuum Variability
ASP Conference Series, Vol. 110, 1996
H. R. Miller, J. R. Webb, and J. C. Noble, eds.

On the Spectral Energy Distributions of Blazars

Rita M. Sambruna

NASA/GSFC, Code 662, Greenbelt, MD 20771

L. Maraschi

Oss. Astronomico di Brera, via Brera 28, Milano, Italy

C. Megan Urry

STScI, 3700 San Martin Dr., Baltimore, MD 21218

Abstract. We investigate the multifrequency spectral properties of complete samples of three different kinds of blazars, all observed with the ROSAT PSPC in pointed mode, including X-ray-selected BL Lacs (XBLs), radio-selected BL Lacs (RBLs), and Flat Spectrum Radio Quasars (FSRQs). We find that different blazar classes have characteristic colors, but forming a continuous spectral sequence from XBLs to RBLs to FSRQs. This can not be reproduced by a change in viewing angle *alone*, but requires a smooth change of physical parameters, with XBLs having larger magnetic fields/electron energies and smaller sizes than RBLs and FSRQs.

1. Introduction

The relationship between X-ray-selected and radio-selected BL Lacs (XBLs and RBLs, respectively), and between BL Lacs and Flat Spectrum Radio Quasars (FSRQs), is still uncertain, but of great interest (e.g., Urry & Padovani 1995). The systematically different properties of XBLs and RBLs have been attributed to different orientations of intrinsically similar relativistic jets, with XBLs being at larger viewing angles than RBLs (Ghisellini & Maraschi 1989). Alternatively, the differences may be intrinsic, with selection effects playing a major role (Padovani & Giommi 1995). While evolutionary links between BL Lacs and FSRQs have been suggested (Vagnetti, Giallongo, & Cavaliere 1991), but not supported by the broad-band and emission line properties of large samples (Padovani 1992), the continuity of the radio and X-ray luminosity functions suggests a continuity of some kind between the two blazar subclasses (Maraschi & Rovetti 1994).

Here we address the connection among XBLs, RBLs, and FSRQs, via the spectral energy distributions of three complete samples: the 1 Jy RBL sample (Stickel et al. 1991), the EMSS XBL sample (Morris et al. 1991), and the S5 FSRQs (Brunner et al. 1994), which were all observed in pointed mode with the ROSAT PSPC. X-ray spectra are crucial for theoretical considerations, since the synchrotron and inverse Compton components, responsible for the radio through UV and for the γ-ray emission, respectively, overlap in the X-ray band.

2. Analysis

We combined the ROSAT data with non-simultaneous fluxes at radio and optical frequencies from the literature, and constructed two point spectral indices, connecting radio to X-rays (α_{rx}), radio to optical (α_{ro}), and optical to X-rays (α_{ox}). We introduce the composite spectral index $\alpha_{ox} - \alpha_{x}$, where α_{x} is the spectral index within the ROSAT band, to describe the shape of the optical-to-X-ray continuum. Positive or negative values of $\alpha_{ox} - \alpha_{x}$ correspond to a concave or convex shape, respectively. The individual rest-frame distributions of the power per decade Log (νL_{ν}) vs. Log ν from radio to X-rays were interpolated with a parabolic function (Landau et al. 1986) to determine the peak frequency and the bolometric luminosity of each source. A full account of this work can be found in Sambruna, Maraschi, and Urry (1996; S96 thereafter).

3. Results

We find that different blazar subclasses are characterized by distinctive α_{rx}, α_{ro}, α_{ox} values, in agreement with previous reports (e.g., Padovani & Giommi 1995). Table 1 lists the average colors of each blazar sample, and shows that XBLs are characterized by flatter spectral distributions than RBLs and FSRQs, while no difference is present between the latter two samples. Also reported in Table 1 are the average values of the bolometric luminosities, L_{bol}, and peak frequencies, ν_p, for the three samples, derived from the parabolic fits (S96).

Table 1: Average Spectral Parameters

Class	$\langle\alpha_{rx}\rangle$	$\langle\alpha_{ro}\rangle$	$\langle\alpha_{ox}\rangle$	$\langle\alpha_{ox} - \alpha_{x}\rangle$	$\langle$ Log L_{bol} $\rangle$ (ergs s^{-1})	$\langle$ Log $\nu_p\rangle$ (Hz)
EMSS XBLs	0.57 ± 0.06	0.36 ± 0.07	0.96 ± 0.15	-0.34 ± 0.46	45.57 ± 0.37	15.93 ± 1.45
1 Jy RBLs	0.86 ± 0.08	0.59 ± 0.10	1.38 ± 0.18	0.26 ± 0.54	46.49 ± 0.80	13.91 ± 0.93
S5 FSRQs	0.89 ± 0.09	0.57 ± 0.06	1.49 ± 0.23	1.01 ± 0.72	47.38 ± 0.76	14.01 ± 0.93

Comparing the $\alpha_{ox} - \alpha_{x}$ distributions, we find that the three classes have significantly different shapes of the optical-to-X-ray continuum, which is typically convex in XBLs and concave in FSRQs. The RBL sample is intermediate, including both XBL-like and FSRQ-like spectra. This suggests that in XBLs the X-rays are the high-energy tail of the lower frequency synchrotron emission, while in FSRQs the presence of a separate, most likely inverse Compton origin of the X-ray emission, is indicated. For RBLs, both situations are possible.

Fig. 1a–b show the plot of α_{ro} versus $\alpha_{ox} - \alpha_{x}$ and versus the bolometric luminosity. A similar diagram is obtained plotting α_{ro} versus the redshift. Different samples occupy different regions in this plane; however, a continuity

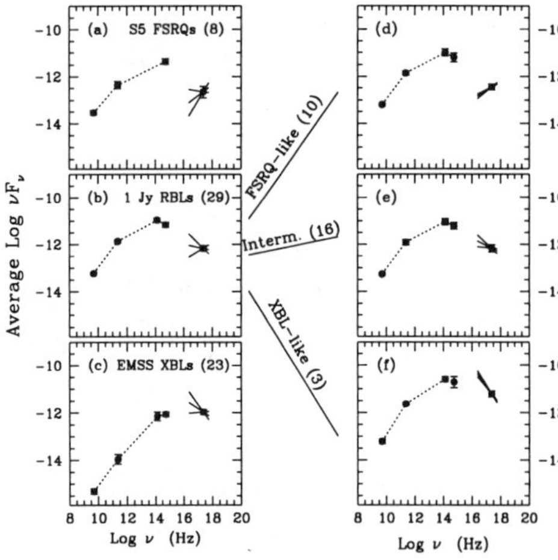

Figure 1. Correlation of $\alpha_{\rm ro}$ with $\alpha_{\rm ox} - \alpha_{\rm x}$ *(a)* and the bolometric luminosity *(b)*. A decrease of $\alpha_{\rm ro}$ from 0.8 to 0.3 corresponds to an increase of the synchrotron peak frequency from 10^{13} to 10^{18} Hz.

is apparent. Considering the three samples jointly there is a strong correlation ($>$ 99.9% significance, from a Spearman-Kendall test) between $\alpha_{\rm ro}$ and $\alpha_{\rm ox} - \alpha_{\rm x}$ and between $\alpha_{\rm ro}$ and bolometric luminosity and/or redshift, in the sense that sources with steeper $\alpha_{\rm ro}$ are more luminous (distant) and have concave optical-to-X-ray shapes. Note that $\alpha_{\rm ro}$ is strongly anticorrelated with the synchrotron peak frequency ν_p, as expected; an increase of $\alpha_{\rm ro}$ from 0.3 to 0.8 corresponds to a decrease in ν_p from 10^{18} to 10^{13} Hz (S96).

Fig. 1 suggests that the shape of the blazar continuum can be characterized by only one spectral parameter, and indicates that the bolometric luminosity and/or redshift plays a major role in determining its shape.

Fig. 2 shows the average radio-to-X-ray spectra for each blazar class. These were constructed using the mean of the logarithmic K-corrected flux densities at different frequencies and the mean spectral indices in the X-ray band, with errors on the mean as uncertainties for the average fluxes, and the rms deviations as the error on the average X-ray spectral indices (1σ, or $\sim 68\%$). The average spectral energy distributions of the S5 FSRQs (Fig. 2a) shows a clear upturn of the X-ray spectrum with respect to the optical-to-X-ray slope and a peak in the emitted power is suggested at or below the infrared/optical range. In contrast, for the EMSS XBLs (Fig. 2c), the X-ray spectrum is steeper than the optical-to-X-ray slope and the peak power should be in the UV/soft-X-ray range. The average spectral energy distribution for the total RBL sample (Fig. 2b) is intermediate between the two.

As mentioned above, there is a significant spread in the spectral shapes of RBLs, overlapping on one end with XBLs and on the other with FSRQs. Splitting the RBL sample into three subgroups defined by the values of $\alpha_{\rm ox} - \alpha_{\rm x}$ (> 0.5, > -0.5 and < 0.5, < -0.5), we derived the corresponding average spectral energy distributions which are shown in Fig. 2d–f, respectively.

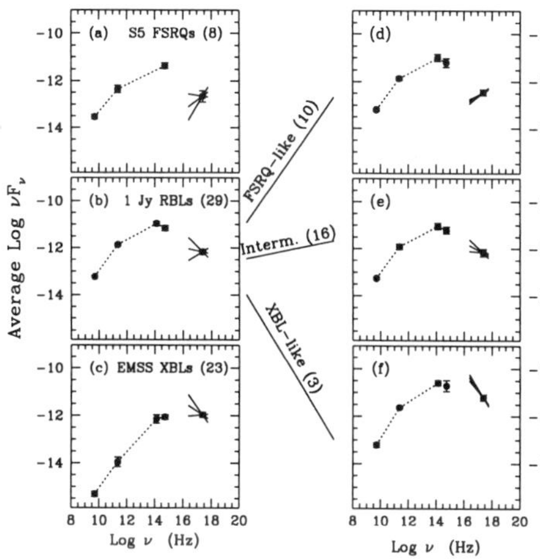

Figure 2. Average spectra of blazar samples (right) and RBL subclasses (left).

The average spectral properties of the three RBL subgroups closely reproduce those of the three original samples confirming that the spectral continuity between blazar subclasses is real.

4. Discussion

The observed spectral sequence formed by XBLs/RBLs/FSRQs can be interpreted in the frame of available emission models. We restrict our considerations to synchrotron-self Compton homogeneous models, where the radiation is emitted by a uniform plasma with constant particle density and magnetic field B. The relativistic electrons have a a broken power law energy distribution with a break point at γ_b, where γ is the Lorentz factor of the electrons (Jones, O'Dell, & Stein 1974). If the energy spectral indices are smaller and larger than 3 below and above the break energy respectively, as is typically the case, the bulk of the synchrotron emission is produced at the frequency

$$\nu_S = 2.8 \times 10^6 \gamma_b^2 B\delta \ \text{Hz}, \qquad (1)$$

where δ is the beaming factor ensuing from bulk relativistic motion of the emitting plasma, and is defined as $\delta = \left(\Gamma - (\Gamma^2-1)^{1/2}\cos\theta\right)^{-1}$, with Γ the bulk Lorentz factor of the plasma and θ the viewing angle to the jet axis (Rybicki & Lightman 1979). This frequency corresponds to the maximum observed power per decade of the synchrotron component ν_p (Table 1; Fig. 2).

Let us now consider different objects along the spectral sequence in Fig. 1 as ν_S increases from 10^{13} to 10^{18} Hz. Assuming that δ is approximately constant in different sources, Eq. (1) shows that the increase in ν_p corresponds to an increase of either the energy of the emitting electrons (by a factor $\approx 10^3$) or of the magnetic field (by a factor 10^5), or both. On the other hand, for fixed B and γ_b, and assuming a constant bulk Lorentz factor, an increase of δ due to a decreasing vewing angle can produce a shift in peak frequency. However, high-power objects would be characterized by *higher* values of the cutoff frequency, contrary to what observed (Fig. 1). To reproduce the observed shift in ν_S from 10^{13} to 10^{18}, the beaming factor would need to vary by a factor 10^5, while estimates of δ based on SSC arguments indicate variations of at most 2 decades (0.1–10; Ghisellini et al. 1993).

These simple considerations based on the homogeneous model show that the observed continuity of spectral properties among blazar classes cannot be easily accounted for in terms of a changing viewing angle but suggests instead a change in intrinsic physical parameters such as magnetic field and/or maximum electron energy, with XBLs having higher magnetic fields/electron energies than radio-selected blazars. Fitting specific multifrequency spectra from radio to γ-rays of representative objects of the three samples, we find similar results, i.e., the transition XBLs/RBLs/FSRQs requires a decrease of magnetic field/electron energies and an increase of jet size (see S96 for a full account).

Although orientation effects *alone* are not sufficient to account for the observed spectral properties of blazars, this does not imply that the average beaming factors and viewing angles of XBLs and RBLs should be the same. The above discussion suggests that the physical properties of the jets change systematically going from X-ray-selected to radio-selected sources, and the beaming factor itself may be an additional *intrinsic* difference between XBLs and RBLs.

References

Brunner, H., Lamer, G., Worrall, D. M., & Staubert, R. 1994, A&A, 287, 436

Landau, R., et al. 1986, ApJ, 308, 78

Ghisellini, G., Padovani, P., Celotti, A., & Maraschi, L. 1993, ApJ, 407, 65

Ghisellini, G., & Maraschi, L. 1989, ApJ, 340, 181

Jones, T. W., O'Dell, S. L., & Stein, W. A. 1974, ApJ, 188, 353

Maraschi, L., & Rovetti, F. 1994, ApJ, 436, 79

Morris, S. L., Stocke, J. T., Gioia, I. M., Schild, R. E., Wolter, A., Maccacaro, T., & Della Ceca, R. 1991, ApJ, 380, 49

Padovani, P. 1992, MNRAS, 257, 404

Padovani, P., & Giommi, P. 1995, ApJ, 444, 567

Rybicki, G., & Lightman, A. 1979, in Radiation Processes in Astrophysics (New York, Wiley)

Sambruna, R. M., Maraschi, L., & Urry, C. M. 1996, ApJ, in press: June 1, 1996 issue (S96)

Stickel, M., Padovani, P., Urry, C. M., Fried, J. W., and Kühr, H. 1991, ApJ, 374, 431

Vagnetti, F., Giallongo, E., & Cavaliere, A. 1991, ApJ, 368, 366

Urry, C. M. & Padovani, P. 1995, PASP, 107, 803

Blazar Continuum Variability
ASP Conference Series, Vol. 110, 1996
H. R. Miller, J. R. Webb, and J. C. Noble, eds.

Principal Component Analysis Applied to the Spectral Energy Distributions of Blazars

Sandra D. Clements

Department of Astronomy, University of Florida, Gainesville, FL 32611

Abstract. The X-ray-to-radio spectral energy distributions (SEDs) of blazars are examined using Principal Component Analysis (PCA). When applied to an $N \times M$ array of data, PCA finds and ranks those components responsible for variations in the data. (Here, the $N \times M$ array consists of fluxes at M frequencies in the SED of each of the N blazars. In this case, PCA finds those components responsible for object-to-object variations in the SEDs.) Fits to the SEDs of individual blazars are constructed using the principal components. Subcategories of blazars are contrasted based on differences in the contribution to their SEDs made by the various principal components.

1. Introduction

Principal Component Analysis (PCA) was applied to the spectral energy distributions (SEDs) of a sample of blazars. Applied to the spectra of a sample of objects, PCA finds and ranks those components of the spectra responsible for object-to-object variations. For example, consider the simulation shown in Figure 1. Here, the results of a simulation described in Mittaz, Penston, and Snijders (1990) were reproduced in order to test the PCA code. Figure 1a shows 5 of the 80 spectra in the simulation. Each of the spectra consists of 230 points. The only thing that varies from spectrum to spectrum is the line strength. PCA finds 230 principal components (PCs), but only the 1st PC (Figure 1b) is required to account for 100% of the spectrum-to-spectrum variation. In this case, the 1st PC even looks like the varying feature (if turned upside down). That is, it looks like an emission line. Each of the original spectra can be reconstructed by simply multiplying this 1st PC by an appropriate (in this case, negative) weight which is also determined by PCA. The results of this first simulation are very straightforward. However, if both the line strength and the line width vary, the results are not so straightforward. In this case, PCA finds not two, but instead, four PCs which together account for 100% of the spectrum-to-spectrum variations. (See Figure 1 in Mittaz, Penston, & Snijders 1990.) The first component again is due to the variation in line strength. Components 2–4 account for the variation in line width.

The goal of this project is to connect the principal components of a sample of blazar SEDs to the emission mechanisms which contribute to the SEDs, with the ultimate goal being to learn something about the physical properties of the emission regions. If a single PC, or perhaps a small number of PCs, can account

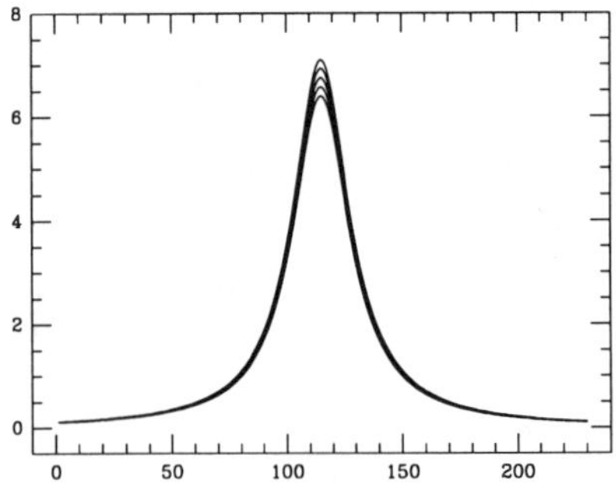

Figure 1a. Simulated data.

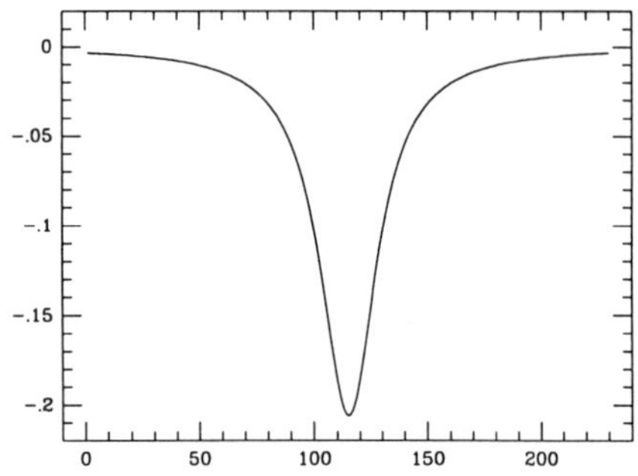

Figure 1b. PC 1.

for variations due to a single emission mechanism, then the physics of that emission mechanism can be investigated independent of other mechanisms. A very simple scenario would be one in which PC 1 accounts for variations due to the synchrotron component, PC 2 accounts for variations due to the accretion disk, and PC 3 accounts for differing contibutions of the galaxy component. Instead of having to fit models with several free parameters to the SED of an object in order to determine its physical characteristics (as was done, for example, by Edelson and Malkan, (1986)), an individual component, isolated from the SED by PCA, can be modelled with fewer free parameters.

Principal component analysis has been used for many years, especially in the social sciences. Murtagh and Heck (1987) cite several references in which PCA was applied to astronomical data. In addition, they include a PCA code. Their PCA code is being used in this research. PCA is often used as a classification tool but can be used in other ways. The two references mentioned below describe alternative applications. Mittaz, Penston, and Snijders (1990) applied PCA to the IUE spectra of NGC 4151 looking for components responsible for epoch-to-epoch variations in the spectra. Francis et al. (1992) applied PCA to the UV spectra of 232 QSOs looking for components responsible for object-to-object variations in the spectra—an application similar to the one used in this research.

An $N \times M$ array is input into PCA. For this project, there are $N(= 14)$ blazars, each with an SED consisting of $M(= 25)$ points. PCA finds the components of the SEDs responsible for the object-to-object variations and ranks them according to the amount of variation they account for. The first principal component accounts for the most object-to-object variation and the last PC accounts for the least object-to-object variation. The original SEDs can be reconstructed from the principal components using the following equation:

$$d_{ik} = \sum_{j=1}^{M'} w_{ij} c_{jk}$$

where d_{ik} = flux at frequency k for object i, w_{ij} = weight for object i and principal component j, c_{jk} = value of principal component j at frequency k, and $M \leq M'$. The SEDs can be reconstructed *exactly* if all M of the principal components are used. However, if fewer PCs can adequately reconstruct the SEDs, then the dimensionality of the data has effectively been reduced.

2. Data

The sample was taken from the Impey and Neugebauer (1988) list of objects which exhibit blazar-like behavior. Objects were included if redshifts could be found for them and if they exhibited 'good coverage' within the three following spectral regions.

$$8.625 \leq \log \nu(Hz) \leq 11.625$$
$$12.625 \leq \log \nu(Hz) \leq 15.125$$
$$17.375 \leq \log \nu(Hz) \leq 17.875$$

Here, 'good coverage' means that there were no gaps greater than $\log \nu = 0.5$. The 14 SEDs included cover roughly 9 decades in frequency from the radio to the x-ray. Eight of the 14 SEDs are composites of 'quasi-simultaneous' observations from Landau et al. (1986) and non-simultaneous observations gleaned from references cited in Impey and Neugebauer (1988). The remaining 6 SEDs consist only of the non-simultaneous observations. The 14 blazars in the final sample are listed in Table 1 along with their redshifts.

Table 1. Sample of Blazars used in PCA Anaylsis

Object	Alias	Type	z	Type of z
0316+413[a]	NGC 1275	cD,pec	0.18	
0420–014	OA 129	QSO	0.915	
0735+178	OI 158	BL Lac	0.424	absorption
0736+017	OI 061	QSO	0.191	
0851+202	OJ 287	BL Lac	0.306	
1219+285	ON 231	BL Lac	0.102	galaxy
1226+023	3CR 273	QSO	0.158	
1308+326	OP 313	BL Lac	0.996	
1514–241	AP Lib	BL Lac	0.0486	galaxy
1641+399	3CR 345	QSO	0.595	
1652+398	Mnk 501	BL Lac	0.034	galaxy
2200+420	BL Lac	BL Lac	0.0688	galaxy
2223–052	3C 446	QSO and BL Lac	1.404	
2251+158	3C 454.3	QSO	0.859	

[a]Information from NASA/IPAC Extragalactic Database (NED).

Figure 2 shows three representative SEDs, with fluxes rescaled for display. The spectral region from $\log \nu =$ 12.625 to 15.125 for 0735+178 is fairly smooth, indicating that the synchrotron component is the only one contributing significantly to the SED in this portion of the spectrum. The SED for 1226+023 (3C 273) exhibits flattening of the spectrum at the ultraviolet freqencies indicating the presence of an accretion disk. A large galaxy component in the SED of 1652+398 (Mkn 501) is evident as a bump centered on $\log \nu =$ 14.375.

3. Preliminary results

Applied to this sample of SEDs (14 blazars with 25 points per SED) PCA finds 25 principal components. (See Figure 3 for three of these.) Only the first 14 PCs are

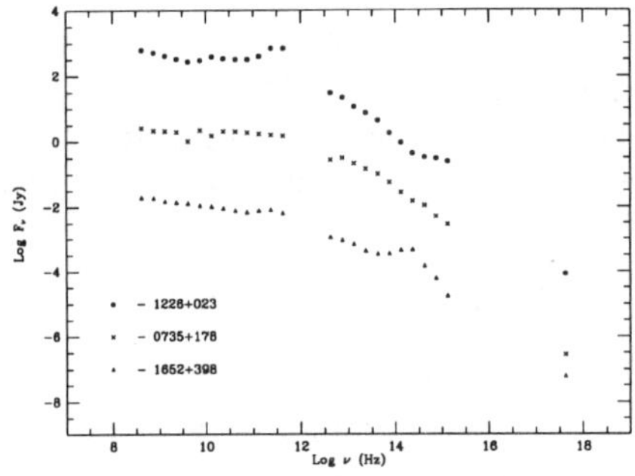

Figure 2. Three representative SEDs.

given any weight for reconstructions. The 1st PC which accounts for 92.9% of the object-to-object variations, is due to the sychrotron component. It even looks like the sychrotron component (if turned upside down). Simulations suggest that PCs 1 and 2 are required to fully reconstruct the synchroton component. Figure 4 shows the reconstruction of 0735+178 using the principal components. Although a better reconstruction is obtained with more PCs, the 1st PC alone provides a fair fit, especially to the $\log \nu =$ 12.625 to 15.125 region. A better fit is obtained with both PCs 1 and 2.

Applying PCA to centered data (data from which the mean SED has been subtracted) prevents some variables from "shouting louder" than others. (Murtagh & Heck (1987) p. 23) When applied to centered data, the 1st principal component no longer corresponds as strongly to the synchrotron component. This is because the synchrotron component looks very much like the mean SED which was subtracted from each object's SED prior to analysis. Figure 5 shows some principal components which result when PCA is applied to centered data. PC 1 obtained with centered data looks like the inverse of PC 2 obtained with uncentered data. Therefore, PC 1 in Figure 5 plays the same role as PC 2 in Figure 3, that is, it acts to improve the slope of the synchrotron component. Here PC 1 accounts for 71.1% of the object-to-object variation and PC 2 accounts for 13.9%. PC 1 looks somewhat like the inversion of PC 2 and vice versa. In PC 1, there is a broad bump centered on $\log \nu =$ 14.375 and a sharp dip centered on $\log \nu =$ 14.875. Premultiplied by a positive weight, this principal component, when added to the mean SED for the sample, will flatten the spectral slope for $\log \nu < 14.375$ and steepen it for $\log \nu > 14.375$. If premultiplied by a negative weight and added to the mean SED of the sample, PC 1 will steepen the spectral slope for $\log \nu < 14.375$ but will flatten the spectral slope for $\log \nu > 14.375$. PC 2 can similarly flatten or steepen the slope above and below $\log \nu = 14.375$ depending on the sign of the weight. For example, a negative weight is applied to PC 2 for 1652+398, contributing to the galaxy component for that object. A positive weight is applied to PC 2 for 1226+023, contributing to that object's accretion disk component. Figure 6 shows the reconstruction of 1652+398 using the principal components. Here, PCs 2 and 3 bring out the galaxy component.

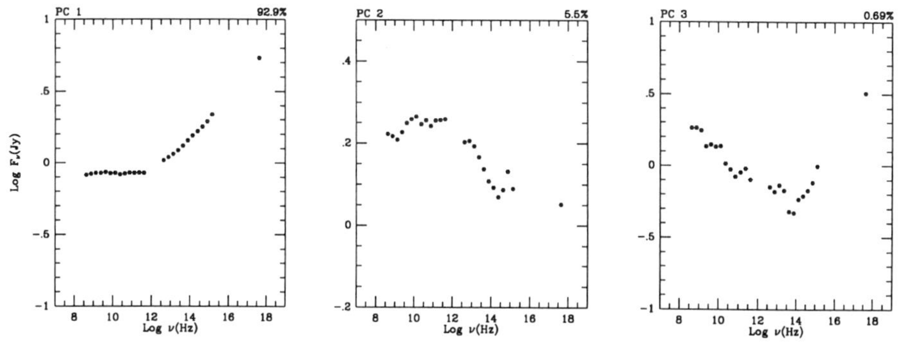

Figure 3. First three principal components obtained using uncentered data.

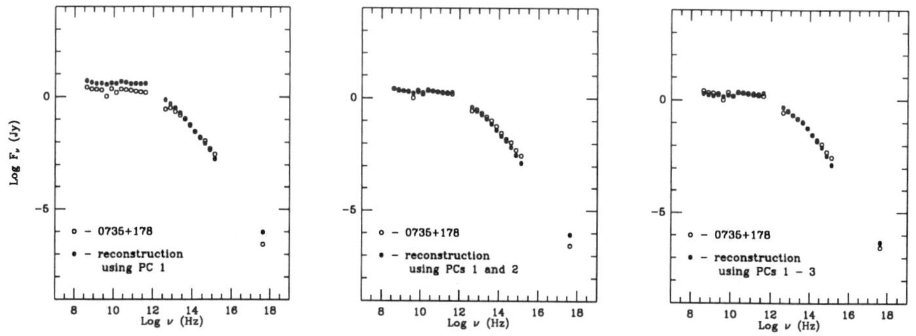

Figure 4. Reconstructions of 0735+178.

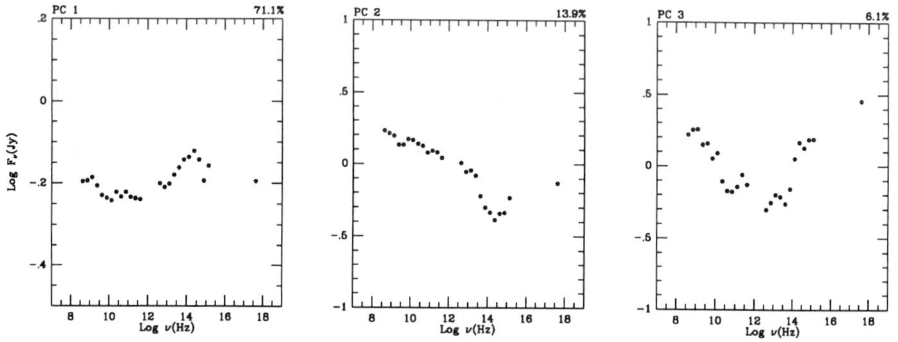

Figure 5. First three principal components obtained using centered data.

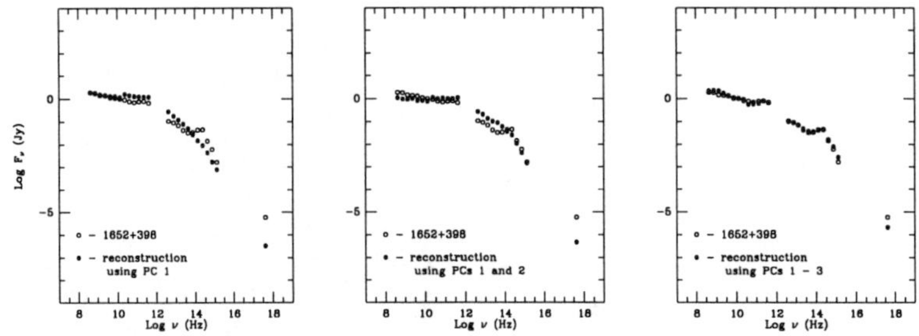

Figure 6. Reconstructions of 1652+398.

4. Summary

To summarize, PC 1 (obtained from analysis of uncentered data) accounts for the majority of the variation due to the synchrotron component. PC 2 also contributes to the synchrotron component. From the analysis of centered data, it is evident that the accretion disk and galaxy components have not been isolated. Not only can more than one PC account for the variation due to a single emission mechanism, but a single PC can account for variations due to more than one emission mechanism.

5. Next steps

More can be done to try to fully decompose the accretion disk and galaxy components. 1) The sample can be improved and enlarged by including more SEDs constructed from quasi-simultaneous observations. Numerous multifrequency campaigns have provided such data for several individual objects (e.g., OJ 287, 2155-304, and 1156+295). 2) Since the accretion disk and galaxy components are evident in the range $\log \nu =$ 12.625 to 15.125, only that range need be included in the SEDs to be analyzed. Many more SEDs (24 instead of 14) satify the 'good coverage' criterion for this smaller spectral range. 3) Subsets can be analyzed. In particular, the sample to be analyzed may consist of objects exhibiting a synchrotron component and/or an accretion disk component, but no galaxy component. Another sample may consist of objects exhibiting a synchrotron component and/or a galaxy component, but no accretion disk component.

Acknowledgments. This research has made use of the NASA/IPAC Extragalactic Database (NED) which is operated by the Jet Propulsion Laboratory, CalTech, under contract with the National Aeronautic and Space Administration. This research has been supported by the National Science Foundation through grant AST-9523318.

References

Edelson, R. A., & Malkan, M. A. 1986, ApJ, 308, 59

Francis, P. J., Hewett, P. C., Foltz, C. B., & Chaffee, F. H. 1992, ApJ, 398, 476

Hewitt, A., & Burbidge, G. 1987, ApJS, 63, 1

Impey & Neugebauer 1988, AJ, 95, 307

Landau, R., et al. 1986, ApJ, 308, 78

Mittaz, J. P. D., Penston, M. V., & Snijders, M. A. J. 1990, MNRAS, 242, 370

Murtagh, F., & Heck, A. 1987, in Multivariate Data Analysis (Dordrecht, Reidel), p. 13

Blazar Continuum Variability
ASP Conference Series, Vol. 110, 1996
H. R. Miller, J. R. Webb, and J. C. Noble, eds.

EUVE Observations of 3C 273

E. Ramos and M. Kafatos

Center for Earth Observing and Space Research, Institute for Computational Sciences and Informatics, George Mason University, Fairfax, VA 22030

A. Fruscione

Harvard-Smithsonian Center for Astrophysics, 60 Garden Street, Cambridge, MA 02138

Abstract. We present the results of our January 1994 and January 1995 observations of the blazar 3C 273 obtained with the *Extreme Ultraviolet Explorer*. Our January 1995 photometric observations with the EUVE Lexan/Boron (Lex/B) Deep Survey (DS) instrument indicate strong evidence for variability, at a 99% confidence level, during the 12 days observing period. We have utilized soft ROSAT PSPC X-ray power law models to correlate with EUVE count rates. These observations were part of a large multiwavelength campaign to observe 3C 273 from radio to γ-rays.

1. Introduction

3C 273 is one of the nearest quasars ($z = 0.158$) and one of the best studied at all frequencies from radio to γ-rays. The first detection in the X-rays was reported by Bowyer et al. (1970). The existence of a soft X-ray/EUV component was first detected by EXOSAT observations (Turner et al. 1985) and subsequently confirmed with other instruments (Courvoisier et al. 1987 and Turner et al. 1991) . A recent detailed study with the ROSAT Position Sensitive Proportional Counter (PSPC) (Leach et al. 1995) indicates that the X-ray spectrum can be decomposed into a hard X-ray (1.5–2.4 keV) power-law with a photon index of ~ 1.5 ($F_\epsilon \propto E^{-\Gamma}$) and a soft X-ray (0.1–0.3 keV) power-law with a photon index of ~ 2.7. 3C 273 was first observed with the EUVE during the EUVE all-sky survey showing a Lex/B filter rate of 40 counts/ksec (Malina et al. 1994). The existence of a blue bump in the UV/EUV range of 3C 273 indicates that is a very important source to observe with the EUVE. The presence of this bump may be indicating an origin from an underlying accretion disk around a supermassive black hole. By this reason observations of AGNs in the EUV band could help to clarify the physical mechanisms operating in these systems. In this contribution we report our January 1994 and January 1995 observations with the EUVE. These observations were part of a big multifrequency campaign to observe 3C 273 from radio to γ-rays.

2. Observations

The reported data was taken with the Lex/B filter in the Deep Survey (DS) instrument. This filter covers the range from 67–178 Å (at 10 % of the peak) and has an effective area of $\sim$ 29 cm^2 at 91 Å (Malina et al. 1994).

2.1. January 1994

3C 273 was observed with the EUVE Deep Survey instrument (DS) from 16:09 UT on 1994 January 8 to 20:47 UT January 14. The source was observed for 96 orbits, corresponding to a total effective exposure of 130,093 seconds spread over around 6 days. This exposure time excludes observations when 1) the background was particularly high (e.g., passage through the South Atlantic Anomaly) and 2) the source crossed the deadspot (e.g., during short jitters due to antenna slews). The average, error weighted count rate from 3C 273 in the Lexan/B band was 0.072 ± 0.00081 counts s^{-1}.

Due to the low signal to noise in our simultaneous spectroscopic observations we were unable to obtain acceptable spectral information. We, therefore, used ROSAT PSPC observations to model the soft X-ray spectral shape and normalization in order to obtain an estimate of the flux density from the photometric data. We modeled the EUV to soft X-ray spectrum with an absorbed power law of the form:

$$f(\lambda) = K\,\lambda^{\alpha-1} \exp[-\sum_x \sigma_{\lambda\,x}\, N_x]\ \text{photons cm}^{-2}\,\text{s}^{-1}\,\text{Å}^{-1} \tag{1}$$

where K is the normalization factor, α is the energy index ($\alpha \equiv \Gamma - 1$), and the interstellar absorption is characterized by a cross-section $\sigma_{\lambda\,x}$ and a column density N_x for each contributing species.

We will adopt the interstellar photoionization cross-sections of Rumph, Bowyer, and Vennes (1994) for H I, He I and He II. The column density $N_{\rm H}$ is assumed fixed at the galactic value of 1.8×10^{20} cm^{-2}(Stark et al. 1992) and the contributions of absorption from HeI and HeII, respectively, are $N_{\rm HeI} = 0.09 \times N({\rm H})$ and $N_{\rm HeII} = 0.01 \times N({\rm H})$. The energy index $\alpha = 1.77 \pm 0.13$ and $K = (7.7 \pm 1.7)\,10^{-4}$ photons cm^{-2} s^{-1} Å^{-1} (at 1 keV) were obtained from ROSAT PSPC observations (Leach et al. 1995). In principle, $f(\lambda)$ is the quantity that we should derive from the EUV spectrum, but since our signal to noise is too low the only quantity that we know from the EUVE observations is the total count rate which is defined as the integral over the DS Lex/B band of the same $f(\lambda)$ convolved with the instrument effective area:

$$\text{count rate}|_{EUVE} = \int_{\lambda_1}^{\lambda_2} f(\lambda)\, A_{\rm eff}\,(\lambda)\, d\lambda, \tag{2}$$

where the DS band goes from $\lambda_1 = 67$ Å to $\lambda_2 = 178$ Å. The EUV total flux is given by the integral over the DS [Lex/B] band of the flux density or:

$$f_{\rm EUV} = \int_{\lambda_1}^{\lambda_2} f_\lambda\, d\lambda \tag{3}$$

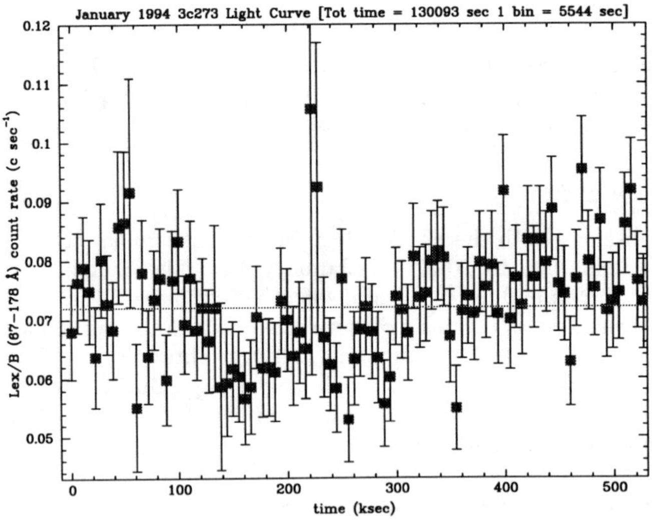

Figure 1. Light curve for January 1994 EUVE observations with the DS Lex/B imaging detector. The average count rate is 0.072 ± 0.00081 counts s^{-1}.

From the ROSAT PSPC model and Equation 2 we can obtain an estimate of the normalization K that corresponds to our measured count rate and from Equation 1 we can obtain an estimate of the flux density at a given wavelength. Taking the spectral model in (1), the unabsorbed flux density S_ν , from 3C 273 at 100 Å (although the EUVE flux is negligible above $\sim$ 85 Å due to absorption) would be 396 μJy during the EUVE observations. The corresponding unabsorbed non-contemporaneous ROSAT PSPC extrapolation at 100 Å was 253 μJy.

2.2. January 1995

3C 273 was observed with the EUVE DS instrument from 12:42 UT on 1995 January 3 to 20:04 UT January 15. The source was observed for 187 orbits, corresponding to a total effective exposure of 277,358 sec. The average, error weighted count rate from 3C 273 in the Lexan/B band was 0.044 ± 0.0043 counts s^{-1}. From this average count rate and the ROSAT PSPC spectral model presented in the previous section we estimated the unabsorbed flux density to be 242 μJy at 100 Å.

3. Light curves

In Figure 1 we present the light curve for the January 1994 EUV/DS observations of 3C 273. A χ^2 test for a constant flux gives a value of 107.7 which is significant at the 80% confidence level. This result indicates weak or no evidence

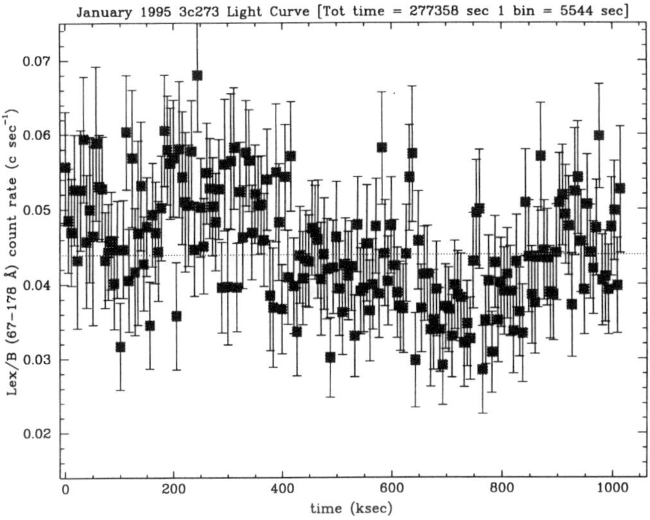

Figure 2. Light curve for January 1994 EUVE observations with the DS Lex/B imaging detector. The average count rate is 0.044 ± 0.0043 counts s^{-1}.

of variability during the January 1994 six-day observing period. In Figure 2 we present the light curve for the January 1995 EUV/DS observations of 3C 273. A similar analysis for the January 1995 observations results in a χ^2 of 241.2 for 187 degrees of freedom which is significant at the 99% confidence level. This result indicates stronger evidence for variability during the January 1995 12-day EUVE/DS observation period. A light curve for a smaller interval was extracted in order to look for possible rapid (200 ksec) intraday variations. The χ^2 test indicates weak or no evidence ($< 68\%$ confidence level) of variations at these smaller time scales. Details about the extraction and reduction procedure of our photometric data will be provided elsewhere.

4. Summary and discussion

Discrepancies between our January 1994 estimated flux densities and ROSAT non-contemporaneous flux density, presented above, can be interpreted from simultaneous observations of 3C 273 using the ROSAT HRI instrument (McHardy 1996). These observations indicated variability during a 30 days observing period with the lowest level at ~ 15 days before our observations and a highest peak at ~ 6 days from our EUVE/DS observations. Our EUV estimate may correspond to an "intermediate" level between the highest count rates and the lowest count rates detected during the ROSAT HRI observing period. As a matter of comparison we can perform a similar calculation to obtain the corresponding flux density from the average count rate for 3C 273 obtained during the EUVE all-sky survey. These observations were carried out between 1992 December 25 to 29 and simultaneous with the ROSAT PSPC observations. From Marshall et al. (1995) the observed average count rate is $0.043 \pm 1.14 \times 10^{-2}$ counts s^{-1} which corresponds to an unabsorbed flux density of 237 μJy. The simultaneity of both observations accounts for the consistency between both estimates.

Besides not being simultaneous, our January 1995 flux density estimate presented above is more consistent with the ROSAT PSPC observations and the all-sky observations presented in the previous sections. From this we can conclude that the January 1995 observations corresponded to the base line or "low" state of the soft-X-ray/EUV variations of 3C 273 and the January 1994 observations corresponded to a "intermediate" state in the level of variations of 3C 273. The implication of our observations and correlations with contemporaneous observation in other frequencies will be discussed in a forthcoming paper.

The observations reported here were part of a big multifrequency campaign to observe 3C 273 from radio to γ-rays with a variety of instruments: UMRAO (radio), IUE (UV), OSSE, COMPTEL, and EGRET. The results and analysis of this multifrequency campaign will be presented elsewhere (von Montigny et al. 1996).

Acknowledgments. This work has been supported by NASA grant NAG-5-2360.

References

Bowyer, C. S., Lampton, M., Mack, J., & de Mendoca, F. 1970, ApJ, 161, l1

Courvoisier, T. J.-L., et al. 1987, A&A, 176, 197

Leach, C. M., McHardy, I. M., & Papadakis, I. E. 1995, MNRAS, 272, 221

Malina, R. F., et al. 1994, ApJ, 107, 751

Marshall, H. L., Carone, T. E., & Fruscione, A. 1993, ApJ, 414, L53

McHardy, I. M. 1996, private communication

von Montigny, C., et al. 1996, in preparation

Rumph, T., Bowyer S., & Vennes, S. 1994, ApJ, 107, 2108

Stark, A. A., et al. 1992, ApJS, 79, 77

Turner, M. J. L., Courvoisier, T., Staubert, R., Molteni, D., & Trümper, J. 1985, in Proc. 18th ESLAB Symp., ed. A. Peackock (Dordrecht, Reidel), p. 623

Turner, T. J., et al. 1991, ApJ, 381, 85

Blazar Continuum Variability
ASP Conference Series, Vol. 110, 1996
H. R. Miller, J. R. Webb, and J. C. Noble, eds.

A Multi-wavelength Test of the Synchrotron Self-Compton Model

Mark W. Sincell

Department of Physics, University of Illinois at Urbana-Champaign, 1110 W. Green Street, Urbana, IL 61801

Abstract. We calculate the relative amplitudes of variations in the synchrotron and self-Compton fluxes (R_ν) caused by fluctuations in the electron density of a synchrotron source. The change in the optically thin synchrotron flux is one-half the change in the X-ray flux, independent of frequency. We find that $R_\nu < 0.5$ over two decades of frequency when the induced Compton scattering optical depth $\tau_{\rm ICS} > 1$. In addition, R_ν increases with frequency. The amplitudes of the correlated radio and X-ray variability of 3C279 are inconsistent with a homogeneous synchrotron self-Compton model. The discrepancy can be removed by including stimulated scattering.

1. Introduction

The emission from 3C279 varies coherently over its entire spectrum and the amplitude of the variations increases steadily from low to high energy (Maraschi et al. 1994, Grandi et al. 1995). The strong correlation of the radio and X-ray emission on time scales < 70 days implies that the flux in both bands comes from physically related regions separated by < 0.06 pc (Grandi et al. 1995).

This correlation is expected if the electrons radiating synchrotron photons scatter a fraction of these photons to higher energies (e.g., Marscher 1987). The relative amplitudes of the synchrotron and self-Compton X-ray flux variations are sensitive to the radio frequency opacity of the source. The ratio of the change in the peak radio flux density to the change in the X-ray flux at 1 kev is ~ 0.4 for a self-absorbed source and can be reduced to ~ 0.2 by induced Compton scattering when the radio brightness temperature $T_{\rm B} > 2 \times 10^{11}$K (Sincell & Krolik 1994: SK94). Unfortunately, the effects of stimulated scattering are unlikely to be detected in this way because it is difficult to define the peak of the flux density distribution (Valtaoja et al. 1988).

In this paper, we calculate the relative amplitudes of the radio and X-ray variability in a synchrotron source, including induced Compton scattering, as a function of radio frequency and compare to the observed variability amplitudes of 3C279.

2. The flux variability ratio

The relative amplitude of the variations in the synchrotron flux at $\nu\,(S_\nu)$ and the self-Compton flux at 1 kev (S_x),

$$R_\nu = \frac{\partial \log S_\nu}{\partial \log S_x}, \tag{1}$$

is sensitive to the dominant source of opacity. Variations in the electron density of an optically thin synchrotron self-Compton source lead to $R_\nu = 0.5$, independent of frequency. This is because $S_\nu \propto \tau_T$ and the self-Compton X-ray flux, $S_x \propto \tau_T^2$ (e.g., Marscher 1987). Self-absorption reduces R_ν below the turnover frequency (Fig. 1).

Induced Compton scattering becomes the dominant source of opacity in a synchrotron source when $\tau_{\rm ICS} \simeq (kT_{\rm B}/m_e c^2)\tau_T \geq 1$ (SK94). Stimulated scattering causes three important changes: the spectrum becomes progressively flatter, the frequency range of the flattened region increases and the turnover frequency rises more rapidly as τ_T increases (SK94). Stimulated scattering reduces the number of synchrotron seed photons and the self-Compton flux is $\propto \tau_T^{1.8}$ for $\tau_T \geq 0.1$ (SK94).

We find that induced Compton scattering reduces R_ν over almost two decades of frequency when $\tau_{\rm ICS} > 1$. In Fig. 1 b, we see that $R_\nu \simeq 0$ below the self-absorption turnover frequency and R_ν increases with frequency. Stimulated scattering has a small effect when $\tau_{\rm ICS} < 1$ (Fig. 1 a). The shape of the FVR is nearly independent of B_{-5}.

The stimulated scattering curves in Fig. 1a,b were computed using the steady state models of SK94 and the approximation

$$\Delta \log S_\nu \simeq 2\frac{S_2 - S_1}{S_2 + S_1}. \tag{2}$$

The change in the electron density between the two models is fairly large $(\Delta\tau_T/\tau_T > 1)$, so these results are only qualitatively correct.

3. 3C279

The strong correlation of the radio and X-ray fluxes from 3C279 is consistent with the synchrotron self-Compton model for the continuum emission. We can estimate

$$R_\nu \simeq \frac{V(\nu)}{V(1\,\text{kev})}, \tag{3}$$

where V are the variability amplitudes defined by Grandi et al. (1995). The estimated FVR for the epochs 1988–1991.4 and 1991.4–1993.2 are plotted in Fig. 2.

The model R_ν for $\tau_T \sim 1.0$ has been shifted up to $\nu' = 250\nu$ and displayed in the same figure. This corresponds to raising the magnetic field strength and/or assuming that the radiating material is relativistically beamed because

$$\nu \propto B_{-5}^{(p+2)/(p+4)} \left(\frac{\delta}{1+z}\right), \tag{4}$$

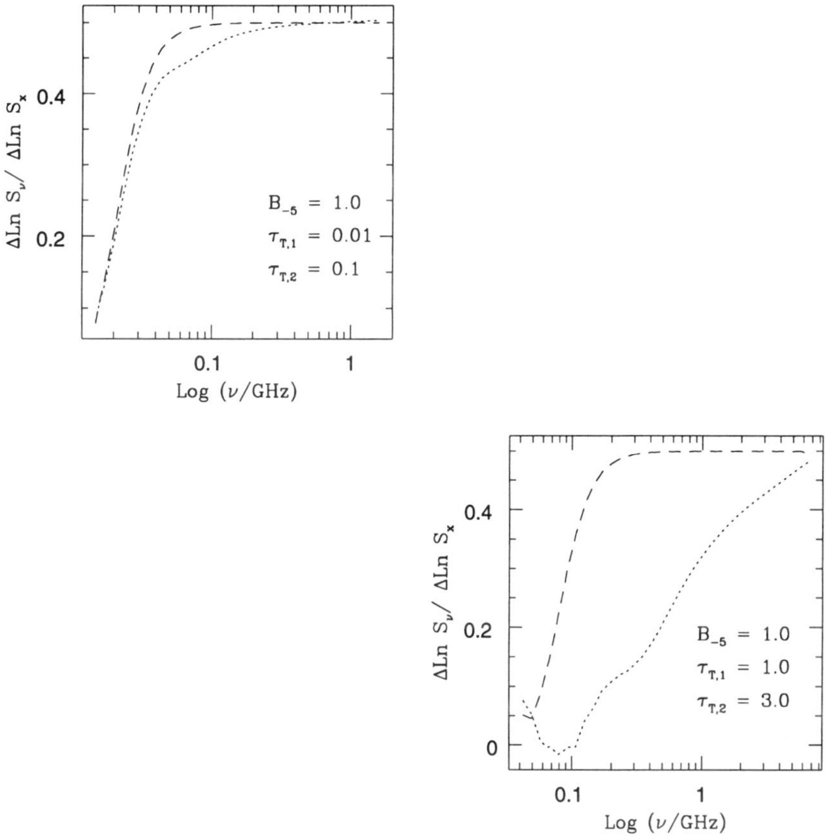

Figure 1. Relative variation in the radio flux density for synchrotron emission (dashed line) and including induced Compton scattering (dotted line) for two values of $\Delta\tau_T$.

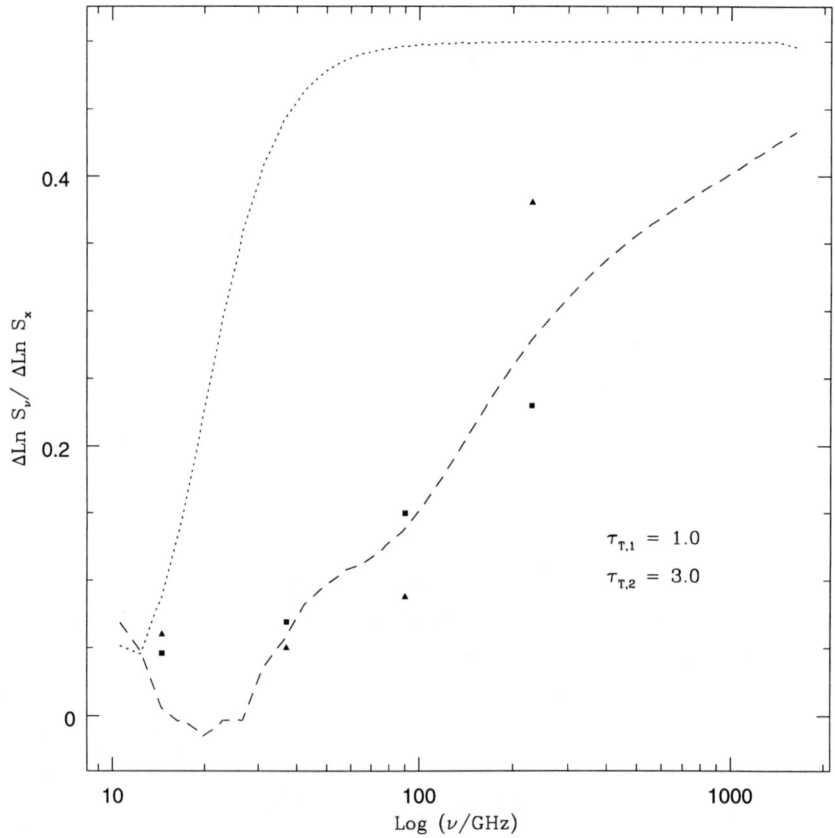

Figure 2. Comparison to the variability amplitudes of 3C279 for the epochs 1988–1991.4 (squares) and 1991.4–1993.2 (triangles). The frequencies of the model FVR have been scaled up by a factor of 250.

where B_{-5} is the magnetic field strength in 10^{-5} G and δ is the Doppler factor of the radiating material.

A homogeneous source of synchrotron self-Compton emission is not consistent with the observed R_ν (Fig. 2). However, the increase of R_ν with frequency can be reproduced by a homogeneous source if $\tau_{\rm ICS} \geq 1$. The radio spectrum will be flattened if this explanation is correct.

4. Conclusions

The ratio of the fractional change in the radio flux density to the fractional change in the X-ray flux at 1 kev (R_ν) is sensitive to the radio opacity of the synchrotron source. Variations in the electron density of an optically thin synchrotron self-Compton source result in $R_\nu = 0.5$, independent of frequency.

Induced Compton scattering reduces R_ν near the turnover frequency by $\sim 10\%$ when $\tau_{\rm ICS} \leq 1$. At larger $\tau_{\rm ICS}$, $R_\nu < 0.5$ over two decades of frequency. The steady state models of SK94 were used to show that R_ν is reduced by more than a factor of 5 below the self-absorption turnover and increases with frequency.

The observed variability amplitudes of 3C279 increase with frequency and are always smaller than the expected value of $R_\nu = 0.5$ for a homogeneous synchrotron self-absorbed source. The observed R_ν can be reproduced if $\tau_{\rm ICS} \geq 1$. The source must be beamed and/or have a magnetic field strength $B_{-5} > 1$.

References

Brown, L. M. J., et al. 1989, ApJ, 340, 129

Grandi, P., et al. 1995, in Advances in Space Research, 15, (5) 23

Maraschi, L., et al. 1994, ApJ, 435, L91

Marscher, A. P. 1987, in Superluminal Radio Sources, ed. J. A. Zensus & T. J. Pearson (Cambridge, Cambridge University Press), p. 280

Sincell, M. W., & Krolik, J. H. 1994, ApJ, 430, 550 (SK94)

Valtaoja, E., et al. 1988, A&A, 203, 1

Part V
Conference Summary

Blazar Continuum Variability
ASP Conference Series, Vol. 110, 1996
H. R. Miller, J. R. Webb, and J. C. Noble, eds.

Present and Future Blazar Variability

R. D. Blandford

130-33 Caltech, Pasadena CA 91125

Abstract. A brief overview of papers presented at a Workshop on Blazar Continuum Variability held in honor of Alex Smith at Florida International University is presented. The importance of multi-wavelength monitoring for elucidating the physics of blazar emission was emphasized although it remains difficult to disentangle a true cross-correlated signal from more random and independent variation. The interpretation of the blazar phenomenon in terms of synchrotron and Compton emission from an ultrarelativistic jet directed towards us is becoming increasingly secure and theoretical models are more ambitious in terms of the observational details that they attempt to accommodate. Although EGRET is coming to the end of its life, there are fresh observational opportunities from radio to X-ray frequencies. The major theoretical challenge is to relate these detailed jet models to the putative central black holes and accretion disks that are generally held to be responsible for their formation.

1. Introduction

I have been asked to to try to summarize the papers we have heard these past three days, to give some personal commentary on what has been learned and then to try to guess how this field might develop in the foreseeable future. This is such a tall order that I hope I can be allowed some license in what I say and some laziness in my bibliography, deferring to the previous contributors who will be identified by name in square brackets.

2. What is a Blazar anyway?

As I am sure almost everyone here realizes, "Blazar" is a neologism invented in jest by Ed Spiegel who happened to be visiting Pittsburgh at the time of the 1978 meeting on BL Lac Objects. He failed to give a precise definition! Not surprisingly, a trail of taxonomic confusion has ensued. I presume that he saw blazar as a hybrid, a contraction of BL Lac object and quasar, the union of both classes. However now we have reason to regard blazar as a "phenomenon" [Robson], what happens when we stand in the line of fire of a much larger set of cosmic objects. Not only have we had to abandon "spectral chauvinism" because, as this meeting epitomizes, the observations encompass almost 60 octaves, we have also in more recent years renounced "orientation chauvinism." What you see depends upon where you stand.

Nevertheless, if we return to an observational definition, then the three common characteristics are flat radio spectrum, rapid variability and high polarization. The BL Lacs have lineless and bump-free spectra and are mostly found at low redshift ($z \sim 0.1$). Their variation is more characterized as a sequence of "flares" rather than a "trend" and the duty cycle for variation is high $\delta \sim 0.8$. Finally when the radio polarization can be measured on VLBI scales, we find that the magnetic field is perpendicular to the jet. (Naturally enough the eponymous object, BL Lac itself, is anomalous [Corbett].). The BL Lac objects can be further subdivided into radio-selected (RBL) and X-ray selected (XBL) objects.

By contrast, flat spectrum radio quasars (FSRQ), that are also included in the blazar class are typically located at $z \sim 2$ and, when they exhibit blazar activity, there are both incipient lines and bumps. Under these circumstances, trend variation dominates, and the duty cycle is lower ($\delta \sim 0.4$) while the polarization direction is parallel [Smith, Aller]. (If you are worried about whether some of these distinctions depend more upon the temporal and spatial resolution of the observations rather than reflect some deep underlying physical distinction, then you are probably not alone!)

There has been much discussion here of both empirical and physical procedures for interpreting the complete radio to UHE γ-ray spectra. There seems to be widespread agreement that the majority of the X- and γ-rays are produced through Compton scattering of synchrotron photons [Maraschi]. Furthermore, [Sambruna, Urry] find evidence that the sequence FSRQ-RBL-XBL can be covered by a single parameter, the radio curvature and concludes that, contrary to prior belief, the RBL-XBL distinction is not due to orientation. The RBL exhibit greater relative γ-ray emission than the XBL. A key problem in all of these spectral studies is to separate the non-thermal from the stellar radiation. One approach is to synthesize the spectra of large numbers of blazars using Principal Component Analysis [Clements].

3. Jets and scaling laws

Blazar imaging, direct and indirect, centers on the jets. We have been reminded, graphically, that jets can be seen over a dynamic range ranging from ~ 300 times the radius of the putative, underlying black hole, to scales some eight or nine orders of magnitude larger [Marscher, Maraschi, Kafatos]. For example, in the case of 3C279, the beamed γ-rays detected by EGRET probably originate from $\sim 10^{15}$ cm, and yet radio emission is traced out to a de-projected distance of $\sim 10^{24}$ cm. In principle, this large range should allow us to establish scaling laws for jet propagation, synchrotron emission and so on, especially if the external medium can be described using a power-law density and pressure radial variation, (e.g., $P, \rho \propto r^{-1.5-2}$, [Marscher]). On the scales where we can resolve the radio emission, it appears that the source is quite inhomogeneous. What remains controversial is whether or not this is true on scales smaller than ~ 10 pc where the radio emission cannot be resolved.

There is a further argument in favor of scaling laws. This is, if one actually performs the gas dynamic experiment of firing a hypersonic jet across a uniform pressure chamber, then one finds that the jet typically propagates a distance of order the product of the Mach number and the jet diameter, the distance it takes rarefraction fronts to cross the jet. Clearly, extragalactic jets are far more persistent than this. I suspect that the power law density variation in the ambient pressure is actually necessary for a jet to propagate. Conversely, if there is structure in the external density variation, then this may well be reflected in the jet structure.

Jet models, especially those applied to VLBI and γ-ray jets, usually adopt a constant Lorentz factor Γ. This is probably too simplistic and if Γ is large then either acceleration [Georganopolos], curvature [Dreissigacker] or velocity shear must be taken into account in interpreting the observations. The beautiful analyses of unbeamed FR 1 radio galaxies, the probable parent population of BL Lac objects by Laing (private communication) substantiates this point.

4. Variability

A proper understanding of the causes of blazar variability is contingent upon an accurate description of the phenomenology. At this meeting a wide variety of approaches have been described. By far the largest observational effort has gone into the major monitoring programs at both optical [Smith, Webb] and radio [Aller] wavelengths. It is common to interpret the variations in terms of flares and trends. However, this separation depends upon one's attention span and one man's flare is another woman's baseline. In order to circumvent this verbal ambiguity, it is increasingly popular to quantify the results in terms of a structure function

$$D(\tau) = < [F(t+\tau) - F(t)]^2 > \tag{1}$$

where the average is over t. In order to derive a reliable statistical estimator of this quantity, large stretches of uninterrupted data must be obtained, especially for the more slowly varying quasars. However, there is no more reason to expect that blazar variability is any more likely to be a Gaussian process than turbulence, for example, and by going to three point correlation functions, or equivalently bispectra, it should be possible to see if the time series exhibits a significant arrow of time. One sees strong suggestions of intermittency in blazar flux time series that are strongly reminiscent of fluid turbulence. In some formal sense, we can think of a blazar as a dynamical system and appropriate some of the modern techniques [Sadun, Nair].

There has been much discussion of trans-spectral flare correlation and various examples have been presented, representing a bewildering variety of source behaviors. 3C279 has been the object of major campaigns [Urry, Wehrle] and remains one of the most striking EGRET variables. PKS 1622−297 was extensively monitored following detection of an EGRET flare in which its flux doubled in 2.5 hr, and a $\sim 3^{\rm m}$ optical variation ensued [Mattox]. There may also have been a correlated mm outburst. In general, variability with a timescale of days, is quite common in the brighter EGRET γ-ray sources [Hartmann]. The γ-ray radio correlation is more problematic, with the bright γ-ray flares being associated with epochs of intense radio activity [Pohl]. It is also possible to claim that the γ rays precede the radio emission as is surely the theoretical expectation. However, in at least one source, the γ-rays follow the optical variation [Marscher]. In the very bright source PKS 2155−30, it appears that the ultraviolet variation leads the X-rays [Pesce, Pian, Urry]. Some of the results are more statistical in character. Finally the most powerful source followed to date is 0528+134 has a γ-ray luminosity for $\sim 3 \times 10^{49}\Omega$ erg s^{-1} where Ω is the solid angle illuminated by the jets. [McHardy and Takahashi] reported on the multifarious X-ray campaigns and attempts to infer the underlying emission mechanisms in the different types of object. In particular the ASCA observations show that at least in MKN421, the lower energy X-ray variation follows that at high energy. In addition, [Valtaoja] reports that the characteristic variability timescale increases with observing wavelength. Several participants [Takahashi, Robson, McHardy, Mukherjee] discussed the nature of spectral variation at various frequencies and demonstrated a form of hysteresis in which the spectrum appeared to harden as the flux increased. This also provides a clear demonstration that the system is not time-reversible.

Despite all of this hard work, a concern remains that the observations are undersampled and that claimed identifications of outbursts at, say, γ-ray wavelengths with outbursts at radio wavelengths may not be significant? Probably the most important ambiguous case is provided by intraday radio variability [Wagner]. Here, the association of optical and radio variation would conclusively rule out association of the latter with refractive interstellar scintillation. Although the observational data are still not completely compelling, there is no doubt that, through much hard work, the evidence for correlation is steadily improving. If the variations can be shown to be intrinsic then simple synchrotron source models are required to have bulk Lorentz factors $\Gamma > 100$. Such models are radiatively quite inefficient and this leads to prodigious jet power estimates [Marscher].

A quite different, and extremely promising approach to source variability was described here by [Helfand] who presented first results from the FIRST project. He and his collaborators detect 100 radio sources per square degree brighter than 1 mJy and they can use repeated observations to perform serious statistical investigations. They have found hundreds of sources varying over successive days and find that these are preferentially identified with bright sources on the sky survey and especially AGN. Clearly, this is only the beginning, and it should provide a very useful complement to the detailed monitoring of selected individual sources.

For many of us, this conference provided a first introduction to microvariability [Miller, Heidt, Noble]. It is surprising that it is possible to perform photometry to such accuracy, but the results are persuasive. The origin of this variation is unknown and both disk and jet models can be entertained. At the other extreme are reports by [Hudec] of enormous optical variations in objects detected on archival plates.

Probably the most exciting discovery that can be envisaged is of unambiguous periodicity. Once more OJ287 provided the best case, [Valtonen, Sillanpao, Marchenko, Takalo]. Although I personally remain unconvinced by the data, there is no question that future searches of this type are well-motivated. The implications for quasar modeling are extremely important. The interpretation of these data in terms of a precessing binary black hole system is probably one of the more conservative possibilities.

5. The physics of blazar emission

All of this hard observational effort has, quite naturally, stimulated many interpretative theoretical approaches. These were ably summarized by [Maraschi]. In this description, the non-negotiable elements are relativistic jets, the synchrotron explanation of the radio spectrum and Compton scattering for the X- and γ-ray emission. However, it seems to me that there are then a list of choices that have to be made.

1. Is the source best described as essentially steady with small perturbations or should it instead be modeled as a serious of discrete, though possibly overlapping flares? In the context of the jet model, we can express this question as "Is the jet flow approximately steady or are we observing the effects of a sequence of traveling shock fronts?"

2. Should we see the source as inhomogeneous with the emission at different frequencies being identified with different jet radii, or is it essentially homogeneous, with all the particle acceleration and consequently all the emission coming from one particular site?

3. Is the plasma a conventional electron-proton plasma or does it comprise relativistic electrons and positrons? This may be answerable observationally through Faraday rotation or circular polarization measurements [Robson]. So far the evidence favors positrons, though, again, it is not overwhelming.

4. Are the Compton seed photons produced externally, or internally? A third possibility has recently been provided by Ghisellini and Madau (1996) who use quasar broad line regions to back-scatter jet photons into the advancing relativistic particles.

5. Are jets accelerating or decelerating?[Georganopolous]

There are additional kinematic embellishments, which, because of the power of the relativistic amplifier, can have major observational consequences. We already know from the example of SS433 that jets can precess. This can also happen in extragalactic jets [Marscher, Dreissigacker]. This is one of the explanations that has been advanced for the large number of perpendicular jet features discussed by [Xu]. It appears that the higher polarization jets are preferentially elongated perpendicular to the large scale structure, while their lower polarization counterparts are extended in a parallel direction.

Now, there are two simple ways to make helical jets. Firstly, there may be genuine dynamical precession of an angular momentum vector associated with a black hole or an accretion disk. This will then produce a helical jet with constant repeat distance. Secondly a helical mode may be the natural way for an unnaturally axisymmetric magnetic structure to break its symmetry. In this case, one expects the repeat distance to scale with the radius. These two possibilities should be distinguishable. Turning to the moving features themselves, [Unwin] reported that these tended to follow separate tracks. This suggests, to me at least, that the jet occupies quite a larger solid angle than is subtended by individual features and that only small regions are lit up at any time.

Radiating relativistic shocks remain the most plausible emitting structures [Marscher, Aller]. They provide a plausibly efficient *in situ* particle accelerator and an equally important magnetic field amplification mechanism. The radio polarization data provides some evidence for their presence. The jet flows themselves do not have to be conical and tapered jets have also been considered [Marscher].

6. Origin of the species

There is no smoke without fire and the jets that we are observing either directly or indirectly are generally held to be the exhaust gases from the internal combustion engine in the center of the galactic nucleus that is polluting the intergalactic environment with noxious fumes of relativistic electrons, magnetic field and heavy metal additives. Theorists dearly want to look under the hood. They ask "Is there a common mechanism which operates around young stellar objects, SS433 and extragalactic radio sources?" There are, I think, some relevant inferences that can already be drawn from blazar observations. Firstly, we can deduce that, even if the jet is comprised of electrons and positrons at large radii, at small enough radii the radiative drag is so severe that the momentum must be carried by some other species. The prime candidates are protons and electromagnetic field. Secondly, we find that the best studied jet to date, that associated with M87, must be collimated within ~ 300 Schwarzschild radii. This places severe restrictions on the confining agent and it appears that pure gas pressure collimation would lead to an overproduction of X-rays. No such difficulty attends magnetic confinement.

7. Diskless nodes?

The general presumption to date has been that disks are essential to jet formation. However, we do not often have direct evidence of this at the radii where the jets are supposed to be formed, although there are more and more examples of rotating gas at larger radii. In addition, [Smith] has argued that we do not see evidence for disks in the wavelength-dependence of the polarization in radio-selected BL Lacs. This may be because the jet flux completely dominates that from the disk (as well as from the lines) as we see in 3C 345, for example. Perhaps FR1 radio galaxies have obscured broad emission line regions as is argued by [Corbett] to be the case in BL Lac itself. Conversely, [Wiita] and [Miller] have interpreted microvariability in terms of flares on an accretion disk.

8. Happy families

No discussion of blazars would be complete without reference to the vexed issue of unification. Many of the arguments pertaining to unification come from polarimetric studies [Smith]. These show evidence for obscured emission lines in radio-quiet quasars. In addition about the most convincing case for a nuclear disk in NGC 4258 also shows dichroic, optical absorption.

The usual supposition is that the parent population of RBL are mostly FR1 radio galaxies. Similarly OVV quasars are identified with regular radio loud quasars. Even if these identifications are correct, we lack a working knowledge of which parameters control the type of AGN we actually observe. If we reduce everything to fundamentals then a plausible set would be the mass of the hole, the spin of the hole and the mass accretion rate. To this set of "intrinsic" parameters, I would add the extrinsic factors, the viewing angle and the nuclear star and gas densities.

9. Observatories of the 21st century

Part of my charge was to anticipate what future observations might yield. The VLBA is already having a major impact [Unwin, Mattox, Wehrle]. If plans to extend its frequency range from 43 GHz to 89 GHz are successful, then it will be used to map the jets down to even smaller radii. For the brighter sources, the launch of VSOP this fall, should bring about even greater improvements in angular resolution. Moving to mm wavelengths, one of the great opportunities that SCUBA will introduce is the ability to perform high frequency polarimetry [Robson]. The immediate future of infrared astronomy is represented by ISO.

At optical wavelengths, I will take for granted all that big telescopes will accomplish and highlight a discussion by [Balonek] and [Boltwood] of the great advantage that will follow from equipping qualified amateurs and small colleges with CCDs and coordinating optical monitoring. It is apparently quite practical to imaging an association of blazar observers contributing in much the same way as supernova hunters and variable star observers.

Moving on to the ultraviolet, one of the successes of EUVE has been the detection of PKS 2155−30. The most recent X-ray satellite is the X-ray timing explorer (XTE). It too has a polarimetric capability [Perlman]. This may allow us to confirm that the X-ray emission is Compton scattering as opposed to thermal or synchrotron radiation. Finally, we come to the γ rays. With the imminent demise of EGRET, proposed missions like GLAST represent the way forward, offering improved sensitivity and angular resolution.

All of these missions represent piecemeal solutions to the problems of multi-wavelength source monitoring. [Urry] and [Maraschi] have a much bolder idea and this is build integrated platforms that take care of all the space-based monitoring needs from a single spacecraft. Unfortunately, it is not clear how likely this is to happen in the present fiscal climate.

10. The theoretical horizon

The theoretical opportunities are also considerable. I think it is clear that we already have enough data to justify yet more detailed phenomenological analyses along the lines presented here by [Maraschi].

If, like me, you see jet formation as essentially a hydromagnetic matter, then you probably also reckon that the future lies in 3D numerical simulations that are not constrained by imposing axisymmetry. Already there is discussion of serious teraflop machines which allow 256^3 grids. Indeed more futuristic plans discuss a computing power which would admit 1024^3 grids, equivalent to magnetic Reynolds' numbers of ~ 300.

A rather different class of problems, represented here by [Sincell] is to discuss coherent processes perhaps associated with jets. If intraday radio variability is also tracked in the optical variation, then I suspect that this is the best way to account for them. However, postulating coherent emission alone is not enough. Whatever its provenance high brightness radio emission is subject to strong propagation effects as it escapes from the galactic nucleus, e.g., through induced Compton or stimulated Raman scattering.

Finally we turn to the black hole itself. I have argued elsewhere that the relativistic plasma that comprises the inner jet, is powered by the electromagnetic extraction of spin energy by the surrounding magnetic field. There is still not a completely satisfactory formal description of how this is supposed to occur. This is the grandest theoretical challenge of all.

Acknowledgments. I thank Alex Smith for being constant in his search for variability and making this meeting possible. I also speak for many in noting Jim Webb's genial hospitality and efficient organization. Support under NSF contract AST 93-23375 is also gratefully acknowledged.

Quotes

1. "The best data available is not enough, but we know where to go."
2. "We should have an optical [UV, X-ray, Gamma-ray, IR, radio] monitor on every telescope."
3. "Those damn Julian Dates....."
4. "*I'm* a γ-ray emitter at *some* level!"
5. "We know we are not moving super-relativistically ourselves, so it must be the source."
6. "Unified schemes!"
7. "My apologies, I know it is difficult to wake up to Stokes' parameters."
8. "When you do polarization you really learn what 1σ is."
9. "What I said in the paper at that meeting has been generally ignored, so I will say it louder this time."
10. "The two or three photons I have seen seem to be very important."
11. "...so we did it again"
12. "3C 273 is not being fully helpful here"
13. "I've smoothed the data in this figure."

    Reply: "Delete data, *then* smooth it."
14. "I hope I won't bore you too long."
15. "My contribution to this paper is insignificant to the rest of yours."
16. "The next step will be to show Julian date in logarithm."
17. "We ask our friends and our friends' friends."